UNITEXT

La Matematica per il 3+2

Volume 166

Editor-in-Chief

Alfio Quarteroni, Politecnico di Milano, Milan, Italy
  École Polytechnique Fédérale de Lausanne (EPFL), Lausanne, Switzerland

Series Editors

Luigi Ambrosio, Scuola Normale Superiore, Pisa, Italy

Paolo Biscari, Politecnico di Milano, Milan, Italy

Ciro Ciliberto, Università di Roma "Tor Vergata", Rome, Italy

Camillo De Lellis, Institute for Advanced Study, Princeton, NJ, USA

Victor Panaretos, Institute of Mathematics, École Polytechnique Fédérale de Lausanne (EPFL), Lausanne, Switzerland

Lorenzo Rosasco, DIBRIS, Università degli Studi di Genova, Genova, Italy
  Center for Brains Mind and Machines, Massachusetts Institute of Technology, Cambridge, Massachusetts, US
  Istituto Italiano di Tecnologia, Genova, Italy

The **UNITEXT - La Matematica per il 3+2** series is designed for undergraduate and graduate academic courses, and also includes books addressed to PhD students in mathematics, presented at a sufficiently general and advanced level so that the student or scholar interested in a more specific theme would get the necessary background to explore it.

Originally released in Italian, the series now publishes textbooks in English addressed to students in mathematics worldwide.

Some of the most successful books in the series have evolved through several editions, adapting to the evolution of teaching curricula.

Submissions must include at least 3 sample chapters, a table of contents, and a preface outlining the aims and scope of the book, how the book fits in with the current literature, and which courses the book is suitable for.

For any further information, please contact the Editor at Springer: francesca.bonadei@springer.com

THE SERIES IS INDEXED IN SCOPUS

***

UNITEXT is glad to announce a new series of free webinars and interviews handled by the Board members, who rotate in order to interview top experts in their field.

Access this link to subscribe to the events:

https://cassyni.com/s/springer-unitext

Andrea Pascucci

Probability Theory II

Stochastic Calculus

 Springer

Andrea Pascucci
Dipartimento di Matematica
Alma Mater Studiorum – Università di Bologna
Bologna, Italy

ISSN 2038-5714   ISSN 2532-3318   (electronic)
UNITEXT
ISSN 2038-5722   ISSN 2038-5757   (electronic)
La Matematica per il 3+2
ISBN 978-3-031-63192-4   ISBN 978-3-031-63193-1   (eBook)
https://doi.org/10.1007/978-3-031-63193-1

This book is a translation of the original Italian edition "Teoria della Probabilità" by Andrea Pascucci, published by Springer-Verlag Italia S.r.l. in 2024. The translation was done with the help of an artificial intelligence machine translation tool. A subsequent human revision was done primarily in terms of content, so that the book will read stylistically differently from a conventional translation. Springer Nature works continuously to further the development of tools for the production of books and on the related technologies to support the authors.

Translation from the Italian language edition: "Teoria della Probabilità. Processi e calcolo stocastico" by Andrea Pascucci, © Springer-Verlag Italia S.r.l., part of Springer Nature 2024. Published by Springer Milano. All Rights Reserved.

© The Editor(s) (if applicable) and The Author(s), under exclusive license to Springer Nature Switzerland AG 2024
This work is subject to copyright. All rights are solely and exclusively licensed by the Publisher, whether the whole or part of the material is concerned, specifically the rights of reprinting, reuse of illustrations, recitation, broadcasting, reproduction on microfilms or in any other physical way, and transmission or information storage and retrieval, electronic adaptation, computer software, or by similar or dissimilar methodology now known or hereafter developed.
The use of general descriptive names, registered names, trademarks, service marks, etc. in this publication does not imply, even in the absence of a specific statement, that such names are exempt from the relevant protective laws and regulations and therefore free for general use.
The publisher, the authors and the editors are safe to assume that the advice and information in this book are believed to be true and accurate at the date of publication. Neither the publisher nor the authors or the editors give a warranty, expressed or implied, with respect to the material contained herein or for any errors or omissions that may have been made. The publisher remains neutral with regard to jurisdictional claims in published maps and institutional affiliations.

Cover illustration: Cino Valentini, Archeologia 2 , 2021, affresco acrilico, private collection

This Springer imprint is published by the registered company Springer Nature Switzerland AG
The registered company address is: Gewerbestrasse 11, 6330 Cham, Switzerland

If disposing of this product, please recycle the paper.

To my students

*E ora, che ne sarà
del mio viaggio?
Troppo accuratamente l'ho studiato
senza saperne nulla. Un imprevisto
è la sola speranza. Ma mi dicono
che è una stoltezza dirselo*[1]

Eugenio Montale, *Prima del viaggio*

---

[1] *And now, what will become of my journey?
I've studied it too meticulously
without knowing anything about it. An unforeseen
event is the only hope. But they tell me
it's foolish to say so.*

Preface

"For over two millennia, Aristotle's logic has ruled over the thinking of western intellectuals. All precise theories, all scientific models, even models of the process of thinking itself, have in principle conformed to the straight-jacket of logic. But from its shady beginnings devising gambling strategies and counting corpses in medieval London, probability theory and statistical inference now emerge as better foundations for scientific models, especially those of the process of thinking and as essential ingredients of theoretical mathematics, even the foundations of mathematics itself. We propose that this sea change in our perspective will affect virtually all of mathematics in the next century."
    David Bryant Mumford, The Dawning of the Age of Stochasticity [99]

"A mathematician is someone who loves philosophy, art, and poetry because they find the profound human need everywhere, against and beyond the often ridiculous oppositions between "hard" and "soft" sciences. Awareness of such an intertwining further enhances (...) the high, inescapable and indestructible moral choice to carry out one's own action as a scientist and as a human being in society towards good. And if good and true come together, they can only produce beauty."
    Rino Caputo, Preface to *Le anime della matematica* [147]

In Volume 1 of Probability Theory [113], we introduced fundamental concepts such as probability space and distribution, random variables, limit theorems, and conditional expectation. This second volume complements the earlier material by delving into more advanced classical topics in stochastic analysis. The primary focus of this book lies in stochastic processes, with particular emphasis on two crucial classes: Markov processes and martingales. The initial chapters provide a general introduction to stochastic processes and explore the analysis of two key examples of Markov processes: Brownian motion and the Poisson process. Historically, two major approaches have been employed for the construction of continuous Markov process, often referred to as "diffusions." The classical approach, pioneered by A. N. Kolmogorov [69] and W. Feller [45], involves constructing a diffusion based on its transition law, which is defined as the distributional solution to the backward and forward Kolmogorov differential equations. This approach relies on intricate analytical results from the theory of partial differential equations. Starting from Chap. 8, we embark on a systematic study of martingales. One of the most significant results in martingale theory is Doob's decomposition theorem, which, under appropriate assumptions, represents a process as the direct sum of a "drift part" and a "martingale part," each with its own regularity properties. This type of result forms the basis for the second approach, proposed by K. Itô, to the construction of continuous Markov processes. Itô builds upon P. Lévy's idea of considering the infinitesimal increment of a diffusion as a Gaussian-type increment with a suitable mean (drift) and covariance matrix (martingale part). In light of these results, Itô develops a theory of stochastic differential calculus and provides a method for constructing diffusions as solutions to stochastic differential equations. The final part of the book is dedicated to an in-depth study of the existence and uniqueness problems for stochastic equations and their connections to elliptic-parabolic partial differential equations.

This book comprises sufficient material to support at least two semester-long courses on stochastic processes and calculus, suitable for graduate or doctoral-level studies. It is designed as a relatively concise compendium, given the intricacy of the subject, providing a solid groundwork for those interested in exploring stochastic models for practical applications and for those beginning their research journey in the field of stochastic analysis.

As emphasized in the introduction of the first volume [113], it is worth restating the quote by David Mumford that initiates the preface: nowadays, probability theory is regarded as an indispensable component for the theoretical advancement of mathematics and the very foundations of mathematics itself. In this context, the noteworthy review article [97] examines the remarkable progress in research on stochastic processes since the mid-twentieth century.

From an applied standpoint, probability theory serves as the fundamental tool used to model and manage risk in all fields where phenomena are studied under conditions of uncertainty:

- **Physics and Engineering** where stochastic numerical methods, such as Monte Carlo methods, are extensively used. These methods were first formalized by Enrico Fermi and John von Neumann.
- **Economics and Finance**, starting with the famous Black-Scholes-Merton formula, for which the authors received the Nobel Prize. Financial modeling generally requires an advanced background in mathematical-probabilistic-numerical methods. The text [112] provides an introduction to the theory of financial derivative valuation, balancing the probabilistic approach (based on martingale theory) and the analytic approach (based on partial differential equations theory).
- **Telecommunications**: NASA utilizes the Kalman-Bucy filter method to filter signals from satellites and probes sent into space. From [102], page 2: "In 1960 Kalman and in 1961 Kalman and Bucy proved what is now known as the Kalman-Bucy filter. Basically the filter gives a procedure for estimating the state of a system which satisfies a 'noisy' linear differential equation, based

on a series of 'noisy' observations. Almost immediately the discovery found applications in aerospace engineering (Ranger, Mariner, Apollo etc.) and it now has a broad range of applications. Thus the Kalman-Bucy filter is an example of a recent mathematical discovery which has already proved to be useful - it is not just 'potentially' useful. It is also a counterexample to the assertion that 'applied mathematics is bad mathematics' and to the assertion that 'the only really useful mathematics is the elementary mathematics'. For the Kalman-Bucy filter—as the whole subject of stochastic differential equations—involves advanced, interesting and first class mathematics."

- **Medicine and Botany**: the most important stochastic process, the Brownian motion, is named after Robert Brown, a botanist who observed the irregular movement of colloidal particles in suspension around 1830. Brownian motion was used by Louis Jean Baptist Bachelier in 1900 in his doctoral thesis to model stock prices and was the subject of one of Albert Einstein's most famous works published in 1905. The first mathematically rigorous definition of the Brownian motion was given by Norbert Wiener in 1923.
- **Genetics**: it is the science that studies the transmission of traits and the mechanisms by which they are inherited. Gregor Johann Mendel (1822–1884), a Czech Augustinian monk considered the precursor of modern genetics, made a fundamental methodological contribution by applying probability calculus to the study of biological inheritance for the first time.
- **Computer Science**: quantum computers exploit the laws of quantum mechanics for data processing. In a "classical" computer, the unit of information is the bit: we can always determine the state of a bit and precisely establish whether it is 0 or 1. However, we cannot determine the state of a quantum bit (qubit), the unit of quantum information, with the same level of precision. We can only determine the probabilities of it assuming the values 0 and 1.
- **Jurisprudence**: the verdict issued by a judge in a court is based on the probability of the defendant's guilt estimated from the information provided by the investigations. In this field, the concept of conditional probability plays a fundamental role, and its misuse can lead to notorious miscarriages of justice, some of which are recounted in [116].
- **Meteorology**: for forecasts beyond the fifth day, it is crucial to have probabilistic meteorological models. These probabilistic models are generally run in major international meteorological centers because they require highly complex statistical-mathematical procedures that are computationally intensive. Starting from 2020, the Data Center of the European Centre for Medium-Range Weather Forecasts (ECMWF) is located in Bologna.
- **Military Applications**: from [127] page 139: "In 1938, Kolmogorov had published a paper that established the basic theorems for smoothing and predicting stationary stochastic processes. An interesting comment on the secrecy of war efforts comes from Norbert Wiener (1894–1964) who, at the Massachusetts Institute of Technology, worked on applications of these methods to military problems during and after the war. These results were considered so important to

America's Cold War efforts that Wiener's work was declared top secret. But all of it, Wiener insisted, could have been deduced from Kolmogorov's early paper."

Finally, probability is at the foundation of the development of the most recent technologies in Machine Learning and all related applications in Artificial Intelligence, such as autonomous driving, speech and image recognition, and more (see, for example, [54] and [122]). Nowadays, an advanced knowledge of Probability Theory is a minimum requirement for anyone interested in pursuing applied mathematics in any of the aforementioned fields.

It should be acknowledged that there are numerous monographs on stochastic analysis: among my favorites I mention, in alphabetical order, Baldi [6], Bass [9], Baudoin [13], Doob [35], Durrett [37], Friedman [50], Kallenberg [66], Karatzas and Shreve [67], Mörters and Peres [98], Revuz and Yor [123], Schilling [129], and Stroock [133]. Other excellent texts that have been major sources of inspiration and ideas include those by Bass [10], Durrett [38], Klenke [68], and Williams [148]. In any case, this list is far from exhaustive.

After more than two decades of teaching experience in this field, this book represents my endeavor to systematically, concisely, and as comprehensively as possible, compile the fundamental concepts of stochastic calculus that, in my view, should constitute the essential knowledge for a modern mathematician, whether pure or applied.

I would like to conclude by expressing my heartfelt gratitude to the exceptional group of probabilists at the Department of Mathematics in Bologna: Stefano Pagliarani, Elena Bandini, Cristina Di Girolami, Salvatore Federico, Antonello Pesce, and Giacomo Lucertini, as well as those whom I hope will join us in the future. A big thank you also goes to Andrea Cosso for his valuable collaboration during the (all too short!) time he was a member of our department. Lastly, I extend a special thank you to all the students who have taken my courses on probability theory and stochastic calculus. This book was created for them, inspired by the passion and energy they have shared with me. It is dedicated to them because I cannot refrain from making it my own, at least as an attempt, the famous phrase of a great scientist "I never teach my pupils; I only attempt to provide the conditions in which they can learn."

Readers who wish to report any errors, typos, or suggestions for improvement can do so at the following address: andrea.pascucci@unibo.it.

The corrections received after publication will be made available on the website at: https://unibo.it/sitoweb/andrea.pascucci/.

Bologna, Italy
April 2024

Andrea Pascucci

Frequently Used Symbols and Notations

- $A := B$ means that A is, *by definition*, equal to B
- $\uplus$ indicates the *disjoint* union
- $A_n \nearrow A$ indicates that $(A_n)_{n \in \mathbb{N}}$ is an *increasing* sequence of sets such that $A = \bigcup_{n \in \mathbb{N}} A_n$
- $A_n \searrow A$ indicates that $(A_n)_{n \in \mathbb{N}}$ is a *decreasing* sequence of sets such that $A = \bigcap_{n \in \mathbb{N}} A_n$
- $\mathscr{B}_d = \mathscr{B}(\mathbb{R}^d)$ is the Borel σ-algebra in $\mathbb{R}^d$; $\mathscr{B} := \mathscr{B}_1$
- $m\mathscr{F}$ is the class of $\mathscr{F}$-measurable functions

$$f : (\Omega, \mathscr{F}) \longrightarrow (E, \mathscr{E});$$

If $(E, \mathscr{E}) = (\mathbb{R}, \mathscr{B})$, $m\mathscr{F}^+$ (resp. $b\mathscr{F}$) denotes the class of $\mathscr{F}$-measurable and non-negative (resp. $\mathscr{F}$-measurable and bounded) functions.
- $\mathscr{N}$ is the family of negligible sets (cf. Definition 1.1.16 in [113])
- Numerical sets:
  - natural numbers: $\mathbb{N} = \{1, 2, 3, \ldots\}$, $\mathbb{N}_0 = \mathbb{N} \cup \{0\}$, $I_n := \{1, \ldots, n\}$ for $n \in \mathbb{N}$
  - real numbers $\mathbb{R}$, extended real numbers $\bar{\mathbb{R}} = \mathbb{R} \cup \{\pm\infty\}$, positive real numbers $\mathbb{R}_{>0} =]0, +\infty[$, non-negative real numbers $\mathbb{R}_{\geq 0} = [0, +\infty[$
- Leb_d indicates the d-dimensional Lebesgue measure; $\mathrm{Leb} := \mathrm{Leb}_1$
- Indicator function of a set A

$$\mathbb{1}_A(x) := \begin{cases} 1 & \text{if } x \in A \\ 0 & \text{otherwise} \end{cases}$$

- Euclidean scalar product:

$$\langle x, y \rangle = x \cdot y = \sum_{i=1}^{d} x_i y_i, \qquad x = (x_1, \ldots, x_d), \; y = (y_1, \ldots, y_d) \in \mathbb{R}^d$$

  In matrix operations, the d-dimensional vector x is identified with the $d \times 1$ column matrix.
- Maximum and minimum of real numbers:

$$x \wedge y = \min\{x, y\}, \qquad x \vee y = \max\{x, y\}$$

- Positive and negative part:

$$x^+ = x \vee 0, \qquad x^- = (-x) \vee 0$$

- Argument of the maximum and minimum of $f : A \longrightarrow \mathbb{R}$:

$$\arg\max_{x \in A} f(x) = \{y \in A \mid f(y) \geq f(x) \text{ for every } x \in A\}$$

$$\arg\min_{x \in A} f(x) = \{y \in A \mid f(y) \leq f(x) \text{ for every } x \in A\}$$

Contents

1 Stochastic Processes 1
  1.1 Stochastic Processes: Law and Finite-Dimensional
      Distributions 1
      1.1.1 Measurable Processes 7
  1.2 Uniqueness 8
  1.3 Existence 10
  1.4 Filtrations and Martingales 14
  1.5 Proof of Kolmogorov's Extension Theorem 18
  1.6 Key Ideas to Remember 23

2 Markov Processes 25
  2.1 Transition Law and Feller Processes 25
  2.2 Markov Property 30
  2.3 Processes with Independent Increments and Martingales 34
  2.4 Finite-Dimensional Laws and Chapman-Kolmogorov Equation ... 36
  2.5 Characteristic Operator and Kolmogorov Equations 41
      2.5.1 The Local Case 44
      2.5.2 Backward Kolmogorov Equation 47
      2.5.3 Forward Kolmogorov (or Fokker-Planck) Equation 51
  2.6 Markov Processes and Diffusions 55
  2.7 Key Ideas to Remember 57

3 Continuous Processes 59
  3.1 Continuity and a.s. Continuity 59
  3.2 Canonical Version of a Continuous Process 61
  3.3 Kolmogorov's Continuity Theorem 64
  3.4 Proof of Kolmogorov's Continuity Theorem 66
  3.5 Key Ideas to Remember 69

4 Brownian Motion 71
  4.1 Definition 71
  4.2 Markov and Feller Properties 75

		4.3	Wiener Space	76
		4.4	Brownian Martingales	77
		4.5	Key Ideas to Remember	80
5	**Poisson Process**			83
		5.1	Definition	83
		5.2	Markov and Feller Properties	88
		5.3	Martingale Properties	90
		5.4	Proof of Theorem 5.2.1	91
		5.5	Key Ideas to Remember	94
6	**Stopping Times**			97
		6.1	The Discrete Case	97
			6.1.1 Optional Sampling, Maximal Inequalities, and Upcrossing Lemma	102
		6.2	The Continuous Case	107
			6.2.1 Usual Conditions and Stopping Times	107
			6.2.2 Filtration Enlargement and Markov Processes	110
			6.2.3 Filtration Enlargement and Lévy Processes	114
			6.2.4 General Results on Stopping Times	118
		6.3	Key Ideas to Remember	120
7	**Strong Markov Property**			123
		7.1	Feller and Strong Markov Properties	123
		7.2	Reflection Principle	126
		7.3	The Homogeneous Case	128
8	**Continuous Martingales**			133
		8.1	Optional Sampling and Maximal Inequalities	134
		8.2	Càdlàg Martingales	139
		8.3	The Space $\mathcal{M}^{c,2}$ of Square-Integrable Continuous Martingales	141
		8.4	The Space $\mathcal{M}^{c,\mathrm{loc}}$ of Continuous Local Martingales	143
		8.5	Uniformly Square-Integrable Martingales	146
		8.6	Key Ideas to Remember	149
9	**Theory of Variation**			151
		9.1	Riemann-Stieltjes Integral	151
		9.2	Lebesgue-Stieltjes Integral	158
		9.3	Semimartingales	160
			9.3.1 Brownian Motion as a Semimartingale	161
			9.3.2 Semimartingales of Bounded Variation	163
		9.4	Doob's Decomposition and Quadratic Variation Process	165
		9.5	Covariation Matrix	166
		9.6	Proof of Doob's Decomposition Theorem	168
		9.7	Key Ideas to Remember	173

10 Stochastic Integral ... 175
- 10.1 Integral with Respect to a Brownian Motion ... 176
  - 10.1.1 Proof of Lemma 10.1.7 ... 182
- 10.2 Integral with Respect to Continuous Square-Integrable Martingales ... 183
  - 10.2.1 Integral of Indicator Processes ... 184
  - 10.2.2 Integral of Simple Processes ... 188
  - 10.2.3 Integral in $\mathbb{L}^2$ ... 190
  - 10.2.4 Integral in $\mathbb{L}^2_{\text{loc}}$ ... 195
  - 10.2.5 Stochastic Integral as a Riemann-Stieltjes Integral ... 200
- 10.3 Integral with Respect to Continuous Semimartingales ... 202
- 10.4 Scalar Itô Processes ... 204
- 10.5 Key Ideas to Remember ... 206

11 Itô's Formula ... 209
- 11.1 Itô's Formula for Continuous Semimartingales ... 209
  - 11.1.1 Itô's Formula for Brownian Motion ... 211
  - 11.1.2 Itô's Formula for Itô Processes ... 213
- 11.2 Some Consequences of Itô's Formula ... 216
  - 11.2.1 Burkholder-Davis-Gundy Inequalities ... 216
  - 11.2.2 Quadratic Variation Process ... 220
- 11.3 Proof of Itô's Formula ... 221
- 11.4 Key Ideas to Remember ... 226

12 Multidimensional Stochastic Calculus ... 227
- 12.1 Multidimensional Brownian Motion ... 227
- 12.2 Multidimensional Itô Processes ... 231
- 12.3 Multidimensional Itô's Formula ... 233
- 12.4 Lévy's Characterization and Correlated Brownian Motion ... 238
- 12.5 Key Ideas to Remember ... 240

13 Changes of Measure and Martingale Representation ... 243
- 13.1 Change of Measure and Itô Processes ... 243
  - 13.1.1 An Application: Risk-Neutral Valuation of Financial Derivatives ... 245
- 13.2 Integrability of Exponential Martingales ... 247
- 13.3 Girsanov Theorem ... 251
- 13.4 Approximation by Exponential Martingales ... 254
- 13.5 Representation of Brownian Martingales ... 257
  - 13.5.1 Proof of Theorem 13.1.1 ... 259
- 13.6 Key Ideas to Remember ... 260

14 Stochastic Differential Equations ... 263
- 14.1 Solving SDEs: Concepts of Existence and Uniqueness ... 264
- 14.2 Weak Existence and Uniqueness via Girsanov Theorem ... 269
- 14.3 Weak vs Strong Solutions: The Yamada-Watanabe Theorem ... 272
- 14.4 Standard Assumptions and Preliminary Estimates ... 277

	14.5	Some A Priori Estimates	280
	14.6	Key Ideas to Remember	284
15	**Feynman-Kac Formulas**		**287**
	15.1	Characteristic Operator of an SDE	288
	15.2	Exit Time from a Bounded Domain	290
	15.3	The Autonomous Case: The Dirichlet Problem	292
	15.4	The Evolutionary Case: The Cauchy Problem	297
	15.5	Key Ideas to Remember	300
16	**Linear Equations**		**303**
	16.1	Solution and Transition Law of a Linear SDE	303
	16.2	Controllability of Linear Systems and Absolute Continuity	308
	16.3	Kalman Rank Condition	310
	16.4	Hörmander's Condition	312
	16.5	Examples and Applications	314
	16.6	Key Ideas to Remember	321
17	**Strong Solutions**		**323**
	17.1	Uniqueness	324
	17.2	Existence	326
	17.3	Markov Property	330
		17.3.1 Forward Kolmogorov Equation	332
	17.4	Continuous Dependence on Parameters	333
18	**Weak Solutions**		**337**
	18.1	The Stroock-Varadhan Martingale Problem	338
	18.2	Equations with Hölder Coefficients	341
	18.3	Other Results for the Martingale Problem	346
	18.4	Strong Uniqueness Through Regularization by Noise	347
	18.5	Key Ideas to Remember	350
19	**Complements**		**353**
	19.1	Markovian Projection and Gyöngy's Lemma	353
	19.2	Backward Stochastic Differential Equations	356
	19.3	Filtering and Stochastic Heat Equation	359
	19.4	Backward Stochastic Integral and Krylov's SPDE	362
20	**A Primer on Parabolic PDEs**		**369**
	20.1	Uniqueness: The Maximum Principle	371
		20.1.1 Cauchy-Dirichlet Problem	372
		20.1.2 Cauchy Problem	375
	20.2	Existence: The Fundamental Solution	379
	20.3	The Parametrix Method	383
		20.3.1 Gaussian Estimates	385

	20.3.2	Proof of Proposition 20.3.2	390
	20.3.3	Potential Estimates	395
	20.3.4	Proof of Theorem 20.2.5	401
	20.3.5	Proof of Proposition 18.4.3	408
20.4	Key Ideas to Remember		410

References 413

Index 421

Abbreviations

r.v. = random variable, a.s. = almost surely. A certain property holds a.s. if there exists $N \in \mathcal{N}$ (negligible set) such that the property is true for every $\omega \in \Omega \setminus N$

a.e. = almost everywhere (with respect to the Lebesgue measure)

We indicate the importance of the results with the following symbols:

[!] means that you should pay close attention and try to understand well, because an important concept, a new idea, or a new technique is being introduced
[!!] means that the result is very important
[!!!] means that the result is fundamental

Certain points of particular significance or relevance will be denoted by the gray shades

Chapter 1
Stochastic Processes

> *Infinite product spaces are the natural habitat of probability theory*
>
> William Feller

Random variables describe the *state* of a random phenomenon: for example, an unobservable position of a particle in a physics model or the price at a future date of a stock in a financial model. Stochastic processes describe the *dynamics*, over time or depending on other parameters, of a random phenomenon. A stochastic process can be defined as a parameterized family of random variables, each of which represents the state of the phenomenon corresponding to a fixed value of the parameters. We have already encountered a simple but notable stochastic process in Volume 1, Example 2.6.4 in [113], in which $(X_n)_{n\in\mathbb{N}}$ represents the evolution over time of the price of a risky asset. From a more abstract perspective, a stochastic process can be defined as a random variable with values in a functional space, typically a space of curves in $\mathbb{R}^N$: each curve represents a *trajectory* or possible evolution of the phenomenon in $\mathbb{R}^N$ as the parameters vary.

The theory of stochastic processes is nowadays one of the richest and most fascinating fields of mathematics: we point out the excellent review article [97] which, with a wealth of insights, tells the story of research on stochastic processes from the middle of the last century onwards.

1.1 Stochastic Processes: Law and Finite-Dimensional Distributions

In this section, we give two *equivalent* definitions of stochastic process. The first definition is quite simple and intuitive; the second is more abstract but essential for the proof of some general results on stochastic processes. We also introduce some accessory notions: the *space of trajectories,* the *law* and the *finite-dimensional distributions.*

Let I be a generic non-empty set. Given $d \in \mathbb{N}$, let $m\mathscr{F}$ be the set of random variables with values in $\mathbb{R}^d$, defined on a probability space $(\Omega, \mathscr{F}, P)$. The concept of a stochastic process extends that of a function from I to $\mathbb{R}^d$, admitting that the values taken may be random: in other words, just as a function

$$f : I \longrightarrow \mathbb{R}^d$$

associates $t \in I$ with the dependent variable $f(t) \in \mathbb{R}^d$, similarly a stochastic process

$$X : I \longrightarrow m\mathscr{F}$$

associates $t \in I$ with the d-dimensional random variable $X_t \in m\mathscr{F}$.

Definition 1.1.1 (Stochastic Process) A stochastic process is a function with d-dimensional random values

$$X : I \longrightarrow m\mathscr{F}$$
$$t \longrightarrow X_t.$$

If $d = 1$ we say that X is a *real* stochastic process. If I is finite or countable then we say that X is a *discrete* stochastic process.

One can equivalently think of the stochastic process X as *an indexed family* $X = (X_t)_{t \in I}$ *of random variables*. To fix ideas, often the domain I will be a subset of $\mathbb{R}$ that represents a set of time indices; for example, if $I = \mathbb{N}$ then a process $(X_n)_{n \in \mathbb{N}}$ is simply a sequence of random variables.

More generally, a stochastic process X can be defined by assuming that X_t, for each $t \in I$, is a random variable with values in a generic measurable space $(E, \mathscr{E})$ instead of $\mathbb{R}^d$.

To give the second definition of a stochastic process, it is necessary to introduce some preliminary notations. We denote by

$$\mathbb{R}^I = \{x : I \longrightarrow \mathbb{R}\}$$

the family of functions from I to $\mathbb{R}$. For each $x \in \mathbb{R}^I$ and $t \in I$, we write x_t instead of $x(t)$ and say that x_t is the *t-th component* of x: in this way we interpret $\mathbb{R}^I$ as the Cartesian product of $\mathbb{R}$ for a number $|I|$ of times (even if I is not finite or countable). For example, if $I = \{1, \ldots, d\}$ then $\mathbb{R}^I$ is identifiable with $\mathbb{R}^d$, while if $I = \mathbb{N}$ then $\mathbb{R}^\mathbb{N}$ is the set of sequences $x = (x_1, x_2, \ldots)$ of real numbers. An element $x \in \mathbb{R}^I$ can be seen as a parameterized curve in $\mathbb{R}$, where I is the set of parameters.

We say that $\mathbb{R}^I$ is the *space of trajectories* from I to $\mathbb{R}$ and $x \in \mathbb{R}^I$ is a *real trajectory*. There is nothing special about considering real trajectories: we could directly consider $\mathbb{R}^d$ or even a generic measurable space $(E, \mathscr{E})$ instead of $\mathbb{R}$. In such a case, the space of trajectories is E^I, the set of functions from I with values

1.1 Stochastic Processes: Law and Finite-Dimensional Distributions

in E. However, at least for the moment, we restrict our attention to $E = \mathbb{R}$ which is involved in the study of one-dimensional (or real) stochastic processes.

We endow the space of trajectories with a measurable space structure. On $\mathbb{R}^I$ we introduce a σ-algebra that generalizes the *product σ-algebra* defined in Section 2.3.2 in [113]. We call a *finite-dimensional cylinder*, or simply *cylinder*, a subset of $\mathbb{R}^I$ of which a finite number of components are "fixed".

Definition 1.1.2 (Finite-Dimensional Cylinder) Given $t \in I$ and $H \in \mathscr{B}$, we say that the set

$$C_t(H) := \{x \in \mathbb{R}^I \mid x_t \in H\}$$

is a one-dimensional cylinder. Given $t_1, \ldots, t_n \in I$ distinct and $H_1, \ldots, H_n \in \mathscr{B}$, we set $H = H_1 \times \cdots \times H_n$ and say that

$$C_{t_1,\ldots,t_n}(H) := \{x \in \mathbb{R}^I \mid (x_{t_1}, \ldots, x_{t_n}) \in H\} = \bigcap_{i=1}^{n} C_{t_i}(H_i) \qquad (1.1.1)$$

is a finite-dimensional cylinder. We denote by $\mathscr{C}$ the family of finite-dimensional cylinders and

$$\mathscr{F}^I := \sigma(\mathscr{C})$$

the σ-algebra generated by such cylinders.

The σ-algebra $\mathscr{F}^I$ is a very abstract object and, at least for the moment, it is not important to try to visualize it concretely or to understand its structure in depth: some additional information about $\mathscr{F}^I$ will be provided in Remark 1.1.10. We introduced $\mathscr{F}^I$ in order to give the following alternative definition.

Definition 1.1.3 (Stochastic Process) A real stochastic process $X = (X_t)_{t \in I}$ on the probability space $(\Omega, \mathscr{F}, P)$ is a random variable with values in the space of trajectories $(\mathbb{R}^I, \mathscr{F}^I)$:

$$X : \Omega \longrightarrow \mathbb{R}^I.$$

Remark 1.1.4 The fact that X is a random variable means that the measurability condition holds

$$(X \in C) \in \mathscr{F} \text{ for every } C \in \mathscr{F}^I. \qquad (1.1.2)$$

In turn, condition (1.1.2) is equivalent[1] to the fact that

$$(X_t \in H) \in \mathscr{F} \text{ for every } H \in \mathscr{B}, \, t \in I, \qquad (1.1.3)$$

and therefore Definitions 1.1.1 and 1.1.3 are equivalent. In summary, one can also say that a real stochastic process X is a function

$$X : I \times \Omega \longrightarrow \mathbb{R}$$
$$(t, \omega) \longrightarrow X_t(\omega)$$

that

- associates to each $t \in I$ the random variable $\omega \mapsto X_t(\omega)$: this is the standpoint of Definition 1.1.1;
- associates to each $\omega \in \Omega$ the trajectory $t \mapsto X_t(\omega)$: this is the standpoint of Definition 1.1.3. Note that each outcome $\omega \in \Omega$ corresponds to (and can be identified with) a trajectory of the process.

Example 1.1.5 Every function $f : I \longrightarrow \mathbb{R}$ can be seen as a stochastic process interpreting, for each fixed $t \in I$, $f(t)$ as a constant random variable. In other words, if $\Omega = \{\omega\}$ is a sample space consisting of a single element, the process defined by $X_t(\omega) = f(t)$ has only one trajectory which is the function f. The measurability condition (1.1.3) is obvious since $\mathscr{F} = \{\emptyset, \Omega\}$. In this sense, the concept of a stochastic process generalizes that of a function because it allows the existence of multiple trajectories.

From the standpoint of Definition 1.1.3 a stochastic process is a random variable and therefore we can define its law.

Definition 1.1.6 (Law) The distribution (or law) of the stochastic process X is the probability measure on $(\mathbb{R}^I, \mathscr{F}^I)$ defined by

$$\mu_X(C) = P(X \in C), \qquad C \in \mathscr{F}^I.$$

Remark 1.1.7 (Finite-Dimensional Distributions) Even the concept of law of a stochastic process is abstract and not very convenient: from an operational perspective, a much more effective tool are the so-called *finite-dimensional distributions*

---

[1] Indeed, $(X_t \in H) = (X \in C)$ where C is the one-dimensional cylinder (i.e., in which only one component is fixed) defined by $\{x \in \mathbb{R}^I \mid x_t \in H\}$: so it is clear that if X is a stochastic process then $X_t \in m\mathscr{F}$ for every $t \in I$. Conversely, the family

$$\mathscr{H} := \{C \in \mathscr{F}^I \mid X^{-1}(C) \in \mathscr{F}\}$$

is a σ-algebra that, by hypothesis, includes one-dimensional cylinders and therefore also $\mathscr{C}$ (cylinders are finite intersections of one-dimensional cylinders). Then $\mathscr{H} \supseteq \sigma(\mathscr{C}) = \mathscr{F}^I$.

1.1 Stochastic Processes: Law and Finite-Dimensional Distributions

which are the distributions $\mu_{(X_{t_1},...,X_{t_n})}$ of the random vectors $(X_{t_1},\ldots,X_{t_n})$ as the choice of a finite number of indices $t_1,\ldots,t_n \in I$ varies. The law of a process is *uniquely determined by the finite-dimensional distributions*: in other words, it is equivalent to knowing the law or the finite-dimensional distributions of a stochastic process.[2]

The *one-dimensional* distributions are not sufficient to identify the law of a process. This is clear when I is finite and therefore the process is simply a random vector: in fact, the one-dimensional distributions are the marginal laws of the vector which obviously do not identify the joint law. Another interesting example is given in Remark 4.1.5.

Example 1.1.8 Let $A, B \sim \mathcal{N}_{0,1}$ be independent random variables. Consider the stochastic process $X = (X_t)_{t \in \mathbb{R}}$ defined by

$$X_t = At + B, \qquad t \in \mathbb{R}.$$

Each trajectory of X is a linear function (a straight line) on $\mathbb{R}$. It is not obvious to specify the distribution of this process but it is easy to calculate the finite-dimensional distributions, in fact given $t_1,\ldots,t_n \in \mathbb{R}$ we have

$$\begin{pmatrix} X_{t_1} \\ \vdots \\ X_{t_n} \end{pmatrix} = \alpha \begin{pmatrix} A \\ B \end{pmatrix}, \qquad \alpha = \begin{pmatrix} t_1 & 1 \\ \vdots & \vdots \\ t_n & 1 \end{pmatrix}$$

and therefore, by Proposition 2.5.15 in [113], $(X_{t_1},\ldots,X_{t_n}) \sim \mathcal{N}_{0,\alpha\alpha^*}$.

Example 1.1.9 (Gaussian Process) We say that a stochastic process is Gaussian if it has normal finite-dimensional distributions. If $X = (X_t)_{t \in I}$ is Gaussian, consider the mean and covariance functions

$$m(t) := E[X_t], \qquad c(s,t) := \mathrm{cov}(X_s, X_t), \qquad s, t \in I.$$

---

[2] The measure of a generic cylinder $C_{t_1,\ldots,t_n}(H)$ is expressed as

$$\mu_X\left(C_{t_1,\ldots,t_n}(H)\right) = \mu_{(X_{t_1},\ldots,X_{t_n})}(H)$$

and therefore the finite-dimensional distributions identify μ_X on $\mathscr{C}$. On the other hand, $\mathscr{C}$ is a ∩-closed family and generates $\mathscr{F}^I$: by Corollary I-cc2 in [113] if two probability measures on $(\mathbb{R}^I, \mathscr{F}^I)$ coincide on $\mathscr{C}$ then they are equal. In other words, if $\mu_1(C) = \mu_2(C)$ for each $C \in \mathscr{C}$ then $\mu_1 \equiv \mu_2$. We will see that, thanks to Carathéodory's theorem, *a probability measure extends uniquely from $\mathscr{C}$ to $\mathscr{F}^I$*: this is the content of one of the first fundamental results on stochastic processes, Kolmogorov's extension theorem, which we will examine in Sect. 1.3.

These functions determine the finite-dimensional distributions (and therefore also the law!) of the process because, for each choice $t_1, \ldots, t_n \in I$, we have

$$(X_{t_1}, \ldots, X_{t_n}) \sim \mathcal{N}_{M,C}$$

where

$$M = (m(t_1), \ldots, m(t_n)) \quad \text{and} \quad C = \big(c(t_i, t_j)\big)_{i,j=1,\ldots,n}. \tag{1.1.4}$$

We observe that $C = \big(c(t_i, t_j)\big)_{i,j=1,\ldots,n}$ is a symmetric and positive semi-definite matrix. Obviously, if I is finite then X is nothing but a random vector with multinormal distribution. The process of Example 1.1.8 is Gaussian with zero mean and covariance function $c(s,t) = st + 1$. The trivial process of Example 1.1.5 is also Gaussian with mean function $f(t)$ and identically zero covariance function: in this case, $X_t \sim \delta_{f(t)}$ for every $t \in I$. Finally, a fundamental example of Gaussian process is the Brownian motion that we will define in Chap. 4.

Remark 1.1.10 ([!]) There are families of trajectories, even very significant ones, that do not belong to the σ-algebra $\mathscr{F}^I$. The idea is that *every element of $\mathscr{F}^I$ is characterized by a countably number of coordinates*[3] and this is highly restrictive when I is uncountable. For example, if $I = [0, 1]$ we have

$$C[0,1] \notin \mathscr{F}^{[0,1]}$$

since the family $C[0,1]$ of continuous functions cannot be characterized, in the space of all functions from $[0,1]$ to $\mathbb{R}$, by imposing conditions on a countable number of coordinates.[4] For the same reason, even the singletons $\{x\}$ with $x \in \mathbb{R}^{[0,1]}$, the subsets of $\mathbb{R}^{[0,1]}$ with a finite number of elements, and other significant

---

[3] More precisely, let us solve Exercise 1.4 in [9]: consider $I = [0, 1]$ (thus the space of trajectories $\mathbb{R}^I$ is the family of functions from $[0, 1]$ to $\mathbb{R}$). Given a sequence $\tau = (t_n)_{n \geq 1} \in [0, 1]^{\mathbb{N}}$, we identify τ with the map

$$\tau : \mathbb{R}^{[0,1]} \longrightarrow \mathbb{R}^{\mathbb{N}}, \qquad \tau(x) := (x_{t_n})_{n \geq 1},$$

and put

$$\mathscr{M} = \{\tau^{-1}(H) \mid \tau \in [0,1]^{\mathbb{N}}, H \in \mathscr{F}^{\mathbb{N}}\}, \qquad \tau^{-1}(H) = \{x \in \mathbb{R}^{[0,1]} \mid \tau(x) \in H\},$$

where $\mathscr{F}^{\mathbb{N}}$ denotes the σ-algebra generated by cylinders in $\mathbb{R}^{\mathbb{N}}$. Then $\mathscr{M} \subseteq \mathscr{F}^{[0,1]}$ and contains the family of finite-dimensional cylinders of $\mathbb{R}^{[0,1]}$, which is a $\cap$-closed family that generates $\mathscr{F}^{[0,1]}$. Moreover, one proves that $\mathscr{M}$ is a monotone family: it follows from Corollary A.0.4 in [113] that $\mathscr{M} = \mathscr{F}^{[0,1]}$ i.e., every element $C \in \mathscr{F}^{[0,1]}$ is of the form $C = \tau^{-1}(H)$ for some sequence τ in $[0, 1]$ and some $H \in \mathscr{F}^{\mathbb{N}}$. In other words, C is characterized by the choice of a *countable number of coordinates* $\tau = (t_n)_{n \geq 1}$ (as well as by $H \in \mathscr{F}^{\mathbb{N}}$).

[4] By contradiction, if $C[0,1] = \tau^{-1}(H)$, for some sequence of coordinates $\tau = (t_n)_{n \geq 1}$ in $[0, 1]$ and $H \in \mathscr{F}^{\mathbb{N}}$, then modifying $x \in C[0, 1]$ at a point $t \notin \tau$ should still result in a continuous function and this is clearly false.

families such as for example

$$\{x \in \mathbb{R}^{[0,1]} \mid \sup_{t \in [0,1]} x_t < 1\}$$

do not belong to $\mathscr{F}^{[0,1]}$.

These examples may raise strong perplexity towards the σ-algebra $\mathscr{F}^I$ which is not wide enough to contain important families of trajectories like those just considered. Actually, the problem is that the sample space $\mathbb{R}^I$, of *all* the functions from I to $\mathbb{R}$, is so large as to be hardly tractable as a measurable space, thus making it difficult to develop a general theory of stochastic processes. For this reason, as soon as possible we will replace $\mathbb{R}^I$ with a state space that, in addition to being "smaller", also possesses a useful metric space structure: this is the case of the space of continuous trajectories that we will examine in Sect. 3.2.

1.1.1 Measurable Processes

We have given two equivalent definitions of stochastic process, each with its own advantages and disadvantages:

(i) **a stochastic process is a function with random values (Definition 1.1.1)**

$$X : I \longrightarrow m\mathscr{F}$$

that associates to each $t \in I$ the random variable X_t defined on the probability space $(\Omega, \mathscr{F}, P)$;

(ii) **a stochastic process is a random variable with values in a space of trajectories (Definition 1.1.3)**: according to this much more abstract definition, a process $X = X(\omega)$ is a random variable

$$X : \Omega \longrightarrow \mathbb{R}^I$$

from the probability space $(\Omega, \mathscr{F}, P)$ to the space of trajectories $\mathbb{R}^I$, equipped with the structure of a measurable space with the σ-algebra $\mathscr{F}^I$. This definition is used in the proof of the most general and theoretical results even if it is a less operational notion and more difficult to apply to the study of concrete examples.

Note that the previous definitions do not require any assumptions about the type of dependence of X with respect to the variable t (for example, measurability or some kind of regularity). Obviously, the problem does not arise if I is a generic set, devoid of any measurable or metric space structure; however, if I is a real interval then it is possible to endow the product space $I \times \Omega$ with a structure of measurable space with the product σ-algebra $\mathscr{B} \otimes \mathscr{F}$.

Definition 1.1.11 (Measurable Process) A measurable stochastic process is a measurable function

$$X : (I \times \Omega, \mathscr{B} \otimes \mathscr{F}) \longrightarrow (\mathbb{R}, \mathscr{B}).$$

By Lemma 2.3.11 in [113], if X is a measurable stochastic process then:

- X_t is a random variable for each $t \in I$;
- the trajectory $t \mapsto X_t(\omega)$ is a Borel measurable function from I to $\mathbb{R}$, for each $\omega \in \Omega$.

If $I \subseteq \mathbb{R}$ it is natural to interpret $t \in I$ as a *time index*: then, as we will see in Sect. 1.4, the space of probability will be enriched with new elements (filtrations) and a predominant role will be assumed by a particular class of stochastic processes, called *martingales*. In that context, we will strengthen the notion of measurability by introducing the concept of *progressively measurable process* (cf. Definition 6.2.27).

The term "General Theory of Stochastic Processes" is usually referred in the literature to the field that deals with the study of the general properties of processes when $I = \mathbb{R}_{\geq 0}$: for a concise introduction see, for example, Chapter 16 in [9] and Chapter 1 in [65].

1.2 Uniqueness

There are various notions of equivalence between stochastic processes. First of all, two processes $X = (X_t)_{t \in I}$ and $Y = (Y_t)_{t \in I}$ are *equal in law* if they have the same distribution (or, equivalently, if they have the same finite-dimensional distributions): in this case X and Y could even be defined on different probability spaces. When X and Y are defined on the same probability space $(\Omega, \mathscr{F}, P)$, we can provide other notions of equivalence expressed in terms of equality of trajectories. We first recall that, in a probability space $(\Omega, \mathscr{F}, P)$, a *subset A of Ω is almost sure* (with respect to P) if there exists an event $C \subseteq A$ such that $P(C) = 1$. If the probability space is *complete*[5] then every almost sure set A is an event and therefore we can simply write $P(A) = 1$.

Definition 1.2.1 (Modifications) Let $X = (X_t)_{t \in I}$ and $Y = (Y_t)_{t \in I}$ be stochastic processes on $(\Omega, \mathscr{F}, P)$. We say that X and Y are *modifications* if $P(X_t = Y_t) = 1$ for every $t \in I$.

Remark 1.2.2 The previous definition can be easily generalized to the case of X, Y *generic functions* from Ω to values in $\mathbb{R}^I$: in this case $(X_t = Y_t)$ is not necessarily

---

[5] We recall the definition given in Remark 2.1.11 in [113]: a probability space $(\Omega, \mathscr{F}, P)$ is complete if $\mathscr{N} \subseteq \mathscr{F}$ where $\mathscr{N}$ denotes the family of negligible sets (cf. Definition 1.1.16 in [113]).

1.2 Uniqueness

an event and therefore we say that X is a *modification* of Y if the set $(X_t = Y_t)$ is almost sure. This can be useful if it is not known a priori that X and/or Y are stochastic processes.

Definition 1.2.3 (Indistinguishable Processes) Let $X = (X_t)_{t \in I}$ and $Y = (Y_t)_{t \in I}$ be stochastic processes on $(\Omega, \mathscr{F}, P)$. We say that X and Y are *indistinguishable* if the set

$$(X = Y) := \{\omega \in \Omega \mid X_t(\omega) = Y_t(\omega) \text{ for every } t \in I\}$$

is almost sure.

Remark 1.2.4 ([!]) Two processes X and Y are indistinguishable if they have almost all the same trajectories. Even if X and Y are stochastic processes, it is not necessarily true that $(X = Y)$ is an event. In fact, $(X = Y) = (X - Y)^{-1}(\{\mathbf{0}\})$ where $\mathbf{0}$ denotes the identically zero trajectory: however, $\{\mathbf{0}\} \notin \mathscr{F}^I$ unless I is finite or countable (cf. Remark 1.1.10).

On the other hand, *if the space $(\Omega, \mathscr{F}, P)$ is complete* then X and Y are indistinguishable if and only if $P(X = Y) = 1$ since the completeness of the space guarantees that $(X = Y) \in \mathscr{F}$ in the case $(X = Y)$ is almost sure. For this and other reasons that we will explain later, from now on we will often assume that $(\Omega, \mathscr{F}, P)$ is complete.

Remark 1.2.5 ([!]) If X and Y are modifications then they have the same finite-dimensional distributions and therefore are equal in law. If X and Y are indistinguishable then they are also modifications since for every $t \in I$ we have $(X = Y) \subseteq (X_t = Y_t)$. Conversely, if X and Y are modifications then they are not necessarily indistinguishable: indeed,

$$(X = Y) = \bigcap_{t \in I}(X_t = Y_t)$$

but if I is uncountable, such intersection might not belong to $\mathscr{F}$ or have probability less than one. If I is finite or countable then X, Y are modifications if and only if they are indistinguishable.

Let us give an explicit example of processes that are modifications but are not indistinguishable.

Example 1.2.6 ([!]) Consider the sample space $\Omega = [0, 1]$ with the Lebesgue measure as the probability measure. Let $I = [0, 1]$, $X = (X_t)_{t \in I}$ be the identically zero process and $Y = (Y_t)_{t \in I}$ be the process defined by

$$Y_t(\omega) = \begin{cases} 1 & \text{if } \omega = t, \\ 0 & \text{if } \omega \in [0, 1] \setminus \{t\}. \end{cases}$$

Then X and Y are modifications since, for every $t \in I$,

$$(X_t = Y_t) = \{\omega \in \Omega \mid \omega \neq t\} = [0,1] \setminus \{t\}$$

has Lebesgue measure equal to one, i.e., it is an almost sure event. On the other hand, *all* the trajectories of X are different from those of Y at one point.

We also note that X and Y are equal in law, but X has all continuous trajectories and Y has all discontinuous trajectories: therefore, *there are important properties of the trajectories of a stochastic process (such as, for example, continuity) that do not depend on the distribution of the process.*

In the case of continuous processes, we have the following particular result.

Proposition 1.2.7 *Let I be a real interval and let $X = (X_t)_{t \in I}$ and $Y = (Y_t)_{t \in I}$ be processes with a.s. continuous trajectories.[6] If X is a modification of Y, then X, Y are indistinguishable.*

Proof By assumption, the trajectories $X(\omega)$ and $Y(\omega)$ are continuous for every $\omega \in A$ with A almost sure. Moreover, $P(X_t = Y_t) = 1$ for every $t \in I$ and consequently the set

$$C := A \cap \bigcap_{t \in I \cap \mathbb{Q}} (X_t = Y_t)$$

is almost sure. For every $t \in I$, there exists an approximating sequence $(t_n)_{n \in \mathbb{N}}$ in $I \cap \mathbb{Q}$: by the continuity hypothesis, for every $\omega \in C$ we have

$$X_t(\omega) = \lim_{n \to \infty} X_{t_n}(\omega) = \lim_{n \to \infty} Y_{t_n}(\omega) = Y_t(\omega)$$

and this proves that X, Y are indistinguishable. □

Remark 1.2.8 The result of Proposition 1.2.7 remains valid for processes that are only continuous from the right or from the left.

1.3 Existence

In this section, we show that *it is "always" possible to construct a stochastic process with assigned finite-dimensional distributions.*

---

[6] The set of $\omega \in \Omega$ such that $t \mapsto X_t(\omega)$ and $t \mapsto Y_t(\omega)$ are continuous functions is almost sure.

1.3 Existence

Let us make a preliminary remark: if $\mu_{t_1,\ldots,t_n}$ are the finite-dimensional distributions of a real stochastic process $(X_t)_{t \in I}$, then we have

$$\mu_{t_1,\ldots,t_n}(H_1 \times \cdots \times H_n)$$
$$= P\left((X_{t_1} \in H_1) \cap \cdots \cap (X_{t_n} \in H_n)\right), \qquad t_1,\ldots,t_n \in I, \; H_1,\ldots,H_n \in \mathscr{B}. \tag{1.3.1}$$

As a consequence, the following *consistency properties* hold: for every finite family of indices $t_1, \ldots, t_n \in I$, for every $H_1, \ldots, H_n \in \mathscr{B}$ and for every permutation ν of the indices $1, 2, \ldots, n$, we have

$$\mu_{t_1,\ldots,t_n}(H_1 \times \cdots \times H_n) = \mu_{t_{\nu(1)},\ldots,t_{\nu(n)}}(H_{\nu(1)} \times \cdots \times H_{\nu(n)}), \tag{1.3.2}$$

$$\mu_{t_1,\ldots,t_n}(H_1 \times \cdots \times H_{n-1} \times \mathbb{R}) = \mu_{t_1,\ldots,t_{n-1}}(H_1 \times \cdots \times H_{n-1}). \tag{1.3.3}$$

A posteriori, it is clear that (1.3.2) and (1.3.3) are *necessary* conditions for the distributions $\mu_{t_1,\ldots,t_n}$ to be the finite-dimensional distributions of a stochastic process. The following result shows that these conditions are also sufficient.

Theorem 1.3.1 (Kolmogorov's Extension Theorem [!!!]) *Let I be a non-empty set. Suppose that, for each finite family of indices $t_1, \ldots, t_n \in I$, a distribution $\mu_{t_1,\ldots,t_n}$ on $\mathbb{R}^n$ is given, and the consistency properties (1.3.2) and (1.3.3) are satisfied. Then there exists a unique probability measure μ on $\left(\mathbb{R}^I, \mathscr{F}^I\right)$ that has $\mu_{t_1,\ldots,t_n}$ as finite-dimensional distributions, i.e., such that*

$$\mu(C_{t_1,\ldots,t_n}(H)) = \mu_{t_1,\ldots,t_n}(H) \tag{1.3.4}$$

for each finite family of indices $t_1, \ldots, t_n \in I$ and $H = H_1 \times \cdots \times H_n \in \mathscr{B}_n$.

Remark 1.3.2 ([!]) Under the hypotheses of the previous theorem, the measure μ extends further to a σ-algebra $\mathscr{F}^I_\mu$ that contains $\mathscr{F}^I$ and such that the probability space $(\mathbb{R}^I, \mathscr{F}^I_\mu, \mu)$ is *complete*: this is a consequence of Corollary 1.5.11 in [113] and the constructive method used in the proof of Carathéodory's theorem. Sometimes, $\mathscr{F}^I_\mu$ is called the μ-completion of $\mathscr{F}^I$.

We postpone the proof of Theorem 1.3.1 to Sect. 1.5 and now examine some remarkable applications.

Corollary 1.3.3 (Existence of Processes with Assigned Finite-Dimensional Distributions [!]) *Let I be a non-empty set. Suppose that, for each finite family of indices $t_1, \ldots, t_n \in I$, a distribution $\mu_{t_1,\ldots,t_n}$ on $\mathbb{R}^n$ is given, and the consistency properties (1.3.2) and (1.3.3) are satisfied. Then there exists a stochastic process $X = (X_t)_{t \in I}$ that is defined on a complete probability space and has $\mu_{t_1,\ldots,t_n}$ as finite-dimensional distributions.*

Proof Proceed in a similar way to the case of real random variables (cf. Remark 2.1.17 in [113]). Let $(\Omega, \mathscr{F}, P) = (\mathbb{R}^I, \mathscr{F}^I_\mu, \mu)$ be the complete probability space defined in Remark 1.3.2. The identity function

$$X : (\mathbb{R}^I, \mathscr{F}_\mu^I) \longrightarrow (\mathbb{R}^I, \mathscr{F}^I)$$

defined by $X(w) = w$ for each $w \in \mathbb{R}^I$, is a stochastic process since $X^{-1}(\mathscr{F}^I) = \mathscr{F}^I \subseteq \mathscr{F}_\mu^I$. Moreover, X has $\mu_{t_1,\ldots,t_n}$ as finite-dimensional distributions since, for each finite-dimensional cylinder $C_{t_1,\ldots,t_n}(H)$ as in (1.1.1), we have

$$\mu_X(C_{t_1,\ldots,t_n}(H)) = \mu(X \in C_{t_1,\ldots,t_n}(H)) =$$

(since X is the identity function)

$$= \mu(C_{t_1,\ldots,t_n}(H)) =$$

(by (1.3.4))

$$= \mu_{t_1,\ldots,t_n}(H).$$

$\square$

Now consider a stochastic process X on the space $(\Omega, \mathscr{F}, P)$. Denote by μ_X the law of X and by $\mathscr{F}_{\mu_X}^I$ the μ_X-completion of $\mathscr{F}^I$ (cf. Remark 1.3.2).

Definition 1.3.4 (Canonical Version of a Stochastic Process [!]) The canonical version (or realization) of a process X is the process $\mathbf{X}$, on the probability space $(\mathbb{R}^I, \mathscr{F}_{\mu_X}^I, \mu_X)$, defined by $\mathbf{X}(w) = w$ for each $w \in \mathbb{R}^I$.

Remark 1.3.5 By Corollary 1.3.3, X and its canonical realization $\mathbf{X}$ are equal in law. Moreover, $\mathbf{X}$ is defined on the complete probability space $(\mathbb{R}^I, \mathscr{F}_{\mu_X}^I, \mu_X)$ in which the sample space is $\mathbb{R}^I$ and *the outcomes are the trajectories of the process*.

Corollary 1.3.6 (Existence of Gaussian Processes [!]) *Let*

$$m : I \longrightarrow \mathbb{R}, \qquad c : I \times I \longrightarrow \mathbb{R}$$

be functions such that, for every finite family of indices $t_1, \ldots, t_n \in I$, the matrix $C = \big(c(t_i, t_j)\big)_{i,j=1,\ldots,n}$ is symmetric and positive semi-definite. Then there exists a Gaussian process, defined on a complete probability space $(\Omega, \mathscr{F}, P)$, with mean function m and covariance function c.

In particular, choosing $I = \mathbb{R}_{\geq 0}$, there exists a Gaussian process with mean function $m \equiv 0$ and covariance function $c(s, t) = t \wedge s \equiv \min\{s, t\}$.

Proof The family of distributions $\mathcal{N}_{M,C}$, with M, C as in (1.1.4), is well defined thanks to the hypothesis on the covariance function c. Moreover, it satisfies the consistency properties (1.3.2) and (1.3.3), as can be verified by applying (1.3.1)

1.3 Existence

with $\mathcal{N}_{M,C}$ instead of $\mu_{t_1,\dots,t_n}$ and $(X_{t_1},\dots,X_{t_n}) \sim \mathcal{N}_{M,C}$. Then the first part of the thesis follows from Corollary 1.3.3.

Now let $t_1,\dots,t_n \in \mathbb{R}_{\geq 0}$: the matrix $C = \big(\min\{t_i,t_j\}\big)_{i,j=1,\dots,n}$ is obviously symmetric and is also positive semi-definite since, for every $\eta_1,\dots,\eta_n \in \mathbb{R}$, we have

$$\sum_{i,j=1}^n \eta_i \eta_j \min\{t_i,t_j\} = \sum_{i,j=1}^n \eta_i \eta_j \int_0^\infty \mathbb{1}_{[0,t_i]}(s) \mathbb{1}_{[0,t_j]}(s)\,ds$$

$$= \int_0^\infty \left(\sum_{i=1}^n \eta_i \mathbb{1}_{[0,t_i]}(s)\right)^2 ds \geq 0.$$

□

Corollary 1.3.7 (Existence of Independent Sequences of Random Variables [!]) *Let $(\mu_n)_{n\in\mathbb{N}}$ be a sequence of real distributions. There exists a sequence $(X_n)_{n\in\mathbb{N}}$ of independent random variables defined on a complete probability space $(\Omega, \mathcal{F}, P)$, such that $X_n \sim \mu_n$ for every $n \in \mathbb{N}$.*

Proof Apply Corollary 1.3.3 with $I = \mathbb{N}$. The family of finite-dimensional distributions defined by

$$\mu_{k_1,\dots,k_n} := \mu_{k_1} \otimes \cdots \otimes \mu_{k_n}, \qquad k_1,\dots,k_n \in \mathbb{N},$$

verifies the consistency properties (1.3.2)–(1.3.3). By Corollary 1.3.3, there exists a process $(X_k)_{k\in\mathbb{N}}$ that has $\mu_{k_1,\dots,k_n}$ as finite-dimensional distributions. Independence follows from Theorem 2.3.25 in [113] and the arbitrariness of the choice of indices $k_1,\dots,k_n \in \mathbb{N}$. □

Corollary 1.3.7 admits the following slightly more general version, whose proof is left as an exercise. The following result requires a simplified version, compared to Corollary 1.3.3, of the consistency property.

Corollary 1.3.8 (Existence of Sequences of Random Variables with Assigned Distribution [!]) *Let a sequence $(\mu_n)_{n\in\mathbb{N}}$ be given, where μ_n is a distribution on $\mathbb{R}^n$ and*

$$\mu_{n+1}(H \times \mathbb{R}) = \mu_n(H), \qquad H \in \mathcal{B}_n,\ n \in \mathbb{N}.$$

Then there exists a sequence $(X_n)_{n\in\mathbb{N}}$ of random variables defined on a complete probability space $(\Omega, \mathcal{F}, P)$, such that $(X_1,\dots,X_n) \sim \mu_n$ for every $n \in \mathbb{N}$.

1.4 Filtrations and Martingales

In this section, we consider the particular case where I is a subset of $\mathbb{R}$, typically

$$I = \mathbb{R}_{\geq 0} \quad \text{or} \quad I = [0,1] \quad \text{or} \quad I = \mathbb{N}.$$

In this case, it is useful to think of t as a parameter denoting a point in time.

Definition 1.4.1 (Filtration) Let $I \subseteq \mathbb{R}$ and $(\Omega, \mathscr{F}, P)$ be a probability space. A filtration $(\mathscr{F}_t)_{t \in I}$ is an increasing family of sub-σ-algebras of $\mathscr{F}$, in the sense that

$$\mathscr{F}_s \subseteq \mathscr{F}_t \subseteq \mathscr{F}, \qquad s,t \in I, \ s \leq t.$$

In many applications, a σ-algebra represents a set of information; as for filtrations, the idea is that

- the σ-algebra $\mathscr{F}_t$ represents *the information available at time t*;
- the filtration $(\mathscr{F}_t)_{t \in I}$ represents *the flow of information that increases over time*.

The concept of information is crucial in probability theory: for example, the very definition of conditional probability is essentially motivated by the problem of describing the effect of information on the probability of events. Filtrations constitute the mathematical tool that dynamically describes (as a function of time) the available information and for this reason play a fundamental role in the theory of stochastic processes. The following definition formalizes the idea that a stochastic process is observable based on the information of some filtration.

Definition 1.4.2 (Adapted Process) Let $X = (X_t)_{t \in I}$ be a stochastic process on the space $(\Omega, \mathscr{F}, P)$. We say that X is *adapted* to the filtration $(\mathscr{F}_t)_{t \in I}$ if $X_t \in m\mathscr{F}_t$ for every $t \in I$.

Definition 1.4.3 (Filtration Generated by a Process) Let $X = (X_t)_{t \in I}$ be a stochastic process on the space $(\Omega, \mathscr{F}, P)$. The *filtration generated by* X, denoted by $\mathscr{G}^X = (\mathscr{G}^X_t)_{t \in I}$, is defined as

$$\mathscr{G}^X_t := \sigma(X_s, \ s \leq t) \equiv \sigma(X_s^{-1}(H), \ s \leq t, \ H \in \mathscr{B}), \qquad t \in I. \tag{1.4.1}$$

Remark 1.4.4 We use the notation $\mathscr{G}^X$ for the filtration generated by X because we want to reserve the symbol $\mathscr{F}^X$ for another filtration that we will define later in Sect. 6.2.2 and call *standard filtration for X*. The filtration generated by X is the "smallest" filtration that includes information about the process X: clearly, X is adapted to $(\mathscr{F}_t)_{t \in I}$ if and only if $\mathscr{G}^X_t \subseteq \mathscr{F}_t$ for every $t \in I$.

Remark 1.4.5 If $\mathbf{X}$ is the canonical version of X (cf. Definition 1.3.4), then

$$\mathscr{G}^\mathbf{X}_t = \sigma(C_s(H) \mid s \in I, \ s \leq t, \ H \in \mathscr{B}), \qquad t \in I,$$

that is, the filtration generated by $\mathbf{X}$ is the one generated by cylinders.

1.4 Filtrations and Martingales

We now introduce a fundamental class of stochastic processes.

Definition 1.4.6 (Martingale [!!!]) Let $X = (X_t)_{t \in I}$, with $I \subseteq \mathbb{R}$, be a stochastic process on the filtered space $(\Omega, \mathscr{F}, P, \mathscr{F}_t)$. We say that X is a *martingale* if:

(i) X is an *absolutely integrable process*, i.e. $X_t \in L^1(\Omega, P)$ for every $t \in I$;
(ii) we have

$$X_t = E[X_T \mid \mathscr{F}_t], \qquad t, T \in I, \, t \leq T. \tag{1.4.2}$$

If I is finite or countable, we say that X is a *discrete martingale*.

The concept of martingale is central to the theory of stochastic processes and in many applications. Equation (1.4.2), called the *martingale property*, means that the current value (at time t) of the process is the best estimate of the future value (at time $T \geq t$) given the currently available information. In economics, for example, the martingale property translates into the fact that if X represents the price of a good, then such price is *fair* in the sense that it is the best estimate of the future value of the good based on the information available at the moment.

Let X be a martingale on the filtered space $(\Omega, \mathscr{F}, P, \mathscr{F}_t)$. As an immediate consequence of Definition 1.4.6 and the properties of conditional expectation, we have:

(i) X is *adapted to* $(\mathscr{F}_t)_{t \in I}$;
(ii) X has *constant expectation* since, applying the expected value to both sides of (1.4.2) we get[7]

$$E[X_t] = E[X_T], \qquad t, T \in I.$$

Remark 1.4.7 The term *martingale* originally referred to a series of strategies used by French gamblers in the 18th century, including the doubling strategy we mentioned in Example 3.2.4 in [113]. The interesting monograph [94] illustrates the history of the concept of martingale through the contribution of many famous historians and mathematicians.

Example 1.4.8 ([!]) The sequence over time of wins and losses in a *fair* gambling game can be represented by a discrete martingale: sometimes we win and sometimes we lose, but if the game is fair, wins and losses balance each other on average.

More precisely, let $(Z_n)_{n \in \mathbb{N}}$ be a sequence of i.i.d. random variables with $Z_n \sim q\delta_1 + (1-q)\delta_{-1}$ and $0 < q < 1$ fixed. Consider the stochastic process

$$X_n := Z_1 + \cdots + Z_n, \qquad n \in \mathbb{N}.$$

---

[7] We recall that $E[E[X_T \mid \mathscr{F}_t]] = E[X_T]$ by definition of conditional expectation.

Here Z_n represents the win or loss at the n-th play, q is the probability of winning, and X_n is the balance after n plays. Consider the filtration $(\mathscr{G}_n^Z)_{n\in\mathbb{N}}$ of information on the outcomes of the plays, $\mathscr{G}_n^Z = \sigma(Z_1, \ldots, Z_n)$. Then we have

$$E\left[X_{n+1} \mid \mathscr{G}_n^Z\right] = E\left[X_n + Z_{n+1} \mid \mathscr{G}_n^Z\right] =$$

(since $X_n \in m\mathscr{G}_n^Z$ and Z_{n+1} is independent of $\mathscr{G}_n^Z$)

$$= X_n + E[Z_{n+1}] = X_n + 2q - 1.$$

So (X_n) is a martingale if $q = \frac{1}{2}$, that is, if the game is fair. If $q \geq \frac{1}{2}$, that is, if the probability of winning a single bet is greater than or equal to the probability of losing, then $X_n \leq E\left[X_{n+1} \mid \mathscr{G}_n^Z\right]$ (and we say that (X_n) is a *sub-martingale*): in this case, we also have $E[X_n] \leq E[X_{n+1}]$, that is, the process is *increasing on average*.

This example shows that the martingale property *is not a property of the trajectories of the process but depends on the probability measure and the filtration considered.*

Example 1.4.9 Let $X \in L^1(\Omega, P)$ and $(\mathscr{F}_t)_{t \in I}$ be a filtration on $(\Omega, \mathscr{F}, P)$. A simple application of the tower property shows that the process defined by $X_t = E[X \mid \mathscr{F}_t]$, $t \in I$, is a martingale, in fact we have

$$E[X_T \mid \mathscr{F}_t] = E[E[X \mid \mathscr{F}_T] \mid \mathscr{F}_t] = E[X \mid \mathscr{F}_t] = X_t, \qquad t, T \in I, \; t \leq T.$$

Remark 1.4.10 ([!]) We will often use the following remarkable identity, valid for a real-valued square-integrable martingale X, i.e. X such that $E\left[X_t^2\right] < \infty$ for $t \in I$:

$$E\left[(X_t - X_s)^2 \mid \mathscr{F}_s\right] = E\left[X_t^2 - X_s^2 \mid \mathscr{F}_s\right], \qquad s \leq t. \tag{1.4.3}$$

It is enough to observe that

$$E\left[(X_t - X_s)^2 \mid \mathscr{F}_s\right] = E\left[X_t^2 - 2X_t X_s + X_s^2 \mid \mathscr{F}_s\right]$$

$$= E\left[X_t^2 \mid \mathscr{F}_s\right] - 2X_s E[X_t \mid \mathscr{F}_s] + X_s^2 =$$

(by the martingale property)

$$= E\left[X_t^2 \mid \mathscr{F}_s\right] - X_s^2$$

from which (1.4.3) follows.

1.4 Filtrations and Martingales

Definition 1.4.11 Let $X = (X_t)_{t \in I}$ be a stochastic process on the filtered space $(\Omega, \mathscr{F}, P, \mathscr{F}_t)$. We say that X is a *sub-martingale* if:

(i) X is an *absolutely integrable* and *adapted* process to $(\mathscr{F}_t)_{t \in I}$;
(ii) we have

$$X_t \leq E[X_T \mid \mathscr{F}_t], \qquad t, T \in I, \ t \leq T.$$

Furthermore, X is a *super-martingale* if $-X$ is a *sub-martingale*.

Proposition 1.4.12 ([!]) *If X is a martingale and $\varphi : \mathbb{R} \longrightarrow \mathbb{R}$ is a convex function such that $\varphi(X_t) \in L^1(\Omega, P)$ for every $t \in I$, then $\varphi(X)$ is a sub-martingale.*

If X is a sub-martingale and $\varphi : \mathbb{R} \longrightarrow \mathbb{R}$ is a convex, increasing function such that $\varphi(X_t) \in L^1(\Omega, P)$ for every $t \in I$, then $\varphi(X)$ is a sub-martingale.

Proof The first part is an immediate consequence of Jensen's inequality. Similarly, if X is a sub-martingale then $X_t \leq E[X_T \mid \mathscr{F}_t]$ for $t \leq T$ and since φ is increasing, we also have

$$\varphi(X_t) \leq \varphi(E[X_T \mid \mathscr{F}_t]) \leq E[\varphi(X_T) \mid \mathscr{F}_t]$$

where for the second inequality we have reapplied Jensen's inequality. □

Remark 1.4.13 If X is a martingale then $|X|$ is a non-negative sub-martingale: however, this is not necessarily true if X is a sub-martingale since $x \mapsto |x|$ is not increasing. Moreover, if X is a sub-martingale then also $X^+ := X \vee 0 = \frac{|X|+X}{2}$ is.

In the last part of this section, we consider the particular case where $I = \mathbb{N} \cup \{0\}$. We give a deep result, valid also in a much more general framework, on the structure of adapted stochastic processes: Doob's decomposition theorem. First, we introduce the following

Definition 1.4.14 (Predictable Process) Let $A = (A_n)_{n \geq 0}$ be a discrete stochastic process, defined on the filtered space $(\Omega, \mathscr{F}, P, (\mathscr{F}_n)_{n \geq 0})$. We say that A is predictable if:

(i) $A_0 = 0$;
(ii) $A_n \in m\mathscr{F}_{n-1}$ for every $n \in \mathbb{N}$.

Theorem 1.4.15 (Doob's Decomposition Theorem) *Let $X = (X_n)_{n \geq 0}$ be an adapted and absolutely integrable stochastic process on the filtered space $(\Omega, \mathscr{F}, P, (\mathscr{F}_n)_{n \geq 0})$. There exist, and are a.s. unique, a martingale M and a predictable process A such that*

$$X_n = M_n + A_n, \qquad n \geq 0. \tag{1.4.4}$$

In particular, if X is a martingale then $M \equiv X$ and $A \equiv 0$; if X is a sub-martingale then the process A has almost surely monotone increasing trajectories.

Proof **Uniqueness** If two processes M and A, with the properties of the statement, exist then we have

$$X_{n+1} - X_n = M_{n+1} - M_n + A_{n+1} - A_n, \qquad n \geq 0. \tag{1.4.5}$$

Conditioning on $\mathscr{F}_n$ and exploiting the fact that X is adapted, M is a martingale and A is predictable, we have

$$E[X_{n+1} \mid \mathscr{F}_n] - X_n = E[M_{n+1} \mid \mathscr{F}_n] - M_n + A_{n+1} - A_n = A_{n+1} - A_n.$$

Consequently, the process A is uniquely determined by the recursive formula

$$\begin{cases} A_{n+1} = A_n + E[X_{n+1} \mid \mathscr{F}_n] - X_n, & \text{if } n \in \mathbb{N}, \\ A_0 = 0. \end{cases} \tag{1.4.6}$$

Note that from (1.4.6) it follows that if X is a sub-martingale then the process A has almost surely monotone increasing trajectories.

Inserting (1.4.6) into (1.4.5) we also find

$$\begin{cases} M_{n+1} = M_n + X_{n+1} - E[X_{n+1} \mid \mathscr{F}_n], & \text{if } n \in \mathbb{N}, \\ M_0 = X_0. \end{cases} \tag{1.4.7}$$

Existence It is enough to prove that the processes M and A, defined respectively by (1.4.7) and (1.4.6), verify the properties of the statement. This is a simple check: for example, it is easy to prove by induction on n that A is predictable. Similarly, we prove that M is a martingale and (1.4.4) holds.

□

Example 1.4.16 ([!]) Let X be as in Example 1.4.8. Then the processes of the Doob's decomposition of X are easily calculated:

$$M_n = X_n - n(2q - 1), \qquad A_n = n(2q - 1).$$

Note that in this case the process A is deterministic; moreover, X is a sub-martingale for $q \geq \frac{1}{2}$ and in this case $(A_n)_{n \geq 0}$ is a monotone increasing sequence.

1.5 Proof of Kolmogorov's Extension Theorem

Lemma 1.5.1 *The family $\mathscr{C}$ of finite-dimensional cylinders is a semi-ring.*

Proof Recalling definition (1.1.1) of finite-dimensional cylinder

$$C_{t_1,\ldots,t_n}(H_1 \times \cdots \times H_n) = \bigcap_{i=1}^{n} C_{t_i}(H_i), \tag{1.5.1}$$

1.5 Proof of Kolmogorov's Extension Theorem

and observing that $C_t(H) \cap C_t(K) = C_t(H \cap K)$ for every $t \in I$ and $H, K \in \mathscr{B}$, it is not difficult to prove that $\mathscr{C}$ is a $\cap$-closed family and $\emptyset \in \mathscr{C}$. It remains to prove that the difference of cylinders is a finite and disjoint union of cylinders: since $C \setminus D = C \cap D^c$, for $C, D \in \mathscr{C}$, it is sufficient to prove that the complement of a cylinder is a disjoint union of cylinders.

For a one-dimensional cylinder we have

$$(C_t(H))^c = C_t(H^c),$$

and therefore, by (1.5.1),

$$\left(C_{t_1,\ldots,t_n}(H_1 \times \cdots \times H_n)\right)^c = \bigcup_{i=1}^n \left(C_{t_i}(H_i)\right)^c = \bigcup_{i=1}^n C_{t_i}(H_i^c)$$

where in general the union is not disjoint: however, we observe that

$$C_{t_1}(H_1) \cup C_{t_2}(H_2) = C_{t_1,t_2}(H_1 \times H_2) \uplus C_{t_1,t_2}(H_1^c \times H_2) \uplus C_{t_1,t_2}(H_1 \times H_2^c),$$

and in general

$$\bigcup_{i=1}^n C_{t_i}(H_i) = \biguplus C_{t_1,\ldots,t_n}(K_1 \times \cdots \times K_n)$$

where the disjoint union is taken among all the different possible combinations of $K_1 \times \cdots \times K_n$ where K_i is H_i or H_i^c, except for the case where $K_i = H_i^c$ for every $i = 1, \ldots, n$. □

We define μ on $\mathscr{C}$ as in (1.3.4), that is

$$\mu(C_{t_1,\ldots,t_n}(H_1 \times \cdots \times H_n))$$
$$:= \mu_{t_1,\ldots,t_n}(H_1 \times \cdots \times H_n), \qquad t_1, \ldots, t_n \in I, \ H_1, \cdots H_n \in \mathscr{B}.$$

If we prove that μ is a pre-measure (i.e., μ is additive, σ-sub-additive, and such that $\mu(\emptyset) = 0$) on $\mathscr{C}$, then by Carathéodory's Theorem 1.5.5 in [113], μ extends uniquely to a probability measure on $\mathscr{F}^I$.

Clearly, $\mu(\emptyset) = 0$ and it is not difficult to prove that μ is finitely additive. To prove that μ is σ-sub-additive, consider a sequence $(C_n)_{n \in \mathbb{N}}$ of disjoint cylinders whose union is a cylinder C and show that[8]

$$\mu(C) = \sum_{n \in \mathbb{N}} \mu(C_n). \qquad (1.5.2)$$

To this end, set

$$D_n = C \setminus \biguplus_{k=1}^{n} C_k, \qquad n \in \mathbb{N}.$$

By Lemma 1.5.1, D_n is a finite and disjoint union of cylinders: therefore, $\mu(D_n)$ is well-defined (by the additivity of μ) and we have

$$\mu(C) = \sum_{k=1}^{n} \mu(C_k) + \mu(D_n).$$

Then it is enough to prove that

$$\lim_{n \to \infty} \mu(D_n) = 0. \qquad (1.5.3)$$

Clearly, $D_n \searrow \emptyset$ as $n \to \infty$. We prove (1.5.3) by contradiction and, without loss of generality, by passing to a subsequence if necessary, suppose there exists $\varepsilon > 0$ such that $\mu(D_n) \geq \varepsilon$ for every $n \in \mathbb{N}$: using a compactness argument, we show that in this case the intersection of D_n is not empty, from which the contradiction.

---

[8] Formula (1.5.2) implies the σ-sub-additivity: if $A \in \mathscr{C}$ and $(A_n)_{n \in \mathbb{N}}$ is a sequence of elements in $\mathscr{C}$ such that

$$A \subseteq \bigcup_{n \in \mathbb{N}} A_n$$

it is enough to set $C_1 = A \cap A_1 \in \mathscr{C}$ and

$$C_n = (A \cap A_n) \setminus \bigcup_{k=1}^{n-1} A_k$$

with C_n which, by Lemma 1.5.1, is a finite and disjoint union of cylinders for each $n \geq 2$. Then from (1.5.2) it follows that

$$\mu(A) \leq \sum_{n \in \mathbb{N}} \mu(A_n).$$

1.5 Proof of Kolmogorov's Extension Theorem

We know that D_n is a finite and disjoint union of cylinders: since $D_n \supseteq D_{n+1}$, possibly repeating[9] the elements of the sequence, we can assume

$$D_n = \biguplus_{k=1}^{N_n} \tilde{C}_k, \qquad \tilde{C}_k = \{x \in \mathbb{R}^I \mid (x_{t_1}, \ldots, x_{t_n}) \in H_{k,1} \times \cdots \times H_{k,n}\}$$

for some sequence $(t_n)_{n \in \mathbb{N}}$ in I and $H_{k,n} \in \mathscr{B}$. Now we use the following fact, the proof of which we postpone to the end: it is possible to construct a sequence $(K_n)_{n \in \mathbb{N}}$ such that

- $K_n \subseteq \mathbb{R}^n$ is a compact subset of

$$B_n := \bigcup_{k=1}^{N_n} (H_{k,1} \times \cdots \times H_{k,n}); \qquad (1.5.4)$$

- $K_{n+1} \subseteq K_n \times \mathbb{R}$;
- $\mu_{t_1,\ldots,t_n}(K_n) \geq \frac{\varepsilon}{2}$.

Thus, we conclude the proof of (1.5.3). Since $K_n \neq \emptyset$, for each $n \in \mathbb{N}$ there exists a vector

$$(y_1^{(n)}, \ldots, y_n^{(n)}) \in K_n.$$

By compactness, the sequence $(y_1^{(n)})_{n \in \mathbb{N}}$ admits a subsequence $(y_1^{(k_n)})_{n \in \mathbb{N}}$ converging to a point $y_1 \in K_1$. Similarly, the sequence $(y_1^{(k_n)}, y_2^{(k_n)})_{n \in \mathbb{N}}$ admits a subsequence converging to $(y_1, y_2) \in K_2$. Repeating the argument, we construct a sequence $(y_n)_{n \in \mathbb{N}}$ such that $(y_1, \ldots, y_n) \in K_n$ for each $n \in \mathbb{N}$. Therefore

$$\{x \in \mathbb{R}^I \mid x_{t_k} = y_k, \ k \in \mathbb{N}\} \subseteq D_n$$

for each $n \in \mathbb{N}$ and this completes the proof by contradiction.

Finally, we prove the existence of the sequence $(K_n)_{n \in \mathbb{N}}$. For each $n \in \mathbb{N}$ there exists[10] a compact subset $\tilde{K}_n$ of B_n in (1.5.4) such that $\mu_{t_1,\ldots,t_n}(B_n \setminus \tilde{K}_n) \leq \frac{\varepsilon}{2^{n+1}}$.

---

[9] Defining a new sequence of the form

$$\mathbb{R}^I, \ldots, \mathbb{R}^I, D_1, \ldots, D_1, D_2, \ldots, D_2, D_3 \ldots$$

in which $\mathbb{R}^I$ and the elements of $(D_n)_{n \in \mathbb{N}}$ are repeated a sufficient number of times.

[10] It is enough to combine the property of internal regularity of $\mu_{t_1,\ldots,t_n}$ (cf. Proposition 1.4.9 in [113]) with the fact that, by the continuity from below, for each $\varepsilon > 0$ there exists a *compact K* such that $\mu_{t_1,\ldots,t_n}(\mathbb{R}^n \setminus K) < \varepsilon$: note that this latter fact is nothing but the tightness property of the distribution $\mu_{t_1,\ldots,t_n}$ (cf. Definition 3.3.5 in [113]).

Setting
$$K_n := \bigcap_{h=1}^{n}(\widetilde{K}_h \times \mathbb{R}^{n-h}), \qquad (1.5.5)$$

we have that K_n is a compact subset of B_n and $K_{n+1} \subseteq K_n \times \mathbb{R}$. Now observe that

$$B_n \setminus K_n \subseteq \bigcup_{h=1}^{n} B_n \setminus (\widetilde{K}_h \times \mathbb{R}^{n-h})$$
$$\subseteq \bigcup_{h=1}^{n} (B_h \setminus \widetilde{K}_h) \times \mathbb{R}^{n-h}$$

and consequently

$$\mu_{t_1,\ldots,t_n}(B_n \setminus K_n) \leq \sum_{h=1}^{n} \mu_{t_1,\ldots,t_n}\left((B_h \setminus \widetilde{K}_h) \times \mathbb{R}^{n-h}\right)$$
$$= \sum_{h=1}^{n} \mu_{t_1,\ldots,t_h}(B_h \setminus \widetilde{K}_h)$$
$$\leq \sum_{h=1}^{n} \frac{\epsilon}{2^{h+1}} \leq \frac{\epsilon}{2}.$$

Then we have

$$\mu_{t_1,\ldots,t_n}(K_n) = \mu_{t_1,\ldots,t_n}(B_n) - \mu_{t_1,\ldots,t_n}(B_n \setminus K_n) \geq \frac{\varepsilon}{2},$$

since $\mu_{t_1,\ldots,t_n}(B_n) = \mu(D_n) \geq \varepsilon$ by hypothesis. This concludes the proof. □

Kolmogorov's extension theorem generalizes, with a substantially identical proof, to the case where the trajectories have values in a separable and complete metric space $(\mathbb{M}, \varrho)$.[11] We recall the notation $\mathscr{B}_\varrho$ for the Borel σ-algebra on $(\mathbb{M}, \varrho)$; moreover, $\mathbb{M}^I$ is the family of functions from I to values in $\mathbb{M}$ and $\mathscr{F}_\varrho^I$ is the σ-algebra generated by finite-dimensional cylinders

$$C_{t_1,\ldots,t_n}(H) := \{x \in \mathbb{M}^I \mid (x_{t_1}, \ldots, x_{t_n}) \in H\}$$

where $t_1, \ldots, t_n \in I$ and $H = H_1 \times \cdots \times H_n$ with $H_1, \ldots, H_n \in \mathscr{B}_\varrho$.

---

[11] The first part of the proof, based on Carathéodory's theorem, is exactly the same. In the second part, and in particular in the construction of the sequence of compact K_n in (1.5.5), the tightness property is crucial: here we exploit the fact that, under the assumption that $(\mathbb{M}, \varrho)$ is separable and complete, every distribution on $\mathscr{B}_\varrho$ is tight (see, for example, Theorem 1.4 in [16]). Kolmogorov's theorem *does not* extend to any measurable space: in this regard, see, for example, [59] page 214.

Theorem 1.5.2 (Kolmogorov's Extension Theorem [!!!]) *Let I be a non-empty set and $(\mathbb{M}, \varrho)$ a separable and complete metric space. Suppose that, for each finite family of indices $t_1, \ldots, t_n \in I$, a distribution $\mu_{t_1,\ldots,t_n}$ is given on $\mathbb{M}^n$, and the following consistency properties are satisfied: for each finite family of indices $t_1, \ldots, t_n \in I$, for each $H_1, \ldots, H_n \in \mathscr{B}_\varrho$ and for each permutation ν of the indices $1, 2, \ldots, n$, we have*

$$\mu_{t_1,\ldots,t_n}(H_1 \times \cdots \times H_n) = \mu_{t_{\nu(1)},\ldots,t_{\nu(n)}}(H_{\nu(1)} \times \cdots \times H_{\nu(n)}),$$

$$\mu_{t_1,\ldots,t_n}(H_1 \times \cdots \times H_{n-1} \times \mathbb{M}) = \mu_{t_1,\ldots,t_{n-1}}(H_1 \times \cdots \times H_{n-1}).$$

Then there exists a unique probability measure μ on $\left(\mathbb{M}^I, \mathscr{F}_\varrho^I\right)$ that has $\mu_{t_1,\ldots,t_n}$ as finite-dimensional distributions, i.e., such that

$$\mu(C_{t_1,\ldots,t_n}(H)) = \mu_{t_1,\ldots,t_n}(H)$$

for each finite family of indices $t_1, \ldots, t_n \in I$ and $H = H_1 \times \cdots \times H_n$ with $H_1, \ldots, H_n \in \mathscr{B}_\varrho$.

1.6 Key Ideas to Remember

We summarize the most significant findings of the chapter and the fundamental concepts to be retained from an initial reading, while excluding overly technical or ancillary details. If you have any doubt about what the following succinct statements mean, please review the corresponding section.

- Section 1.1: we define a *stochastic process* as a function taking random values or equivalently, albeit in a more abstract way, as a random variable with values in the functional space of trajectories. The *finite-dimensional distributions* of a process determine its law, playing the same role as the distribution of a random variable.
- Section 1.2: we compare the different notions of equality between stochastic processes, introducing the definitions of *equivalence in law, indistinguishable processes and modifications*.
- Section 1.3: the main existence result for processes is Kolmogorov's extension Theorem 1.3.1. It states that it is possible to construct a stochastic process starting from given finite-dimensional distributions that satisfy natural consistency properties: it is a corollary of Carathéodory's Theorem 1.4.29 in [113], and the proof, being somewhat technical, can be safely skipped at a first reading.

- Section 1.4: *martingales* constitute a fundamental class of stochastic processes that, together with Markov processes, will be the main object of study in the following chapters. Martingales have constant expectation and originate as models for fair gambling games. The martingale property depends on the fixed probability measure and filtration: a *filtration* describes the increasing flow of observable information as the temporal index varies.

Main notations introduced in this chapter:

Symbol	Description	Page
$\mathbb{R}^I = \{x : I \longrightarrow \mathbb{R}\}$	Space of trajectories, I is the family of parameters	2
$C_{t_1,\ldots,t_n}(H) = \{x \in \mathbb{R}^I \mid x_{t_i} \in H_i, i = 1, \ldots, n\}$	Finite-dimensional cylinder with $t_i \in I$ and $H_i \in \mathscr{B}$	3
$\mathscr{C}$	Family of finite-dimensional cylinders	3
$\mathscr{F}^I = \sigma(\mathscr{C})$	σ-algebra generated by finite-dimensional cylinders	3
$\mathscr{F}^I_\mu$	Completion of $\mathscr{F}^I$ with respect to the measure μ	11
$\mathscr{G}^X_t = \sigma(X_s, s \leq t)$	Filtration generated by the process X	14

Chapter 2
Markov Processes

> *World is stochastic.*
>
> From "Students' opinions on educational activities", A.Y. 2022/23 University of Bologna

Markov processes constitute a fundamental class of stochastic processes, characterized by a memoryless property which renders them highly tractable and beneficial in practical applications. In this chapter the set of indices is $I = \mathbb{R}_{\geq 0}$, where $t \in I$ is interpreted as a time instant.

2.1 Transition Law and Feller Processes

Definition 2.1.1 (Transition Law) A transition law on $\mathbb{R}^N$ is a function

$$p = p(t, x; T, H), \qquad 0 \leq t \leq T, \ x \in \mathbb{R}^N, \ H \in \mathscr{B}_N,$$

that satisfies the following conditions:

(i) for every $0 \leq t \leq T$ and $x \in \mathbb{R}^N$, $p(t, x; T, \cdot)$ is a distribution, i.e., a probability measure on $\mathscr{B}_N$, and $p(t, x; t, \cdot) = \delta_x$;
(ii) for every $0 \leq t \leq T$ and $H \in \mathscr{B}_N$, the function $x \mapsto p(t, x; T, H)$ is $\mathscr{B}_N$-measurable.

Let $X = (X_t)_{t \geq 0}$ be a stochastic process taking values in $\mathbb{R}^N$, defined on the probability space $(\Omega, \mathscr{F}, P)$. We say that X has transition law p if:

(i) p is a transition law;
(ii) we have[1]

$$p(t, X_t; T, H) = P(X_T \in H \mid X_t), \qquad 0 \leq t \leq T, \ H \in \mathscr{B}_N.$$

---
[1] We recall the convention where $P(X_T \in H \mid X_t)$ denotes the usual conditional expectation $E\left[\mathbb{1}_H(X_T) \mid X_t\right]$, as in Remark 4.3.5 in [113].

Remark 2.1.2 By properties (i) and (ii) of Definition 2.1.1, if X has transition law p then $p(t, X_t; T, \cdot)$ is a *regular version*[2] of the conditional law of X_T given X_t. Hence, we have

$$\int_{\mathbb{R}^N} p(t, X_t; T, dy)\varphi(y) = E\left[\varphi(X_T) \mid X_t\right], \qquad \varphi \in b\mathscr{B}_N, \tag{2.1.1}$$

by Theorem 4.3.8 in [113]. Analogously, $p(t, x; T, \cdot)$ is a regular version of the conditional distribution *function*[3] of X_T given X_t and we have

$$\int_{\mathbb{R}^N} p(t, x; T, dy)\varphi(y) = E\left[\varphi(X_T) \mid X_t = x\right], \qquad x \in \mathbb{R}^N, \ \varphi \in b\mathscr{B}_N, \tag{2.1.2}$$

by Theorem 4.3.19 in [113]. Notice that

$$u(x) := \int_{\mathbb{R}^N} p(t, x; T, dy)\varphi(y), \qquad x \in \mathbb{R}^N,$$

is a $\mathscr{B}_N$-measurable, bounded function: indeed, by (ii) of Definition 2.1.1, $u \in b\mathscr{B}_N$ if $\varphi = \mathbb{1}_H$ and by approximation, thanks to Lemma 2.2.3 in [113] and Beppo Levi's theorem, so is for every $\varphi \in b\mathscr{B}_N$. In accordance with notation (4.2.10) in [113], formula (2.1.2) indicates that u is a version of the conditional expectation *function* of $\varphi(X_T)$ given X_t.

Remark 2.1.3 Definition 2.1.1 extends in an obvious way to the case where, instead of $(\mathbb{R}^N, \mathscr{B}_N)$, a generic metric space $(\mathbb{M}, \varrho)$ is considered, equipped with the Borel σ-algebra $\mathscr{B}_\varrho$ (cf. Definition 1.4.4 in [113]).

Example 2.1.4 ([!]) Consider the "trivial" case of the deterministic process $X_t = \gamma(t)$ with $\gamma : \mathbb{R}_{\geq 0} \longrightarrow \mathbb{R}^N$ which is interpreted as a parametrized curve in $\mathbb{R}^N$. We have

$$E\left[\varphi(X_T) \mid X_t\right] = \varphi(\gamma(T)) = \varphi(\gamma(t) + \gamma(T) - \gamma(t))$$

and therefore a regular version of the conditional expectation function of $\varphi(X_T)$ given X_t equals

$$E\left[\varphi(X_T) \mid X_t = x\right] = \varphi(x + \gamma(T) - \gamma(t)) = \int_{\mathbb{R}} \delta_{x+\gamma(T)-\gamma(t)}(dy)\varphi(y).$$

In other words,

$$p(t, x; T, \cdot) = \delta_{x+\gamma(T)-\gamma(t)}$$

---

[2] Definition 4.3.1 in [113].
[3] Theorem 4.3.16 in [113].

2.1 Transition Law and Feller Processes

is a transition law of X: this result is a very particular case of Proposition 2.3.2 which we will prove later. Notice that the transition law *is not unique*: for example, if for every $0 \leq t \leq T$ we set

$$\widetilde{p}(t, x; T, \cdot) = \begin{cases} \delta_{x+\gamma(T)-\gamma(t)} & \text{if } x = \gamma(t), \\ \delta_x & \text{if } x \neq \gamma(t), \end{cases}$$

then even $\widetilde{p}$ is a transition law for X.

Remark 2.1.5 (Time-Homogeneous Transition Law) A transition law p is said to be *time-homogeneous* if

$$p(t, x; T, H) = p(0, x; T - t, H), \qquad 0 \leq t \leq T, \ x \in \mathbb{R}, \ H \in \mathscr{B}.$$

If X has a time-homogeneous transition law p, then

$$E[\varphi(X_T) \mid X_t = x] = \int_{\mathbb{R}} p(t, x; T, dy) \varphi(y)$$
$$= \int_{\mathbb{R}} p(0, x; T - t, dy) \varphi(y) = E[\varphi(X_{T-t}) \mid X_0 = x]. \tag{2.1.3}$$

Equation (2.1.3) means that the conditional expectation function of $\varphi(X_T)$ given X_t is equal to the conditional expectation function of the temporally translated process at the initial time.[4]

Example 2.1.6 (Poisson Transition Law [!]) Recall that $\text{Poisson}_{x,\lambda}$ denotes the Poisson distribution with parameter $\lambda > 0$ and centered at $x \in \mathbb{R}$, defined in Example 1.4.17 in [113]. The *Poisson transition law with parameter* $\lambda > 0$ is defined by

$$p(t, x; T, \cdot) = \text{Poisson}_{x, \lambda(T-t)}$$
$$= e^{-\lambda(T-t)} \sum_{n=0}^{+\infty} \frac{(\lambda(T-t))^n}{n!} \delta_{x+n}, \qquad 0 \leq t \leq T, \ x \in \mathbb{R}.$$

---

[4] If, for simplicity, we denote

$$E_x[Y] = E[Y \mid X_0 = x],$$

Eq. (2.1.3) can be written in the more compact form

$$E[\varphi(X_T) \mid X_t] = E_{X_t}[\varphi(X_{T-t})]. \tag{2.1.4}$$

For clarity: the right-hand side of (2.1.4) is the conditional expectation of $\varphi(X_{T-t})$ given X_0, evaluated at X_t.

Properties (i) and (ii) of Definition 2.1.1 are obvious. The Poisson transition law is time-homogeneous and *invariant under translations* in the sense that

$$p(t, x; T, H) = p(0, 0; T - t, H - x), \qquad 0 \leq t \leq T, \ x \in \mathbb{R}, \ H \in \mathscr{B}.$$

Definition 2.1.7 (Transition Density) A transition law p is absolutely continuous if, for every $0 \leq t < T$ and $x \in \mathbb{R}^N$, there exists a density $\Gamma = \Gamma(t, x; T, \cdot)$ such that

$$p(t, x; T, H) = \int_H \Gamma(t, x; T, y) dy, \qquad H \in \mathscr{B}_N.$$

We say that Γ is a *transition density* of p (or, of X, if p is the transition law of a process X).

Remark 2.1.8 The transition density $\Gamma = \Gamma(t, x; T, y)$ of a process X is a function of four variables: the first pair (t, x) represents the time and starting point of X; the second pair (T, y) represents the time and random position of *arrival* of X. For any $\varphi \in b\mathscr{B}_N$, we have

$$\int_{\mathbb{R}^N} \Gamma(t, X_t; T, y) \varphi(y) dy = E\left[\varphi(X_T) \mid X_t\right],$$

or, in terms of conditional expectation function,

$$\int_{\mathbb{R}^N} \Gamma(t, x; T, y) \varphi(y) dy = E\left[\varphi(X_T) \mid X_t = x\right], \qquad x \in \mathbb{R}^N.$$

Example 2.1.9 (Gaussian Transition Law [!]) The Gaussian transition law is defined by $p(t, x; T, \cdot) = \mathcal{N}_{x, T-t}$ for every $0 \leq t \leq T$ and $x \in \mathbb{R}$. It is an absolutely continuous transition law since

$$p(t, x; T, H) := \mathcal{N}_{x, T-t}(H)$$

$$= \int_H \Gamma(t, x; T, y) dy, \qquad 0 \leq t < T, \ x \in \mathbb{R}, \ H \in \mathscr{B},$$

where

$$\Gamma(t, x; T, y) = \frac{1}{\sqrt{2\pi(T - t)}} e^{-\frac{(x-y)^2}{2(T-t)}}, \qquad 0 \leq t < T, \ x, y \in \mathbb{R},$$

is the Gaussian transition density. It is clear that p satisfies properties (i) and (ii) of Definition 2.1.1.

We now introduce a notion of "continuous dependence" of the transition law with respect to the initial datum (t, x).

2.1 Transition Law and Feller Processes

Definition 2.1.10 (Feller Property) A transition law p has the Feller property if for every $h > 0$ and $\varphi \in bC(\mathbb{R}^N)$ the function

$$(t, x) \longmapsto \int_{\mathbb{R}^N} p(t, x; t + h, dy)\varphi(y)$$

is continuous. A Feller process is a process with a transition law that satisfies the Feller property.

The Feller property is equivalent to the *continuity under weak convergence* of the transition law $p = p(t, x; t + h, \cdot)$ with respect to the pair (t, x) of initial time and position: more precisely, recalling the definition of weak convergence of distributions (cf. Remark 3.1.1 in [113]), the fact that X is a Feller process with transition law p means that

$$p(t_n, x_n; t_n + h, \cdot) \xrightarrow{d} p(t, x; t + h, \cdot)$$

for every sequence (t_n, x_n) converging to (t, x) as $n \to +\infty$.

When p is homogeneous in time, the Feller property reduces to continuity with respect to x: precisely, p has the Feller property if for every $h > 0$ and $\varphi \in bC(\mathbb{R}^N)$ the function

$$x \longmapsto \int_{\mathbb{R}^N} p(0, x; h, dy)\varphi(y)$$

is continuous. The Feller property plays an important role in the study of Markov processes (cf. Chap. 7) and the regularity properties of continuous-time filtrations (cf. Sect. 6.2.1).

Example 2.1.11 Poisson and Gaussian transition laws satisfy the Feller property (cf. Examples 2.4.5 and 2.4.6): therefore, we say that the related stochastic processes that we will introduce later, respectively the Poisson process and the Brownian motion, are Feller processes.

We conclude the section with a technical result. Recall Definition 1.3.4 of the canonical version of a stochastic process.

Proposition 2.1.12 *If p is a transition law for the process X, defined on the space $(\Omega, \mathscr{F}, P)$, then it is also a transition law for its canonical version $\mathbf{X}$.*

Proof Recall that $\mathbf{X}$ is defined on the probability space $(\mathbb{R}^I, \mathscr{F}^I_{\mu_X}, \mu_X)$, where $\mathscr{F}^I_{\mu_X}$ denotes the μ_X-completion of $\mathscr{F}^I$, and $\mathbf{X}(w) = w$ for every $w \in \mathbb{R}^I$. Given $0 \leq t \leq T$ and $H \in \mathscr{B}$, let $Z := p(t, \mathbf{X}_t, T, H)$: we have to verify that

$$Z = E^{\mu_X}\left[\mathbb{1}_H(\mathbf{X}_T) \mid \mathbf{X}_t\right] \tag{2.1.5}$$

where $E^{\mu_X}[\cdot]$ denotes the expected value under the probability measure μ_X. Clearly $Z \in m\sigma(\mathbf{X}_t)$. Moreover, if $W \in b\sigma(\mathbf{X}_t)$ then by Doob's theorem $W = \varphi(\mathbf{X}_t)$ with $\varphi \in b\mathscr{B}$ and we have

$$E^{\mu_X}[ZW] = E^{\mu_X}[p(t, \mathbf{X}_t, T, H)\varphi(\mathbf{X}_t)] = \qquad (2.1.6)$$

(since X and $\mathbf{X}$ have the same law)

$$= E^P[p(t, X_t, T, H)\varphi(X_t)] =$$

(since p is a transition law of X)

$$= E^P[\mathbb{1}_H(X_T)\varphi(X_t)] =$$

(again by the equality in law of X and $\mathbf{X}$)

$$= E^{\mu_X}[\mathbb{1}_H(\mathbf{X}_T)\varphi(\mathbf{X}_t)].$$

This proves (2.1.5) and concludes the proof. $\square$

2.2 Markov Property

We consider for simplicity the scalar case, $N = 1$.

Definition 2.2.1 (Markov Process) *Let $X = (X_t)_{t \geq 0}$ be an* adapted *stochastic process on the filtered space $(\Omega, \mathscr{F}, P, \mathscr{F}_t)$. We say that X is a Markov process if it has a transition law p such that*[5]

$$p(t, X_t; T, H) = P(X_T \in H \mid \mathscr{F}_t), \qquad 0 \leq t \leq T, \ H \in \mathscr{B}. \qquad (2.2.1)$$

Formula (2.2.1) is a memoryless property: intuitively, it expresses the fact that the knowledge of $\mathscr{F}_t$ (and, in particular, of the entire trajectory of X up to time t) or just the value X_t provide the same information regarding the distribution of the future value X_T.

Proposition 2.2.2 (Markov Property) *Let $X = (X_t)_{t \geq 0}$ be an adapted stochastic process on the filtered space $(\Omega, \mathscr{F}, P, \mathscr{F}_t)$, with transition law p. Then X is a Markov process if and only if*

$$\int_{\mathbb{R}} p(t, X_t; T, dy)\varphi(y) = E[\varphi(X_T) \mid \mathscr{F}_t], \qquad 0 \leq t \leq T, \ \varphi \in b\mathscr{B}. \qquad (2.2.2)$$

---

[5] Here, as in Remark 4.3.5 in [113], $P(X_T \in \cdot \mid \mathscr{F}_t)$ indicates a regular version of the conditional distribution of X_T given $\mathscr{F}_t$. Formula (2.2.1) is equivalent to $p(t, X_t; T, H) = E[\mathbb{1}_H(X_T) \mid \mathscr{F}_t]$, that is, $p(t, X_t; T, H)$ is a version of the conditional expectation of $\mathbb{1}_H(X_T)$ given $\mathscr{F}_t$.

2.2 Markov Property

Proof If X is a Markov process then $p(t, X_t; T, \cdot)$ is a regular version of the conditional law of X_T given $\mathscr{F}_t$ and (2.2.2) follows from Theorem I-cc20 in [113]. The converse is obvious, with the choice $\varphi = \mathbb{1}_H$, $H \in \mathscr{B}$. □

Remark 2.2.3 By density, we also have that X is a Markov process if and only if (2.2.2) holds for every $\varphi \in C_0^\infty$. Combining (2.1.1) with (2.2.2), it is customary to write[6]

$$E\left[\varphi(X_T) \mid X_t\right] = E\left[\varphi(X_T) \mid \mathscr{F}_t\right]. \tag{2.2.3}$$

The Markov property can be generalized in the following way. Observe that if $t \le t_1 < t_2$ and $\varphi_1, \varphi_2 \in b\mathscr{B}$ then, by the tower property, we have

$$E\left[\varphi_1(X_{t_1})\varphi_2(X_{t_2}) \mid X_t\right] = E\left[E\left[\varphi_1(X_{t_1})\varphi_2(X_{t_2}) \mid \mathscr{F}_{t_1}\right] \mid X_t\right]$$
$$= E\left[\varphi_1(X_{t_1}) E\left[\varphi_2(X_{t_2}) \mid \mathscr{F}_{t_1}\right] \mid X_t\right] =$$

(by the Markov property)

$$= E\left[\varphi_1(X_{t_1}) E\left[\varphi_2(X_{t_2}) \mid X_{t_1}\right] \mid X_t\right] =$$

(by the Markov property applied to the external conditional expectation, being $\varphi_1(X_{t_1}) E\left[\varphi_2(X_{t_2}) \mid X_{t_1}\right]$ a bounded and Borel-measurable function of X_{t_1} by Doob's theorem)

$$= E\left[\varphi_1(X_{t_1}) E\left[\varphi_2(X_{t_2}) \mid X_{t_1}\right] \mid \mathscr{F}_t\right] =$$

(by the Markov property applied to the internal conditional expectation)

$$= E\left[\varphi_1(X_{t_1}) E\left[\varphi_2(X_{t_2}) \mid \mathscr{F}_{t_1}\right] \mid \mathscr{F}_t\right]$$
$$= E\left[E\left[\varphi_1(X_{t_1})\varphi_2(X_{t_2}) \mid \mathscr{F}_{t_1}\right] \mid \mathscr{F}_t\right]$$
$$= E\left[\varphi_1(X_{t_1})\varphi_2(X_{t_2}) \mid \mathscr{F}_t\right].$$

---

[6] Formula (2.2.3) is not an equality but a notation that must be interpreted in the sense of Convention 4.2.5 in [113]: precisely, (2.2.3) means that if $Z = E[\varphi(X_T) \mid X_t]$ then $Z = E[\varphi(X_T) \mid \mathscr{F}_t]$. However, there may exist a version Z' of $E[\varphi(X_T) \mid \mathscr{F}_t]$ that is not $\sigma(X_t)$-measurable (It is enough to modify Z on a negligible event that belongs to $\mathscr{F}_t$ but not to $\sigma(X_t)$.) and therefore is not the expectation of $\varphi(X_T)$ conditioned on X_t. On the other hand, if (2.2.3) holds and $Z' = E[\varphi(X_T) \mid \mathscr{F}_t]$ then $Z' = f(X_t)$ a.s. for some $f \in m\mathscr{B}$: indeed, taking a version Z of $E[\varphi(X_T) \mid X_t]$, by Doob's theorem, $Z = f(X_t)$ and by (2.2.3) (and the uniqueness of the conditional expectation) $Z = Z'$ a.s. These subtleties are relevant when one has to verify in practice the validity of the Markov property: Example 11.1.10 is illuminating in this sense.

Hence, we have[7]

$$E[Y \mid X_t] = E[Y \mid \mathscr{F}_t] \qquad (2.2.4)$$

for $Y = \varphi_1(X_{t_1})\varphi_2(X_{t_2})$ with $t \leq t_1 < t_2$ and $\varphi_1, \varphi_2 \in b\mathscr{B}$. By induction, it is not difficult to prove that (2.2.4) also holds if

$$Y = \prod_{k=1}^{n} \varphi_k(X_{t_k}) \qquad (2.2.5)$$

for every $t \leq t_1 < \cdots < t_n$ and $\varphi_1, \ldots, \varphi_n \in b\mathscr{B}$. Finally, by[8] Dynkin's Theorem A.0.8 in [113], (2.2.4) is valid for every bounded r.v. that is measurable with respect to the σ-algebra generated by the random variables of the type X_s with $s \geq t$, that is

$$\mathscr{G}_{t,\infty}^X := \sigma(X_s, s \geq t). \qquad (2.2.6)$$

The σ-algebra $\mathscr{G}_{t,\infty}^X$ represents *the future information on X starting from time t*, by analogy with Definition 1.4.3. In conclusion, we have proven the following generalized Markov property.

Theorem 2.2.4 (Extended Markov Property) *Let X be a Markov process on $(\Omega, \mathscr{F}, P, \mathscr{F}_t)$. We have*

$$E[Y \mid X_t] = E[Y \mid \mathscr{F}_t], \qquad Y \in b\mathscr{G}_{t,\infty}^X. \qquad (2.2.7)$$

The following corollary expresses the essence of the Markov property: *the past (i.e., $\mathscr{F}_t$) and the future (i.e., $\mathscr{G}_{t,\infty}^X$) are conditionally independent*[9] *given the present (i.e., $\sigma(X_t)$)*.

---

[7] In accordance with convention (2.2.3).

[8] We use Dynkin's Theorem A.0.8 in [113] in the following way: let $\mathscr{A}$ be the family of cylinders of the form $C = \bigcap_{k=1}^{n}(X_{t_k} \in H_k)$ as $t \leq t_1 \leq \cdots \leq t_n$ and $H_1, \ldots, H_n \in \mathscr{B}$. Then $\mathscr{A}$ is a $\cap$-closed family of events. Let $\mathscr{H}$ be the family of bounded random variables for which (2.2.4) holds: by Beppo Levi's theorem for conditional expectation, $\mathscr{H}$ is a monotone family; moreover, choosing $\varphi_k = \mathbb{1}_{H_k}$ in (2.2.5), we have that $\mathscr{H}$ contains the indicator functions of elements of $\mathscr{A}$. Then Theorem A.0.8 in [113] ensures that $\mathscr{H}$ also contains the bounded and $\sigma(\mathscr{A})$-measurable random variables.

[9] More precisely: if there exists a regular version of the conditional probability $P(\cdot \mid X_t)$ (this is guaranteed if Ω is a Polish space) then (2.2.8) with $Y = \mathbb{1}_A$, $A \in \mathscr{G}_{t,\infty}^X$, and $Z = \mathbb{1}_B$, $B \in \mathscr{F}_t$, becomes

$$P(A \mid X_t) P(B \mid X_t) = P(A \cap B \mid X_t).$$

2.2 Markov Property

Corollary 2.2.5 ([!]) *Let X be a Markov process on $(\Omega, \mathscr{F}, P, \mathscr{F}_t)$. Then we have*

$$E[Y \mid X_t]\, E[Z \mid X_t] = E[YZ \mid X_t], \qquad Y \in b\mathscr{G}^X_{t,\infty},\ Z \in b\mathscr{F}_t. \qquad (2.2.8)$$

Proof We verify that $E[Y \mid X_t]\, E[Z \mid X_t]$ is a version of the expectation of YZ conditioned on X_t: the measurability property $E[Y \mid X_t]\, E[Z \mid X_t] \in m\sigma(X_t)$ is obvious. Given $W \in b\sigma(X_t)$, we have

$$E[W E[Y \mid X_t]\, E[Z \mid X_t]] =$$

(since $W E[Y \mid X_t] \in b\sigma(X_t)$ and by property (ii) of the definition of conditional expectation $E[Z \mid X_t]$)

$$= E[W E[Y \mid X_t]\, Z] =$$

(by the extended Markov property (2.2.7))

$$= E[W E[Y \mid \mathscr{F}_t]\, Z]$$
$$= E[E[WYZ \mid \mathscr{F}_t]] = E[WYZ]$$

which proves the second property of the definition of conditional expectation. $\square$

Finally, we introduce the canonical version of a Markov process. The insistence on prioritizing the canonical version (cf. Definition 1.3.4) of a process is justified by the importance of the completeness property of the space and the fact that we can identify the *outcomes* with the *trajectories* of the process: this will be even clearer when, in Chap. 7, we will express the Markov property using an appropriate time translation operator.

Proposition 2.2.6 (Canonical Version of a Markov Process) *Let X be a Markov process on the space $(\Omega, \mathscr{F}, P, \mathscr{F}_t)$ with transition law p and let $\mathbf{X}$ be its canonical version. Then $\mathbf{X}$ is a Markov process with transition law p on $(\mathbb{R}^I, \mathscr{F}^I_{\mu_X}, \mu_X, \mathscr{G}^{\mathbf{X}})$ where, as usual, $\mathscr{G}^{\mathbf{X}}$ denotes the filtration generated by $\mathbf{X}$ (cf. (1.4.1) and Remark 1.4.5).*

Proof By Proposition 2.1.12, p is also a transition law of $\mathbf{X}$, so it suffices to prove that, for every $0 \le t \le T$ and $H \in \mathscr{B}$, setting $Z := p(t, \mathbf{X}_t, T, H)$, we have

$$Z = E^{\mu_X}\!\left[\mathbb{1}_H(\mathbf{X}_T) \mid \mathscr{G}^{\mathbf{X}}_t\right]$$

where $E^{\mu_X}[\cdot]$ denotes the expected value under the probability measure μ_X. Obviously, $Z \in m\mathscr{G}^{\mathbf{X}}_t$ and therefore it remains to verify that

$$E^{\mu_X}[ZW] = E^{\mu_X}[\mathbb{1}_H(\mathbf{X}_T)W], \qquad W \in b\mathscr{G}^{\mathbf{X}}_t.$$

Actually, thanks to[10] Dynkin's Theorem A.0.8 in [113], it is sufficient to consider W of the form

$$W = \varphi(\mathbf{X}_{t_1}, \ldots, \mathbf{X}_{t_n})$$

with $0 \leq t_1 < \cdots < t_n \leq t$ and $\varphi \in b\mathscr{B}_n$. Now, it is enough to proceed as in the proof of Proposition 2.1.12:

$$E^{\mu_X}[ZW] = E^{\mu_X}\left[p(t, \mathbf{X}_t, T, H)\varphi(\mathbf{X}_{t_1}, \ldots, \mathbf{X}_{t_n})\right] =$$

(since X and $\mathbf{X}$ have the same distribution)

$$= E^P\left[p(t, X_t, T, H)\varphi(X_{t_1}, \ldots, X_{t_n})\right] =$$

(by the Markov property of X)

$$= E^P\left[\mathbb{1}_H(X_T)\varphi(X_{t_1}, \ldots, X_{t_n})\right] =$$

(again by the equality in distribution of X and $\mathbf{X}$)

$$= E^{\mu_X}\left[\mathbb{1}_H(\mathbf{X}_T)\varphi(\mathbf{X}_{t_1}, \ldots, \mathbf{X}_{t_n})\right].$$

□

2.3 Processes with Independent Increments and Martingales

Let $X = (X_t)_{t \geq 0}$ be a stochastic process on the filtered space $(\Omega, \mathscr{F}, P, \mathscr{F}_t)$.

Definition 2.3.1 (Process with Independent Increments) We say that X has *independent increments* if:

(i) X is adapted to $(\mathscr{F}_t)_{t \geq 0}$;
(ii) the increment $X_T - X_t$ is independent of $\mathscr{F}_t$ for every $0 \leq t < T$.

Proposition 2.3.2 ([!]) *Let $X = (X_t)_{t \geq 0}$ be a process with independent increments, then X is a Markov process with transition law $p = p(t, x; T, \cdot)$ equal to the law of*

$$X_T^{t,x} := X_T - X_t + x, \qquad 0 \leq t \leq T, \; x \in \mathbb{R}. \tag{2.3.1}$$

---

[10] We use Dynkin's Theorem A.0.8 in [113] in a similar way to what was done in the proof of Theorem 2.2.4.

2.3 Processes with Independent Increments and Martingales

Proof Let us prove that p in (2.3.1) is a transition law for X. Clearly, $p(t, x; T, \cdot)$ is a distribution and $p(t, x; t, \cdot) = \delta_x$. Moreover, if $\mu_{X_T - X_t}$ denotes the law of $X_T - X_t$, then by the Fubini's theorem, for any $H \in \mathscr{B}$ the function

$$x \longmapsto p(t, x; T, H) = \mu_{X_T - X_t}(H - x)$$

is $\mathscr{B}$-measurable. Finally, fixed $H \in \mathscr{B}$, $p(t, X_t; T, H) = P(X_T \in H \mid X_t)$ as a consequence of the fact that for every function $\varphi \in b\mathscr{B}$ we have

$$E[\varphi(X_T) \mid X_t] = E[\varphi(X_T - X_t + X_t) \mid X_t] =$$

(by the freezing Lemma 4.2.11 in [113], since $X_T - X_t$ is independent of X_t and obviously X_t is $\sigma(X_t)$-measurable)

$$= E\left[\varphi(X_T^{t,x})\right]\Big|_{x=X_t} = \int_{\mathbb{R}} p(t, X_t; T, dy) \varphi(y).$$

Similarly, the Markov property (2.2.2) (and consequently (2.2.1)) is established, conditioning on $\mathscr{F}_t$ rather than X_t. □

It is interesting to compare the definitions of a process with independent increments and a martingale. We begin by observing that if X has independent increments, then for every $n \in \mathbb{N}$ and $0 \leq t_0 < t_1 < \cdots < t_n$, the increments $X_{t_k} - X_{t_{k-1}}$ are indeed independent; in particular, if X is square-integrable, i.e., $X_t \in L^2(\Omega, P)$ for any t, then the increments are uncorrelated:

$$\operatorname{cov}(X_{t_k} - X_{t_{k-1}}, X_{t_h} - X_{t_{h-1}}) = 0, \quad 1 \leq k < h \leq n.$$

Even a martingale has uncorrelated (but not necessarily independent) increments.

Proposition 2.3.3 *Let X be a square-integrable martingale. Then X has uncorrelated increments.*

Proof Let $t_0 \leq t_1 \leq t_2 \leq t_3$. We have

$$\begin{aligned}\operatorname{cov}(X_{t_1} - X_{t_0}, X_{t_3} - X_{t_2}) &= E\left[(X_{t_1} - X_{t_0})(X_{t_3} - X_{t_2})\right] \\&= E\left[E\left[(X_{t_1} - X_{t_0})(X_{t_3} - X_{t_2}) \mid \mathscr{F}_{t_2}\right]\right] \\&= E\left[(X_{t_1} - X_{t_0}) E\left[X_{t_3} - X_{t_2} \mid \mathscr{F}_{t_2}\right]\right] = 0.\end{aligned}$$

□

A process with independent increments is not necessarily integrable, nor constant in mean, and therefore not necessarily a martingale. However, we have the following

Proposition 2.3.4 *Let X be an absolutely integrable process with independent increments. Then the "compensated" process defined by $\widetilde{X}_t := X_t - E[X_t]$ is a martingale.*

Proof It is enough to observe that for every $t \leq T$ we have

$$E[\widetilde{X}_T \mid \mathscr{F}_t] = E[\widetilde{X}_T - \widetilde{X}_t \mid \mathscr{F}_t] + \widetilde{X}_t =$$

(since also $\widetilde{X}$ has independent increments)

$$= E[\widetilde{X}_T - \widetilde{X}_t] + \widetilde{X}_t = \widetilde{X}_t$$

since $\widetilde{X}$ has zero mean. □

Remark 2.3.5 [!] Proposition 2.3.4 provides the Doob's decomposition $X = \widetilde{X} + A$ of the process X: in this case the drift process $A_t = E[X_t]$ is deterministic.

2.4 Finite-Dimensional Laws and Chapman-Kolmogorov Equation

Let X be a Markov process with initial distribution μ (i.e., $X_0 \sim \mu$) and transition law p. The following result shows that, starting from the knowledge of μ and p, it is possible to determine the finite-dimensional distributions (and therefore the law!) of X.

Proposition 2.4.1 (Finite-Dimensional Distributions [!]) *Let $X = (X_t)_{t \geq 0}$ be a Markov process with transition law p and such that $X_0 \sim \mu$. For every $t_0, t_1, \ldots, t_n \in \mathbb{R}$ with $0 = t_0 < t_1 < t_2 < \cdots < t_n$, and $H \in \mathscr{B}_{n+1}$ we have*

$$P((X_{t_0}, X_{t_1}, \ldots, X_{t_n}) \in H) = \int_H \mu(dx_0) \prod_{i=1}^n p(t_{i-1}, x_{i-1}; t_i, dx_i). \quad (2.4.1)$$

Proof By Corollary A.0.5 in [113] it is sufficient to prove the thesis when $H = H_0 \times \cdots \times H_n$ with $H_i \in \mathscr{B}$. We proceed by induction: in the case $n = 1$ we have

$$P((X_{t_0}, X_{t_1}) \in H_0 \times H_1) = E\left[\mathbb{1}_{H_0}(X_{t_0})\mathbb{1}_{H_1}(X_{t_1})\right]$$

$$= E\left[\mathbb{1}_{H_0}(X_{t_0})E\left[\mathbb{1}_{H_1}(X_{t_1}) \mid X_{t_0}\right]\right]$$

$$= E\left[\mathbb{1}_{H_0}(X_{t_0})\int_{H_1} p(t_0, X_{t_0}; t_1, dx_1)\right] =$$

2.4 Finite-Dimensional Laws and Chapman-Kolmogorov Equation

(by Fubini's theorem)

$$= \int_{H_0 \times H_1} \mu(dx_0) p(t_0, x_0; t_1, dx_1).$$

Now suppose (2.4.1) is true for n and prove it for $n+1$: for $H \in \mathscr{B}_{n+1}$ and $K \in \mathscr{B}$ we have

$$P((X_{t_0}, \ldots, X_{t_{n+1}}) \in H \times K) = E\left[\mathbb{1}_H(X_{t_0}, \ldots, X_{t_n}) E\left[\mathbb{1}_K(X_{t_{n+1}}) \mid \mathscr{F}_{t_n}\right]\right] =$$

(by the Markov property)

$$= E\left[\mathbb{1}_H(X_{t_0}, \ldots, X_{t_n}) E\left[\mathbb{1}_K(X_{t_{n+1}}) \mid X_{t_n}\right]\right]$$

$$= E\left[\mathbb{1}_H(X_{t_0}, \ldots, X_{t_n}) \int_K p(t_n, X_{t_n}; t_{n+1}, dx_{n+1})\right] =$$

(by inductive hypothesis and Fubini's theorem)

$$= \int_{H \times K} \mu(dx_0) \prod_{i=1}^{n+1} p(t_{i-1}, x_{i-1}; t_i, dx_i).$$

$\square$

Remark 2.4.2 In the particular case $\mu = \delta_{x_0}$, (2.4.1) becomes

$$P((X_{t_1}, \ldots, X_{t_n}) \in H) = \int_H \prod_{i=1}^{n} p(t_{i-1}, x_{i-1}; t_i, dx_i), \qquad H \in \mathscr{B}_n. \qquad (2.4.2)$$

The following remarkable result provides a *necessary condition* for a transition law to be the transition law of a Markov process.

Proposition 2.4.3 (Chapman-Kolmogorov Equation [!!]) *Let X be a Markov process with transition law p. For every $0 \leq t_1 < t_2 < t_3$ and $H \in \mathscr{B}$, we have*

$$p(t_1, X_{t_1}; t_3, H) = \int_{\mathbb{R}} p(t_1, X_{t_1}; t_2, dx_2) p(t_2, x_2; t_3, H). \qquad (2.4.3)$$

Proof Intuitively, the Chapman-Kolmogorov equation expresses the fact that the probability of moving from position x_1 at time t_1 to a position in H at time t_3 is equal to the probability of transitioning to a position x_2 at an interim time t_2, followed by a transition from x_2 to H, integrated over all possible values of x_2.

We have

$$p(t_1, X_{t_1}; t_3, H) = E\left[\mathbb{1}_H(X_{t_3}) \mid X_{t_1}\right] =$$

(by the tower property)

$$= E\left[E\left[\mathbb{1}_H(X_{t_3}) \mid \mathscr{F}_{t_2}\right] \mid X_{t_1}\right] =$$

(by the Markov property (2.2.1))

$$= E\left[p(t_2, X_{t_2}; t_3, H) \mid X_{t_1}\right] =$$

(by (2.1.1))

$$= \int_{\mathbb{R}} p(t_1, X_{t_1}; t_2, dx_2) p(t_2, x_2; t_3, H).$$

□

We now show that the Chapman-Kolmogorov equation is actually a *necessary and sufficient condition*, in the sense that it is always possible to construct a Markov process from an initial law and a transition law p provided that it verifies (2.4.3).

Theorem 2.4.4 ([!]) *Let μ be a distribution on $\mathbb{R}$ and let $p = p(t, x; T, H)$ be a transition law*[11] *that verifies the Chapman-Kolmogorov equation*

$$p(t_1, x; t_3, H) = \int_{\mathbb{R}} p(t_1, x; t_2, dy) p(t_2, y; t_3, H), \tag{2.4.4}$$

for every $0 \leq t_1 < t_2 < t_3$, $x \in \mathbb{R}$ and $H \in \mathscr{B}$. Then there exists a Markov process $X = (X_t)_{t \geq 0}$ with transition law p and such that $X_0 \sim \mu$.

Proof Consider the family of finite-dimensional distributions defined by (2.4.1): specifically, if $0 = t_0 < t_1 < t_2 < \cdots < t_n$ we set

$$\mu_{t_0,\ldots,t_n}(H) = \int_H \mu(dx_0) \prod_{i=1}^n p(t_{i-1}, x_{i-1}; t_i, dx_i), \qquad H \in \mathscr{B}_{n+1},$$

and if $t_0, \ldots, t_n$ are not ordered in increasing order, we define $\mu_{t_0,\ldots,t_n}$ by (1.3.2) by reordering the times. In this way, the consistency property (1.3.2) is automatically satisfied by construction. On the other hand, the Chapman-Kolmogorov equation guarantees the validity of the second consistency property (1.3.3) since, after ordering the times in increasing order, we have

$$\mu_{t_0,\ldots,t_{k-1},t_k,t_{k+1},\ldots,t_n}(H_0 \times \cdots \times H_{k-1} \times \mathbb{R} \times H_{k+1} \times \cdots \times H_n)$$
$$= \mu_{t_0,\ldots,t_{k-1},t_{k+1},\ldots,t_n}(H_0 \times \cdots \times H_{k-1} \times H_{k+1} \times \cdots \times H_n).$$

---

[11] That is, p verifies properties (i) and (ii) of Definition 2.1.1.

2.4 Finite-Dimensional Laws and Chapman-Kolmogorov Equation

Since the assumptions of Kolmogorov's extension theorem are satisfied, we consider the stochastic process $X = (X_t)_{t \geq 0}$ constructed canonically as in Corollary 1.3.3: X has the finite-dimensional distributions in (2.4.1) and is defined on the filtered space $(\Omega, \mathscr{F}, P, (\mathscr{G}^X_t)_{t \geq 0})$ with $\Omega = \mathbb{R}^{[0,+\infty)}$: we recall that, by Remark 1.4.4, the filtration $(\mathscr{G}^X_t)_{t \geq 0}$ is the one generated by finite-dimensional cylinders.

It remains to prove that X is a Markov process with transition distribution p. Fixing $0 \leq t < T$ and $\varphi \in b\mathscr{B}$, we prove that the following formula, equivalent to (2.2.2), holds

$$\int_{\mathbb{R}} p(t, X_t; T, dy) \varphi(y) = E\left[\varphi(X_T) \mid \mathscr{G}^X_t\right],$$

by directly verifying the properties of conditional expectation. Setting

$$Z = \int_{\mathbb{R}} p(t, X_t; T, dy) \varphi(y)$$

clearly $Z \in m\mathscr{G}^X_t$. By Remark 4.2.2 in [113], to conclude it is sufficient to prove that

$$E[\mathbb{1}_C \varphi(X_T)] = E[\mathbb{1}_C Z]$$

where C is a finite-dimensional cylinder in $\mathscr{G}^X_t$ of the form in (1.1.1): in particular, it is not restrictive to assume $C = C_{t_0, t_1, \ldots, t_n}(H)$ with $H \in \mathscr{B}_{n+1}$ and $t_n = t$. This allows us to use the finite-dimensional distributions in (2.4.1): in fact, we have

$$\begin{aligned} & E\left[\mathbb{1}_{C_{t_0, \ldots, t_n}(H)} \varphi(X_T)\right] \\ &= E\left[\mathbb{1}_H(X_{t_0}, X_{t_1}, \ldots, X_{t_n}) \varphi(X_T)\right] \\ &= \int_H \mu(dx_0) \prod_{i=1}^n p(t_{i-1}, x_{i-1}; t_i, dx_i) \int_{\mathbb{R}} p(t_n, x_n; T, dy) \varphi(y) \\ &= E\left[\mathbb{1}_H(X_{t_0}, \ldots, X_{t_n}) \int_{\mathbb{R}} p(t_n, X_{t_n}; T, dy) \varphi(y)\right] \\ &= E\left[\mathbb{1}_{C_{t_0, \ldots, t_n}(H)} Z\right]. \end{aligned}$$

□

Example 2.4.5 (Poisson Transition Law [!]) The Poisson transition law with parameter $\lambda > 0$ (cf. Example 2.1.6)

$$p(t, x; T, \cdot) = \text{Poisson}_{x, \lambda(T-t)}$$

$$= e^{-\lambda(T-t)} \sum_{n=0}^{+\infty} \frac{(\lambda(T-t))^n}{n!} \delta_{x+n}, \qquad 0 \leq t \leq T, \ x \in \mathbb{R},$$

satisfies the Chapman-Kolmogorov equation: this can be proved proceeding as[12] in Example 2.6.5 in [113] on the sum of independent Poisson random variables. The Markov process associated with p is called the *Poisson process* and will be studied in Chap. 5. For any $\varphi \in bC$ and $t > 0$ the function

$$x \longmapsto \int_{\mathbb{R}} \text{Poisson}_{x,\lambda t}(dy)\varphi(y) = e^{-\lambda t} \sum_{n=0}^{+\infty} \frac{(\lambda t)^n}{n!} \varphi(x+n)$$

is continuous and therefore *the Poisson process is a Feller process*.

Example 2.4.6 (Gaussian Transition Law [!]) Consider the Gaussian transition law of Example 2.1.9:

$$p(t,x;T,H) := \int_H \Gamma(t,x;T,y)dy, \qquad 0 \le t < T, \ x \in \mathbb{R}, \ H \in \mathscr{B},$$

where

$$\Gamma(t,x;T,y) = \frac{1}{\sqrt{2\pi(T-t)}} e^{-\frac{(x-y)^2}{2(T-t)}}, \qquad 0 \le t < T, \ x,y \in \mathbb{R},$$

is the Gaussian transition density. The Gaussian transition law satisfies the Chapman-Kolmogorov equation as it is verified directly by calculating the convolution of two Gaussians or, more easily, the product of their characteristic functions. We will study later, in Chap. 4, the Markov process associated with p,

---

[12] For $0 \le t < s < T$, we have

$$\int_{\mathbb{R}} p(t,x;s,dy) p(s,y;T,H) = e^{-\lambda(s-t)} \sum_{n=0}^{+\infty} \frac{(\lambda(s-t))^n}{n!} p(s, x+n; T, H)$$

$$= e^{-\lambda(T-t)} \sum_{n,m=0}^{+\infty} \frac{(\lambda(s-t))^n}{n!} \frac{(\lambda(T-s))^m}{m!} \delta_{x+n+m}(H) =$$

(by the change of indices $i = n + m$ and $j = n$)

$$= e^{-\lambda(T-t)} \sum_{i=0}^{+\infty} \sum_{j=0}^{i} \lambda^i \frac{(s-t)^j (T-s)^{i-j}}{j! (i-j)!} \delta_{x+i}(H)$$

$$= e^{-\lambda(T-t)} \sum_{i=0}^{+\infty} \frac{\lambda^i}{i!} \delta_{x+i}(H) \sum_{j=0}^{i} \binom{i}{j} (s-t)^j (T-s)^{i-j}$$

$$= p(t,x;T,H).$$

the so-called *Brownian motion*. For any $\varphi \in bC$ and $T > 0$ the function

$$x \longmapsto \int_{\mathbb{R}} \Gamma(0, x; T, y)\varphi(y)dy \qquad (2.4.5)$$

is continuous and therefore the *Brownian motion is a Feller process*. Actually, one verifies that the function in (2.4.5) is C^{∞} for each $T > 0$ and $\varphi \in b\mathscr{B}$ (not just for $\varphi \in bC$): for this reason we say that Brownian motion verifies the *strong Feller property*.

Remark 2.4.7 (Transition Law and Semigroups) For each transition law $p = p(t, x; T, \cdot)$, there exists a corresponding family $\mathbf{p} = (\mathbf{p}_{t,T})_{0 \leq t \leq T}$ of linear and bounded operators

$$\mathbf{p}_{t,T} : b\mathscr{B} \longrightarrow b\mathscr{B}$$

defined by

$$\mathbf{p}_{t,T}\varphi := \int_{\mathbb{R}} p(t, \cdot; T, dy)\varphi(y), \qquad \varphi \in b\mathscr{B}.$$

Note that $\mathbf{p}_{t,T}\varphi \in b\mathscr{B}$ for every $\varphi \in b\mathscr{B}$ and by Jensen's inequality we have

$$\|\mathbf{p}_{t,T}\varphi\|_{\infty} \leq \|\varphi\|_{\infty}.$$

The Chapman-Kolmogorov equation (2.4.4) corresponds to the so-called *semigroup property* of $\mathbf{p}$:

$$\mathbf{p}_{t,s} \circ \mathbf{p}_{s,T} = \mathbf{p}_{t,T}, \qquad t \leq s \leq T.$$

The family $\mathbf{p} = (\mathbf{p}_{t,T})_{0 \leq t \leq T}$ is called the *semigroup of operators* associated with the transition law p. Moreover, we say that $\mathbf{p}$ is a homogeneous semigroup if $\mathbf{p}_{t,T} = \mathbf{p}_{0,T-t}$ for every $t \leq T$: in this case, we simply write $\mathbf{p}_t$ instead of $\mathbf{p}_{0,t}$. There are many monographs on Markov processes and semigroup theory: among the most recent, we mention [71, 142] and [138].

2.5 Characteristic Operator and Kolmogorov Equations

Let X be a stochastic process on the space $(\Omega, \mathscr{F}, P, \mathscr{F}_t)$. In various applications, there is a notable interest in calculating the conditional expectation

$$E\left[\varphi(X_T) \mid \mathscr{F}_t\right], \qquad 0 \leq t < T,$$

where $\varphi \in b\mathscr{B}$ is a given function. The problem is not trivial, even from a computational standpoint, because such a conditional expectation is an $\mathscr{F}_t$-measurable random variable, i.e., it depends on the information up to time t, which in mathematical terms translates into a *functional* dependency. However, if X is a Markov process with transition law p then, by the memoryless property, we have

$$E\left[\varphi(X_T) \mid \mathscr{F}_t\right] = u(t, X_t) \qquad (2.5.1)$$

where

$$u(t, x) := \int_{\mathbb{R}^N} p(t, x; T, dy)\varphi(y), \qquad 0 \le t \le T, \; x \in \mathbb{R}^N. \qquad (2.5.2)$$

Thus, the problem reduces to determining u as a function of real variables: this is a significant advantage of Markov processes.

In this section, we show that, as a consequence of the Chapman-Kolmogorov equation, the function u in (2.5.2) solves a Cauchy problem for which theoretical results and efficient numerical computation methods are available. More generally, we prove that, under appropriate assumptions, the transition law $p = p(t, x; T, dy)$ solves the so-called *Kolmogorov backward and forward equations*: these are integro-differential equations solved by $p(t, x; T, dy)$ in the *backward variables* (t, x) (corresponding to the *initial* time and value of the process X) and in the *forward variables* (T, y) (corresponding to the *final* time and value of the process X), respectively.

Notation 2.5.1 Given a function $f = f(t, T)$, with $t < T$, we use the notation

$$\lim_{T-t \to 0^+} f(t, T) := \lim_{T \to t^+} f(t, T) = \lim_{t \to T^-} f(t, T)$$

when the second and third limits exist and coincide.

Definition 2.5.2 (Characteristic Operator) Let p be a transition law on $\mathbb{R}^N$. Suppose that the limit

$$\mathscr{A}_t\varphi(x) := \lim_{T-t \to 0^+} \int_{\mathbb{R}^N} \frac{p(t, x; T, dy) - p(t, x; t, dy)}{T - t} \varphi(y)$$

exists for every $(t, x) \in \mathbb{R}_{>0} \times \mathbb{R}^N$ and $\varphi \in \mathscr{D}$ where $\mathscr{D}$ is a suitable subspace of $b\mathscr{B}_N$, the space of measurable and bounded functions from $\mathbb{R}^N$ to $\mathbb{R}$. Then we say that $\mathscr{A}_t$ is *the characteristic operator (or infinitesimal generator) of p*. If p is the transition law of a Markov process X, then we also say that $\mathscr{A}_t$ is the characteristic operator of X.

Note that $\mathscr{A}_t$ is a linear operator on $\mathscr{D}$. The "domain" $\mathscr{D}$ on which the characteristic operator is defined depends on the transition law p: in the following

2.5 Characteristic Operator and Kolmogorov Equations

sections we present some particular cases in which $\mathscr{D}$ can be explicitly determined. Let us start with the following simple

Example 2.5.3 ([!]) Consider the deterministic Markov process $X_t = \gamma(t)$ from Example 2.1.4. A transition law of X is

$$p(t, x; T, \cdot) = \delta_{x+\gamma(T)-\gamma(t)} \qquad (2.5.3)$$

and therefore

$$\mathscr{A}_t \varphi(x) = \lim_{T-t \to 0^+} \frac{\varphi(x + \gamma(T) - \gamma(t)) - \varphi(x)}{T - t} =$$

(assuming $\varphi \in \mathscr{D} := bC^1(\mathbb{R}^N)$, the vector space of bounded and C^1 functions, and expanding in a first-order Taylor series)

$$= \lim_{T-t \to 0^+} \frac{1}{T - t} \left(\nabla \varphi(x) \cdot (\gamma(T) - \gamma(t)) + o(|\gamma(T) - \gamma(t)|) \right).$$

Such a limit exists only if the function γ is sufficiently regular: in particular, if γ is differentiable then we have

$$\mathscr{A}_t \varphi(x) = \gamma'(t) \cdot \nabla \varphi(x).$$

In this case, the characteristic operator is simply the directional derivative of φ along the curve γ: precisely, $\mathscr{A}_t$ is the first-order differential operator with constant coefficients

$$\mathscr{A}_t = \gamma'(t) \cdot \nabla = \sum_{j=1}^{N} \gamma_j'(t) \partial_{x_j}.$$

Remark 2.5.4 ([!]) Since $p(t, x; t, \cdot) = \delta_x$ for every $t \geq 0$, we have

$$\mathscr{A}_t \varphi(x) = \lim_{T-t \to 0^+} \int_{\mathbb{R}^N} p(t, x; T, dy) \frac{\varphi(y) - \varphi(x)}{T - t}. \qquad (2.5.4)$$

Hence, if p is the transition law of a Markov process X, we have

$$\mathscr{A}_t \varphi(x) = \lim_{T-t \to 0^+} E\left[\frac{\varphi(X_T) - \varphi(X_t)}{T - t} \mid X_t = x \right]. \qquad (2.5.5)$$

Notice, in particular, that the characteristic operator $\mathscr{A}_t$ depends on the process X and not on the specific version of its transition law. By (2.5.5), in analogy with Example 2.5.3, we can interpret $\mathscr{A}_t \varphi(x)$ as an "average directional derivative" (or average infinitesimal increment) of φ along the trajectories of X starting at time t from x. Let us also note that

$$\mathscr{A}_t\varphi(x) = - \lim_{T-t\to 0^+} \int_{\mathbb{R}^N} \frac{p(T, x; T, dy) - p(t, x; T, dy)}{T - t} \varphi(y). \qquad (2.5.6)$$

In the following section, we show that for a wide class of transition laws, it is possible to give a more detailed representation of the characteristic operator.

2.5.1 The Local Case

Definition 2.5.5 Let $x_0 \in \mathbb{R}^N$. We say that a linear operator $\mathscr{A} : C^2(\mathbb{R}^N) \longrightarrow \mathbb{R}$

- *satisfies the maximum principle at x_0* if $\mathscr{A}\varphi \leq 0$ for any $\varphi \in C^2(\mathbb{R}^N)$ such that $\varphi(x_0) = \max_{x \in \mathbb{R}^N} \varphi(x)$;
- is *local at x_0* if $\mathscr{A}\varphi = 0$ for every $\varphi \in C^2(\mathbb{R}^N)$ that vanishes in a neighborhood of x_0.

Remark 2.5.6 We note that:

(i) if $\mathscr{A}$ satisfies the maximum principle at x_0 then $\mathscr{A}\varphi = 0$ for every constant function φ;
(ii) if $\mathscr{A}$ is a local operator at x_0 then $\mathscr{A}\varphi = \mathscr{A}\psi$ for every φ, ψ that are equal in a neighborhood of x_0;
(iii) combining (i) and (ii) we have that if $\mathscr{A}$ satisfies the maximum principle and is local at x_0 then $\mathscr{A}\varphi = 0$ for every φ that is constant in a neighborhood of x_0;
(iv) if $\mathscr{A}$ satisfies the maximum principle and is local at x_0 then $\mathscr{A}\varphi = \mathscr{A}\mathbf{T}_{2,x_0}(\varphi)$ where $\mathbf{T}_{2,x_0}(\varphi)$ is the second-order Taylor polynomial of φ with initial point x_0.

Indeed, since $\mathscr{A}$ is a linear operator, it is enough to prove that $\mathscr{A}\varphi = 0$ for every $\varphi \in C^2(\mathbb{R}^N)$ whose second-order Taylor polynomial with initial point x_0 is null. Moreover, it is not restrictive to assume $x_0 = 0$. Consider a "cut-off" function $\chi \in C_0^\infty(\mathbb{R}^N; \mathbb{R})$ such that $0 \leq \chi \leq 1$, $\chi(x) \equiv 1$ for $|x| \leq 1$ and $\chi(x) \equiv 0$ for $|x| \geq 2$. Letting $\varphi_\delta(x) = \varphi(x)\chi\left(\frac{x}{\delta}\right)$ for $\delta > 0$, there exists[13] a

---

[13] By assumption, $|\varphi(x)| \leq |x|^2 g(|x|)$ for $|x| \leq 1$ with g going to zero as $|x| \to 0^+$ and it is not restrictive to assume g monotonically increasing. Then (2.5.7) follows from the fact that

$$g(|x|)\chi\left(\frac{x}{\delta}\right) \leq \chi(x)g(\delta), \qquad x \in \mathbb{R}^N, \ 0 < \delta \leq \frac{1}{2}.$$

2.5 Characteristic Operator and Kolmogorov Equations

function g such that $g(\delta) \to 0$ as $\delta \to 0^+$ and

$$|\varphi_\delta(x)| \le g(\delta)|x|^2 \chi(x), \qquad x \in \mathbb{R}^N,\ 0 < \delta \le \frac{1}{2}. \tag{2.5.7}$$

Then, applying the maximum principle at 0 to the functions $\psi_\delta^\pm(x) = -g(\delta)|x|^2\chi(x) \pm \varphi_\delta(x)$, we obtain $\mathscr{A}\psi_\delta^\pm \le 0$ or equivalently, by point (i),

$$\pm \mathscr{A}\varphi = \pm \mathscr{A}\varphi_\delta \le g(\delta)\mathscr{A}\psi, \qquad \psi(x) := |x|^2\chi(x).$$

The thesis is follows since $\delta > 0$ is arbitrarily small.

The following result, which is a particular case of Courrège's theorem [26], provides an interesting characterization of local linear operators that satisfy the maximum principle.

Theorem 2.5.7 (Courrège's Theorem) *A linear operator $\mathscr{A}$ on $C^2(\mathbb{R}^N)$ satisfies the maximum principle and is local at $x_0 \in \mathbb{R}^N$ if and only if there exist $b \in \mathbb{R}^N$ and a symmetric and positive semidefinite $\mathscr{C} = (c_{ij})_{1 \le i,j \le N}$ such that*

$$\mathscr{A}\varphi = \frac{1}{2}\sum_{i,j=1}^N c_{ij}\partial_{x_i x_j}\varphi(x_0) + \sum_{i=1}^N b_i \partial_{x_i}\varphi(x_0), \qquad \varphi \in C^2(\mathbb{R}^N). \tag{2.5.8}$$

Proof By Remark 2.5.6 we have

$$\mathscr{A}\varphi = \mathscr{A}\mathbf{T}_{2,x_0}(\varphi) =$$

(by the linearity of $\mathscr{A}$)

$$= \frac{1}{2}\sum_{i,j=1}^N c_{ij}\partial_{x_i x_j}\varphi(x_0) + \sum_{i=1}^N b_i \partial_{x_i}\varphi(x_0)$$

where $c_{ij} := \mathscr{A}\varphi_{ij}$ and $b_j := \mathscr{A}\varphi_j$ with

$$\varphi_{ij}(x) = (x - x_0)_i (x - x_0)_j, \qquad \varphi_j(x) = (x - x_0)_j, \qquad x \in \mathbb{R}^N. \tag{2.5.9}$$

To check that $\mathscr{C} = (c_{ij}) \ge 0$, consider $\eta \in \mathbb{R}^N$ and set

$$\varphi_\eta(x) = -\langle x - x_0, \eta \rangle^2 = -\sum_{i,j=1}^N \eta_i \eta_j \varphi_{ij}(x);$$

then by linearity and by the maximum principle at x_0 we have

$$\mathscr{A}\varphi_\eta = -2\langle \mathscr{C}\eta, \eta \rangle \le 0.$$

Conversely, if $\mathscr{A}$ is of the form (2.5.8) then it is clearly local at x_0. Moreover, there exists a symmetric and positive semi-definite matrix $M = (m_{ij})$ such that

$$\mathscr{C} = M^2 = \left(\sum_{h=1}^{N} m_{ih} m_{hj}\right)_{i,j} = \left(\sum_{h=1}^{N} m_{ih} m_{jh}\right)_{i,j}.$$

If x_0 is a maximum point for φ then $\nabla\varphi(x_0) = 0$ and the Hessian matrix of φ in x_0 is negative semi-definite, so we have

$$\mathscr{A}\varphi = \frac{1}{2}\sum_{i,j=1}^{N} \partial_{x_i x_j}\varphi(x_0) \sum_{h=1}^{N} m_{ih} m_{jh} = \frac{1}{2}\sum_{h=1}^{N}\sum_{i,j=1}^{N} \partial_{x_i x_j}\varphi(x_0) m_{ih} m_{jh} \leq 0,$$

that is, $\mathscr{A}$ satisfies the maximum principle at x_0. □

Remark 2.5.8 ([!]) For every $x \in \mathbb{R}^N$, the characteristic operator $\mathscr{A}_t$ of a transition law p satisfies the maximum principle at x: this follows immediately from (2.5.4). Then, under the further assumption that $\mathscr{A}_t$ is local[14] at x, Theorem 2.5.7 provides the representation

$$\mathscr{A}_t\varphi(x) = \frac{1}{2}\sum_{i,j=1}^{N} c_{ij}(t,x)\partial_{x_i x_j}\varphi(x) + \sum_{i=1}^{N} b_i(t,x)\partial_{x_i}\varphi(x), \qquad (t,x) \in \mathbb{R}_{>0} \times \mathbb{R}^N,$$
(2.5.10)

where $\mathscr{C}(t,x) = (c_{ij}(t,x))$ is an $N \times N$ symmetric, positive semi-definite matrix and $b(t,x) = (b_j(t,x)) \in \mathbb{R}^N$. In other words, $\mathscr{A}_t$ *is a second-order partial differential operator of elliptic-parabolic type*.

Combining (2.5.4) with the expression of the coefficients of $\mathscr{A}_t$ given by the functions in (2.5.9), we obtain the formulas[15]

$$b_i(t,x) = \lim_{T-t\to 0^+} \int_{\mathbb{R}^N} \frac{p(t,x;T,dy)}{T-t}(y-x)_i$$

$$= \lim_{T-t\to 0^+} E\left[\frac{(X_T - X_t)_i}{T-t} \mid X_t = x\right], \qquad (2.5.11)$$

---

[14] It can be shown that the property of being local corresponds to the *continuity* of the trajectories of the associated Markov process. For the characterization of the characteristic operator of a generic Markov process, see, for example, [132].

[15] If $\mathscr{A}_t$ is local at x then the integration domain in (2.5.11) and (2.5.12) can be restricted to $|x - y| < 1$.

2.5 Characteristic Operator and Kolmogorov Equations

$$c_{ij}(t,x) = \lim_{T-t\to 0^+} \int_{\mathbb{R}^N} \frac{p(t,x;T,dy)}{T-t}(y-x)_i(y-x)_j$$

$$= \lim_{T-t\to 0^+} E\left[\frac{(X_T-X_t)_i(X_T-X_t)_j}{T-t} \;\Big|\; X_t = x\right], \qquad (2.5.12)$$

for $i, j = 1, \ldots, N$. Hence, *the coefficients of $\mathscr{A}_t$ represent the infinitesimal increments of the mean and covariance matrix*[16] *of the process X as it starts from (t, x)*. From formulas (2.5.11) and (2.5.12) it also follows that $c_{ij} = c_{ij}(t, x)$ and $b_j = b_j(t, x)$ are Borel measurable functions on $\mathbb{R}_{>0} \times \mathbb{R}^N$.

2.5.2 Backward Kolmogorov Equation

Let p be the transition law of a Markov process X. We exploit the Chapman-Kolmogorov equation to study the conditional expectation function in (2.5.2), defined by

$$u(t,x) := \int_{\mathbb{R}^N} p(t,x;T,dy)\varphi(y) = E[\varphi(X_T) \mid X_t = x], \quad 0 \le t \le T, \; x \in \mathbb{R}^N, \tag{2.5.13}$$

for $\varphi \in b\mathscr{B}$. If it exists, the derivative $\partial_t u(t,x)$ is given by

$$\partial_t u(t,x) = \lim_{h\to 0^+} \int_{\mathbb{R}^N} \frac{p(t,x;T,dy) - p(t-h,x;T,dy)}{h}\varphi(y) =$$

---

[16] Notice that

$$c_{ij}(t,x) = \lim_{T-t\to 0^+} \int_{\mathbb{R}^N} \frac{p(t,x;T,dy)}{T-t}(y-x-(T-t)b(t,x))_i(y-x-(T-t)b(t,x))_j$$

$$= \lim_{T-t\to 0^+} E\left[\frac{(X_T-X_t-(T-t)b(t,X_t))_i(X_T-X_t-(T-t)b(t,X_t))_j}{T-t} \;\Big|\; X_t = x\right]$$

as can be verified by expanding the product inside the integral and observing that

$$\lim_{T-t\to 0^+}(T-t)\int_{\mathbb{R}^N} p(t,x;T,dy)b_i(t,x)b_j(t,x) = \lim_{T-t\to 0^+}\int_{\mathbb{R}^N} p(t,x;T,dy)(y-x)_i b_j(t,x) = 0.$$

(by the Chapman-Kolmogorov equation)

$$= \lim_{h \to 0^+} \int_{\mathbb{R}^N} \frac{p(t, x; t, dz) - p(t - h, x; t, dz)}{h} \underbrace{\int_{\mathbb{R}^N} p(t, z; T, dy)\varphi(y)}_{=u(t,z)}$$

$$= -\mathscr{A}_t u(t, x) \tag{2.5.14}$$

based on the definition of the characteristic operator in the form (2.5.6). The previous steps are justified rigorously under the assumption that $u(t, \cdot) \in \mathscr{D}$: in Example 2.5.12 this assumption is satisfied if $\varphi \in C^1(\mathbb{R}^N)$ since $x \mapsto u(t, x) = \varphi(x + \gamma(T) - \gamma(t))$ inherits the regularity properties of φ. We will examine later other significant examples in which $u(t, \cdot) \in bC^2(\mathbb{R}^N)$ thanks to the regularizing properties of the kernel $p(t, x; T, dy)$.

Therefore, at least formally, the function u in (2.5.13) solves the Cauchy problem for the backward Kolmogorov equation[17] (with final datum)

$$\begin{cases} \partial_t u(t, x) + \mathscr{A}_t u(t, x) = 0, & (t, x) \in [0, T[\times \mathbb{R}^N, \\ u(T, x) = \varphi(x), & x \in \mathbb{R}^N, \end{cases} \tag{2.5.15}$$

or in integral form

$$u(t, x) = \varphi(x) + \int_t^T \mathscr{A}_s u(s, x) ds, \qquad (t, x) \in [0, T] \times \mathbb{R}^N.$$

We emphasize that problem (2.5.15) is written in the *backward variables* (t, x) assuming the *forward time* T fixed.

Example 2.5.9 ([!]) Consider the Gaussian transition law $p(t, x; T, dy) = \Gamma(t, x; T, y) dy$ of Example 2.1.9 with transition density defined by

$$\Gamma(t, x; T, y) = \frac{1}{\sqrt{2\pi(T-t)}} e^{-\frac{(x-y)^2}{2(T-t)}}, \qquad 0 \le t < T, \; x, y \in \mathbb{R}. \tag{2.5.16}$$

---

[17] Being $u(t, x) = \int_{\mathbb{R}^N} p(t, x; T, dy) \varphi(y)$, it is also customary to say that the transition law $(t, x) \mapsto p(t, x; T, dy)$ solves the backward problem

$$\begin{cases} \partial_t p(t, x; T, dy) + \mathscr{A}_t p(t, x; T, dy) = 0, & (t, x) \in [0, T[\times \mathbb{R}^N, \\ p(T, x; T, \cdot) = \delta_x, & x \in \mathbb{R}^N, \end{cases}$$

in the backward variables (t, x).

2.5 Characteristic Operator and Kolmogorov Equations

The Markov process associated with p is the Brownian motion that will be introduced in Chap. 4. A direct calculation shows that

$$\partial_t \Gamma(t, x; T, y) = -\partial_T \Gamma(t, x; T, y) = \frac{T - t - (x - y)^2}{2(T - t)^2} \Gamma(t, x; T, y),$$

$$\partial_x \Gamma(t, x; T, y) = -\partial_y \Gamma(t, x; T, y) = \frac{y - x}{T - t} \Gamma(t, x; T, y),$$

$$\partial_{xx} \Gamma(t, x; T, y) = \partial_{yy} \Gamma(t, x; T, y) = -\frac{T - t - (x - y)^2}{(T - t)^2} \Gamma(t, x; T, y),$$

from which we obtain the backward Kolmogorov equation

$$\left(\partial_t + \frac{1}{2}\partial_{xx}\right) \Gamma(t, x; T, y) = 0, \qquad t < T, \ x, y \in \mathbb{R} \tag{2.5.17}$$

and also

$$\left(\partial_T - \frac{1}{2}\partial_{yy}\right) \Gamma(t, x; T, y) = 0, \qquad t < T, \ x, y \in \mathbb{R} \tag{2.5.18}$$

which is called *forward Kolmogorov equation* and will be studied in Sect. 2.5.3. The characteristic operator of p is the Laplace operator

$$\mathscr{A}_t = \frac{1}{2}\partial_{xx}$$

as can also be verified using formulas (2.5.11) and (2.5.12) which here become

$$b(t, x) = \lim_{T-t \to 0^+} \int_{\mathbb{R}^N} \frac{\Gamma(t, x; T, y)}{T - t}(y - x) dy = 0,$$

$$c(t, x) = \lim_{T-t \to 0^+} \int_{\mathbb{R}^N} \frac{\Gamma(t, x; T, y)}{T - t}(y - x)^2 dy = 1.$$

Obviously, $\mathscr{A}_t$ is a local operator at every $x \in \mathbb{R}$.

Equations (2.5.17) and (2.5.18) are well known for their importance in physics and economics:

- (2.5.18) is also called *forward heat equation* and intervenes in models that describe the physical phenomenon of heat diffusion in a body. Precisely, the solution $v = v(T, y)$ of the forward Cauchy problem

$$\begin{cases} \partial_T v(T, y) = \frac{1}{2}\partial_{yy} v(T, y), & (T, y) \in]t, +\infty[\times \mathbb{R}, \\ v(t, y) = \varphi(y), & y \in \mathbb{R}, \end{cases} \tag{2.5.19}$$

represents the temperature, at time T and position y, of an infinitely long body with assigned temperature φ at the initial time t;
- (2.5.17) is called *backward heat equation* and intervenes naturally in mathematical finance, in the valuation of certain complex financial instruments, called *derivatives*, of which the value φ is known at the future time T: the price at time $t < T$ is given by the solution $u = u(t, x)$ of the backward Cauchy problem

$$\begin{cases} \partial_t u(t, x) + \frac{1}{2} \partial_{xx} u(t, x) = 0, & (t, x) \in [0, T[\times \mathbb{R}, \\ u(T, x) = \varphi(x), & x \in \mathbb{R}. \end{cases} \quad (2.5.20)$$

Note that, if v denotes the solution of the forward problem (2.5.19) with initial time $t = 0$, then $u(t, x) := v(T - t, x)$ solves the backward problem (2.5.20); moreover, u is given by formula (2.5.13) which here becomes

$$u(t, x) = \int_{\mathbb{R}} \Gamma(t, x; T, y) \varphi(y) dy, \qquad (t, x) \in [0, T] \times \mathbb{R}. \quad (2.5.21)$$

By exchanging signs of derivative and integral, one can prove that $u \in C^\infty([0, T[\times \mathbb{R})$ and $\|u\|_\infty \leq \|\varphi\|_\infty$ for every $\varphi \in b\mathscr{B}$ and this justifies the validity of (2.5.14).

Remark 2.5.10 In the theory of differential equations, Γ in (2.5.16) is called *fundamental solution of the heat operator* since, through the resolutive formula (2.5.21), it provides the solution of the backward problem (2.5.20) *for every* final datum $\varphi \in bC$ (and similarly of the forward problem (2.5.19) *for every* initial datum $\varphi \in bC$). We refer to Sect. 20.2 for the general definition of fundamental solution.

A deep connection between the theory of stochastic processes and that of partial differential equations is given by the fact that, if it exists, *the transition density of a Markov process (for example, the Gaussian density in the case of a Brownian motion) is the fundamental solution of the Kolmogorov equations (corresponding to the heat equations in the case of a Brownian motion)*. A general treatment on the existence and uniqueness of the solution of the Cauchy problem for partial differential equations of parabolic type is given in Chap. 20, while in Chap. 15 we deepen the connection with stochastic differential equations.

Example 2.5.11 ([!]) Consider the Poisson transition law with parameter $\lambda > 0$ of Example 2.4.5:

$$p(t, x; T, \cdot) = \text{Poisson}_{x, \lambda(T-t)}$$

$$:= e^{-\lambda(T-t)} \sum_{n=0}^{+\infty} \frac{(\lambda(T-t))^n}{n!} \delta_{x+n}, \qquad 0 \leq t \leq T, \ x \in \mathbb{R}.$$

2.5 Characteristic Operator and Kolmogorov Equations

For u as in (2.5.13) we have

$$\partial_t u(t,x) = \partial_t \left(e^{-\lambda(T-t)} \sum_{n \geq 0} \varphi(x+n) \frac{(\lambda(T-t))^n}{n!} \right)$$

$$= \lambda e^{-\lambda(T-t)} \sum_{n \geq 0} \varphi(x+n) \frac{(\lambda(T-t))^n}{n!}$$

$$+ e^{-\lambda(T-t)} \partial_t \sum_{n \geq 0} \varphi(x+n) \frac{(\lambda(T-t))^n}{n!} =$$

(the exchange of series-derivative is justified by the fact that it is a series of powers with infinite convergence radius if $\varphi \in b\mathscr{B}$)

$$= \lambda u(t,x) - \lambda e^{-\lambda(T-t)} \sum_{n \geq 1} \varphi(x+n) \frac{(\lambda(T-t))^{n-1}}{(n-1)!}$$

$$= \lambda u(t,x) - \lambda e^{-\lambda(T-t)} \sum_{n \geq 0} \varphi(x+n+1) \frac{(\lambda(T-t))^n}{n!}$$

$$= -\lambda \left(u(t, x+1) - u(t,x) \right).$$

Hence $\mathscr{A}_t$ is defined by

$$\mathscr{A}_t \varphi(x) = \lambda \left(\varphi(x+1) - \varphi(x) \right), \qquad \varphi \in \mathscr{D} := b\mathscr{B}.$$

In this case, $\mathscr{A}_t$ is a *non-local operator* at any $x \in \mathbb{R}$.

2.5.3 Forward Kolmogorov (or Fokker-Planck) Equation

Assume that p is the transition law of a Markov process X. By definition of characteristic operator and assuming the existence of the derivative $\partial_T p(t,x;T,dz)$, for every $\varphi \in \mathscr{D}$ we have

$$\int_{\mathbb{R}^N} \partial_T p(t,x;T,dz) \varphi(z) = \int_{\mathbb{R}^N} \lim_{h \to 0^+} \frac{p(t,x;T+h,dz) - p(t,x;T,dz)}{h} \varphi(z) =$$

(by the Chapman-Kolmogorov equation)

$$= \int_{\mathbb{R}^N} p(t,x;T,dy) \lim_{h \to 0^+} \int_{\mathbb{R}^N} \frac{p(T,y;T+h,dz) - p(T,y;T,dz)}{h} \varphi(z)$$

$$= \int_{\mathbb{R}^N} p(t,x;T,dy) \mathscr{A}_T \varphi(y).$$

In conclusion, we have

$$\int_{\mathbb{R}^N} \partial_T p(t, x; T, dy)\varphi(y) = \int_{\mathbb{R}^N} p(t, x; T, dy)\mathscr{A}_T\varphi(y), \qquad \varphi \in \mathscr{D}, \tag{2.5.22}$$

which is called the *forward Kolmogorov equation* or also the *Fokker-Planck equation*. Here φ must be interpreted as a test function and (2.5.22) as the weak (or distributional) form of the equation

$$\partial_T p(t, x; T, \cdot) = \mathscr{A}_T^* p(t, x; T, \cdot)$$

where $\mathscr{A}_T^*$ denotes the adjoint operator of $\mathscr{A}_T$. For example, if $\mathscr{A}_T$ is a differential operator of the form (2.5.10) then $\mathscr{A}_T^*$ is obtained formally by integration by parts:

$$\int_{\mathbb{R}^N} \left(\mathscr{A}_T^* u(y) \right) v(y) dy = \int_{\mathbb{R}^N} u(y) \mathscr{A}_T v(y) dy,$$

for any pair of test functions u, v. If the coefficients are sufficiently regular, it is possible to write the forward operator more explicitly:

$$\mathscr{A}_T^* u = \frac{1}{2} \sum_{i,j=1}^N c_{ij} \partial_{y_i y_j} u + \sum_{j=1}^N b_j^* \partial_{y_j} + a^*, \tag{2.5.23}$$

where

$$b_j^* := -b_j + \sum_{i=1}^N \partial_{y_i} c_{ij}, \qquad a^* := -\sum_{i=1}^N \partial_{y_i} b_i + \frac{1}{2} \sum_{i,j=1}^N \partial_{y_i y_j} c_{ij}. \tag{2.5.24}$$

Formula (2.5.22) is also expressed by stating that $p(t, x; \cdot, \cdot)$ is a *distributional solution* of the forward Cauchy problem (with initial datum)

$$\begin{cases} \partial_T p(t, x; T, \cdot) = \mathscr{A}_T^* p(t, x; T, \cdot), & T > t, \\ p(t, x; t, \cdot) = \delta_x. \end{cases} \tag{2.5.25}$$

The term "distributional solution" is used to indicate the fact that $p(t, x; T, \cdot)$, being a distribution, does not generally have the regularity required to support the operator $\mathscr{A}_T$ which in fact appears in (2.5.22) applied to the test function φ. Note that the problem (2.5.25) is written in the *forward variables* (T, y) on $]t, +\infty[\times\mathbb{R}^N$, assuming fixed the *backward variables* (t, x).

The existence of the distributional solution of (2.5.25) can be proved under very general assumptions (see, for example, Theorem 1.1.9 in [133]): although the notion of distributional solution is very weak, this is the best result one can hope to obtain without assuming further hypotheses, as shown by the following

2.5 Characteristic Operator and Kolmogorov Equations

Example 2.5.12 ([!]) Let us resume Example 2.5.3. The operator $\mathscr{A}_t = \gamma'(t) \cdot \nabla_x$, with $\nabla_x = (\partial_{x_1}, \ldots, \partial_{x_N})$, is obviously local at every $x \in \mathbb{R}^N$: it can also be determined using formulas (2.5.11) and (2.5.12) which, for p as in (2.5.3) with γ differentiable, give

$$b(t,x) = \lim_{T-t \to 0^+} \frac{1}{T-t} \int_{\mathbb{R}^N} \delta_{x+\gamma(T)-\gamma(t)}(dy)(y-x) = \gamma'(t),$$

$$c_{ij}(t,x) = \lim_{T-t \to 0^+} \frac{1}{T-t} \int_{\mathbb{R}^N} \delta_{x+\gamma(T)-\gamma(t)}(dy)(y-x)_i(y-x)_j = 0.$$

The Cauchy problem (2.5.25) for the forward Kolmogorov equation is

$$\begin{cases} \partial_T p(t,x;T,\cdot) = -\gamma'(T) \cdot \nabla_y p(t,x;T,\cdot), & T > t, \\ p(t,x;t,\cdot) = \delta_x. \end{cases} \quad (2.5.26)$$

Clearly, since $p(t,x;T,\cdot)$ is a measure, the gradient $\nabla_y p(t,x;T,\cdot)$ is not defined in the classical sense but in the sense of distributions. Therefore, problem (2.5.26) should be understood as in (2.5.22), that is, as an integral equation where the gradient is applied to the function φ:

$$\varphi(x+\gamma(T)-\gamma(t)) = \varphi(x) + \int_t^T \gamma'(s) \cdot (\nabla \varphi)(x+\gamma(s)-\gamma(t)) ds, \qquad \varphi \in C^1(\mathbb{R}^N);$$

by differentiating, we find

$$\frac{d}{dT} \varphi(x+\gamma(T)-\gamma(t)) = \gamma'(T) \cdot (\nabla \varphi)(x+\gamma(T)-\gamma(t)).$$

Intuitively, the characteristic operator provides the infinitesimal increment (also called, the *drift*) of a process: *by removing the drift, we get a martingale*. This fact is made rigorous by the following remarkable result, which shows how to compensate a process to make it a martingale, by means of the characteristic operator.

Theorem 2.5.13 ([!]) *Let X be a Markov process with characteristic operator $\mathscr{A}_t$ defined on a domain $\mathscr{D}$. If $\psi \in \mathscr{D}$ is such that $\mathscr{A}_t \psi(X_t) \in L^1([0,T] \times \Omega)$, then the process*

$$M_t := \psi(X_t) - \int_0^t \mathscr{A}_s \psi(X_s) ds, \qquad t \in [0,T],$$

is a martingale.

Proof We have $M_t \in L^1(\Omega, P)$, for any $t \in [0, T]$, thanks to the assumptions[18] on ψ. It remains to prove that

$$E[M_t - M_s \mid \mathscr{F}_t] = 0, \qquad 0 \leq s \leq t \leq T,$$

that is

$$E\left[\psi(X_t) - \psi(X_s) - \int_s^t \mathscr{A}_r \psi(X_r) dr \mid \mathscr{F}_s\right] = 0, \qquad 0 \leq s \leq t \leq T.$$

Integrating the forward Kolmogorov equation (2.5.22) over time with $x = X_s$, we have

$$0 = \int_{\mathbb{R}^N} p(s, X_s; t, dy)\psi(y) - \psi(X_s) - \int_s^t \int_{\mathbb{R}^N} p(s, X_s; r, dy) \mathscr{A}_r \psi(y) dr =$$

(by the Markov property (2.5.1) applied to the first and last term)

$$= E[\psi(X_t) \mid \mathscr{F}_s] - \psi(X_s) - \int_s^t E[\mathscr{A}_r \psi(X_r) \mid \mathscr{F}_s] dr =$$

(since, as we will prove shortly, it is possible to exchange the time integral with the conditional expectation)

$$= E\left[\psi(X_t) - \psi(X_s) - \int_s^t \mathscr{A}_r \psi(X_r) dr \mid \mathscr{F}_s\right]$$

which proves the thesis.

To justify the exchange between the integral and the conditional expectation, we verify that the random variable

$$Z := \int_s^t E[\mathscr{A}_r \psi(X_r) \mid \mathscr{F}_s] dr$$

is a version of the conditional expectation of $\int_s^t \mathscr{A}_r \psi(X_r) dr$ given $\mathscr{F}_s$. First of all, from the fact that $E[\mathscr{A}_r \psi(X_r) \mid \mathscr{F}_s] \in m\mathscr{F}_s$ it follows that also $Z \in m\mathscr{F}_s$. Then, for every $G \in \mathscr{F}_s$, we have

$$E[Z \mathbb{1}_G] = E\left[\int_s^t E[\mathscr{A}_r \psi(X_r) \mid \mathscr{F}_s] dr \, \mathbb{1}_G\right] =$$

---

[18] We also recall that ψ is bounded since $\mathscr{D} \subseteq b\mathscr{B}_N$: this assumption is not restrictive and can be significantly weakened.

(by Fubini's theorem, given the integrability assumption on $\mathscr{A}_r\psi(X_r)$)

$$= \int_s^t E\left[E\left[\mathscr{A}_r\psi(X_r) \mid \mathscr{F}_s\right] \mathbb{1}_G\right] dr =$$

(by the properties of conditional expectation)

$$= \int_s^t E\left[\mathscr{A}_r\psi(X_r)\mathbb{1}_G\right] dr =$$

(reapplying Fubini's theorem)

$$= E\left[\int_s^t \mathscr{A}_r\psi(X_r) dr \, \mathbb{1}_G\right].$$

$\square$

2.6 Markov Processes and Diffusions

Continuous Markov processes are sometimes called *diffusions*, although it should be noted that there is no unanimous agreement on this definition in the literature. Associated with each N-dimensional diffusion are the measurable functions $b = (b_i)_{1\le i\le N}$ and $\mathscr{C} = (c_{ij})_{1\le i,j\le N}$ defined in (2.5.11) and (2.5.12); these functions are the coefficients of the characteristic operator (2.5.10):

$$\mathscr{A}_t = \frac{1}{2}\sum_{i,j=1}^N c_{ij}(t,x)\partial_{x_i x_j} + \sum_{i=1}^N b_i(t,x)\partial_{x_i}, \qquad (t,x)\in\mathbb{R}\times\mathbb{R}^N.$$

We recall that $\mathscr{C}$ is an $N\times N$ symmetric and positive semi-definite matrix.

Historically, there are two main approaches to the construction of diffusions. The first and more classical one is based on Kolmogorov's equations: specifically, the idea of A. N. Kolmogorov [69] and W. Feller [45] is to determine a transition law $p(t,x;T,dy)$ as the solution of the forward Kolmogorov equation

$$\partial_T p(t,x;T,dy) = \mathscr{A}_T^* \partial_T p(t,x;T,dy) \qquad (2.6.1)$$

associated with the initial datum $p(t,x;t,\cdot) = \delta_x$ as in (2.5.25). Equation (2.6.1) is the starting point for the study of the existence and regularity properties of a density of p through *analytical*[19] and *probabilistic*[20] techniques. Although it seems the most natural approach, Eq. (2.6.1) presents some technical difficulties due to being interpreted in a distributional sense in the forward variables and the presence of the *adjoint* operator of $\mathscr{A}_t$ whose precise definition requires appropriate regularity assumptions on the coefficients (cf. (2.5.23) and (2.5.24)). For this reason, attention has subsequently shifted to the Kolmogorov *backward* equation. The study of diffusions using the backward equation has been one of the most effective and successful approaches: Sect. 18.2 is dedicated to a summary of the main results in this regard. The main objection to the use of Kolmogorov's equations for the study of diffusions is that the tools used are predominantly analytical in nature and rely on technically complex results from the theory of partial differential equations: among these, first and foremost, the construction of the fundamental solution of parabolic equations that we will present in a synthetic way in Chap. 20.

The second approach to the construction of diffusions is the one initiated by K. Itô: it is inspired by P. Lévy's idea of considering the infinitesimal increment $X_{t+dt} - X_t$ of a diffusion as a Gaussian increment with drift $b(t, X_t)$ and covariance matrix $\mathscr{C}(t, X_t)$, consistently with Eqs. (2.5.11) and (2.5.12). Itô developed a theory of stochastic calculus based on which the previous idea can be formalized in terms of the stochastic differential equation

$$dX_t = b(t, X_t)dt + \sigma(t, X_t)dW_t, \qquad (2.6.2)$$

where W denotes a stochastic process with independent and Gaussian increments (a Brownian motion, cf. Chap. 4) and $\mathscr{C} = \sigma\sigma^*$. The primary challenge with this approach lies in defining the stochastic differential (or integral) of processes whose trajectories, while continuous, exhibit such irregularity that traditional mathematical analysis tools prove inadequate: Chap. 10 is entirely dedicated to the theory of stochastic integration in the Itô sense. Secondly, in order to construct a diffusion X as a solution of Eq. (2.6.2), existence and uniqueness results are required for such an equation: this problem has also been solved by Itô under standard assumptions of local Lipschitz continuity and linear growth of the coefficients in perfect analogy with the theory of ordinary differential equations. Subsequently, a significant step forward was made by Stroock and Varadhan [134, 135] who built a bridge between the theory of diffusions and that of martingales: Stroock and Varadhan showed that the problem of the existence of a diffusion, as a solution of (2.6.2), is equivalent to the so-called "martingale problem", i.e., the problem of the existence of a probability measure, on the canonical space of trajectories, with respect to which the compensated process of Theorem 2.5.13 is a martingale. A concise presentation of the main results by Stroock and Varadhan is provided in Chap. 18.

---

[19] The most important result in this regard is the famous Hörmander's theorem [62].

[20] Malliavin's calculus extends the mathematical field of calculus of variations from deterministic functions to stochastic processes. For a general reference see, e.g., [101].

2.7 Key Ideas to Remember

We summarize the core concepts and key insights from the chapter to facilitate comprehension, omitting the more technical or less significant details. As usual, if you have any doubt about what the following succinct statements mean, please review the corresponding section.

- Section 2.1: the *transition law* of a stochastic process $X = (X_t)_{t \geq 0}$ is the family of the conditional distributions of X_T given X_t, indexed by t, T with $t \leq T$. Two notable examples of transition laws are the Gaussian and Poisson ones.
- Section 2.2: for a *Markov process*, conditioning on $\mathscr{F}_t$ (the σ-algebra of information *up to time t*) is equivalent to conditioning on X_t: in this sense, the Markov property is a "memoryless" property.
- Section 2.3: *processes with independent increments* are Markov processes.
- Section 2.4: starting from the initial distribution and the transition law of a Markov process, it is possible to derive the finite-dimensional distributions, and therefore the law of the process: moreover, the transition law of a Markov process verifies an important identity, the *Chapman-Kolmogorov equation* (2.4.3), which expresses a consistency property between the distributions that make up the transition law.
- Section 2.5: if it exists, the average directional derivative along the trajectories of X, i.e.

$$\lim_{T-t \to 0^+} E\left[\frac{\varphi(X_T) - \varphi(X_t)}{T-t} \mid X_t = x\right] =: \mathscr{A}_t \varphi(x),$$

defines the *characteristic operator* $\mathscr{A}_t$ of the Markov process X, at least for φ in an appropriate space of functions.
- Section 2.5.1: for continuous Markov processes, $\mathscr{A}_t$ is a second-order elliptic-parabolic partial differential operator whose prototype is the Laplace operator. The coefficients of $\mathscr{A}_t$ are the infinitesimal increments of the mean and covariance matrix of X (cf. formulas (2.5.11) and (2.5.12)).
- Sections 2.5.2 and 2.5.3: the transition law is the solution of the *backward and forward Kolmogorov equations*. The prototypes of such equations are the backward and forward versions of the heat equation.
- Section 2.6: we call *diffusion* a continuous Markov process. A classical approach to the construction of diffusions consists in determining their transition law as fundamental solutions of the backward or forward Kolmogorov equation. Alternatively, diffusions are constructed as solutions of stochastic differential equations, the theory of which will be developed starting from Chap. 14.

Main notations introduced in this chapter:

Symbol	Description	Page
$p = p(t, x; T, H)$	Transition law	25
$\text{Poisson}_{x, \lambda(T-t)}$	Poisson transition law	27
$\Gamma(t, x; T, y)$	Gaussian transition density	28
X	Canonical version of the process X	29
$\mathscr{G}^X_{t,\infty} = \sigma(X_s, s \geq t)$	σ-algebra of future information on X	32
$X^{t,x}_T = X_T - X_t + x$	Translated process	34
$\mathscr{A}_t$	Characteristic operator	42
$\mathscr{A}^*_t$	Adjoint operator	52

Chapter 3
Continuous Processes

> *As far as the laws of mathematics refer to reality, they are not certain; and as far as they are certain, they do not refer to reality.*
>
> Albert Einstein

The notion of continuity for stochastic processes, although intuitive, hides some small pitfalls and must therefore be analyzed carefully.

In this chapter, I denotes a real interval of the form $I = [0, T]$ or $I = [0, +\infty[$. Moreover, $C(I)$ is the set of continuous functions mapping I to real values. In the first part of the chapter, we confirm a natural and unsurprising fact: a continuous process can be defined as a random variable with values in the space of continuous functions $C(I)$, rather than in the space $\mathbb{R}^I$ of *all* trajectories, as seen in the broader definition of a stochastic process (cf. Definition 1.1.3). Then we prove the fundamental *Kolmogorov's continuity theorem* according to which, up to modifications, one can deduce the continuity of a process from a condition on its law: this is a deep result because it allows to deduce a "pointwise" property (of individual trajectories) from a condition "in the average" (i.e. on the law of the process).

3.1 Continuity and a.s. Continuity

Definition 3.1.1 (Continuous Process) A stochastic process $X = (X_t)_{t \in I}$ on the space $(\Omega, \mathscr{F}, P)$ is almost surely (a.s.) continuous if the family of continuous trajectories

$$(X \in C(I)) := \{\omega \in \Omega \mid X(\omega) \in C(I)\}$$

is an almost sure set, i.e., it includes a certain event: $(X \in C(I)) \supseteq A$ with $A \in \mathscr{F}$ such that $P(A) = 1$.

Remark 3.1.2 (Continuity and Completeness) If the space $(\Omega, \mathscr{F}, P)$ is complete, then X is a.s. continuous if and only if $P(X \in C(I)) = 1$. If $(\Omega, \mathscr{F}, P)$ is not complete, then it is not necessarily true that $(X \in C(I))$ is an event. In fact, recall that, denoting by $\mathscr{F}^I$ the σ-algebra on $\mathbb{R}^I$ generated by cylinders, by the Definition 1.1.3 of stochastic process, we have $X^{-1}(H) \in \mathscr{F}$ for every $H \in \mathscr{F}^I$: however, by Remark 1.1.10, $C(I) \notin \mathscr{F}^I$ and therefore it is not necessarily true that $(X \in C(I)) \in \mathscr{F}$. Similarly, in an incomplete space, even if X is a.s. continuous, it is not necessarily the case that quantities such as

$$M := \sup_{t \in I} X_t, \quad J := \int_I X_t dt, \quad T := \begin{cases} \inf I^+ & \text{if } I^+ := \{t \in I \mid X_t > 0\} \neq \emptyset, \\ 0 & \text{otherwise}, \end{cases}$$
(3.1.1)

are random variables.

Remark 3.1.3 (Continuity and Almost Sure Continuity) Let X be an a.s. continuous process defined on the space $(\Omega, \mathscr{F}, P)$ and let A be as in Definition 3.1.1. Then X is indistinguishable from $\bar{X} := X \mathbb{1}_A$ which has *all continuous trajectories*.[1] More explicitly, $\bar{X}$ is defined by

$$\bar{X}(\omega) = \begin{cases} X(\omega) & \text{if } \omega \in A, \\ 0 & \text{otherwise}. \end{cases}$$

We say that $\bar{X}$ is a *continuous version* of X. Hence, provided that we switch to a continuous version, we can eliminate the term "almost surely" and consider *continuous* processes instead of *a.s. continuous* ones.

Now, one might wonder why the definition of a.s. continuous process was introduced and not directly that of a continuous process. The fact is that a stochastic process, such as the Brownian motion, is usually constructed from a given law, using Kolmogorov's extension theorem: in this way, one can only prove[2] the almost sure continuity of the trajectories and only later switch to a continuous version.

Remark 3.1.4 If $X = (X_t)_{t \in I}$, with $I = [0, 1]$, is a continuous process then M, J and T in (3.1.1) are well-defined and are random variables. In fact, it is enough to observe that

$$M = \sup_{t \in [0,1] \cap \mathbb{Q}} X_t.$$

---

[1] We cannot use $(X \in C(I))$ instead of A because if $(\Omega, \mathscr{F}, P)$ is not complete then $X \mathbb{1}_{(X \in C(I))}$ would not necessarily be a stochastic process.

[2] Actually, the argument is more subtle and will be clarified in Sect. 3.3.

Moreover, $J(\omega)$ is well-defined for each $\omega \in \Omega$, since all trajectories of X are continuous, and equals

$$J(\omega) = \lim_{n\to\infty} \frac{1}{n} \sum_{k=1}^{n} X_{\frac{k}{n}}(\omega)$$

since the integral of a continuous function is equal to the limit of Riemann sums. Finally, $(I^+ = \emptyset) = (M \le 0) \in \mathscr{F}$ and thus also

$$(T < t) = (I^+ = \emptyset) \cup \bigcup_{s \in \mathbb{Q} \cap [0,t[} (X_s > 0)$$

belongs to $\mathscr{F}$ for every $0 < t \le 1$: this is enough to prove that $T \in m\mathscr{F}$.

3.2 Canonical Version of a Continuous Process

In this section, we focus on the case $I = [0, 1]$. We recall that $C([0, 1])$ (we also write, more simply, $C[0, 1]$) is a *separable and complete metric space*, i.e., a Polish space, with the uniform metric

$$\rho_{\max}(v, w) = \max_{t \in [0,1]} |v(t) - w(t)|, \qquad v, w \in C[0, 1].$$

We consider $I = [0, 1]$ only for simplicity: the results of this section can be easily extended to the case where $I = [0, T]$ or even $I = \mathbb{R}_{\ge 0}$ considering the distance

$$\rho_{\max}(v, w) = \sum_{n \ge 1} \frac{1}{2^n} \min\left\{1, \max_{t \in [0,n]} |v(t) - w(t)|\right\}, \qquad v, w \in C(\mathbb{R}_{\ge 0}).$$

We denote by $\mathscr{B}_{\rho_{\max}}$ the Borel σ-algebra on $C[0, 1]$ (cf. Section 1.4.2 in [113]).

According to the general Definition 1.1.3, a stochastic process $X = (X_t)_{t \in I}$ is a measurable function from $(\Omega, \mathscr{F})$ to $(\mathbb{R}^I, \mathscr{F}^I)$. We now show that if X is continuous then it is possible to replace the codomain $(\mathbb{R}^I, \mathscr{F}^I)$ with $(C(I), \mathscr{B}_{\rho_{\max}})$, maintaining the measurability property with respect to the σ-algebra $\mathscr{B}_{\rho_{\max}}$. This fact is not trivial and deserves to be proven rigorously. In fact, based on Remark 1.1.10, $C[0, 1]$ itself does not belong to $\mathscr{F}^{[0,1]}$ and therefore in general $(X \in C[0, 1])$ is not an event. Similarly, the singletons $\{w\}$ are not elements of $\mathscr{F}^{[0,1]}$ and therefore even if

$$X : (\Omega, \mathscr{F}) \longrightarrow (\mathbb{R}^{[0,1]}, \mathscr{F}^{[0,1]})$$

is a stochastic process, it is not necessarily true that $(X = w)$ is an event. On the contrary, in the space $(C[0, 1], \mathscr{B}_{\varrho_{\max}})$ singletons are measurable (they are disks of radius zero in the uniform metric), that is, $\{w\} \in \mathscr{B}_{\varrho_{\max}}$ for each $w \in C[0, 1]$.

Proposition 3.2.1 *Let $X = (X_t)_{t \in [0,1]}$ be a continuous stochastic process on the space $(\Omega, \mathscr{F}, P)$. Then the map*

$$X : (\Omega, \mathscr{F}) \longrightarrow (C[0, 1], \mathscr{B}_{\varrho_{\max}})$$

is measurable.

Proof First, we show that $\mathscr{B}_{\varrho_{\max}}$ is the σ-algebra generated by the family $\widetilde{\mathscr{C}}$ of cylinders of the form[3]

$$\widetilde{C}_t(H) := \{w \in C[0, 1] \mid w(t) \in H\}, \qquad t \in [0, 1], \ H \in \mathscr{B}. \tag{3.2.1}$$

In fact, cylinders of the type (3.2.1) with H open in $\mathbb{R}$ generate $\sigma(\widetilde{\mathscr{C}})$ and are open with respect to $\varrho_{\max}$: therefore $\mathscr{B}_{\varrho_{\max}} \supseteq \sigma(\widetilde{\mathscr{C}})$.

Conversely, since $(C[0, 1], \varrho_{\max})$ is separable, every open set is a countable union of open disks. Therefore, $\mathscr{B}_{\varrho_{\max}}$ is generated by the family of open disks that are sets of the form

$$D(w, r) = \{v \in C[0, 1] \mid \varrho_{\max}(v, w) < r\},$$

where $w \in C[0, 1]$ is the center and $r > 0$ is the radius of the disk. On the other hand, each disk is obtained by countable operations of union and intersection of cylinders of $\widetilde{\mathscr{C}}$ in the following way

$$D(w, r) = \bigcup_{n \in \mathbb{N}} \bigcap_{t \in [0,1] \cap \mathbb{Q}} \{v \in C[0, 1] \mid |v(t) - w(t)| < r - \tfrac{1}{n}\}.$$

Thus, each disk belongs to $\sigma(\widetilde{\mathscr{C}})$ and this proves the opposite inclusion.

Now we prove the thesis: as just proven, we have

$$X^{-1}\left(\mathscr{B}_{\varrho_{\max}}\right) = X^{-1}\left(\sigma(\widetilde{\mathscr{C}})\right) =$$

(since X is continuous)

$$= X^{-1}\left(\sigma(\mathscr{C})\right) \subseteq \mathscr{F}$$

where the last inclusion is due to the fact that X is a stochastic process. □

---

[3] We use the "tilde" to distinguish the cylinders of continuous functions from the cylinders of $\mathbb{R}^{[0,1]}$ defined in (1.1.1).

3.2 Canonical Version of a Continuous Process

Proposition 3.2.1 allows us to give the following

Definition 3.2.2 (Law of an a.s. Continuous Process) Let $X = (X_t)_{t \in I}$ be a continuous process[4] on the space $(\Omega, \mathscr{F}, P)$. The law of X is the distribution μ_X defined on $(C(I), \mathscr{B}_{\varrho_{\max}})$ by

$$\mu_X(H) = P(X \in H), \qquad H \in \mathscr{B}_{\varrho_{\max}}.$$

Two continuous processes X and Y are equal in law (or in distribution) if $\mu_X = \mu_Y$: in this case we write $X \stackrel{d}{=} Y$.

In analogy with Definition 1.3.4 we give the following

Definition 3.2.3 (Canonical Version of an a.s. Continuous Process [!]) Let $X = (X_t)_{t \in I}$ be an a.s. continuous process defined on the space $(\Omega, \mathscr{F}, P)$ and with law μ_X. The canonical version of X is the stochastic process defined as the identity function $\mathbf{X}(w) = w$, $w \in C(I)$, on the probability space $(C(I), \mathscr{B}_{\varrho_{\max}}, \mu_X)$.

Remark 3.2.4 The main properties of the canonical version $\mathbf{X}$ are:

(i) $\mathbf{X}$ is a continuous process equal in law to X;
(ii) $\mathbf{X}$ is defined on the *Polish metric space* $(C(I), \varrho_{\max})$: this fact is relevant for the existence of the regular version of conditional probability (cf. Theorem 4.3.2 in [113]) and is crucial in the study of stochastic differential equations. In Chap. 14 we will make extensive use of the canonical version of continuous processes;
(iii) $\mathbf{X}$ is defined on a sample space in which *the outcomes are the trajectories*: $t \mapsto \mathbf{X}_t(w) \equiv w(t)$, $t \in I$. This fact allows, for example, to give an intuitive characterization of the strong Markov property (cf. Sect. 7.3).

Furthermore, the space $(C(I), \mathscr{B}_{\varrho_{\max}}, \mu_X)$ can be completed by considering as σ-algebra of events the completion of $\mathscr{B}_{\varrho_{\max}}$ with respect to μ_X (cf. Remark 1.4.3 in [113]).

Remark 3.2.5 (Skorokhod Space) The *Skorokhod space* is an extension of the space of continuous trajectories that intervenes in the study of discontinuous stochastic processes (such as, for example, the Poisson process). The Skorokhod space $\mathscr{D}(I)$ is formed by càdlàg functions (cf. Definition 5.2.2) from I to $\mathbb{R}$ or, more generally, with values in a metric space. All the results of this section extend to the case of a.s. processes with càdlàg trajectories. In particular, it is possible to define on $\mathscr{D}(I)$ a metric, the Skorokhod distance, equipped with which $\mathscr{D}(I)$ is a Polish space. Obviously $C(I)$ is a subspace of $\mathscr{D}(I)$ and it can be proved that the uniform and Skorokhod distances are equivalent on $C(I)$. The monograph [16] provides a complete treatment of the Skorokhod space and the compactness

---

[4] By Remark 3.1.3, the definition extends to the case of X a.s. continuous in an obvious way.

properties (tightness) of families of probability measures on $\mathscr{D}(I)$, in analogy with what was seen in Section 3.3.2 in [113].

3.3 Kolmogorov's Continuity Theorem

Kolmogorov's extension theorem establishes the existence of a process with a given law but does not provide information on the regularity of its trajectories. In fact, Example 1.2.6 shows that nothing can be said about the continuity of a process's trajectories based on its distribution: modifying[5] a continuous process can make it discontinuous without changing its law. For this reason, the construction of a process using Kolmogorov's extension theorem takes place in the space $\mathbb{R}^I$ of *all* the trajectories.

On the other hand, if the law of a process X satisfies suitable conditions, then there exists a continuous modification of X: the fundamental result in this regard is the classical *Kolmogorov's continuity theorem*, of which we offer various versions, with the simplest being the following

Theorem 3.3.1 (Kolmogorov's Continuity Theorem [!!!]) *Let* $X = (X_t)_{t \in [0,1]}$ *be a real stochastic process defined on a probability space* $(\Omega, \mathscr{F}, P)$. *If there exist three positive constants* c, ε, p, *with* $p > \varepsilon$, *such that*

$$E\left[|X_t - X_s|^p\right] \leq c|t-s|^{1+\varepsilon}, \qquad t, s \in [0, 1], \tag{3.3.1}$$

then X admits a modification $\widetilde{X}$ with α-Hölder continuous trajectories for every $\alpha \in [0, \frac{\varepsilon}{p}[$: *precisely, for every* $\alpha \in [0, \frac{\varepsilon}{p}[$ *and* $\omega \in \Omega$ *there exists a positive constant* $c_{\alpha,\omega}$, *which depends only on α and ω, such that*

$$|\widetilde{X}_t(\omega) - \widetilde{X}_s(\omega)| \leq c_{\alpha,\omega}|t-s|^\alpha, \qquad t, s \in [0, 1].$$

In Sect. 3.4 we give a proof of Theorem 3.3.1, inspired by the original ideas of Kolmogorov. Let us consider some examples.

Example 3.3.2 ([!]) We resume Corollary 1.3.6 and consider a Gaussian process $(X_t)_{t \in [0,1]}$ with mean function $m \equiv 0$ and covariance $c(s, t) = s \wedge t$. By definition, $(X_t, X_s) \sim \mathcal{N}_{0, C_{t,s}}$ where

$$C_{t,s} = \begin{pmatrix} t & s \wedge t \\ s \wedge t & s \end{pmatrix}$$

and therefore $X_t - X_s \sim \mathcal{N}_{0, t+s-2s \wedge t}$. It is easy to prove an estimate of the type (3.3.1): first of all, it is not restrictive to assume $s < t$ so that $X_t - X_s =$

---

[5] Here "modifying a process" means taking a modification of it.

3.3 Kolmogorov's Continuity Theorem

$\sqrt{t-s}\,Z$ with $Z \sim \mathcal{N}_{0,1}$; then, for every $p > 0$ we have

$$E\left[|X_t - X_s|^p\right] = |t-s|^{\frac{p}{2}} E\left[|Z|^p\right]$$

where $E\left[|Z|^p\right]$ is a finite constant. By Kolmogorov's continuity theorem, X admits a modification $\widetilde{X}$ which is α-Hölder for every $\alpha < \frac{p/2-1}{p} = \frac{1}{2} - \frac{1}{p}$. Given the arbitrariness of p, it follows that $\widetilde{X}$ is α-Hölder for every $\alpha < \frac{1}{2}$.

Example 3.3.3 ([!]) Let us verify the Kolmogorov's criterion (3.3.1) for the Poisson transition law. If $N_t - N_s \sim \text{Poisson}_{\lambda(t-s)}$, then for $p > 0$ we have

$$E\left[|N_t - N_s|^p\right] = e^{-\lambda(t-s)} \sum_{n=0}^{\infty} n^p \frac{(\lambda(t-s))^n}{n!} =$$

(since the first term of the series is zero)

$$= e^{-\lambda(t-s)} \sum_{n=1}^{\infty} n^p \frac{(\lambda(t-s))^n}{n!}$$

$$\geq e^{-\lambda(t-s)} \sum_{n=1}^{\infty} \frac{(\lambda(t-s))^n}{n!}$$

$$= e^{-\lambda(t-s)} \left(e^{\lambda(t-s)} - 1\right) \approx \lambda(t-s) + o(t-s)$$

for $t - s \to 0$. Thus, condition (3.3.1) is not satisfied for any value of $\varepsilon > 0$. Indeed, in Chap. 5 we will discover that the Poisson law corresponds to a process N with discontinuous trajectories.

Theorem 3.3.1 can be extended in several directions: the most interesting ones concern higher-order regularity, the extension to the case of multidimensional I, and the case of processes with values in Banach spaces. In relatively recent times, it has been observed that Kolmogorov's continuity theorem is essentially an *analytical* result that can be proved as a corollary of the Sobolev embedding theorem, in a very general version for the so-called Besov spaces. We provide here the statement given in [128].

Theorem 3.3.4 (Kolmogorov's Continuity Theorem) *[[!!!]] Let $X = (X_t)_{t \in \mathbb{R}^d}$ be a real stochastic process. If there exist $k \in \mathbb{N}_0$, $0 < \varepsilon < p$, and $\delta > 0$ such that*

$$E\left[|X_t - X_s|^p\right] \leq c|t-s|^{d+\varepsilon+kp}$$

for every $t, s \in \mathbb{R}^d$ with $|t - s| < \delta$, then X admits a modification $\widetilde{X}$ whose trajectories are differentiable up to order k, with locally α-Hölder derivatives for every $\alpha \in [0, \frac{\varepsilon}{p}[$.

Theorem 3.3.4 also extends to processes with values in a Banach space: the following example is particularly relevant in the study of stochastic differential equations.

Example 3.3.5 Let $(X_t^x)_{t \in [0,1]}$ be a family of continuous stochastic processes, indexed by $x \in \mathbb{R}^d$: as in Sect. 3.2, we consider X^x as a r.v. with values in $(C[0,1], \mathscr{B}_{\varrho_{\max}})$ which is a Banach space with the norm

$$\|X\|_\infty := \max_{t \in [0,1]} |X_t|.$$

If

$$E\left[\|X^x - X^y\|_\infty^p\right] \le c|x-y|^{d+\varepsilon}, \qquad x, y \in \mathbb{R}^d,$$

then there exists a modification $\widetilde{X}$ (i.e., we have[6] $\widetilde{X}^x = X^x$ a.s. for each $x \in \mathbb{R}^d$) such that

$$\|\widetilde{X}_t^x(\omega) - \widetilde{X}_t^y(\omega))\|_\infty \le c|x-y|^\alpha, \qquad x, y \in K,$$

for every compact subset K of $\mathbb{R}^d$ and $\alpha < \frac{\varepsilon}{p}$, with $c > 0$ depending only on ω, α and K.

3.4 Proof of Kolmogorov's Continuity Theorem

We have to prove that, if $X = (X_t)_{t \in [0,1]}$ is a real stochastic process and there exist three constants $p, \varepsilon, c > 0$ such that

$$E\left[|X_t - X_s|^p\right] \le c|t-s|^{1+\varepsilon}, \qquad t, s \in [0,1], \tag{3.4.1}$$

then X admits a modification $\widetilde{X}$ with α-Hölder continuous trajectories for every $\alpha \in [0, \frac{\varepsilon}{p}[$.

We divide the proof into four steps, of which the third is the most technical and can be skipped at a first reading.

First Step We combine Markov's inequality (3.1.2) in [113] with (3.4.1) to obtain the estimate

$$P(|X_t - X_s| \ge \lambda) \le \frac{E[|X_t - X_s|^p]}{\lambda^p} \le \frac{c|t-s|^{1+\varepsilon}}{\lambda^p}, \qquad \lambda > 0. \tag{3.4.2}$$

---

[6] In the sense that $P\left(\widetilde{X}_t^x = X_t^x, \ t \in [0,1]\right) = 1$.

3.4 Proof of Kolmogorov's Continuity Theorem

We observe that from (3.4.2) it follows that, fixing $t \in [0, 1]$, there exists the limit in probability

$$\lim_{s \to t} X_s = X_t$$

and consequently, there is also almost sure convergence. However, this is not enough to prove the thesis: in fact, the same result holds, for example, for the Poisson process which has all discontinuous trajectories (cf. (5.1.5)). Indeed, Kolmogorov realized that from (3.4.2) it is not possible to directly obtain an estimate of the increment $X_t - X_s$ *for every t, s* since $[0, 1]$ is uncountable. Thus, his idea was to first restrict t, s to the *countable* family of dyadic rationals of $[0, 1]$ defined by

$$\mathscr{D} = \bigcup_{n \geq 1} \mathscr{D}_n, \qquad \mathscr{D}_n = \left\{ \tfrac{k}{2^n} \mid k = 0, 1, \ldots, 2^n \right\}.$$

We observe that $\mathscr{D}_n \subseteq \mathscr{D}_{n+1}$ for every $n \in \mathbb{N}$. Two elements $t, s \in \mathscr{D}_n$ are called *consecutive* if $|t - s| = 2^{-n}$.

Second Step We estimate the increment $X_t - X_s$ assuming that t, s are consecutive in $\mathscr{D}_n$: by (3.4.2) we have

$$P\left(|X_{\frac{k}{2^n}} - X_{\frac{k-1}{2^n}}| \geq 2^{-n\alpha} \right) \leq c\, 2^{n(\alpha p - 1 - \varepsilon)}.$$

Then, setting

$$A_n = \left(\max_{1 \leq k \leq 2^n} |X_{\frac{k}{2^n}} - X_{\frac{k-1}{2^n}}| \geq 2^{-n\alpha} \right) = \bigcup_{1 \leq k \leq 2^n} \left(|X_{\frac{k}{2^n}} - X_{\frac{k-1}{2^n}}| \geq 2^{-n\alpha} \right),$$

by the sub-additivity of P, we have

$$P(A_n) \leq \sum_{k=1}^{2^n} P\left(|X_{\frac{k}{2^n}} - X_{\frac{k-1}{2^n}}| \geq 2^{-n\alpha} \right) \leq \sum_{k=1}^{2^n} c\, 2^{n(\alpha p - 1 - \varepsilon)} = c\, 2^{n(\alpha p - \varepsilon)}.$$

Hence, if $\alpha < \frac{\varepsilon}{p}$, we have

$$\sum_{n \geq 1} P(A_n) < \infty$$

and by Borel-Cantelli's Lemma 1.3.28 in [113] $P(A_n \text{ i.o.}) = 0$: this means that there exists $N \in \mathscr{F}$, with $P(N) = 0$, such that for every $\omega \in \Omega \setminus N$ there exists $n_{\alpha, \omega} \in \mathbb{N}$ for which

$$\max_{1 \leq k \leq 2^n} |X_{\frac{k}{2^n}}(\omega) - X_{\frac{k-1}{2^n}}(\omega)| \leq 2^{-n\alpha}, \qquad n \geq n_{\alpha, \omega}.$$

As a consequence, we also have that for every $\omega \in \Omega \setminus N$ there exists $c_{\alpha,\omega} > 0$ such that

$$\max_{1 \leq k \leq 2^n} |X_{\frac{k}{2^n}}(\omega) - X_{\frac{k-1}{2^n}}(\omega)| \leq c_{\alpha,\omega} 2^{-n\alpha}, \qquad n \in \mathbb{N}.$$

Third Step We estimate the increment $X_t - X_s$ with $t, s \in \mathcal{D}$, constructing an appropriate chain of consecutive points connecting s to t, and then using, through the triangle inequality, the estimate obtained in the previous step. Let $t, s \in \mathcal{D}$ with $s < t$: we set

$$\bar{n} = \min\{k \mid t, s \in \mathcal{D}_k\}, \qquad n = \max\{k \mid t - s < 2^{-k}\},$$

so that $n < \bar{n}$. Moreover, for $k = n + 1, \ldots, \bar{n}$, we recursively define the sequence

$$s_n = \max\{\tau \in \mathcal{D}_n \mid \tau \leq s\}, \qquad s_k = s_{k-1} + 2^{-k}\operatorname{sgn}(s - s_{k-1})$$

where $\operatorname{sgn}(x) = \frac{x}{|x|}$ if $x \neq 0$ and $\operatorname{sgn}(0) = 0$. We define $(t_k)_{n \leq k \leq \bar{n}}$ in an analogous way. Then $s_k, t_k \in \mathcal{D}_k$ and we have

$$|s_k - s_{k-1}| \leq 2^{-k}, \qquad |t_k - t_{k-1}| \leq 2^{-k}, \qquad k = n+1, \ldots, \bar{n}.$$

Furthermore, we prove that $|t_n - s_n| \leq 2^{-n}$ and we have

$$|s - s_k| < 2^{-k}, \qquad |t - t_k| < 2^{-k}, \qquad k = n, \ldots, \bar{n},$$

from which $s_{\bar{n}} = s$ and $t_{\bar{n}} = t$. Then we have

$$X_t - X_s = X_{t_n} - X_{s_n} + \sum_{k=n+1}^{\bar{n}} (X_{t_k} - X_{t_{k-1}}) - \sum_{k=n+1}^{\bar{n}} (X_{s_k} - X_{s_{k-1}})$$

and therefore, for every $\omega \in \Omega \setminus N$,

$$|X_t(\omega) - X_s(\omega)| \leq c_{\alpha,\omega} 2^{-n\alpha} + 2 \sum_{k=n+1}^{\bar{n}} c_{\alpha,\omega} 2^{-k\alpha}$$

$$\leq 2 c_{\alpha,\omega} \sum_{k=n}^{\infty} 2^{-k\alpha}$$

$$= \frac{2 c_{\alpha,\omega}}{1 - 2^{-\alpha}} 2^{-n\alpha},$$

so that $|X_t - X_s| \leq c'_{\alpha,\omega} |t - s|^\alpha$ for some positive constant $c'_{\alpha,\omega}$.

Fourth Step We proved that for every $\omega \in \Omega \setminus N$ the trajectory $X(\omega)$ is α-Hölder continuous on $\mathscr{D}$ and therefore extends uniquely to an α-Hölder continuous function on $[0, 1]$, which we denote by $\widetilde{X}(\omega)$. Now we define the process $\widetilde{X}$ whose trajectories are equal to $\widetilde{X}(\omega)$ if $\omega \in \Omega \setminus N$ and are identically zero on N. We prove that $\widetilde{X}$ is a modification of X, that is, $P(X_t = \widetilde{X}_t) = 1$ for every fixed $t \in [0, 1]$: this is obvious if $t \in \mathscr{D}$. On the other hand, if $t \in [0, 1] \setminus \mathscr{D}$, we consider a sequence $(t_n)_{n \in \mathbb{N}}$ in $\mathscr{D}$ that approximates t. We already have observed that, by (3.4.2), X_{t_n} converges to X_t in probability and thus also pointwise a.s., up to a subsequence: since $X_{t_n} = \widetilde{X}_{t_n}$ a.s. then also $X_t = \widetilde{X}_t$ a.s. and this concludes the proof.

3.5 Key Ideas to Remember

We provide a summary of the chapter's major findings and essential concepts for initial comprehension, focusing on omitting technical or secondary details. As usual, if you have any doubt about what the following succinct statements mean, please review the corresponding section.

- Sections 3.1 and 3.2: a continuous stochastic process X can be regarded as a random variable with values in the Polish metric space of continuous trajectories, $(C(I), \mathscr{B}_{\varrho_{\max}})$. The law of X is therefore a distribution on the Borel σ-algebra $\mathscr{B}_{\varrho_{\max}}$.
- Section 3.3: Kolmogorov's continuity theorem provides a condition on the law of a process so that it admits a modification with locally Hölder continuous trajectories. This is the case of the Gaussian transition law of Example 3.3.2 but not of the Poisson transition law of Example 3.3.3.
- Section 3.4: the first two steps of the proof of Kolmogorov's continuity theorem are based on Markov's inequality and Borel-Cantelli's lemma: they contain the key ideas of the proof of this deep and fundamental result.

Main notations used or introduced in this chapter:

Symbol	Description	Page
$C(I)$	Continuous functions on the interval I	59
$\mathscr{F}^I$	σ-algebra on $\mathbb{R}^I$ generated by finite-dimensional cylinders	3
$\varrho_{\max}$	Uniform distance on $C(I)$	61
$\mathscr{B}_{\varrho_{\max}}$	Borel σ-algebra on $C(I)$	61

Chapter 4
Brownian Motion

> *In this section we will define Brownian motion and construct it. This event, like the birth of a child, is messy and painful, but after a while we will be able to have fun with our new arrival.*
>
> Richard Durrett

Brownian motion stands out as one of the paramount stochastic processes. It owes its name to the botanist Robert Brown, who, circa 1820, documented the erratic motion exhibited by pollen grains suspended within a solution. This phenomenon, characterized by the seemingly random movement of particles due to collisions with surrounding molecules, has since found widespread applications in various fields, ranging from physics and chemistry to finance and biology. Brownian motion was used by Louis Bachelier in 1900 in his doctoral thesis as a model for the price of stocks and was studied by Albert Einstein in one of his famous papers in 1905. The first rigorous mathematical definition of a Brownian motion is due to Norbert Wiener in 1923.

4.1 Definition

Definition 4.1.1 (Brownian Motion [!!!]) Let $W = (W_t)_{t \geq 0}$ be a real stochastic process defined on a filtered probability space $(\Omega, \mathscr{F}, P, \mathscr{F}_t)$. We say that W is a Brownian motion if it satisfies the following properties:

(i) $W_0 = 0$ a.s.;
(ii) W is a.s. continuous;
(iii) W is adapted to $(\mathscr{F}_t)_{t \geq 0}$, i.e., $W_t \in m\mathscr{F}_t$ for every $t \geq 0$;
(iv) $W_t - W_s$ is independent of $\mathscr{F}_s$ for every $t \geq s \geq 0$;
(v) $W_t - W_s \sim \mathscr{N}_{0, t-s}$ for every $t \geq s \geq 0$.

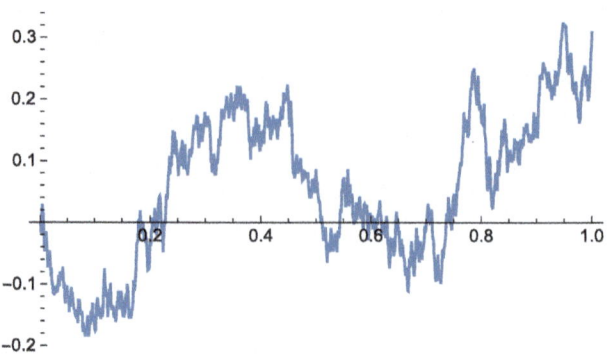

Fig. 4.1 A trajectory of a Brownian motion

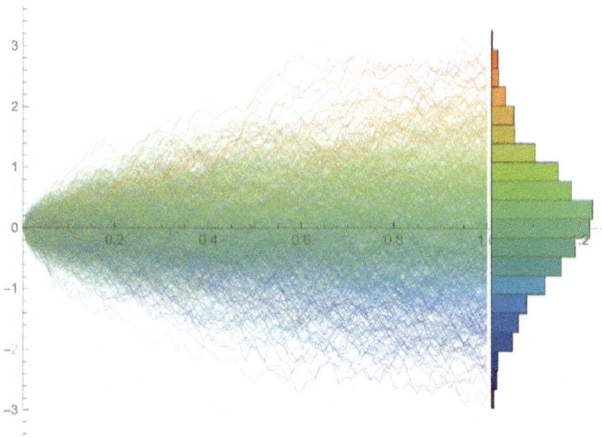

Fig. 4.2 1000 trajectories of a Brownian motion and histogram of its sample distribution at time $t = 1$

Remark 4.1.2 Let us briefly comment on the properties of Definition 4.1.1: by (i) a Brownian motion starts from the origin, just as a convention. Property (ii) ensures that almost all trajectories of W are continuous. Moreover, W is *adapted to the filtration* $(\mathscr{F}_t)_{t\geq 0}$: this means that, at any fixed time t, the information in $\mathscr{F}_t$ is sufficient to observe the entire trajectory of W up to time t. Properties (iv) and (v) are less intuitive but can be justified by some notable features, observable at the statistical level, of random motions: we call (iv) and (v) the *independence* and *stationarity* properties of the increments, respectively (cf. Definition 2.3.1). Notice that $W_t - W_s$ is equal in law to W_{t-s}. Figures 4.1 and 4.2 show the plot of trajectories of a Brownian motion.

Remark 4.1.3 In Definition 4.1.1 the filtration $(\mathscr{F}_t)$ is not necessarily the one generated by W: the latter was denoted by $(\mathscr{G}_t^W)_{t\geq 0}$ in Definition 1.4.3. Clearly, property (iii) of a Brownian motion implies that $\mathscr{G}_t^W \subseteq \mathscr{F}_t$ for every $t \geq 0$. We

4.1 Definition

will see in Sect. 6.2 that it is generally preferable to work with filtrations strictly larger than $\mathscr{G}^W$ in order to satisfy appropriate technical assumptions, including, for example, completeness.

We give a useful characterization of a Brownian motion.

Proposition 4.1.4 ([!]) *An a.s. continuous stochastic process $W = (W_t)_{t \geq 0}$ is a Brownian motion with respect to its own filtration $(\mathscr{G}_t^W)_{t \geq 0}$ if and only if it is a Gaussian process with zero mean function, $E[W_t] = 0$, and covariance function $\mathrm{cov}(W_s, W_t) = s \wedge t$.*

Proof Let W be a Brownian motion on $(\Omega, \mathscr{F}, P, (\mathscr{G}_t^W)_{t \geq 0})$. For each $0 = t_0 < t_1 < \cdots < t_n$, the random variables $Z_k := W_{t_k} - W_{t_{k-1}}$, have normal distribution; moreover, by properties (iii) and (v) of a Brownian motion, Z_k is independent of $\mathscr{G}_{t_{k-1}}^W$ and therefore of $Z_1, \ldots, Z_{k-1} \in m\mathscr{G}_{t_{k-1}}^W$. This proves that $(Z_1, \ldots, Z_n)$ is a multi-normal vector with independent components. Also $(W_{t_1}, \ldots, W_{t_n})$ is multi-normal because it is obtained from $(Z_1, \ldots, Z_n)$ by the linear transformation

$$W_{t_h} = \sum_{k=1}^{h} Z_k, \qquad h = 1, \ldots, n,$$

and this proves that W is a Gaussian process. We also observe that, assuming $s < t$, we have

$$\mathrm{cov}(W_s, W_t) = \mathrm{cov}(W_s, W_t - W_s + W_s) = \mathrm{cov}(W_s, W_t - W_s) + \mathrm{var}(W_s) = s$$

by the independence of W_s and $W_t - W_s$: this proves that $\mathrm{cov}(W_s, W_t) = s \wedge t$.

Conversely, let W be a Gaussian process with zero mean function and covariance function $\mathrm{cov}(W_s, W_t) = s \wedge t$. Since $E[W_0] = \mathrm{var}(W_0) = 0$, we have $W_0 = 0$ a.s. Properties (ii) and (iii) of the definition of a Brownian motion are obvious. To prove (v), it is enough to observe that, if $s < t$, we have

$$\mathrm{var}(W_t - W_s) = \mathrm{var}(W_t) + \mathrm{var}(W_s) - 2\mathrm{cov}(W_t, W_s) = t + s - 2(s \wedge t) = t - s.$$

Finally, given $\tau \leq s < t$, the vector $(W_t - W_s, W_\tau)$ has a normal distribution because it is a linear combination of (W_τ, W_s, W_t) and

$$\mathrm{cov}(W_t - W_s, W_\tau) = \mathrm{cov}(W_t, W_\tau) - \mathrm{cov}(W_s, W_\tau) = \tau - \tau = 0.$$

Consequently, $W_t - W_s$ and W_τ are independent: since W is a Gaussian process, it follows that $W_t - W_s$ is independent of $(W_{\tau_1}, \ldots, W_{\tau_n})$ for every $\tau_1, \ldots, \tau_n \leq s$. Then, by Lemma 2.3.20 in [113], $W_t - W_s$ is independent of $\mathscr{G}_s^W$ and this proves the validity of property (iv). □

Remark 4.1.5 ([!]) Proposition 4.1.4 states that the finite-dimensional distributions of a Brownian motion are uniquely determined: hence the Brownian motion is unique in law. Notice that, if W is a Brownian motion then the process $\widetilde{W}_t := \sqrt{t} W_1$ has the same *one-dimensional* distributions as W but is obviously not a Brownian motion.

There are numerous proofs of the existence of a Brownian motion: some of them can be found, for example, in the monographs by Schilling [129] and Bass [9]. Here we see the result as a corollary of Kolmogorov's extension and continuity theorems.

Theorem 4.1.6 *A Brownian motion exists.*

Proof The main step is the construction of a Brownian motion on the bounded time interval [0, 1]. By Kolmogorov's extension theorem (in particular, by Corollary 1.3.6) there exists a Gaussian process $W^{(0)} = (W_t^{(0)})_{t \in [0,1]}$ with zero mean function and covariance function $\text{cov}(W_s^{(0)}, W_t^{(0)}) = s \wedge t$. By Kolmogorov's continuity theorem and Example 3.3.2, $W^{(0)}$ admits a continuous modification that, by Proposition 4.1.4, satisfies the properties of a Brownian motion on [0, 1].

Now take a sequence $(W^{(n)})_{n \in \mathbb{N}}$ of independent copies of $W^{(0)}$. We "glue" these processes together by defining $W_t = W_t^{(0)}$ for $t \in [0, 1]$ and

$$W_t = \sum_{k=0}^{[t]-1} W_1^{(k)} + W_{t-[t]}^{[t]}, \qquad t > 1,$$

where $[t]$ denotes the integer part of t. Then it is easy to prove that W is a Brownian motion. □

Remark 4.1.7 As seen in Example 3.3.2, a Brownian motion admits a modification with trajectories that are not only continuous but also locally α-Hölder continuous for every $\alpha < \frac{1}{2}$. The exponent α is strictly less than $\frac{1}{2}$, and this result cannot be improved: for more details, we refer, for example, to Chapter 7 in [9]. A classic result, the *Law of the iterated logarithm*, precisely describes the asymptotic behavior of Brownian increments:

$$\limsup_{t \to 0^+} \frac{|W_t|}{\sqrt{2t \log \log \frac{1}{t}}} = 1 \qquad \text{a.s.}$$

Consequently, the trajectories of a Brownian motion are almost surely not differentiable at any point: precisely, there exists $N \in \mathscr{F}$, with $P(N) = 0$, such that for every $\omega \in \Omega \setminus N$, the function $t \mapsto W_t(\omega)$ is not differentiable at any point in $[0, +\infty[$.

4.2 Markov and Feller Properties

Let $W = (W_t)_{t\geq 0}$ be a Brownian motion on $(\Omega, \mathscr{F}, P, \mathscr{F}_t)$. Given $t \geq 0$ and $x \in \mathbb{R}$, we set

$$W_T^{t,x} := W_T - W_t + x, \qquad T \geq t.$$

Definition 4.2.1 The process $W^{t,x} = (W_T^{t,x})_{T \geq t}$ is called *Brownian motion with initial point x at time t* and has the following properties:

(i) $W_t^{t,x} = x$;
(ii) the trajectories $T \mapsto W_T^{t,x}$ are a.s. continuous;
(iii) $W_T^{t,x} \in m\mathscr{F}_T$ for every $T \geq t$;
(iv) $W_T^{t,x} - W_s^{t,x} = W_T - W_s$ is independent of $\mathscr{F}_s$ for every $T \geq s \geq t$;
(v) $W_T^{t,x} - W_s^{t,x} \sim \mathscr{N}_{0,T-s}$ for every $T \geq s \geq t$.

Remark 4.2.2 The process $W^{t,x}$ is also a Brownian motion with respect to its generated filtration, defined by

$$\mathscr{G}_T^{t,x} := \sigma(W_s^{t,x}, s \in [t, T]), \qquad T \geq t.$$

Note that $\mathscr{G}_T^{t,x} \subseteq \mathscr{F}_T$ and there is a strict inclusion $\mathscr{G}_t^{t,x} = \{\emptyset, \Omega\} \subset \mathscr{F}_t$ if $t > 0$.

By Proposition 2.3.2, we have

Theorem 4.2.3 (Markov Property [!]) *Let $W = (W_t)_{t\geq 0}$ be a Brownian motion on $(\Omega, \mathscr{F}, P, \mathscr{F}_t)$. Then W is a Markov process with Gaussian transition density*

$$\Gamma(t, x; T, y) = \frac{1}{\sqrt{2\pi(T-t)}} e^{-\frac{(x-y)^2}{2(T-t)}}, \qquad 0 \leq t < T, \ x, y \in \mathbb{R}. \qquad (4.2.1)$$

Consequently, for every $\varphi \in b\mathscr{B}$, we have

$$u(t, W_t) = E[\varphi(W_T) \mid \mathscr{F}_t]$$

where

$$u(t, x) := \int_{\mathbb{R}} \Gamma(t, x; T, y) \varphi(y) dy. \qquad (4.2.2)$$

We have proven in Example 2.4.6 the following

Proposition 4.2.4 (Feller Property) *A Brownian motion satisfies the strong Feller property.*

Remark 4.2.5 The function u in (4.2.2) belongs to $C^\infty\left([0, T[\times\mathbb{R}\right)$; moreover, if $\varphi \in bC(\mathbb{R})$, proceeding as in Example 3.1.3 in [113], we get

$$\lim_{\substack{(t,x)\to(T,y) \\ t<T}} u(t, x) = \varphi(y)$$

so that $u \in C\left([0, T] \times \mathbb{R}\right)$ and $u(0, \cdot) \equiv \varphi$. Thus, u is a classical solution (cf. Definition 18.2.5) of the backward Cauchy problem

$$\begin{cases} \partial_t u(t, x) + \frac{1}{2}\partial_{xx} u(t, x) = 0, & t \in [0, T[, \ x \in \mathbb{R}, \\ u(T, x) = \varphi(x), & x \in \mathbb{R}. \end{cases}$$

This is in agreement with Example 2.5.9, being $\mathscr{A}_t = \frac{1}{2}\partial_{xx}$ the characteristic operator of the Gaussian transition distribution. Note that the hypothesis $\varphi \in bC(\mathbb{R})$ is only[1] used to prove the continuity of $u(t, x)$ up to $t = T$.

4.3 Wiener Space

By Proposition 4.1.4, a Brownian motion has finite-dimensional Gaussian distributions. More precisely, by Proposition 2.4.1 (in particular, by formula (2.4.2)) we have the following

Theorem 4.3.1 (Finite-Dimensional Densities) *Let $W = (W_t)_{t\geq 0}$ be a real Brownian motion. For every $0 < t_1 < \cdots < t_n$, the vector $(W_{t_1}, \ldots, W_{t_n})$ is absolutely continuous with density*

$$\gamma_{(W_{t_1}, \ldots, W_{t_n})}(x_1, \ldots, x_n) = \Gamma(0, 0; t_1, x_1)\Gamma(t_1, x_1; t_2, x_2)\cdots\Gamma(t_{n-1}, x_{n-1}; t_n, x_n)$$

with Γ as in (4.2.1). The law[2] of W is called Wiener measure.

Definition 4.3.2 (Wiener Space) The probability space $(C(\mathbb{R}_{\geq 0}), \mathscr{B}_{\mu_W}, \mu_W)$, where μ_W is the Wiener measure and $\mathscr{B}_{\mu_W}$ is the μ_W-completion[3] of the Borel σ-algebra, is called Wiener space.

Recall Definition 3.2.3, which defines the canonical version of an a.s. continuous process. An immediate consequence of Proposition 4.1.4 is the following

Corollary 4.3.3 *Given a Brownian motion W, its canonical version $\mathbf{W}$ is a Brownian motion on the Wiener space equipped with the filtration $\mathscr{G}^{\mathbf{W}}$ generated by $\mathbf{W}$.*

---

[1] $u \in C^\infty\left([0, T[\times\mathbb{R}\right)$ for every $\varphi \in b\mathscr{B}$.
[2] Definition 3.2.2.
[3] Cf. Remark 1.4.3 in [113].

4.4 Brownian Martingales

Given a Brownian motion W, we will later introduce (cf. Sect. 6.2.3) a filtration larger than $\mathscr{G}^W$ so that some useful regularity properties hold.

Example 4.3.4 Let W be a real Brownian motion and $0 < t < T$. We have the following expressions for the joint densities of W_t and W_T:

$$\gamma_{(W_t, W_T)}(t, x; T, y) = \gamma_{(W_T, W_t)}(T, y; t, x) = \frac{1}{2\pi\sqrt{t(T-t)}} e^{-\frac{(Tx^2 - 2txy + ty^2)}{2t(T-t)}}.$$

By Proposition 4.3.20 in [113] we also have the conditional densities

$$\gamma_{W_T | W_t}(T, y; t, x) = \frac{\gamma_{(W_T, W_t)}(T, y; t, x)}{\gamma_{W_t}(t, x)} = \Gamma(t, x; T, y),$$

$$\gamma_{W_t | W_T}(t, x; T, y) = \frac{\gamma_{(W_t, W_T)}(t, x; T, y)}{\gamma_{W_T}(T, y)} = \frac{1}{\sqrt{2\pi \frac{t(T-t)}{T}}} e^{-\frac{T\left(x - \frac{t}{T} y\right)^2}{2t(T-t)}}.$$

Thus, in accordance with Theorem 4.2.3, we have

$$\mu_{W_T | W_t} = \mathscr{N}_{W_t, T-t}$$

and

$$\mu_{W_t | W_T} = \mathscr{N}_{\frac{t}{T} W_T, \frac{t(T-t)}{T}}.$$

4.4 Brownian Martingales

Let W be a Brownian motion on the filtered space $(\Omega, \mathscr{F}, P, \mathscr{F}_t)$.

Proposition 4.4.1 *The following processes are martingales:*

(i) *the Brownian motion W;*
(ii) *the quadratic martingale*

$$X_t := W_t^2 - t;$$

(iii) *the exponential martingale*

$$Y_t = e^{\sigma W_t - \frac{\sigma^2}{2} t}$$

for every $\sigma \in \mathbb{C}$.

Proof By Hölder's inequality, we have

$$E[|W_t|] \le E\left[W_t^2\right]^{\frac{1}{2}} = \sqrt{t}$$

and therefore W is an absolutely integrable process. Part (i) follows from Proposition 2.3.4, being W a process with constant zero mean and independent increments.

Similarly, (ii) and (iii) are proven: for example, we have

$$E[X_T \mid \mathscr{F}_t] = E\left[(W_T - W_t + W_t)^2 \mid \mathscr{F}_t\right] - T$$

$$= \underbrace{E\left[(W_T - W_t)^2 \mid \mathscr{F}_t\right]}_{=T-t} + 2W_t \underbrace{E[W_T - W_t \mid \mathscr{F}_t]}_{=0}$$

$$+ W_t^2 - T = W_t^2 - t.$$

□

We give a useful characterization of a Brownian motion in terms of exponential martingales.

Proposition 4.4.2 ([!]) *A continuous and adapted process W, defined on the space $(\Omega, \mathscr{F}, P, \mathscr{F}_t)$ and such that $W_0 = 0$ a.s., is a Brownian motion if and only if*

$$M_t^\eta := e^{i\eta W_t + \frac{\eta^2}{2} t}$$

is a martingale for every $\eta \in \mathbb{R}$.

Proof If W is a Brownian motion then M^η is a martingale by Proposition 4.4.1-(iii). Conversely, it is sufficient to verify that for $0 \le s \le t$:

(i) $W_t - W_s$ has normal distribution $\mathscr{N}_{0,t-s}$;
(ii) $W_t - W_s$ is independent of $\mathscr{F}_s$.

The martingale property of M_t^η is equivalent to

$$E\left[e^{i\eta(W_t - W_s)} \mid \mathscr{F}_s\right] = e^{-\frac{\eta^2}{2}(t-s)}, \qquad \eta \in \mathbb{R}.$$

Applying the expected value, we obtain the characteristic function of $W_t - W_s$:

$$E\left[e^{i\eta(W_t - W_s)}\right] = e^{-\frac{\eta^2}{2}(t-s)}, \qquad \eta \in \mathbb{R},$$

from which the thesis follows: in particular, the independence property follows from 14) of Theorem 4.2.10 in [113]. □

4.4 Brownian Martingales

The following version of Theorem 2.5.13 provides a general method for constructing a martingale by composing a Brownian motion W with a sufficiently regular function $f = f(t, x)$. We also assume on f a growth condition of the type

$$|f(t,x)| \leq c_T e^{c_T |x|^\alpha}, \qquad (t,x) \in [0,T] \times \mathbb{R}, \tag{4.4.1}$$

with c_T a positive constant dependent on T and $\alpha \in [0, 2[$: this ensures the integrability of the process $f(t, W_t)$ for $t \in [0, T]$.

Theorem 4.4.3 ([!]) *Let $f = f(t,x) \in C^{1,2}(\mathbb{R}_{\geq 0} \times \mathbb{R})$ be a function that verifies, together with its first and second derivatives, the growth condition (4.4.1). Then the process*

$$M_t := f(t, W_t) - f(0, W_0) - \int_0^t \left(\partial_s f + \frac{1}{2} \partial_{xx} f \right)(s, W_s) ds, \qquad t \in [0, T],$$

is a martingale. In particular, if f solves the backward heat equation, then $f(t, W_t)$ is a martingale.

Proof The proof is entirely analogous to that of Theorem 2.5.13. For each $s > t$ and $x \in \mathbb{R}$, we have

$$\partial_s \int_\mathbb{R} \Gamma(t, x; s, y) f(s, y) dy = \int_\mathbb{R} \partial_s \big(\Gamma(t, x; s, y) f(s, y) \big) dy =$$

(since $\partial_s \Gamma(t, x; s, y) = \frac{1}{2} \partial_{yy} \Gamma(t, x; s, y)$)

$$= \int_\mathbb{R} \Gamma(t, x; s, y) \partial_s f(s, y) dy + \int_\mathbb{R} \frac{1}{2} \partial_{yy} \Gamma(t, x; s, y) f(s, y) dy =$$

(integrating by parts in the second integral)

$$= \int_\mathbb{R} \Gamma(t, x; s, y) \left(\partial_s f + \frac{1}{2} \partial_{yy} f \right)(s, y) dy.$$

Setting $x = W_t$ in the previous formula, by the Markov property we have

$$\partial_s E[f(s, W_s) \mid \mathscr{F}_t] = E\left[\left(\partial_s f + \frac{1}{2} \partial_{xx} f \right)(s, W_s) \mid \mathscr{F}_t \right].$$

Now we integrate in s between t and T to obtain

$$E[f(T, W_T) \mid \mathscr{F}_t] - f(t, W_t) = \int_t^T E\left[\left(\partial_s f + \frac{1}{2} \partial_{xx} f \right)(s, W_s) \mid \mathscr{F}_t \right] ds =$$

(exchanging the signs of integral and conditional expectation as in the proof of Theorem 2.5.13)

$$= E\left[\int_t^T \left(\partial_s f + \frac{1}{2}\partial_{xx} f\right)(s, W_s)ds \mid \mathscr{F}_t\right].$$

In conclusion, we have

$E[M_T - M_t \mid \mathscr{F}_t]$

$$= E\left[f(T, W_T) - f(t, W_t) - \int_t^T \left(\partial_s f + \frac{1}{2}\partial_{xx} f\right)(s, W_s)ds \mid \mathscr{F}_t\right] = 0$$

and this wraps up the proof. □

4.5 Key Ideas to Remember

We summarize the most significant findings of the chapter and the fundamental concepts to be retained from an initial reading, while disregarding the more technical or secondary matters. As usual, if you have any doubt about what the following succinct statements mean, please review the corresponding section.

- Section 4.1: a Brownian motion W is a continuous and adapted process, with independent and stationary increments having normal distribution. It is characterized by being a Gaussian process with zero mean function and covariance function $\text{cov}(W_s, W_t) = s \wedge t$.
- Section 4.2: W is a Markov process with transition law equal to the law of $W_T^{t,x}$. Moreover, W is a strong Feller process.
- Section 4.3: the finite-dimensional densities of W are uniquely determined and the law of W is called *Wiener measure*.
- Section 4.4: W is a martingale and other notable examples of martingales can be constructed as functions of W: for instance, the quadratic and the exponential martingales. The latter provides a characterization of the Brownian motion (cf. Proposition 4.4.2). Theorem 4.4.3 shows how to "compensate" a function of W to make it a martingale and indicates the connection with the heat equation that will be further explored in the following chapters.

4.5 Key Ideas to Remember

Main notations used or introduced in this chapter:

Symbol	Description	Page
$\mathscr{G}^W$	Filtration generated by W	14
$W^{t,x}$	Brownian motion with initial point x at time t	75
$\mathscr{G}^{t,x}$	Filtration generated by $W^{t,x}$	75
$\Gamma(t, x; T, y)$	Gaussian transition density	75
μ_W	Wiener measure	76

Chapter 5
Poisson Process

> We are too small and the universe too large and too interrelated for thoroughly deterministic thinking.
>
> Don S. Lemons, [88]

The Poisson process, denoted as $(N_t)_{t \geq 0}$, serves as the prototype of what are known as "pure jump processes". Intuitively, N_t indicates the number of times within the time interval $[0, t]$ that a specific event (referred to as an *episode*) occurs: for example, if the single episode consists of the *arrival of a spam email* in a mailbox, then N_t represents the number of spam emails that arrive in the period $[0, t]$; similarly, N_t can indicate the number of children born in some country or the number of earthquakes that occur in some geographical area in the period $[0, t]$.

5.1 Definition

Referring to the general notation of Definition 1.1.3, we assume $I = \mathbb{R}_{\geq 0}$. To construct the Poisson process, we consider a sequence $(\tau_n)_{n \in \mathbb{N}}$ of independent and identically distributed random variables[1] with exponential distribution, $\tau_n \sim \text{Exp}_\lambda$, with parameter $\lambda > 0$, defined on a complete probability space $(\Omega, \mathscr{F}, P)$: here τ_n *represents the time that elapses* between the $(n-1)$-th episode and the next one. Then we define the sequence

$$T_0 := 0, \qquad T_n := \tau_1 + \cdots + \tau_n, \qquad n \in \mathbb{N},$$

in which T_n represents the instant at which the n-th episode occurs.

---

[1] Such a sequence exists by Corollary 1.3.7.

Lemma 5.1.1 *We have[2]*

$$T_n \sim Gamma_{n,\lambda} \qquad n \in \mathbb{N}. \tag{5.1.1}$$

Moreover, almost surely[3] the sequence $(T_n)_{n \geq 0}$ is monotonically increasing and

$$\lim_{n \to \infty} T_n = +\infty. \tag{5.1.2}$$

Proof Formula (5.1.1) follows from (2.6.7) in [113]. The monotonicity follows from the fact that $\tau_n \geq 0$ a.s. for every $n \in \mathbb{N}$. Finally, (5.1.2) follows from Borel-Cantelli's Lemma 1.3.28 in [113]: in fact, for every $\varepsilon > 0$, we have

$$\left(\lim_{n \to \infty} T_n = +\infty \right) \supseteq ((\tau_n > \varepsilon) \text{ i.o.}) = \bigcap_{n \geq 1} \bigcup_{k \geq n} (\tau_k > \varepsilon)$$

and the events $(\tau_k > \varepsilon)$ are independent and such that

$$\sum_{n \geq 1} P(\tau_n > \varepsilon) = +\infty.$$

□

Definition 5.1.2 (Poisson Process, I) The Poisson process $(N_t)_{t \geq 0}$ with parameter $\lambda > 0$ is defined by

$$N_t = \sum_{n=1}^{\infty} n \mathbb{1}_{[T_n, T_{n+1}[}(t), \qquad t \geq 0. \tag{5.1.3}$$

By definition, N_t takes *non-negative integer* values and precisely $N_t = n$ if and only if t belongs to the interval with random endpoints $[T_n, T_{n+1}[$; hence we have the equality of events

$$(N_t = n) = (T_n \leq t < T_{n+1}), \qquad n \in \mathbb{N} \cup \{0\}. \tag{5.1.4}$$

At the random time T_n, when the n-th episode occurs, the process makes a jump of size 1: Fig. 5.1 shows the plot of a Poisson process trajectory in the time interval

---

[2] Thus T_n is absolutely continuous with density

$$\gamma_{n,\lambda}(t) := \lambda e^{-\lambda t} \frac{(\lambda t)^{n-1}}{(n-1)!} \mathbb{1}_{\mathbb{R}_{\geq 0}}(t), \qquad n \in \mathbb{N}.$$

[3] The set of $\omega \in \Omega$ such that $T_n(\omega) \leq T_{n+1}(\omega)$ for every $n \in \mathbb{N}$ and $\lim_{n \to \infty} T_n(\omega) = +\infty$, is a certain event.

5.1 Definition

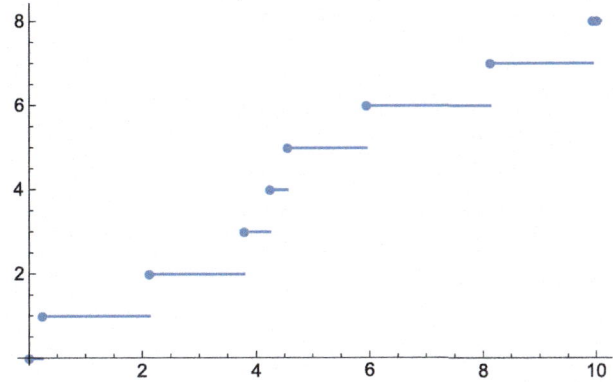

Fig. 5.1 Plot of a Poisson process trajectory

[0, 10]. We recall that a trajectory of N is a function of the form $t \mapsto N_t(\omega)$, defined from $\mathbb{R}_{\geq 0}$ to $\mathbb{N} \cup \{0\}$, and each $\omega \in \Omega$ corresponds to a different trajectory.

In conclusion, the random value N_t is equal to the number of jumps (or the number of episodes) between 0 and t:

$$N_t = \#\{n \in \mathbb{N} \mid T_n \leq t\}.$$

We will later give a more general characterization of the Poisson process, in Definition 5.2.3.

Proposition 5.1.3 *The Poisson process $(N_t)_{t \geq 0}$ has the following properties:*

(i) almost surely the trajectories are right-continuous and monotonically increasing. Moreover, for every $t > 0$, we have[4]

$$P\left(\lim_{s \to t} N_s = N_t\right) = 1; \tag{5.1.5}$$

(ii) $N_t \sim \text{Poisson}_{\lambda t}$, that is

$$P(N_t = n) = e^{-\lambda t} \frac{(\lambda t)^n}{n!}, \qquad t \geq 0, \ n \in \mathbb{N} \cup \{0\}. \tag{5.1.6}$$

---

[4] In other words, every fixed t is *almost surely* (i.e., for almost all trajectories) a point of continuity for the Poisson process. This apparent paradox is explained by the fact that almost every trajectory has at most countably infinite discontinuities, since it is monotonically increasing, and such discontinuities are arranged on the entire interval $[0, +\infty[$ which has the cardinality of the continuum. Thus, all trajectories are discontinuous but every single t is a point of discontinuity only for a negligible family of trajectories.

As a consequence, $N_0 = 0$ a.s. and we have

$$E[N_t] = var(N_t) = \lambda t.$$

In particular, the parameter λ, called *intensity of the Poisson process*, is equal to the expected number of jumps in the unit time interval $[0, 1]$;
(iii) the characteristic function of N_t is given by

$$\varphi_{N_t}(\eta) = e^{\lambda t (e^{i\eta} - 1)}, \qquad t \geq 0, \ \eta \in \mathbb{R}. \tag{5.1.7}$$

Proof

((i) Right-continuity and monotonicity follow from the definition. For every $t > 0$, let $N_{t-} = \lim_{s \nearrow t} N_s$ and $\Delta N_t = N_t - N_{t-}$. We note that $\Delta N_t \in \{0, 1\}$ a.s. and, for a fixed $t > 0$, the set of trajectories that are discontinuous at t is given by

$$(\Delta N_t = 1) = \bigcup_{n=1}^{\infty} (T_n = t)$$

which is a negligible event since the random variables T_n are absolutely continuous. This proves (5.1.5).
(ii) By (5.1.4) we have

$$P(N_t = n) = P(T_n \leq t < T_{n+1}) =$$

(since $(t \geq T_{n+1}) \subseteq (t \geq T_n)$)

$$= P(T_n \leq t) - P(T_{n+1} \leq t) =$$

(since $T_n \sim \text{Gamma}_{n,\lambda}$)

$$= \int_0^t \lambda e^{-\lambda s} \frac{(\lambda s)^{n-1}}{(n-1)!} ds - \int_0^t \lambda e^{-\lambda s} \frac{(\lambda s)^n}{n!} ds$$

from which, integrating by parts the second integral, (5.1.6) follows.
(iii) It is a simple calculation: by (ii) we have

$$E\left[e^{i\eta N_t}\right] = \sum_{n \geq 0} e^{-\lambda t} \frac{(\lambda t)^n}{n!} e^{i\eta n} = e^{-\lambda t} \sum_{n \geq 0} \frac{(\lambda t e^{i\eta})^n}{n!}$$

which concludes the proof. □

5.1 Definition

Remark 5.1.4 (Characteristic Exponent) The characteristic function of the Poisson process has an interesting property of homogeneity with respect to time: in fact, by (5.1.7) the CHF of N_t is of the form $\varphi_{N_t}(\eta) = e^{t\psi(\eta)}$ where

$$\psi(\eta) = \lambda(e^{i\eta} - 1) \qquad (5.1.8)$$

is a function that depends on η but not on t. Consequently, the function ψ determines the CHF of N_t for every t and for this reason is called *characteristic exponent of the Poisson process*.

Example 5.1.5 (Compound Poisson Process [!]) The Poisson process N is the starting point for the construction of stochastic processes even more interesting and useful in applications. The first generalization consists in making the size of the jumps random, as opposed to N where they are all fixed equal to 1.

Consider a probability space on which a Poisson process N is defined and a sequence $(Z_n)_{n \in \mathbb{N}}$ of identically distributed real random variables. Suppose that the family formed by $(Z_n)_{n \in \mathbb{N}}$ and $(\tau_n)_{n \in \mathbb{N}}$ (the exponential random variables that define N) is a family of *independent* random variables: this construction is possible thanks to Corollary 1.3.7. We set by convention $Z_0 = 0$ and define the compound Poisson process in the following way:

$$X_t = \sum_{n=0}^{N_t} Z_n, \qquad t \geq 0.$$

Note that the Poisson process is a particular case of X in which $Z_n \equiv 1$ for $n \in \mathbb{N}$. In Fig. 5.2 two trajectories of the compound Poisson process with normal jumps and different choices of the intensity parameter are represented.

Taking advantage of the independence assumption, it is easy to calculate the CHF of X_t: actually, it is a calculation already carried out in Exercise 2.5.4 in [113] where we proved that

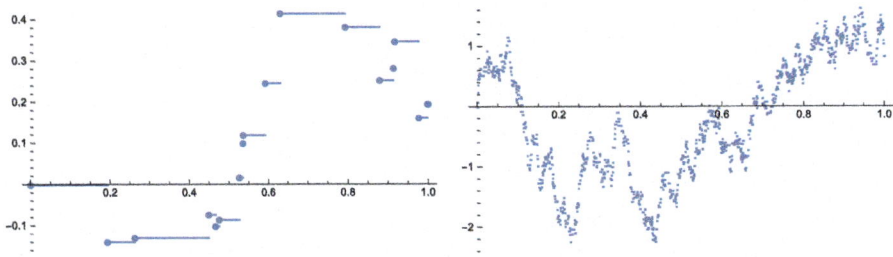

Fig. 5.2 On the left: plot of a trajectory of the compound Poisson process with $\lambda = 10$ and $Z_n \sim \mathcal{N}_{0,10^{-2}}$. **On the right:** plot of a trajectory of the compound Poisson process with $\lambda = 1000$ and $Z_n \sim \mathcal{N}_{0,10^{-2}}$

$$\varphi_{X_t}(\eta) = e^{t\psi(\eta)}, \qquad \psi(\eta) = \lambda\,(\varphi_Z(\eta) - 1)$$

where $\varphi_Z(\eta)$ is the CHF of Z_1. Also in this case, the CHF of X_t is homogeneous in time and ψ is called the *characteristic exponent of the compound Poisson process*. As a particular case, we find (5.1.8) for $Z_n \sim \delta_1$, that is, for unitary jumps as in the Poisson process.

5.2 Markov and Feller Properties

The following theorem provides two crucial properties of the increments $N_t - N_s$ of the Poisson process. As usual (cf. (1.4.1)), $\mathscr{G}^N = (\mathscr{G}_t^N)_{t \geq 0}$ denotes the filtration generated by N.

Theorem 5.2.1 ([!]) *For every $0 \leq s < t$ we have:*

(i) $N_t - N_s \sim \text{Poisson}_{\lambda(t-s)}$;
(ii) $N_t - N_s$ is independent of $\mathscr{G}_s^N$.

Property (i) implies that the r.v. $N_t - N_s$ and N_{t-s} are equal in law and for this reason, we say that N has stationary increments. Property (ii) states that N is a process with independent increments according to Definition 2.3.1.

The proof of Theorem 5.2.1 is postponed to Sect. 5.4.

Definition 5.2.2 (Càdlàg Function) We say that a function f, from a real interval I to $\mathbb{R}$, is *càdlàg* (from the French "continue à droite, limite à gauche") if at every point it is continuous from the right and has a finite limit from the left.[5]

The definition of Poisson process can be generalized as follows.

Definition 5.2.3 (Poisson Process, II) A Poisson process with intensity $\lambda > 0$, defined on a filtered probability space $(\Omega, \mathscr{F}, P, \mathscr{F}_t)$, is a stochastic process $(N_t)_{t \geq 0}$ such that:

(i) $N_0 = 0$ a.s.;
(ii) N is a.s. càdlàg;
(iii) N is adapted to $(\mathscr{F}_t)_{t \geq 0}$, i.e., $N_t \in m\mathscr{F}_t$ for every $t \geq 0$;
(iv) $N_t - N_s$ is independent of $\mathscr{F}_s$ for $s < t$;
(v) $N_t - N_s \sim \text{Poisson}_{\lambda(t-s)}$ for $s < t$.

By Theorem 5.2.1, the process N defined in (5.1.3) is a Poisson process according to Definition 5.2.3 with respect to the filtration $\mathscr{G}^N$ generated by N. Conversely, it can be shown that if N is a Poisson process according to Definition 5.2.3 then the

---

[5] If $I = [a, b]$, at the endpoints we assume by definition that $\lim\limits_{x \searrow a} f(x) = f(a)$ and the limit $\lim\limits_{x \nearrow b} f(x)$ exists and is finite.

5.2 Markov and Feller Properties

r.v. T_n, defined recursively by

$$T_1 = \inf\{t \geq 0 \mid \Delta N_t = 1\}, \qquad T_{n+1} := \inf\{t > T_n \mid \Delta N_t = 1\},$$

are independent and with distribution Exp_λ: for more details see, for example, Chapter 5 in [9]. Note that in Definition 5.2.3 the filtration is not necessarily the one generated by the process.

Theorem 5.2.4 (Markov Property [!]) *The Poisson process N is a Markov and Feller process with transition law*

$$p(t, x; T, \cdot) = \text{Poisson}_{x, \lambda(T-t)}$$

and characteristic operator defined by

$$\mathscr{A}_t \varphi(x) = \lambda \left(\varphi(x+1) - \varphi(x) \right), \qquad x \in \mathbb{R}.$$

If $\varphi \in b\mathscr{B}$ and u is a solution of the backward Cauchy problem

$$\begin{cases} \partial_t u(t, x) + \mathscr{A}_t u(t, x) = 0, & (t, x) \in [0, T[\times \mathbb{R}, \\ u(T, x) = \varphi(x), & x \in \mathbb{R}, \end{cases}$$

then

$$u(t, N_t) = E\left[\varphi(N_T) \mid \mathscr{F}_t \right].$$

Proof The thesis is an immediate consequence of Proposition 2.3.2 and the results of Sect. 2.5.2 for the backward Kolmogorov equation: see in particular Example 2.5.11. The Feller property was proven in Example 2.4.5. □

We give a useful characterization of the Poisson process.

Proposition 5.2.5 ([!]) *Let $N = (N_t)_{t \geq 0}$ be a stochastic process on the space $(\Omega, \mathscr{F}, P, \mathscr{F}_t)$, which satisfies properties (i), (ii) and (iii) of Definition 5.2.3. Then N is a Poisson process of parameter $\lambda > 0$ if and only if*

$$E\left[e^{i\eta(N_t - N_s)} \mid \mathscr{F}_s \right] = e^{\lambda(e^{i\eta} - 1)(t-s)}, \qquad 0 \leq s \leq t, \; \eta \in \mathbb{R}. \tag{5.2.1}$$

Proof If N is a Poisson process, then by the independence and stationarity of increments and (5.1.7), we have

$$E\left[e^{i\eta(N_t - N_s)} \mid \mathscr{F}_s \right] = E\left[e^{i\eta(N_t - N_s)} \right] = E\left[e^{i\eta N_{t-s}} \right] = e^{\lambda(e^{i\eta} - 1)(t-s)}.$$

Conversely, if N satisfies (5.2.1) and properties (i), (ii) and (iii) of Definition 5.2.3, properties (iv) and (v) remain to be proven. Applying the expected value to

(5.2.1), we get

$$E\left[e^{i\eta(N_t-N_s)}\right] = e^{\lambda(e^{i\eta}-1)(t-s)}, \qquad 0 \leq s \leq t, \; \eta \in \mathbb{R}.$$

Then (v) is an obvious consequence of the fact that the characteristic function determines the distribution; property (iv) of independent increments follows from point 14) of Theorem 4.2.10 in [113]. □

Remark 5.2.6 (Poisson Process with Stochastic Intensity) The characterization given in Proposition 5.2.5 enables the definition of a broad range of processes, with the Poisson process being just one specific example. In a space $(\Omega, \mathscr{F}, P, \mathscr{F}_t)$ consider a process $N = (N_t)_{t \geq 0}$ that satisfies properties (i), (ii) and (iii) of Definition 5.2.3 and a non-negative valued process $\lambda = (\lambda_t)_{t \geq 0}$ such that for each $t \geq 0$,

$$\lambda_t \in m\mathscr{F}_0 \quad \text{and} \quad \int_0^t \lambda_s ds < \infty \text{ a.s.}$$

If

$$E\left[e^{i\eta(N_t-N_s)} \mid \mathscr{F}_s\right] = e^{(e^{i\eta}-1)\int_s^t \lambda_r dr}$$

for each $0 \leq s \leq t$ and $\eta \in \mathbb{R}$, then N is called *Poisson process with stochastic intensity* λ. For further insights into stochastic intensity processes and their significant applications, refer to, for instance, [21].

5.3 Martingale Properties

Consider a Poisson process $N = (N_t)_{t \geq 0}$ on the space $(\Omega, \mathscr{F}, P, \mathscr{F}_t)$. Note that N is not a martingale since $E[N_t] = \lambda t$ is a strictly increasing function and therefore the process is not constant in mean. However, being a process with independent increments, from Proposition 2.3.4 we have the following

Proposition 5.3.1 (Compensated Poisson Process) *The compensated Poisson process, defined by*

$$\widetilde{N}_t := N_t - \lambda t, \qquad t \geq 0,$$

is a martingale.

We explicitly observe that $\widetilde{N}$ takes real values, unlike N which takes only integer values: in Fig. 5.3 a trajectory of a compensated Poisson process is depicted.

5.4 Proof of Theorem 5.2.1

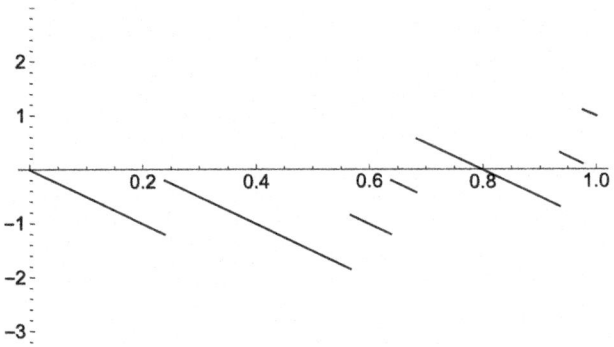

Fig. 5.3 A trajectory of the compensated Poisson process

Remark 5.3.2 The fact that $\tilde{N}$ is a martingale also follows by applying Theorem 2.5.13 with $\varphi(x) = x$. More generally, Theorem 2.5.13 shows how it is possible to "compensate" a process that is a function of N_t in order to obtain a martingale.

5.4 Proof of Theorem 5.2.1

We prove that if N is a Poisson process then for every $0 \leq s < t$:

(i) $N_t - N_s \sim \text{Poisson}_{\lambda(t-s)}$;
(ii) $N_t - N_s$ is independent of $\mathscr{G}_s^N$.

We divide the proof into two steps.

First Step We prove that, given $s > 0$ and $k \in \mathbb{N} \cup \{0\}$, the process defined by

$$N_h^{(s)} = N_{s+h} - N_s, \qquad h \in \mathbb{R}_{\geq 0}, \tag{5.4.1}$$

is a Poisson process with respect to the conditional probability given the event $(N_s = k)$, i.e. $N^{(s)}$ is a Poisson process on the space $(\Omega, \mathscr{F}, P(\cdot \mid N_s = k))$.

To this end, we define the "translated" jumps

$$T_0^{(s)} = 0, \qquad T_n^{(s)} = T_{k+n} - s, \qquad n \in \mathbb{N},$$

which, on the event $A := (N_s = k) \equiv (T_k \leq s < T_{k+1})$, form an increasing sequence a.s. (see Fig. 5.4). We observe that

$$(N_h^{(s)} = n) \cap A = (N_{s+h} = n + k) \cap A = (T_{n+k} \leq s + h < T_{n+k+1}) \cap A$$
$$= (T_n^{(s)} \leq h < T_{n+1}^{(s)}) \cap A$$

Fig. 5.4 Jump times T_n and "translated" jump times $T_n^{(s)}$

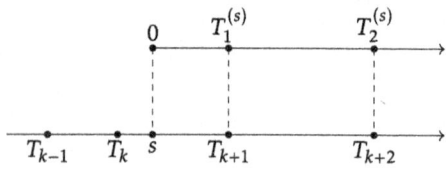

that is, in accordance with the definition of the Poisson process in the form (5.1.4), *on the event A* we have

$$(N_h^{(s)} = n) = (T_n^{(s)} \leq h < T_{n+1}^{(s)}), \qquad n \in \mathbb{N} \cup \{0\}.$$

Thus, it is sufficient to verify that the times

$$\tau_1^{(s)} := T_{k+1} - s, \qquad \tau_n^{(s)} := T_n^{(s)} - T_{n-1}^{(s)} \equiv \tau_{k+n}, \qquad n \geq 2,$$

form a sequence of random variables that, with respect to $P(\cdot \mid N_s = k)$, have distribution Exp_λ and are independent: therefore, we need to prove that

$$P\left(\bigcap_{j=1}^J (\tau_j^{(s)} \in H_j) \mid N_s = k\right) = \prod_{j=1}^J \mathrm{Exp}_\lambda(H_j) \tag{5.4.2}$$

for every $J \in \mathbb{N}$ and $H_1, \ldots, H_J \in \mathcal{B}(\mathbb{R}_{\geq 0})$. Formula (5.4.2) is equivalent to

$$P\left((N_s = k) \cap (T_{k+1} - s \in H_1) \cap \bigcap_{j=2}^J (\tau_{k+j} \in H_j)\right)$$

$$= P(N_s = k) \prod_{j=1}^J \mathrm{Exp}_\lambda(H_j). \tag{5.4.3}$$

Taking advantage of the fact that $(N_s = k) \cap (T_{k+1} - s \in H_1) = (T_k \leq s) \cap (T_{k+1} - s \in H_1)$, $T_{k+1} = T_k + \tau_{k+1}$ and the random variables $T_k, \tau_{k+1}, \ldots, \tau_{k+J}$ are independent under P, (5.4.3) reduces to

$$P\left((T_k \leq s) \cap (T_k + \tau_{k+1} - s \in H_1)\right) = P(N_s = k)\mathrm{Exp}_\lambda(H_1). \tag{5.4.4}$$

Now it is sufficient to consider the case where H_1 is an interval, $H_1 = [0, c]$: since T_k and τ_{k+1} are independent under P, the joint density is given by the product of

5.4 Proof of Theorem 5.2.1

the marginals and, recalling Lemma 5.1.1, we have

$$P((T_k \leq s) \cap (\tau_{k+1} \in [s - T_k, c + s - T_k]))$$

$$= \int_0^s \left(\int_{s-x}^{c+s-x} \lambda e^{-\lambda y} dy \right) \text{Gamma}_{k,\lambda}(dx)$$

$$= \int_0^s e^{-\lambda(c+s-x)} (e^{\lambda c} - 1) \text{Gamma}_{k,\lambda}(dx)$$

$$= \frac{(s\lambda)^k}{k!} e^{-\lambda(c+s)} (e^{\lambda c} - 1) = \text{Poisson}_{\lambda s}(\{k\}) \text{Exp}_\lambda([0, c])$$

which proves (5.4.4) with $H_1 = [0, c]$.

Second Step By the first step, $N_t - N_s$ is a Poisson process conditionally on $(N_s = k)$ and therefore we have

$$P(N_t - N_s = n \mid N_s = k) = \text{Poisson}_{\lambda(t-s)}(\{n\}) \qquad (5.4.5)$$

for every $s < t$ and $n, k \in \mathbb{N} \cup \{0\}$. By the law of total probability, we have

$$P(N_t - N_s = n) = \sum_{k \geq 0} P(N_t - N_s = n \mid N_s = k) P(N_s = k) =$$

(by (5.4.5))

$$= \sum_{k \geq 0} \text{Poisson}_{\lambda(t-s)}(\{n\}) P(N_s = k) = \text{Poisson}_{\lambda(t-s)}(\{n\}), \qquad (5.4.6)$$

and this proves property (i). Moreover, as a consequence of (5.4.6), formula (5.4.5) is equivalent to

$$P((N_t - N_s = n) \cap (N_s = k)) = P(N_s = k) P(N_t - N_s = n)$$

which proves that the consecutive increments $N_t - N_s$ and $N_s = N_s - N_0$ are independent under P.

More generally, we verify that $N_t - N_r$ and $N_r - N_s$, with $0 \leq s < r < t$, are independent under P. Recalling the notation (5.4.1), we have

$$P((N_t - N_r = n) \cap (N_r - N_s = k)) = P((N_{t-s}^{(s)} - N_{r-s}^{(s)} = n) \cap (N_{r-s}^{(s)} = k)) =$$

(by the law of total probability)

$$= \sum_{j \geq 0} P((N_{t-s}^{(s)} - N_{r-s}^{(s)} = n) \cap (N_{r-s}^{(s)} = k) \mid N_s = j) P(N_s = j) =$$

(here we use the fact that $N^{(s)}$ is a Poisson process conditionally on $(N_s = j)$ and therefore, as just proved, the increments $N_{t-s}^{(s)} - N_{r-s}^{(s)}$ and $N_{r-s}^{(s)}$ are independent under $P(\cdot \mid N_s = j)$. Moreover, $N_{r-s}^{(s)} = N_r - N_s$ and N_s are independent under P and therefore $P(N_{r-s}^{(s)} = k \mid N_s = j) = P(N_{r-s}^{(s)} = k))$

$$\begin{aligned}
&= \sum_{j \geq 0} P(N_{t-s}^{(s)} - N_{r-s}^{(s)} = n \mid N_s = j) P(N_{r-s}^{(s)} = k) P(N_s = j) \\
&= P(N_{t-s}^{(s)} - N_{r-s}^{(s)} = n) P(N_{r-s}^{(s)} = k) \\
&= P(N_t - N_r = n) P(N_r - N_s = k).
\end{aligned}$$

Thus, we have proved that, for $0 \leq s < r < t$, the increment $N_t - N_r$ is independent of $X := N_r$ and $Y := N_r - N_s$: consequently, $N_t - N_r$ is also independent of $N_s = X - Y$ and this proves property (ii). □

5.5 Key Ideas to Remember

We summarize the most significant findings of the chapter and the fundamental concepts to be retained from an initial reading, leaving out the more technical or less crucial details. As usual, if you have any doubt about what the following succinct statements mean, please review the corresponding section.

- Section 5.1: the Poisson process N is the prototype of jump processes. Sometimes called a "counting process", N_t indicates the number of times in the interval $[0, t]$ in which an episode occurs. The discontinuities of N are jumps of unit size; in various applications, the *compound* Poisson process is used, which has jumps whose size is random. The CHF of a (compound) Poisson process is homogeneous in time and can be expressed in explicit form in terms of the characteristic exponent.
- Section 5.2: N is a process with independent increments and enjoys the Markov and Feller properties.
- Section 5.3: the compensated process $\tilde{N}_t = N_t - \lambda t$ is a martingale.
- Section 5.4: from the constructive definition of the Poisson process given in Sect. 5.2, one can deduce some remarkable properties, namely the fact that $N_t - N_s \sim \text{Poisson}_{\lambda(t-s)}$ and $N_t - N_s$ is independent of $\mathscr{G}_s^N$ (cf. Theorem 5.2.1); however, this requires some work and the proof can be skipped at a first reading.

5.5 Key Ideas to Remember

Main notations used or introduced in this chapter:

Symbol	Description	Page
$N = (N_t)_{t \geq 0}$	Poisson process	84
τ_n	Time elapsed between two jumps (or episodes)	84
T_n	n-th jump time	84
$\mathscr{G}^N$	Filtration generated by N	88
$\widetilde{N}_t = N_t - \lambda t$	compensated Poisson process	90

Chapter 6
Stopping Times

> *Passion glows within your heart Like a furnace burning bright Until you struggle through the dark You'll never know the joy in life Dream Theater,*
>
> *Illumination theory*

Stopping times are a fundamental tool in the study of stochastic processes: they are particular random times that satisfy a consistency property with respect to the assigned filtration of information. The concept of stopping time is at the basis of some deep results on the structure of martingales: the optional sampling theorem, the maximal inequalities, and the upcrossing lemma. The inherent challenges in establishing these results become apparent even within the discrete framework. To move to continuous time, it will be necessary to introduce further assumptions on filtrations, the so-called *usual conditions*. The second part of the chapter collects some technical results: it shows how to extend the filtrations of Markov processes and other important classes of stochastic processes, in order to guarantee the usual conditions while maintaining the properties of the processes.

6.1 The Discrete Case

In this section, we consider the case of a *finite* number of time instants, within a filtered probability space $(\Omega, \mathscr{F}, P, (\mathscr{F}_n)_{n=0,1,\ldots,N})$ with $N \in \mathbb{N}$.

Definition 6.1.1 (Discrete Stopping Time) A discrete stopping time is a random variable

$$\tau : \Omega \longrightarrow \{0, 1, \ldots, N, \infty\}$$

such that

$$(\tau = n) \in \mathscr{F}_n, \qquad n = 0, \ldots, N. \tag{6.1.1}$$

We employ the symbol ∞ to represent a constant number that is not part of the set $\{0, 1, \ldots, N\}$ of the specified time instances: the reason for using such a symbol will be clearer later, e.g. in Example 6.1.3. We assume $N < \infty$ so that

$$(\tau \geq n) := (\tau = n) \cup \cdots \cup (\tau = N) \cup (\tau = \infty)$$

for every $n = 0, \ldots, N$.

Remark 6.1.2 Note that:

(i) condition (6.1.1) is equivalent to

$$(\tau \leq n) \in \mathscr{F}_n, \qquad n = 0, 1, \ldots, N;$$

(ii) we have

$$(\tau \geq n+1) = (\tau \leq n)^c \in \mathscr{F}_n, \qquad n = 0, \ldots, N, \qquad (6.1.2)$$

and in particular $(\tau = \infty) \in \mathscr{F}_N$;

(iii) if τ, σ are stopping times then $\tau \wedge \sigma$ and $\tau \vee \sigma$ are stopping times because

$$(\tau \wedge \sigma \leq n) = (\tau \leq n) \cup (\sigma \leq n),$$
$$(\tau \vee \sigma \leq n) = (\tau \leq n) \cap (\sigma \leq n), \qquad n = 0, \ldots, N;$$

(iv) constant times are stopping times: precisely, if $\tau \equiv k$ for some $k \in \{0, \ldots, N, \infty\}$, then τ is a stopping time.

Example 6.1.3 (Exit Time [!]) Given an *adapted* real-valued process $X = (X_n)_{n=0,1,\ldots,N}$ and $H \in \mathscr{B}$, we set

$$J(\omega) = \{n \mid X_n(\omega) \notin H\}, \qquad \omega \in \Omega.$$

The *first* exit time of X from H is defined as

$$\tau(\omega) = \begin{cases} \min J(\omega) & \text{if } J(\omega) \neq \emptyset, \\ \infty & \text{otherwise.} \end{cases}$$

From now on, we adopt the convention $\min \emptyset = \infty$ and therefore write more concisely

$$\tau = \min\{n \mid X_n \notin H\}.$$

It is easy to see that τ is a stopping time: in fact, $(\tau = 0) = (X_0 \notin H) \in \mathscr{F}_0$ and we have

$$(\tau = n) = (X_0 \in H) \cap \cdots \cap (X_{n-1} \in H) \cap (X_n \notin H) \in \mathscr{F}_n, \qquad n = 1, \ldots, N.$$

6.1 The Discrete Case

A simple example of random time that is *not* a stopping time is the *last* exit time of X from H:

$$\bar{\tau}(\omega) = \begin{cases} \max J(\omega) & \text{if } J(\omega) \neq \emptyset, \\ \infty & \text{otherwise.} \end{cases}$$

Notation 6.1.4 Given a discrete stopping time τ and a stochastic process $X = (X_n)_{n=0,1,\ldots,N}$, we set

$$(X_\tau)(\omega) := \begin{cases} X_{\tau(\omega)}(\omega) & \text{if } \tau(\omega) \in \{0, \ldots, N\}, \\ X_N(\omega) & \text{if } \tau(\omega) = \infty, \end{cases}$$

that is, $X_\tau := X_{\tau \wedge N}$, and

$$\mathscr{F}_\tau := \{ A \in \mathscr{F} \mid A \cap (\tau = n) \in \mathscr{F}_n \text{ for every } n = 0, \ldots, N \}. \tag{6.1.3}$$

It is easy to prove that $\mathscr{F}_\tau$ is a σ-algebra: in fact, for example, if $A \in \mathscr{F}_\tau$ then $A^c \cap (\tau = n) = (\tau = n) \setminus (A \cap (\tau = n)) \in \mathscr{F}_n$ and therefore $A^c \in \mathscr{F}_\tau$. We note that $\mathscr{F}_\tau = \{ A \in \mathscr{F} \mid A \cap (\tau \leq n) \in \mathscr{F}_n \text{ for every } n = 0, \ldots, N \}$. Moreover $\mathscr{F}_\infty$ (that is, $\mathscr{F}_\tau$ with $\tau \equiv \infty$) is equal to $\mathscr{F}$.

The following proposition collects other useful properties of $\mathscr{F}_\tau$.

Proposition 6.1.5 *Given τ, σ discrete stopping times, we have:*

(i) *if $\tau \equiv k$ for some $k \in \{0, \ldots, N\}$ then $\mathscr{F}_\tau = \mathscr{F}_k$;*
(ii) *if $\tau \leq \sigma$ then $\mathscr{F}_\tau \subseteq \mathscr{F}_\sigma$;*
(iii) *$(\tau \leq \sigma) \in \mathscr{F}_\tau \cap \mathscr{F}_\sigma \equiv \mathscr{F}_{\tau \wedge \sigma}$;*
(iv) *if $X = (X_n)_{n=0,\ldots,N}$ is a process adapted to the filtration then $X_\tau \in m\mathscr{F}_\tau$.*

Proof Part (i) follows from the fact that if $\tau \equiv k$ then

$$A \cap (\tau = n) = \begin{cases} A & \text{if } k = n, \\ \emptyset & \text{if } k \neq n. \end{cases}$$

Regarding (ii), it is enough to observe that, given $n \in \{0, \ldots, N\}$, if $\tau \leq \sigma$ then $(\sigma = n) \subseteq (\tau \leq n)$ and consequently for every $A \in \mathscr{F}_\tau$ we have

$$A \cap (\sigma = n) = \underbrace{A \cap (\tau \leq n)}_{\in \mathscr{F}_n} \cap \underbrace{(\sigma = n)}_{\in \mathscr{F}_n}.$$

As for (iii), recalling (6.1.2) we have

$$(\tau \leq \sigma) \cap (\tau = n) = (\sigma \geq n) \cap (\tau = n) \in \mathscr{F}_n,$$
$$(\tau \leq \sigma) \cap (\sigma = n) = (\tau \leq n) \cap (\sigma = n) \in \mathscr{F}_n,$$

and therefore $(\tau \leq \sigma) \in \mathscr{F}_\tau \cap \mathscr{F}_\sigma$. Now, if $A \in \mathscr{F}_\tau \cap \mathscr{F}_\sigma$ we have

$$A \cap (\tau \wedge \sigma \leq n) = A \cap ((\tau \leq n) \cup (\sigma \leq n))$$
$$= (A \cap (\tau \leq n)) \cup (A \cap (\sigma \leq n)) \in \mathscr{F}_n, \qquad n = 0, \ldots, N,$$

so that $\mathscr{F}_\tau \cap \mathscr{F}_\sigma \subseteq \mathscr{F}_{\tau \wedge \sigma}$. Conversely, if $A \in \mathscr{F}_{\tau \wedge \sigma}$, since $(\tau = n) \subseteq (\tau \wedge \sigma = n)$, we have

$$A \cap (\tau = n) = (A \cap (\tau \wedge \sigma = n)) \cap (\tau = n) \in \mathscr{F}_n$$

which proves the opposite inclusion.

Finally, consider $H \in \mathscr{B}$: to prove that $(X_\tau \in H) \in \mathscr{F}_\tau$ it is enough to observe that

$$(X_\tau \in H) \cap (\tau = n) = (X_n \in H) \cap (\tau = n) \in \mathscr{F}_n, \qquad n = 0, \ldots, N.$$

This proves (iv). □

Definition 6.1.6 (Stopped Process) Given a process $X = (X_n)_{n=0,\ldots,N}$ and a stopping time τ, the *stopped process* $X^\tau = (X_n^\tau)_{n=0,\ldots,N}$ is defined by

$$X_n^\tau = X_{n \wedge \tau}, \qquad n = 0, \ldots, N.$$

Proposition 6.1.7

(i) If X is adapted, then X^τ is adapted;
(ii) if X is a sub-martingale, then X^τ is a sub-martingale as well.

Proof Part (i) follows from the fact that, for $n = 0, \ldots, N$, we have[1]

$$X_{\tau \wedge n} = X_0 + \sum_{k=1}^{\tau \wedge n} (X_k - X_{k-1})$$

$$= X_0 + \sum_{k=1}^{n} (X_k - X_{k-1}) \mathbb{1}_{(k \leq \tau)}$$

and, by (6.1.2), $(k \leq \tau) \in \mathscr{F}_{k-1}$. Part (ii) follows by applying the conditional expectation given $\mathscr{F}_{n-1}$ to the identity

$$X_n^\tau - X_{n-1}^\tau = (X_n - X_{n-1}) \mathbb{1}_{(\tau \geq n)}, \qquad n = 1, \ldots, N,$$

and remembering that $(\tau \geq n) \in \mathscr{F}_{n-1}$. □

---

[1] With the convention $\sum_{k=1}^{0} \cdots = 0$.

From Proposition 6.1.7 it also follows that if X is a martingale (or a super-martingale) then even X^τ is a martingale (or a super-martingale). We leave as an exercise the proof of the following

Lemma 6.1.8 *Let $X \in L^1(\Omega, \mathscr{F}, P)$ and $Z \in L^1(\Omega, \mathscr{G}, P)$, where $\mathscr{G}$ is a sub-σ-algebra of $\mathscr{F}$. Then*[2] $Z \leq E[X \mid \mathscr{G}]$ *if and only if*

$$E[Z\mathbb{1}_G] \leq E[X\mathbb{1}_G] \quad \text{for every } G \in \mathscr{G}.$$

Proposition 6.1.9 *Let $X = (X_n)_{n=0,1,\ldots,N}$ be an absolutely integrable and adapted process on the filtered space $(\Omega, \mathscr{F}, P, (\mathscr{F}_n)_{n=0,1,\ldots,N})$. The following properties are equivalent:*

(i) X is a sub-martingale;
(ii) for every pair of stopping times σ, τ we have

$$X_{\tau \wedge \sigma} \leq E[X_\tau \mid \mathscr{F}_\sigma];$$

(iii) for every stopping time τ_0, the stopped process X^{τ_0} is a sub-martingale.

Proof [(i) $\Longrightarrow$ (ii)] Observe that

$$X_\tau = X_{\tau \wedge \sigma} + \sum_{\sigma < k \leq \tau} (X_k - X_{k-1}) = \quad (6.1.4)$$

(recalling that, by Notation 6.1.4, $X_\tau = X_{\tau \wedge N}$)

$$= X_{\tau \wedge \sigma} + \sum_{k=1}^{N} (X_k - X_{k-1})\mathbb{1}_{(\sigma < k \leq \tau)}.$$

Now, by points (ii) and (iv) of Proposition 6.1.5, $X_{\tau \wedge \sigma} \in m\mathscr{F}_{\tau \wedge \sigma} \subseteq m\mathscr{F}_\sigma$ and therefore conditioning (6.1.4) to $\mathscr{F}_\sigma$ we have

$$E[X_\tau \mid \mathscr{F}_\sigma] = X_{\tau \wedge \sigma} + \sum_{k=1}^{N} E\left[(X_k - X_{k-1})\mathbb{1}_{(\sigma < k \leq \tau)} \mid \mathscr{F}_\sigma\right].$$

To conclude, it is sufficient to prove that $E\left[(X_k - X_{k-1})\mathbb{1}_{(\sigma < k \leq \tau)} \mid \mathscr{F}_\sigma\right] \geq 0$ for $k = 1, \ldots, N$ or equivalently, thanks to Lemma 6.1.8,

$$E\left[X_{k-1}\mathbb{1}_{(\sigma < k \leq \tau)}\mathbb{1}_G\right] \leq E\left[X_k \mathbb{1}_{(\sigma < k \leq \tau)}\mathbb{1}_G\right], \qquad G \in \mathscr{F}_\sigma, \; k = 1, \ldots, N. \quad (6.1.5)$$

---

[2] $Z \leq E[X \mid \mathscr{G}]$ means $Z \leq Y$ a.s. if $Y = E[X \mid \mathscr{G}]$.

Formula (6.1.5) follows from the sub-martingale property of X once observed that, by definition of $\mathscr{F}_\sigma$ and by Remark 6.1.2-(ii), we have

$$(\sigma < k \leq \tau) \cap G = \underbrace{(\sigma < k) \cap G}_{\in \mathscr{F}_{k-1}} \cap \underbrace{(\tau \geq k)}_{\in \mathscr{F}_{k-1}}.$$

[(ii) $\Longrightarrow$ (iii)] From point (ii) with $\tau = \tau_0 \wedge n$ and $\sigma = n - 1$ we get

$$X_{\tau_0 \wedge (n-1)} \leq E\left[X_{\tau_0 \wedge n} \mid \mathscr{F}_{n-1}\right], \qquad n = 1, \ldots, N,$$

which implies the sub-martingale property of X^{τ_0}.

[(iii) $\Longrightarrow$ (i)] The claim follows by choosing $\tau_0 \equiv \infty$. □

6.1.1 Optional Sampling, Maximal Inequalities, and Upcrossing Lemma

The following result is an immediate consequence of Proposition 6.1.9 (see also Notation 6.1.4).

Theorem 6.1.10 (Optional Sampling Theorem [!!!]) *Let $X = (X_n)_{n=0,\ldots,N}$ be a sub-martingale on the space $(\Omega, \mathscr{F}, P, (\mathscr{F}_n)_{n=0,\ldots,N})$. If τ, σ are discrete stopping times such that $\sigma \leq \tau$ then*

$$X_\sigma \leq E[X_\tau \mid \mathscr{F}_\sigma]. \tag{6.1.6}$$

If X is a martingale (respectively, a super-martingale) then formula (6.1.6) becomes an equality (respectively, the direction of the inequality is reversed).

We now prove two important consequences of the optional sampling theorem:

- the *Doob's maximal inequalities* which provide an estimate of the maximum of a martingale;
- the *Upcrossing lemma* which provides an estimate on the local behavior of a martingale and in particular on "how many times it can oscillate around an interval".

A fundamental characteristic of both results is to provide estimates that depend only on the final value of the martingale and *not on the number N of time instants considered*: this crucial fact will allow us to easily move from the discrete case to the continuous one as we will see in Chap. 8.

6.1 The Discrete Case

Theorem 6.1.11 (Doob's Maximal Inequalities [!!!]) *Let $M = (M_n)_{n=0,1,\ldots,N}$ be a martingale or a non-negative sub-martingale on the space $(\Omega, \mathcal{F}, P, (\mathcal{F}_n)_{n=0,1,\ldots,N})$. Then:*

(i) for every $\lambda > 0$ we have

$$P\left(\max_{0 \le n \le N} |M_n| \ge \lambda\right) \le \frac{E[|M_N|]}{\lambda}; \tag{6.1.7}$$

(ii) for every $p > 1$ we have

$$E\left[\max_{0 \le n \le N} |M_n|^p\right] \le \left(\frac{p}{p-1}\right)^p E[|M_N|^p]. \tag{6.1.8}$$

Proof Formula (6.1.7) is a sort of Markov inequality (cf. (3.12) in [113]) for discrete martingales. If M is a martingale then, by Proposition 1.4.12, $|M|$ is a non-negative sub-martingale: therefore it is enough to prove the thesis under the assumption that M is a non-negative sub-martingale. In this case, we denote by τ the first instant in which M exceeds the level λ,

$$\tau = \min\{n \mid M_n \ge \lambda\},$$

and we set

$$\bar{M} = \max_{0 \le n \le N} M_n.$$

By Example 6.1.3 τ is a stopping time and by Proposition 6.1.5-(iii) we have

$$(\bar{M} \ge \lambda) = (\tau \le N) \in \mathcal{F}_{\tau \wedge N}.$$

Then we have

$$P(\bar{M} \ge \lambda) = E\left[\lambda \mathbb{1}_{(\bar{M} \ge \lambda)}\right] \le E\left[M_{\tau \wedge N} \mathbb{1}_{(\bar{M} \ge \lambda)}\right] \le$$

(by the optional sampling theorem)

$$\le E\left[E[M_N \mid \mathcal{F}_{\tau \wedge N}] \mathbb{1}_{(\bar{M} \ge \lambda)}\right] =$$

(since $(\bar{M} \ge \lambda) \in \mathcal{F}_{\tau \wedge N}$)

$$= E\left[E\left[M_N \mathbb{1}_{(\bar{M} \ge \lambda)} \mid \mathcal{F}_{\tau \wedge N}\right]\right] = E\left[M_N \mathbb{1}_{(\bar{M} \ge \lambda)}\right] \tag{6.1.9}$$

which proves (6.1.7).

Now observe that $\bar{M}^p = \max_{0 \leq n \leq N} M_n^p$. From (3.1.7) in [113] we have

$$E[\bar{M}^p] = p \int_0^{+\infty} \lambda^{p-1} P(\bar{M} \geq \lambda) \, d\lambda \leq$$

(by (6.1.9))

$$\leq p \int_0^{+\infty} \lambda^{p-2} E[M_N \mathbb{1}_{(\bar{M} \geq \lambda)}] \, d\lambda \leq$$

(by Fubini's theorem)

$$\leq pE\left[M_N \int_0^{\bar{M}} \lambda^{p-2} d\lambda\right] = \frac{p}{p-1} E[M_N \bar{M}^{p-1}] \leq$$

(by Hölder's inequality, with $\frac{p}{p-1}$ being the conjugate exponent of p)

$$\leq \frac{p}{p-1} E[M_N^p]^{\frac{1}{p}} E[\bar{M}^p]^{1-\frac{1}{p}},$$

hence (6.1.8) follows by dividing by $E[\bar{M}^p]^{1-\frac{1}{p}}$ and raising to the power of p. □

Corollary 6.1.12 (Doob's Maximal Inequalities) *Let $M = (M_n)_{n=0,1,\ldots,N}$ be a martingale or a non-negative sub-martingale on the space $(\Omega, \mathscr{F}, P, (\mathscr{F}_n)_{n=0,1,\ldots,N})$, and let τ be a discrete stopping time. Then:*

(i) for every $\lambda > 0$

$$P\left(\max_{0 \leq n \leq \tau \wedge N} |M_n| \geq \lambda\right) \leq \frac{E[|M_\tau|]}{\lambda};$$

(ii) for every $p > 1$

$$E\left[\max_{0 \leq n \leq \tau \wedge N} |M_n|^p\right] \leq \left(\frac{p}{p-1}\right)^p E[|M_\tau|^p].$$

Proof It is sufficient to apply Theorem 6.1.11 to the stopped martingale M^τ (cf. Definition 6.1.6 and Proposition 6.1.7). □

We now prove a rather bizarre and surprising result, which will play a crucial role in the study of the regularity and convergence properties of martingales: the Upcrossing lemma. It shows that the number of "oscillations" of a martingale is controlled by its final expectation. This result is unexpected and goes against the

6.1 The Discrete Case

idea that we might have of a martingale as a process whose trajectories are strongly "oscillating" (think, for example, of a Brownian motion).

To formalize the result, let us fix $a, b \in \mathbb{R}$ with $a < b$. The upcrossing lemma provides an estimate of the number of times a martingale "rises" from a value *less* than a to a value *greater* than b. More precisely, given a martingale $M = (M_n)_{n=0,\ldots,N}$ on the space $(\Omega, \mathscr{F}, P, (\mathscr{F}_n)_{n=0,\ldots,N})$, let $\tau_0 := 0$ and, recursively for $k \in \mathbb{N}$,

$$\sigma_k := \min\{n \in \{\tau_{k-1}, \ldots, N\} \mid M_n \leq a\}, \quad \tau_k := \min\{n \in \{\sigma_k, \ldots, N\} \mid M_n \geq b\},$$

assuming as usual the convention $\min \emptyset = \infty$. By definition, $\tau_k \geq \sigma_k \geq \tau_{k-1}$ and σ_k, τ_k are stopping times with values in $\{0, \ldots, N, \infty\}$. If $\tau_k(\omega) \leq N$ then $\tau_k(\omega)$ is *the time of the k-th upcrossing of the trajectory* $M(\omega)$; instead, if $\tau_k(\omega) = \infty$ then the total number of upcrossings of the trajectory $M(\omega)$ is less than k. Ultimately, the *number of upcrossings of M on* $[a,b]$ is given by

$$\nu_{a,b} := \max\{k \in \mathbb{N} \cup \{0\} \mid \tau_k \leq N\}. \tag{6.1.10}$$

A fundamental ingredient of the proof of the upcrossing lemma is the optional sampling theorem, according to which, for every sub-martingale M, we have

$$E\left[M_{\tau_k}\right] \leq E\left[M_{\sigma_{k+1}}\right], \quad k \in \mathbb{N}. \tag{6.1.11}$$

Now it is good to remember that, by definition (cf. Notation 6.1.4), $M_{\tau_k} \equiv M_{\tau_k \wedge N}$ so that $M_{\tau_k} = M_N$ on $(\tau_k = \infty)$: in particular, it is not necessarily true that $M_{\tau_k}(\omega) \geq b$ if $\tau_k(\omega) = \infty$. This remark is important because, between an upcrossing time $\tau_k(\omega) \leq N$ and the next one, the trajectory $M(\omega)$ must "descend" from $M_{\tau_k}(\omega) \geq b$ to $M_{\sigma_{k+1}}(\omega) \leq a$. The optional sampling theorem says that this cannot happen "too often": if $\sigma_{k+1} \leq N$, by (6.1.11) we would have $b \leq E\left[M_{\tau_k}\right] \leq E\left[M_{\sigma_{k+1}}\right] \leq a$ and this is absurd by the assumption $a < b$. Therefore, for every $k \in \mathbb{N}$, the event $(\tau_k = \infty)$ cannot be negligible and, as already mentioned, such an event is identifiable with the set of trajectories that have fewer than k upcrossings. In this sense, the martingale property and the optional sampling theorem limit the number of possible upcrossings, and thus oscillations, of M on $[a, b]$. Now it is obvious that $\nu_{a,b} \leq N$, indeed more precisely $\nu_{a,b} \leq \frac{N}{2}$ if $N \geq 2$: the surprising fact of the upcrossing lemma is that it provides an estimate of $\nu_{a,b}$ *independent of N*.

Lemma 6.1.13 (Upcrossing Lemma [!!]) *For every sub-martingale* $M = (M_n)_{n=0,\ldots,N}$ *and* $a < b$, *we have*

$$E\left[\nu_{a,b}\right] \leq \frac{E\left[(M_N - a)^+\right]}{b - a}$$

where $\nu_{a,b}$ *in* (6.1.10) *indicates the number of upcrossings of M on* $[a, b]$.

Proof Since a, b are fixed, during the proof we denote $\nu_{a,b}$ simply by ν. By definition, $\tau_k \leq N$ on $(k \leq \nu)$ and $\tau_k = \infty$ on $(k > \nu)$: therefore, recalling again that $M_\tau \equiv M_{\tau \wedge N}$ for every stopping time τ, we have

$$\sum_{k=1}^{N}(M_{\tau_k} - M_{\sigma_k}) = \sum_{k=1}^{\nu}(M_{\tau_k} - M_{\sigma_k}) + M_{\tau_{\nu+1}} - M_{\sigma_{\nu+1}}. \qquad (6.1.12)$$

Now there is a small problem: the last term $M_{\tau_{\nu+1}} - M_{\sigma_{\nu+1}} = M_N - M_{\sigma_{\nu+1}}$ may have a negative sign (since M_N could also be less than a). To solve this problem (we will see shortly what the advantage will be) we introduce the process Y defined by $Y_n = (M_n - a)^+$. We recall that Y is a non-negative sub-martingale (Proposition 1.4.12) and the number of upcrossings of M on $[a, b]$ is equal to the number of upcrossings of Y on $[0, b - a]$ since

$$\sigma_k = \min\{n \in \{\tau_{k-1}, \ldots, N\} \mid Y_n = 0\}, \quad \tau_k = \min\{n \in \{\sigma_k, \ldots, N\} \mid Y_n \geq b - a\}.$$

Rewriting (6.1.12) for Y, now we have

$$\sum_{k=1}^{N}(Y_{\tau_k} - Y_{\sigma_k}) = \sum_{k=1}^{\nu}(Y_{\tau_k} - Y_{\sigma_k}) + Y_{\tau_{\nu+1}} - Y_{\sigma_{\nu+1}} \geq \sum_{k=1}^{\nu}(Y_{\tau_k} - Y_{\sigma_k}) \geq (b-a)\nu, \qquad (6.1.13)$$

since[3] $Y_{\tau_{\nu+1}} - Y_{\sigma_{\nu+1}} \geq 0$. To conclude, we observe that $Y_N = Y_{\sigma_{N+1}}$ and

$$Y_N \geq Y_{\sigma_{N+1}} - Y_{\sigma_1} = \sum_{k=1}^{N}(Y_{\sigma_{k+1}} - Y_{\sigma_k})$$

$$= \sum_{k=1}^{N}(Y_{\sigma_{k+1}} - Y_{\tau_k}) + \sum_{k=1}^{N}(Y_{\tau_k} - Y_{\sigma_k}) \geq$$

(by (6.1.13))

$$\geq \sum_{k=1}^{N}(Y_{\sigma_{k+1}} - Y_{\tau_k}) + (b-a)\nu.$$

Applying the expected value and the optional sampling theorem ((6.1.11) with $M = Y$) we finally have the thesis

$$E[Y_N] \geq E[(b-a)\nu].$$

$\square$

---

[3] We have $Y_{\tau_{\nu+1}} - Y_{\sigma_{\nu+1}} = Y_N \geq 0$ on $(\sigma_{\nu+1} \leq N)$ and $Y_{\tau_{\nu+1}} - Y_{\sigma_{\nu+1}} = 0$ on $(\sigma_{\nu+1} = \infty)$.

Exercise 6.1.14 Prove that, for every $a < b$, a continuous function $f : [0, 1] \longrightarrow \mathbb{R}$ can have only a finite number of upcrossings on $[a, b]$.

6.2 The Continuous Case

The analysis of stopping times in the continuous case, where $I = \mathbb{R}_{\geq 0}$, requires additional technical assumptions on filtrations, commonly referred to as the "usual conditions". We will delve into these conditions in the subsequent sections.

6.2.1 Usual Conditions and Stopping Times

Definition 6.2.1 (Usual Conditions) We say that a filtration $(\mathscr{F}_t)_{t \geq 0}$ in the complete space $(\Omega, \mathscr{F}, P)$ satisfies the usual conditions if:

(i) it is *complete*, i.e., $\mathscr{F}_0$ (and therefore also $\mathscr{F}_t$ for every $t > 0$) contains the family $\mathscr{N}$ of negligible events;[4]
(ii) it is *right-continuous*, i.e., for every $t \geq 0$ we have $\mathscr{F}_t = \mathscr{F}_{t+}$ where

$$\mathscr{F}_{t+} := \bigcap_{\varepsilon > 0} \mathscr{F}_{t+\varepsilon}. \tag{6.2.1}$$

If X is adapted to a filtration $(\mathscr{F}_t)$ that satisfies the usual conditions, then every modification of X is adapted to $(\mathscr{F}_t)$ as well: without the completeness assumption on the filtration, this statement is false. The right-continuity assumption is more subtle: it means that the knowledge of information up to time t, represented by $\mathscr{F}_t$, allows us to know what happens "immediately after" t, i.e., $\mathscr{F}_{t+}$. To better understand this fact, which may now appear obscure, we introduce the concepts of stopping time in $\mathbb{R}_{\geq 0}$ and exit time of an adapted process.

Definition 6.2.2 (Stopping Time) In a filtered space $(\Omega, \mathscr{F}, P, \mathscr{F}_t)$, a stopping time is a random variable[5]

$$\tau : \Omega \longrightarrow \mathbb{R}_{\geq 0} \cup \{\infty\}$$

such that

$$(\tau \leq t) \in \mathscr{F}_t, \qquad t \geq 0. \tag{6.2.2}$$

---

[4] By assumption $(\Omega, \mathscr{F}, P)$ is complete and therefore every negligible set is an event.
[5] That is, $(\tau \in H) \in \mathscr{F}$ for every $H \in \mathscr{B}$. Consequently, also $(\tau = \infty) = (\tau \in [0, \infty))^c \in \mathscr{F}$.

Example 6.2.3 (First Exit Time [!]) Given a process $X = (X_t)_{t \geq 0}$ and $H \subseteq \mathbb{R}$, we set

$$\tau(\omega) = \begin{cases} \inf J(\omega) & \text{if } J(\omega) \neq \emptyset, \\ \infty & \text{if } J(\omega) = \emptyset, \end{cases} \qquad \text{where } J(\omega) = \{t \geq 0 \mid X_t(\omega) \notin H\}.$$

Hereafter, we will also write

$$\tau = \inf\{t \geq 0 \mid X_t \notin H\}$$

assuming by convention that the infimum of the empty set is ∞ so that $\tau(\omega) = \infty$ if $X_t(\omega) \in H$ for every $t \geq 0$. We say that τ is the *first exit time of X from H*.

Proposition 6.2.4 (Exit Time from an Open Set [!]) *Let X be an adapted and continuous process on the space $(\Omega, \mathscr{F}, P, \mathscr{F}_t)$. The first exit time of X from an open set H is a stopping time.*

Proof The thesis is a consequence of the equality

$$(\tau > t) = \bigcup_{n \in \mathbb{N}} \bigcap_{s \in \mathbb{Q} \cap [0,t)} \left(\operatorname{dist}(X_s, H^c) \geq \tfrac{1}{n} \right) \tag{6.2.3}$$

since $\left(\operatorname{dist}(X_s, H^c) \geq \tfrac{1}{n} \right) \in \mathscr{F}_s$ for $s \leq t$ and therefore $(\tau \leq t) = (\tau > t)^c \in \mathscr{F}_t$. Let us prove (6.2.3): if ω belongs to the right-hand side then there exists $n \in \mathbb{N}$ such that $\operatorname{dist}(X_s(\omega), H^c) \geq \tfrac{1}{n}$ for every $s \in \mathbb{Q} \cap [0, t)$; since X has continuous trajectories, it follows that $\operatorname{dist}(X_s(\omega), H^c) \geq \tfrac{1}{n}$ for every $s \in [0, t]$ and therefore, again by the continuity of X, it must be $\tau(\omega) > t$.

Conversely, if $\tau(\omega) > t$ then $K := \{X_s(\omega) \mid s \in [0, t]\}$ is a compact subset of H: since H is open, it follows that $\operatorname{dist}(K, H^c) > 0$ and this is enough to conclude. $\square$

In the next lemma, we prove that for every stopping time τ we have

$$(\tau < t) \in \mathscr{F}_t, \qquad t > 0. \tag{6.2.4}$$

In general, (6.2.4) is weaker than (6.2.2) but, under the usual conditions on the filtration, the two properties are equivalent.

Lemma 6.2.5 ([!]) *Every stopping time τ satisfies (6.2.4). Conversely, if (6.2.4) holds and the filtration $(\mathscr{F}_t)_{t \geq 0}$ is right-continuous, then τ is a stopping time.*

Proof We have

$$(\tau < t) = \bigcup_{n \in \mathbb{N}} \left(\tau \leq t - \tfrac{1}{n} \right).$$

6.2 The Continuous Case

If τ is a stopping time, then $\left(\tau \leq t - \frac{1}{n}\right) \in \mathscr{F}_{t-\frac{1}{n}} \subseteq \mathscr{F}_t$ for every $n \in \mathbb{N}$, and this proves the first part of the thesis.

Conversely, if (6.2.4) holds, then for every $\varepsilon > 0$ we have

$$(\tau \leq t) = \bigcap_{\substack{n \in \mathbb{N} \\ \frac{1}{n} < \varepsilon}} \left(\tau < t + \frac{1}{n}\right) \in \mathscr{F}_{t+\varepsilon}.$$

Therefore

$$(\tau \leq t) \in \bigcap_{\varepsilon > 0} \mathscr{F}_{t+\varepsilon} = \mathscr{F}_t$$

thanks to the right-continuity assumption on the filtration. $\square$

Remark 6.2.6 If τ is a stopping time then

$$(\tau = t) = (\tau \leq t) \setminus (\tau < t) \in \mathscr{F}_t.$$

Moreover

$$(\tau = \infty) = \bigcap_{t \geq 0} (\tau \geq t) \in \bigcup_{t \geq 0} \mathscr{F}_t.$$

Since the union of σ-algebras is not generally a σ-algebra, we denote by

$$\mathscr{F}_\infty := \sigma\left(\bigcup_{t \geq 0} \mathscr{F}_t\right) \qquad (6.2.5)$$

the smallest σ-algebra that contains $\mathscr{F}_t$ for each $t \geq 0$. Clearly $(\tau = \infty) \in \mathscr{F}_\infty$.

Proposition 6.2.7 (Exit Time from a Closed Set) *Let X be an adapted and continuous process on the space $(\Omega, \mathscr{F}, P, \mathscr{F}_t)$. The first exit time τ of X from a closed set H satisfies (6.2.4). If the filtration is right-continuous then τ is a stopping time.*

Proof Since H^c is open and X is continuous, for each $t > 0$ we have

$$(\tau < t) = \bigcup_{s \in \mathbb{Q} \cap [0,t)} (X_s \in H^c)$$

and the thesis follows from the fact that $(X_s \in H^c) \in \mathscr{F}_t$ for $s \leq t$ since X is adapted to $(\mathscr{F}_t)$. The second part of the thesis follows directly from Lemma 6.2.5. $\square$

Fig. 6.1 A trajectory of a continuous process X and its first exit time from a closed set H

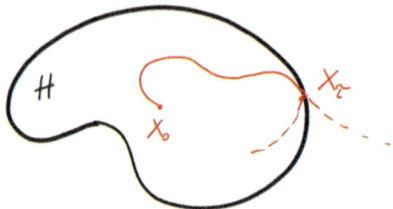

Remark 6.2.8 Under the usual conditions, also the exit time from a Borel set is a stopping time. However, establishing this fact demands a substantially more challenging proof: see, for example, Section I.10 in [20].

Remark 6.2.9 ([!]) Let us comment on Proposition 6.2.7 by observing Fig. 6.1 where the first exit time τ of X from the closed set H is represented. Up to time τ, including τ, the trajectory of X is in H. Now note the difference between the events

$$(\tau < t) = \text{``}X \text{ exits } H \text{ before time } t\text{''},$$

$$(\tau \leq t) = \text{``}X \text{ exits } H \text{ before or immediately after } t\text{''}.$$

Intuitively, it is plausible that, without the need to impose conditions on the filtration, one can prove (this is what we did in Proposition 6.2.7) that $(\tau < t) \in \mathscr{F}_t$, i.e., that *the fact that X exits H before time t is observable based on the knowledge of what happened up to time t* (i.e., $\mathscr{F}_t$, in particular knowing the trajectory of the process up to time t). On the contrary, it is only thanks to the right-continuity of the filtration that one can prove that $(\tau \leq t) \in \mathscr{F}_t$. Indeed, if $t = \tau(\omega)$ then $X_t(\omega) \in \partial H$ and based on the observation of the trajectory of X up to time t (i.e., having the information in $\mathscr{F}_t$) it is not possible to know whether $X(\omega)$ will continue to remain inside H or exit H immediately after t. In fact, for a generic filtration $(\tau \leq t) \notin \mathscr{F}_t$, i.e., as already observed, the condition $(\tau < t) \in \mathscr{F}_t$ is weaker than $(\tau \leq t) \in \mathscr{F}_t$. On the other hand, if $(\mathscr{F}_t)_{t \geq 0}$ satisfies the usual conditions (in particular, the right-continuity property) then the two conditions $(\tau < t) \in \mathscr{F}_t$ and $(\tau \leq t) \in \mathscr{F}_t$ are equivalent (Lemma 6.2.5). As we anticipated, this means that the right-continuity of the filtration ensures that knowing $\mathscr{F}_t$ we can also see what happens "immediately after" time t.

6.2.2 Filtration Enlargement and Markov Processes

We have explained the importance of the usual conditions on filtrations and the reasons why it is it is preferable to assume the validity of such hypotheses. In this section, we prove that it is always possible to modify a filtration so that it satisfies the usual conditions and, under appropriate conditions, it is also possible

6.2 The Continuous Case

to preserve some fundamental properties of the considered processes, such as the Markov property.

> The results of this section and the rest of the chapter are useful but have quite technical and less informative proofs: at a first reading, it is therefore recommended to read the statements but skip the proofs.

Consider a complete space $(\Omega, \mathscr{F}, P)$ equipped with a generic filtration $(\mathscr{F}_t)_{t \geq 0}$ and denote by $\mathscr{N}$ the family of negligible events. It is always possible to expand $(\mathscr{F}_t)_{t \geq 0}$ so that the usual conditions are satisfied:

(i) by setting

$$\bar{\mathscr{F}}_t := \sigma(\mathscr{F}_t \cup \mathscr{N}), \qquad t \geq 0, \tag{6.2.6}$$

we define the smallest filtration[6] in $(\Omega, \mathscr{F}, P)$, which completes and extends $(\mathscr{F}_t)_{t \geq 0}$;

(ii) the filtration $(\mathscr{F}_{t+})_{t \geq 0}$ defined by (6.2.1) is right-continuous.

Combining points (i) and (ii) (in any order), we obtain the filtration $\left(\bar{\mathscr{F}}_{t+}\right)_{t \geq 0}$ which is the smallest filtration that extends $(\mathscr{F}_t)_{t \geq 0}$ and verifies the usual conditions.

Definition 6.2.10 (Standard Enlargement of a Filtration) The filtration $\left(\bar{\mathscr{F}}_{t+}\right)_{t \geq 0}$ is called the *standard enlargement* of the filtration $(\mathscr{F}_t)_{t \geq 0}$.

Now consider a stochastic process $X = (X_t)_{t \geq 0}$ on $(\Omega, \mathscr{F}, P)$ and its associated filtration

$$\mathscr{G}_t^X := \sigma(X_s, s \leq t), \qquad t \geq 0,$$

that is, the filtration *generated* by X.

Definition 6.2.11 (Standard Filtration of a Process) The *standard filtration* of a process X, hereafter denoted by $\mathscr{F}^X = \left(\mathscr{F}_t^X\right)_{t \geq 0}$, is the standard enlargement of $\mathscr{G}^X$.

Suppose that $X = (X_t)_{t \geq 0}$ is a Markov process with transition law p on the complete filtered space $(\Omega, \mathscr{F}, P, \mathscr{F}_t)$. In general, it is not a problem to "shrink" the filtration: more precisely, if $(\mathscr{G}_t)_{t \geq 0}$ is a filtration such that $\mathscr{G}_t^X \subseteq \mathscr{G}_t \subseteq \mathscr{F}_t$ for every $t \geq 0$, i.e., $(\mathscr{G}_t)_{t \geq 0}$ is smaller than $(\mathscr{F}_t)_{t \geq 0}$ but larger than $(\mathscr{G}_t^X)_{t \geq 0}$, then it is immediate to verify that X is a Markov process also on the space $(\Omega, \mathscr{F}, P, \mathscr{G}_t)$.

---

[6] Obviously, we have $\bar{\mathscr{F}}_t \subseteq \bar{\mathscr{F}}_T$ if $0 \leq t \leq T$. Moreover, $\bar{\mathscr{F}}_t \subseteq \mathscr{F}$ for every $t \geq 0$ thanks to the completeness assumption of $(\Omega, \mathscr{F}, P)$.

The problem is not obvious when we want to *enlarge* the filtration. The following results provide conditions under which it is possible to enlarge the filtration of a Markov process so that it verifies the usual conditions, without affecting the Markov property.

Proposition 6.2.12 *Let $X = (X_t)_{t \geq 0}$ be a Markov process with transition law p on the complete filtered space $(\Omega, \mathscr{F}, P, \mathscr{F}_t)$. Then X is a Markov process with transition law p on $(\Omega, \mathscr{F}, P)$ with respect to the completed filtration $(\bar{\mathscr{F}}_t)_{t \geq 0}$ in (6.2.6).*

Proof Clearly, X is adapted to $\bar{\mathscr{F}}$ so we only need to prove that

$$p(t, X_t; T, H) = P(X_T \in H \mid \bar{\mathscr{F}}_t), \qquad 0 \leq t \leq T, \ H \in \mathscr{B}.$$

Let $Z = p(t, X_t; T, H)$, then $Z \in m\sigma(X_t) \subseteq m\bar{\mathscr{F}}_t$; based on the definition of conditional expectation, it remains to verify that for every $G \in \bar{\mathscr{F}}_t$ we have

$$E[Z \mathbb{1}_G] = E\left[\mathbb{1}_{(X_T \in H)} \mathbb{1}_G\right]. \tag{6.2.7}$$

Formula (6.2.7) is true if $G \in \mathscr{F}_t$: on the other hand (see Remark 1.4.3 in [113]) $G \in \bar{\mathscr{F}}_t = \sigma(\mathscr{F}_t \cup \mathscr{N})$ if and only if $G = A \cup N$ for some $A \in \mathscr{F}_t$ and $N \in \mathscr{N}$. Therefore, we have

$$E[Z \mathbb{1}_G] = E[Z \mathbb{1}_A] = E\left[\mathbb{1}_{(X_T \in H)} \mathbb{1}_A\right] = E\left[\mathbb{1}_{(X_T \in H)} \mathbb{1}_G\right].$$

□

It is possible to enlarge the filtration to make it right-continuous and maintain the Markov property, assuming additional continuity assumptions for the process trajectories (e.g., a.s. right-continuity) and for the process transition law (the Feller property, Definition 2.1.10).

Proposition 6.2.13 *Let $X = (X_t)_{t \geq 0}$ be a Markov process with transition law p on the complete filtered space $(\Omega, \mathscr{F}, P, \mathscr{F}_t)$. Suppose that X is a Feller process with a.s. right-continuous trajectories. Then X is a Markov process with transition law p on $(\Omega, \mathscr{F}, P, \mathscr{F}_{t+})$.*

Proof Clearly, X is adapted to $(\mathscr{F}_{t+})_{t \geq 0}$ so there is only to prove the Markov property, namely that for every $0 \leq t < T$ and $\varphi \in b\mathscr{B}$ we have

$$Z = E[\varphi(X_T) \mid \mathscr{F}_{t+}] \quad \text{where} \quad Z := \int_{\mathbb{R}} p(t, X_t; T, dy)\varphi(y).$$

By Fubini's theorem, $Z \in m\mathscr{F}_t \subseteq m\mathscr{F}_{t+}$. Therefore, by definition of conditional expectation, it remains to verify that for every $G \in \mathscr{F}_{t+}$ we have

$$E[\varphi(X_T) \mathbb{1}_G] = E[Z \mathbb{1}_G]. \tag{6.2.8}$$

6.2 The Continuous Case

Now, let $h > 0$ such that $t + h < T$: we have $G \in \mathscr{F}_{t+h}$ and therefore, by the Markov property of X with respect to $(\mathscr{F}_t)_{t \geq 0}$, we have

$$E[\varphi(X_T)\mathbb{1}_G] = E\left[\int_{\mathbb{R}} p(t+h, X_{t+h}; T, dy)\varphi(y)\mathbb{1}_G\right]. \tag{6.2.9}$$

Utilizing the a.s. right-continuity of the trajectories of X and the Feller property of p, we can take the limit as h tends to 0^+ in (6.2.9). Applying the dominated convergence theorem yields (6.2.8). □

Remark 6.2.14 ([!]) Combining Propositions 6.2.12 and 6.2.13 we have the following result: *if X is an a.s. right-continuous, Markov and Feller process on the complete space $(\Omega, \mathscr{F}, P, \mathscr{F}_t)$ then X is a Markov process also on the complete space $(\Omega, \mathscr{F}, P, (\bar{\mathscr{F}}_{t+})_{t \geq 0})$ where the usual conditions hold.*

Next, we show that for a Markov process X with respect to its own standard filtration $\mathscr{F}^X$, we simply have

$$\mathscr{F}^X_t = \sigma(\mathscr{G}^X_t \cup \mathscr{N}), \qquad t \geq 0. \tag{6.2.10}$$

In other words, $\mathscr{F}^X$ is obtained by completing the filtration generated by X and the property of right-continuity is automatically satisfied.

Proposition 6.2.15 ([!]) *If X is a Markov process with respect to its standard filtration $\mathscr{F}^X$ then (6.2.10) holds.*

Proof The proof is based on the extended Markov property of Theorem 2.2.4 according to which we have[7]

$$ZE[Y \mid X_t] = E\left[ZY \mid \mathscr{F}^X_t\right], \qquad Z \in b\sigma(\mathscr{G}^X_t \cup \mathscr{N}), \ Y \in b\mathscr{G}^X_{t,\infty}.$$

Since every version of $E[Y \mid X_t]$ is $\sigma(X_t)$-measurable and given the uniqueness of the conditional expectation up to negligible events, it follows that every version of $E\left[ZY \mid \mathscr{F}^X_t\right]$ is $\sigma(\mathscr{G}^X_t \cup \mathscr{N})$-measurable: given the assumptions on Y and Z, this measurability property also holds if instead of ZY we put any random variable in $b\sigma(\mathscr{G}^X_\infty \cup \mathscr{N})$. In particular, for $A \in \mathscr{F}^X_t \subseteq \sigma(\mathscr{G}^X_\infty \cup \mathscr{N})$ we obtain

$$\mathbb{1}_A = E\left[\mathbb{1}_A \mid \mathscr{F}^X_t\right] \in b\sigma(\mathscr{G}^X_t \cup \mathscr{N}).$$

□

Remark 6.2.16 ([!]) Combining Propositions 6.2.12, 6.2.13, and 6.2.15, we have the following result: *let X be a Markov and Feller right-continuous process with*

---

[7] In the sense of Convention 4.2.5 in [113]. Note that $Z \in b\sigma(\mathscr{G}^X_t \cup \mathscr{N}) \subseteq b\mathscr{F}^X_t$.

respect to $\mathscr{G}^X$; then the standard filtration is $\mathscr{F}_t^X = \sigma(\mathscr{G}_t^X \cup \mathscr{N})$, $t \geq 0$, and X is a Markov process also with respect to $\mathscr{F}^X$.

We now consider a Markov process X on the space $(\Omega, \mathscr{F}, P, \mathscr{F}_t)$ in which the usual conditions hold and recall definition (2.2.6) of the σ-algebra $\mathscr{G}_{t,\infty}^X$ of future information on X starting from time t.

Theorem 6.2.17 (Blumenthal's 0-1 Law) *Let X be a Markov process on $(\Omega, \mathscr{F}, P, \mathscr{F}_t)$. If $A \in \mathscr{F}_t \cap \mathscr{G}_{t,\infty}^X$ then $P(A \mid X_t) = 1$ or $P(A \mid X_t) = 0$.*

Proof We explicitly note that A is not necessarily $\sigma(X_t)$-measurable. In other words, in general $\sigma(X_t)$ is strictly included in $\mathscr{F}_t \cap \mathscr{F}_{t,\infty}^X$ since, by the right continuity of $\mathscr{F}^X$, we have

$$\sigma(X_t) \subseteq \bigcap_{\varepsilon > 0} \sigma(X_s, t \leq s \leq t + \varepsilon) \subseteq \mathscr{F}_t \cap \mathscr{F}_{t,\infty}^X.$$

if this were the case, the thesis would be an obvious consequence of Example 4.3.3 in [113]. On the other hand, by Corollary 2.2.5, $\mathscr{F}_t$ and $\mathscr{G}_{t,\infty}^X$ are, conditionally on X_t, independent: it follows that A is independent of itself (conditionally on X_t) and therefore we have

$$P(A \mid X_t) = P(A \cap A \mid X_t) = P(A \mid X_t)^2.$$

Hence, $P(A \mid X_t)$ can only take the values 0 or 1. □

Example 6.2.18 ([!]) We resume Example 6.2.3 and suppose that τ is the exit time from a closed set H, of a continuous Markov process X on the space $(\Omega, \mathscr{F}, P, \mathscr{F}^X)$. We apply Blumenthal's 0-1 law with $t = 0$: clearly $(\tau = 0) \in \mathscr{F}_0^X = \mathscr{F}_0^X \cap \mathscr{F}_{0,\infty}^X$ since τ is a stopping time; here $(\tau = 0)$ indicates the event according to which the process X exits immediately from H. Then we have $P(\tau = 0 \mid X_0) = 0$ or $P(\tau = 0 \mid X_0) = 1$, that is almost all trajectories of X exit immediately from H or almost none. This fact is particularly interesting when X_0 belongs to the boundary of H.

6.2.3 Filtration Enlargement and Lévy Processes

We now study the filtration enlargement for the Poisson process and the Brownian motion. To treat the subject in a unified way, we introduce a class of processes of which Poisson and Brownian are particular cases.

Definition 6.2.19 (Lévy Process) Let $X = (X_t)_{t \geq 0}$ be a real stochastic process defined on a complete filtered probability space $(\Omega, \mathscr{F}, P, \mathscr{F}_t)$. We say that X is a Lévy process if it satisfies the following properties:

(i) $X_0 = 0$ a.s.;
(ii) the trajectories of X are a.s. càdlàg;

6.2 The Continuous Case

(iii) X is adapted to $(\mathscr{F}_t)$;
(iv) $X_t - X_s$ is independent of $\mathscr{F}_s$ for every $0 \le s \le t$;
(v) the increments $X_t - X_s$ and $X_{t+h} - X_{s+h}$ have the same law for every $0 \le s \le t$ and $h \ge 0$.

Remark 6.2.20 ([!!]) Properties (iv) and (v) are expressed by saying that X has independent and stationary increments. By Proposition 2.3.2, a Lévy process X is a Markov process with transition law $p(t, x; T, \cdot)$ equal to the distribution of $X_T - X_t + x$: such law is homogeneous in time thanks to the stationarity of the increments. It follows in particular that every Lévy process is a Feller process: indeed, for every $\varphi \in bC(\mathbb{R})$ and $h > 0$ we have

$$(t, x) \longmapsto \int_{\mathbb{R}} p(t, x; t+h, dy)\varphi(y) =$$

(since $p(t, x; t+h, \cdot)$ is the distribution of $X_{t+h} - X_t + x$ which is equal in law to $X_h + x$ by the stationarity of the increments)

$$= \int_{\mathbb{R}} p(0, x; h, dy)\varphi(y) = E\left[\varphi(X_h + x)\right]$$

and the continuity in (t, x) follows from Lebesgue's dominated convergence theorem.

Moreover, one can prove that the CHF of a Lévy process X is of the form

$$\varphi_{X_T}(\eta) = e^{T\psi(\eta)}$$

where ψ is called the *characteristic exponent* of X: for example, $\psi(\eta) = -\frac{\eta^2}{2}$ for Brownian motion and $\psi(\eta) = \lambda(e^{i\eta} - 1)$ for the Poisson process (cf. Remark 5.1.4). Then, setting for simplicity $p(T, \cdot) = p(0, 0; T, \cdot)$, we have the following remarkable relation:

$$\psi(\eta) e^{T\psi(\eta)} = \partial_T e^{T\psi(\eta)}$$

$$= \partial_T \int_{\mathbb{R}} e^{i\eta y} p(T, dy) =$$

(assuming we can exchange the signs of derivative and integral)

$$= \int_{\mathbb{R}} e^{i\eta y} \partial_T p(T, dy) =$$

(since $p(T, dy)$ solves the forward Kolmogorov equation (2.5.25), $\partial_T p(T, \cdot) = \mathscr{A}_T^* p(T, \cdot)$ where $\mathscr{A}_T^*$ is the adjoint of the infinitesimal generator or characteristic operator of X)

$$= \int_{\mathbb{R}} e^{i\eta y} \mathscr{A}_T^* p(T, dy).$$

In the language of pseudo-differential calculus, this fact is expressed by stating that ψ *is the symbol of the operator* $\mathscr{A}_T^*$ and is denoted as

$$\mathscr{A}_T^* = \psi(i\partial_y).$$

For example, for the Brownian motion we have $\psi(\eta) = -\frac{\eta^2}{2}$ and

$$\mathscr{A}_T^* = \psi(i\partial_y) = \frac{1}{2}\partial_{yy},$$

while for the Poisson process, since $\psi(\eta) = \lambda(e^{i\eta} - 1)$, we have

$$\mathscr{A}_T^* \varphi(y) = \psi(i\partial_y)\varphi(y) = \lambda(\varphi(y-1) - \varphi(y)). \tag{6.2.11}$$

The representation (6.2.11) of $\mathscr{A}_T^*$ as a pseudo-differential operator is also justified by the formal expression

$$e^{\alpha \partial_y}\varphi(y) = \sum_{n=0}^{\infty} \frac{(\alpha \partial_y)^n}{n!}\varphi(y) = \varphi(y+\alpha)$$

as a Taylor series expansion valid for every analytic function φ. The general expression of the characteristic exponent of a Lévy process is given by the famous *Lévy-Khintchine formula*

$$\psi(\eta) = i\mu\eta - \frac{\sigma^2 \eta^2}{2} + \int_{\mathbb{R}}\left(e^{i\eta x} - 1 - i\eta x \mathbb{1}_{|x|\leq 1}\right)\nu(dx)$$

where $\mu, \sigma \in \mathbb{R}$ and ν is a measure on $\mathbb{R}$ such that $\nu(\{0\}) = 0$ and

$$\int_{\mathbb{R}}(1 \wedge |x|^2)\nu(dx) < \infty.$$

For each $H \in \mathscr{B}$, $\nu(H)$ indicates the expected number of jumps of the process trajectories in a unit time period, with size $\Delta_t X \in H$: for example, for the Poisson process, we have $\nu = \lambda \delta_1$ and for the compound Poisson process of Example 5.1.5, we have $\nu = \lambda \mu_Z$ where μ_Z is the law of the variables Z_n, i.e., the individual jumps of the process.

If a Lévy process X is *a.s. continuous* then $\nu \equiv 0$ and therefore necessarily X is a Brownian motion with drift, i.e., a process of the form $X_t = \mu t + \sigma W_t$ with $\mu, \sigma \in \mathbb{R}$ and W Brownian motion. Among the reference texts for the general theory of Lévy processes, we indicate the monograph [4].

Proposition 6.2.21 *Let* $X = (X_t)_{t\geq 0}$ *be a Lévy process on the complete space* $(\Omega, \mathscr{F}, P, \mathscr{F}_t)$. *Then* X *is a Lévy process also on* $(\Omega, \mathscr{F}, P, (\bar{\mathscr{F}}_t)_{t\geq 0})$ *and on* $(\Omega, \mathscr{F}, P, (\mathscr{F}_{t+})_{t\geq 0})$.

6.2 The Continuous Case

Proof It suffices to verify that, for each $0 \leq s < t$, the increment $X_t - X_s$ is independent of $\bar{\mathscr{F}}_s$ and of $\mathscr{F}_{s+}$, i.e., we have

$$P(X_t - X_s \in H \mid G) = P(X_t - X_s \in H), \qquad H \in \mathscr{B}, \qquad (6.2.12)$$

if $G \in \bar{\mathscr{F}}_s \cup \mathscr{F}_{s+}$ with $P(G) > 0$. Let us first consider the case $G \in \bar{\mathscr{F}}_s$ (always assuming $P(G) > 0$). Equation (6.2.12) is true if $G \in \mathscr{F}_s$: on the other hand (cf. Remark 1.4.3 in [113]) $G \in \bar{\mathscr{F}}_s = \sigma(\mathscr{F}_s \cup \mathscr{N})$ if and only if $G = A \cup N$ for some $A \in \mathscr{F}_s$ and $N \in \mathscr{N}$ (and necessarily $P(A) > 0$ since $P(G) > 0$). Hence we have

$$P(X_t - X_s \in H \mid G) = P(X_t - X_s \in H \mid A) = P(X_t - X_s \in H).$$

Now let us consider the case $G \in \mathscr{F}_{s+}$ with $P(G) > 0$. Here we use the fact that, by Corollary 2.5.8 in [113], Eq. (6.2.12) is true if and only if we have

$$E[\varphi(X_t - X_s) \mid G] = E[\varphi(X_t - X_s)],$$

for every $\varphi \in bC$. We observe that, for every $h > 0$, $G \in \mathscr{F}_{s+h}$ and therefore G is independent from $X_{t+h} - X_{s+h}$: then we have

$$E[\varphi(X_{t+h} - X_{s+h}) \mid G] = E[\varphi(X_{t+h} - X_{s+h})]$$

and we conclude by taking the limit as $h \to 0^+$, by the dominated convergence theorem thanks to the right-continuity of the trajectories of X and the continuity and boundedness of φ. □

Combining the previous results with Remark 6.2.16 we have the following

Theorem 6.2.22 ([!]) *Let X be a Lévy process on the complete space $(\Omega, \mathscr{F}, P)$ equipped with the filtration $\mathscr{G}^X$ generated by X. Then $\mathscr{F}_t^X = \sigma(\mathscr{G}_t^X \cup \mathscr{N})$, for $t \geq 0$, and X is a Lévy process also with respect to the standard filtration $\mathscr{F}^X$.*

As a consequence of Blumenthal's 0-1 law of Theorem 6.2.17, we have

Corollary 6.2.23 (Blumenthal's 0-1 Law) *Let $X = (X_t)_{t \geq 0}$ be a Lévy process. For every $A \in \mathscr{F}_0^X$ we have $P(A) = 0$ or $P(A) = 1$.*

Let $(C(\mathbb{R}_{\geq 0}), \mathscr{B}_{\mu_W}, \mu_W)$ be the Wiener space (cf. Definition 4.3.2): here μ_W is the Wiener measure (i.e., the law of a Brownian motion) defined on the μ_W-completion $\mathscr{B}_{\mu_W}$ of the Borel σ-algebra.

Definition 6.2.24 (Canonical Brownian Motion) The canonical Brownian motion $\mathbf{W}$ is the identity process[8] on the Wiener space equipped with the standard filtration $\mathscr{F}^{\mathbf{W}}$.

---

[8] That is, $\mathbf{W}_t(w) = w(t)$ for every $w \in C(\mathbb{R}_{\geq 0})$ and $t \geq 0$.

Remark 6.2.25 ([!]) By Corollary 4.3.3 and Theorem 6.2.22, the canonical Brownian motion is a Brownian motion, according to Definition 4.1.1, on the space $(C(\mathbb{R}_{\geq 0}), \mathscr{B}_{\mu_W}, \mu_W, \mathscr{F}^{\mathbf{W}})$. Moreover, the Wiener space is a Polish metric space and a complete probability space in which the standard filtration $\mathscr{F}^{\mathbf{W}}$ satisfies the usual conditions: due to these important properties, the Wiener space and the canonical Brownian motion constitute respectively the canonical space and process of reference in the study of stochastic differential equations.

6.2.4 General Results on Stopping Times

We resume the study of stopping times with values in $\mathbb{R}_{\geq 0} \cup \{\infty\}$ (cf. Definition 6.2.2), on a filtered space $(\Omega, \mathscr{F}, P, \mathscr{F}_t)$ *satisfying the usual conditions*. We leave as an exercise the proof of the following

Proposition 6.2.26

(i) *If* $\tau = t$ *a.s. then* τ *is a stopping time;*
(ii) *if* τ, σ *are stopping times then also* $\tau \wedge \sigma$ *and* $\tau \vee \sigma$ *are stopping times;*
(iii) *if* $(\tau_n)_{n \geq 1}$ *is an increasing sequence (i.e.,* $\tau_n \leq \tau_{n+1}$ *a.s. for every* $n \in \mathbb{N}$) *then* $\sup_{n \in \mathbb{N}} \tau_n$ *is a stopping time;*
(iv) *if* $(\tau_n)_{n \geq 1}$ *is a decreasing sequence (i.e.,* $\tau_n \geq \tau_{n+1}$ *a.s. for every* $n \in \mathbb{N}$) *then* $\inf_{n \in \mathbb{N}} \tau_n$ *is a stopping time;*
(v) *if* τ *is a stopping time then for every* $\varepsilon \geq 0$ *also* $\tau + \varepsilon$ *is a stopping time.*

Now consider a stochastic process $X = (X_t)_{t \geq 0}$ on the filtered space $(\Omega, \mathscr{F}, P, \mathscr{F}_t)$ that verifies the usual conditions. In the analysis of stopping times (and later, stochastic integration), it becomes necessary to impose a minimal measurability condition on X concerning the time variable. This condition enhances the notion of adapted process.

Definition 6.2.27 (Progressively Measurable Process) A process $X = (X_t)_{t \geq 0}$ is progressively measurable if, for every $t > 0$, the function $(s, \omega) \mapsto X_s(\omega)$ on $[0, t] \times \Omega$ to $\mathbb{R}^d$ is measurable with respect to the product σ-algebra $\mathscr{B} \otimes \mathscr{F}_t$.

In other words, X is progressively measurable if, for every fixed $t > 0$, the function $g := X|_{[0,t] \times \Omega}$, defined by

$$g : ([0, t] \times \Omega, \mathscr{B} \otimes \mathscr{F}_t) \longrightarrow (\mathbb{R}, \mathscr{B}), \qquad g(s, \omega) = X_s(\omega), \qquad (6.2.13)$$

is $(\mathscr{B} \otimes \mathscr{F}_t)$-measurable. If X is progressively measurable then, by Lemma 2.3.11 in [113], it is adapted to $(\mathscr{F}_t)$. Conversely, a result by Chung and Doob [25] shows that *if X is adapted and measurable*[9] *then it possesses a progressively measurable*

---

[9] That is, $(t, \omega) \mapsto X_t(\omega)$ is $\mathscr{B} \otimes \mathscr{F}$-measurable.

6.2 The Continuous Case

modification (for a proof of this fact see, for example [96], Theorem T46 on p. 68). We will only need the following much simpler result:

Proposition 6.2.28 *If X is adapted to $(\mathscr{F}_t)$ and has a.s. right-continuous trajectories (or has a.s. left-continuous trajectories) then it is progressively measurable.*

Proof Consider the sequences

$$\vec{X}_t^{(n)} := \sum_{k=1}^{\infty} X_{\frac{k-1}{2^n}} \mathbb{1}_{[\frac{k-1}{2^n}, \frac{k}{2^n})}(t), \quad \overleftarrow{X}_t^{(n)} := \sum_{k=1}^{\infty} X_{\frac{k}{2^n}} \mathbb{1}_{[\frac{k-1}{2^n}, \frac{k}{2^n})}(t), \quad t \in [0,T], \, n \in \mathbb{N}.$$

Since X is adapted, it follows from Corollary 2.3.9 in [113] that $\vec{X}^{(n)} \in m(\mathscr{B} \otimes \mathscr{F}_T)$ and $\overleftarrow{X}^{(n)} \in m(\mathscr{B} \otimes \mathscr{F}_{T+\frac{1}{2^n}})$. If X has a.s. left-continuous trajectories then $\vec{X}^{(n)}$ converges pointwise (Leb $\otimes$ P)-a.s. to X on $[0,T] \times \Omega$ as $n \to \infty$: given the arbitrariness of T, it follows that X is progressively measurable.

Similarly, if X has a.s. right-continuous trajectories then $\overleftarrow{X}^{(n)}$ converges pointwise (Leb $\otimes$ P)-a.s. to X on $[0,T] \times \Omega$ as $n \to \infty$: it follows that, for every $\varepsilon > 0$, the map $(t, \omega) \mapsto X_t(\omega)$ is $(\mathscr{B} \otimes \mathscr{F}_{T+\varepsilon})$-measurable on $[0,T] \times \Omega$. Due to the right-continuity of the filtration, we conclude that X is progressively measurable. □

Given a stopping time τ, we recall definition (6.2.5) of $\mathscr{F}_\infty$ and, in analogy with (6.1.3), we define

$$\mathscr{F}_\tau := \{A \in \mathscr{F}_\infty \mid A \cap (\tau \leq t) \in \mathscr{F}_t \text{ for every } t \geq 0\}.$$

Note that $\mathscr{F}_\tau$ is a σ-algebra and $\mathscr{F}_\tau = \mathscr{F}_t$ if τ is the constant stopping time equal to t. Moreover, given a process $X = (X_t)_{t \geq 0}$ we define

$$(X_\tau)(\omega) := \begin{cases} X_{\tau(\omega)}(\omega) & \text{if } \tau(\omega) < \infty, \\ 0 & \text{if } \tau(\omega) = \infty. \end{cases}$$

Proposition 6.2.29 *In a filtered probability space where the usual conditions are in force, we have:*

(i) $\tau \in m\mathscr{F}_\tau$;
(ii) *if* $\tau \leq \sigma$ *then* $\mathscr{F}_\tau \subseteq \mathscr{F}_\sigma$;
(iii) $\mathscr{F}_\tau \cap \mathscr{F}_\sigma = \mathscr{F}_{\tau \wedge \sigma}$;
(iv) *if X is progressively measurable then* $X_\tau \in m\mathscr{F}_\tau$;
(v) $\mathscr{F}_\tau = \mathscr{F}_{\tau+} := \bigcap_{\varepsilon > 0} \mathscr{F}_{\tau+\varepsilon}$;

Proof

(i) We have to show that $(\tau \in H) \cap (\tau \leq t) \in \mathscr{F}_t$ for every $t \geq 0$ and $H \in \mathscr{B}$: the thesis follows easily since by Lemma 2.1.5 in [113] it is sufficient to consider H of the type $(-\infty, s]$ with $s \in \mathbb{R}$.

(ii) If $\tau \leq \sigma$ then $(\sigma \leq t) \subseteq (\tau \leq t)$: hence for every $A \in \mathscr{F}_\tau$ we have

$$A \cap (\sigma \leq t) = \underbrace{A \cap (\tau \leq t)}_{\in \mathscr{F}_t} \cap \underbrace{(\sigma \leq t)}_{\in \mathscr{F}_t}.$$

(iii) By point (ii) the inclusion $\mathscr{F}_\tau \cap \mathscr{F}_\sigma \supseteq \mathscr{F}_{\tau \wedge \sigma}$ holds. Conversely, if $A \in \mathscr{F}_\tau \cap \mathscr{F}_\sigma$ then

$$A \cap (\tau \wedge \sigma \leq t) = A \cap ((\tau \leq t) \cup (\sigma \leq t)) = \underbrace{(A \cap (\tau \leq t))}_{\in \mathscr{F}_t} \cup \underbrace{(A \cap (\sigma \leq t))}_{\in \mathscr{F}_t}.$$

(iv) we have to prove that $(X_\tau \in H) \cap (\tau \leq t) = (X_{\tau \wedge t} \in H) \cap (\tau \leq t) \in \mathscr{F}_t$ for every $t \geq 0$ and $H \in \mathscr{B}$. Since $(\tau \leq t) \in \mathscr{F}_t$ it is sufficient to prove that $X_{\tau \wedge t} \in m\mathscr{F}_t$: this is a consequence of the fact that $X_{\tau \wedge t}(\omega) = (f \circ g)(t, \omega)$ with f and g measurable functions defined by

$$f : (\Omega, \mathscr{F}_t) \longrightarrow ([0, t] \times \Omega, \mathscr{B} \otimes \mathscr{F}_t), \qquad f(t, \omega) := (\tau(\omega) \wedge t, \omega),$$

and g as in (6.2.13). The measurability of f follows from Corollary 2.3.9 in [113] and the fact that, by (i), $(\tau \wedge t) \in m\mathscr{F}_{\tau \wedge t} \subseteq m\mathscr{F}_t$; g is measurable since X is progressively measurable.

(v) The inclusion $\mathscr{F}_\tau \subseteq \mathscr{F}_{\tau+}$ is obvious by (ii). Conversely, if $A \in \mathscr{F}_{\tau+}$ then by definition $A \cap (\tau + \epsilon \leq t) \in \mathscr{F}_t$ for every $t \geq 0$ and $\epsilon > 0$: therefore $A \cap (\tau \leq t - \epsilon) \in \mathscr{F}_t$ for every $t \geq 0$ and $\epsilon > 0$, or equivalently $A \cap (\tau \leq t) \in \mathscr{F}_{t+\epsilon}$ for every $t \geq 0$ and $\epsilon > 0$. Due to the right-continuity hypothesis of the filtration, we have $A \cap (\tau \leq t) \in \mathscr{F}_t$ for every $t \geq 0$ which means $A \in \mathscr{F}_\tau$. □

6.3 Key Ideas to Remember

We summarize the most significant findings of the chapter and the fundamental concepts to be retained from an initial reading, while disregarding the more technical or secondary matters. As usual, if you have any doubt about what the following succinct statements mean, please review the corresponding section.

- Section 6.1: stopping times are random times that comply with the information structure of the assigned filtration. They are a useful tool in various fields and in particular for the study of the fundamental properties of martingales. Even in the

6.3 Key Ideas to Remember

discrete case, many of the main ideas and techniques related to stopping times emerge: the proofs, although using elementary tools, can be quite challenging. Stopping a process maintains its essential properties such as being adapted and the martingale property.

- Section 6.1.1: the optional sampling theorem and Doob's maximal inequalities are crucial results that will be systematically used in the following chapters: so it is useful to dwell on the details of the proofs. The upcrossing lemma is a rather unusual and subtle result, whose use will be limited to proving the continuity of martingale trajectories: its proof can be skipped at a first reading.
- Section 6.2.1: the study of stopping times in the continuous case involves some technical difficulties. First of all, it is necessary to assume the so-called usual conditions on the filtration: these are crucial, for example, in the study of exit times of a process from a closed set.
- Sections 6.2.2 and 6.2.3: every filtration can be enlarged in such a way that it satisfies the usual conditions, but in that case, it is necessary to prove that certain properties of the processes remain valid: for instance, the Markov property or the independence properties of the increments of a Lévy process. It is useful to grasp the statements in these sections, but one can gloss over the technical aspects of the proofs.
- Section 6.2.4: the notion of progressively measurable process strengthens that of an adapted process as it requires a joint measurability property in (t, ω). In particular, a progressively measurable process is also measurable as a function of the time variable: this is relevant in the context of stochastic integration theory.

Main notations used or introduced in this chapter:

Symbol	Description	Page
τ	Typical letter used to indicate a stopping time	97
X_τ	Process X evaluated at (stopping) time τ	99
$\mathscr{F}_\tau$	σ-algebra of information at (stopping) time τ	99
X^τ	Stopped process	100
$\bar{M} = \max_{0 \leq n \leq N} M_n$	Maximum process	103
$\mathscr{N}$	Negligible sets	107
$\bar{\mathscr{F}}_t$	Completed σ-algebra	111
$\mathscr{F}_{t+}$	"Right-augmented" σ-algebra	111
$\mathscr{F}^X$	Standard filtration of a process X	111

Chapter 7
Strong Markov Property

> *L'appartenenza*
> *è assai di più della salvezza personale*
> *è la speranza di ogni uomo che sta male*
> *e non gli basta esser civile.*
> *È quel vigore che si sente se fai parte di qualcosa*
> *che in sé travolge ogni egoismo personale*
> *on quell'aria più vitale che è davvero contagiosa.*[1]
>
> Giorgio Gaber

In this chapter, $X = (X_t)_{t \geq 0}$ denotes a Markov process with transition law p on a filtered probability space $(\Omega, \mathscr{F}, P, \mathscr{F}_t)$ satisfying the usual conditions. The strong Markov property is an extension of the Markov property in which the initial time is a stopping time.

7.1 Feller and Strong Markov Properties

Definition 7.1.1 (Strong Markov property) We say that X satisfies the strong Markov property if for any $h > 0$, $\varphi \in b\mathscr{B}$ and τ being an almost surely finite stopping time, we have

$$\int_{\mathbb{R}} p(\tau, X_\tau; \tau + h, dy)\varphi(y) = E\left[\varphi(X_{\tau+h}) \mid \mathscr{F}_\tau\right]. \tag{7.1.1}$$

---

[1] *Belonging*
is much more than personal salvation
it's the hope of every man who's struggling
and being civil isn't enough for him.

It's that strength you feel when you're part of something
that overwhelms every personal selfishness
with that more vital air that is truly contagious.

Theorem 7.1.2 *Let X be a Markov process. If X is a right-continuous Feller process, then it satisfies the strong Markov property.*

Proof Recall from Definition 2.1.10 that the transition law p of a Feller process is such that, for every $h > 0$ and $\varphi \in bC(\mathbb{R})$, the function

$$(t, x) \longmapsto \int_{\mathbb{R}} p(t, x; t + h, dy)\varphi(y)$$

is continuous. Given $h > 0$ and $\varphi \in bC$, we prove that, setting

$$Z := \int_{\mathbb{R}} p(\tau, X_\tau; \tau + h, dy)\varphi(y),$$

then $Z = E[\varphi(X_{\tau+h}) \mid \mathscr{F}_\tau]$. We verify the properties of conditional expectation. First of all, $Z \in m\mathscr{F}_\tau$ since:

- $Z = f(\tau, X_\tau)$ with $f(t, x) := \int_{\mathbb{R}} p(t, x; t + h, dy)\varphi(y)$ that is a continuous function by the Feller property;
- $X_\tau \in m\mathscr{F}_\tau$ by Proposition 6.2.29-(iv), being X adapted and right-continuous (thus progressively measurable by Proposition 6.2.28).

Secondly, we prove that for every $A \in \mathscr{F}_\tau$ we have

$$E[Z \mathbb{1}_A] = E[\varphi(X_{\tau+h}) \mathbb{1}_A]. \tag{7.1.2}$$

First, consider the case where τ takes only a countable infinity of values t_k, $k \in \mathbb{N}$: in this case, (7.1.2) follows from the fact that

$$E[Z \mathbb{1}_A] = \sum_{k=1}^{\infty} E[Z \mathbb{1}_{A \cap (\tau = t_k)}]$$

$$= \sum_{k=1}^{\infty} E\left[\int_{\mathbb{R}} p(t_k, X_{t_k}; t_k + h, dy)\varphi(y) \mathbb{1}_{A \cap (\tau = t_k)}\right] =$$

(by the Markov property (2.2.2), since $A \cap (\tau = t_k) \in \mathscr{F}_{t_k}$)

$$= \sum_{k=1}^{\infty} E[\varphi(X_{t_k+h}) \mathbb{1}_{A \cap (\tau = t_k)}] = E[\varphi(X_{\tau+h}) \mathbb{1}_A].$$

7.1 Feller and Strong Markov Properties

In the general case, consider the approximating sequence of stopping times defined as

$$\tau_n(\omega) = \begin{cases} \frac{k}{2^n} & \text{if } \frac{k-1}{2^n} \leq \tau(\omega) < \frac{k}{2^n} \text{ for } k \in \mathbb{N}, \\ \infty & \text{if } \tau(\omega) = \infty. \end{cases}$$

For every $n \in \mathbb{N}$, τ_n takes only a countably infinite number of values. Moreover, $\tau_n \geq \tau$ and thus if $A \in \mathscr{F}_\tau$ then also $A \in \mathscr{F}_{\tau_n}$ and we have

$$E\left[\int_\mathbb{R} p(\tau_n, X_{\tau_n}; \tau_n + h, dy)\varphi(y)\mathbb{1}_A\right] = E\left[\varphi\left(X_{\tau_n+h}\right)\mathbb{1}_A\right].$$

By taking the limit as $n \to \infty$, we obtain (7.1.2). This limit is justified by the dominated convergence theorem, given that the integrands are bounded and converge pointwise almost surely. On the right-hand side, the convergence is ensured by the right-continuity of X and the continuity of φ; on the left-hand side, by the right-continuity of X and the Feller property. □

Remark 7.1.3 [!] By Theorem 7.1.2, the Brownian motion, the Poisson process, and more generally Lévy processes (cf. Definition 6.2.19) enjoy the strong Markov property: so we say that they are *strong Markov processes*.

In analogy with the results of Sect. 4.2, we have

Proposition 7.1.4 Let $W = (W_t)_{t \geq 0}$ be a Brownian motion on $(\Omega, \mathscr{F}, P, \mathscr{F}_t)$ and τ an a.s. finite stopping time. Then the process

$$W_t^\tau := W_{t+\tau} - W_\tau, \qquad t \geq 0, \tag{7.1.3}$$

is a Brownian motion on $(\Omega, \mathscr{F}, P, (\mathscr{F}_{t+\tau})_{t \geq 0})$. In particular, W^τ is independent of $\mathscr{F}_\tau$.

Proof For every $\eta \in \mathbb{R}$, we have

$$E\left[e^{i\eta W_t^\tau} \mid \mathscr{F}_\tau\right] = E\left[e^{i\eta(W_{t+\tau} - W_\tau)} \mid \mathscr{F}_\tau\right]$$
$$= e^{i\eta W_\tau} E\left[e^{i\eta W_{t+\tau}} \mid \mathscr{F}_\tau\right]$$
$$= e^{i\eta W_\tau} E\left[e^{i\eta W_{t+\tau}} \mid W_\tau\right] = e^{-\frac{\eta^2 t^2}{2}}$$

thanks to the strong Markov property in the form (7.1.1). From Theorem 4.2.10 in [113] it follows that $W_t^\tau \sim \mathscr{N}_{0,t}$ and is independent of $\mathscr{F}_\tau$. Similarly, we prove that $W_t^\tau - W_s^\tau \sim \mathscr{N}_{0,t-s}$ and is independent of $\mathscr{F}_{\tau+s}$ for every $0 \leq s \leq t$. □

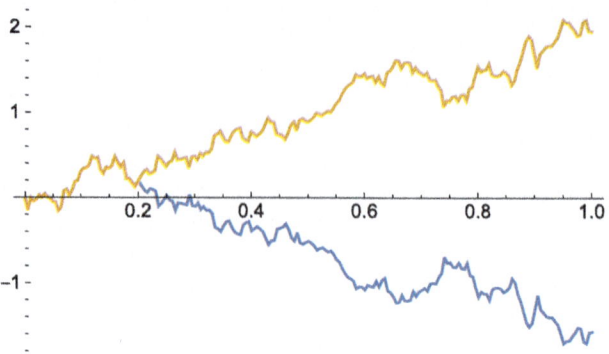

Fig. 7.1 Trajectories of a Brownian and its reflected process starting from $t_0 = 0.2$

7.2 Reflection Principle

Consider a Brownian motion W defined on the filtered space $(\Omega, \mathscr{F}, P, \mathscr{F}_t)$ and fix $t_0 \geq 0$. We say that

$$\widetilde{W}_t := W_{t \wedge t_0} - (W_t - W_{t \wedge t_0}), \qquad t \geq 0,$$

is the *reflected process of W starting from t_0*. Figure 7.1 represents a trajectory of W and its reflected process $\widetilde{W}$ starting from $t_0 = 0.2$.

It is not difficult to check[2] that $\widetilde{W}$ also is a Brownian motion on $(\Omega, \mathscr{F}, P, \mathscr{F}_t)$. It is noteworthy that this result generalizes to the case where t_0 is a stopping time.

Theorem 7.2.1 (Reflection Principle) *[!] Let $W = (W_t)_{t \geq 0}$ be a Brownian motion on the filtered space $(\Omega, \mathscr{F}, P, \mathscr{F}_t)$ and τ a stopping time. Then the reflected process starting from τ, defined as*

$$\widetilde{W}_t := W_{t \wedge \tau} - (W_t - W_{t \wedge \tau}), \qquad t \geq 0,$$

is a Brownian motion on $(\Omega, \mathscr{F}, P, \mathscr{F}_t)$.

---

[2] For $s \leq t$ we have

$$\widetilde{W}_t = \begin{cases} W_t & \text{if } t \leq t_0, \\ 2W_{t_0} - W_t & \text{if } t > t_0, \end{cases}$$

so that $\widetilde{W}_t \in m\mathscr{F}_t$. Moreover,

$$\widetilde{W}_t - \widetilde{W}_s = \begin{cases} W_t - W_s & \text{if } s, t \leq t_0, \\ W_{t_0} - W_s - (W_t - W_{t_0}) & \text{if } s < t_0 < t, \\ -(W_t - W_s) & \text{if } t_0 \leq s, t, \end{cases}$$

and therefore $\widetilde{W}_t - \widetilde{W}_s$ is independent of $\mathscr{F}_s$ and has distribution $\mathscr{N}_{0, t-s}$.

7.2 Reflection Principle

Proof It is enough to prove the thesis on a time interval $[0, T]$ for a fixed $T > 0$ and therefore it is not restrictive to assume $\tau < \infty$ so that the Brownian motion W^τ in (7.1.3) is well defined. We observe that

$$W_t = W_{t \wedge \tau} + W^\tau_{t-\tau} \mathbb{1}_{(t \geq \tau)}, \qquad \widetilde{W}_t = W_{t \wedge \tau} - W^\tau_{t-\tau} \mathbb{1}_{(t \geq \tau)}.$$

The thesis follows from the fact that, being a Brownian motion, W^τ is equal in law to $-W^\tau$ and is independent of $\mathscr{F}_\tau$ and therefore of $W_{t \wedge \tau}$ and of τ: it follows that W and $\widetilde{W}$ are equal in law. □

Consider the process of the *maximum of W*, defined by

$$\bar{W}_t := \max_{s \in [0,t]} W_s, \qquad t \geq 0.$$

Corollary 7.2.2 *For every $a > 0$ we have*

$$P(\bar{W}_t \geq a) = 2P(W_t \geq a), \qquad t \geq 0. \tag{7.2.1}$$

Proof We decompose $(\bar{W}_t \geq a)$ into the disjoint union

$$(\bar{W}_t \geq a) = (W_t > a) \cup (W_t \leq a, \ \bar{W}_t \geq a).$$

We introduce the stopping time

$$\tau_a := \inf\{t \geq 0 \mid W_t \geq a\}$$

and the reflected process $\widetilde{W}$ of W starting from τ_a. Then we have[3]

$$(W_t \leq a, \ \bar{W}_t \geq a) = (\widetilde{W}_t \geq a)$$

and the thesis follows from the reflection principle. □

Remark 7.2.3 [!] Some notable consequences of Corollary 7.2.2 are:

(i) since $P(|W_t| \geq a) = 2P(W_t \geq a)$, from (7.2.1) it follows that $\bar{W}_t$ and $|W_t|$ are equal in law;
(ii) since $(\tau_a \leq t) = (\bar{W}_t \geq a)$, from (7.2.1) we have

$$P(\tau_a \leq t) = 2P(W_t \geq a) = \frac{2}{\sqrt{\pi}} \int_{\frac{a}{\sqrt{2t}}}^{\infty} e^{-y^2} dy, \tag{7.2.2}$$

---

[3] We set $A = (W_t \leq a, \ \bar{W}_t \geq a)$ and $B = (\widetilde{W}_t \geq a)$. If $\omega \in A$ then $\tau_a(\omega) \leq t$ and therefore $\widetilde{W}_t(\omega) = 2W_{\tau_a(\omega)}(\omega) - W_t = 2a - W_t \geq a$ from which $\omega \in B$. Conversely, assume $\widetilde{W}_t(\omega) \geq a$: if $\tau_a(\omega) > t$ we would have $a \leq \widetilde{W}_t(\omega) = W_t(\omega)$ which is absurd. Then it must be $\tau_a(\omega) \leq t$ and therefore obviously $\bar{W}_t(\omega) \geq a$ and also $a \leq \widetilde{W}_t(\omega) = 2a - W_t(\omega)$ so that $W_t(\omega) \geq a$.

so that
$$P(\tau_a < +\infty) = \lim_{n \to +\infty} P(\tau_a \leq n) = 1$$

and, by differentiating (7.2.2), we obtain the expression of a density of τ_a:
$$\gamma_{\tau_a}(t) = \frac{a e^{-\frac{a^2}{2t}}}{\sqrt{2\pi} t^{3/2}} \mathbb{1}_{]0,+\infty[}(t);$$

(iii) for every $\varepsilon > 0$
$$P(W_t \leq 0 \,\forall t \in [0, \varepsilon]) = P(\bar{W}_\varepsilon \leq 0) = P(|W_\varepsilon| \leq 0) = 0.$$

7.3 The Homogeneous Case

We set $I = \mathbb{R}_{\geq 0}$ and suppose that X is the *canonical version* (cf. Proposition 2.2.6) of a Markov process with *time-homogeneous* transition law p: thus, X is defined on the complete space $(\mathbb{R}^I, \mathscr{F}^I_\mu, \mu, \mathscr{F}^X)$ where μ is the law of the process X and $\mathscr{F}^X$ is the standard filtration of X (cf. Definition 6.2.11). Moreover $X_t(\omega) = \omega(t)$ for every $t \geq 0$ and $\omega \in \mathbb{R}^I$.

To express the Markov property more effectively, we introduce the family of *translations* $(\theta_t)_{t \geq 0}$ defined by
$$\theta_t : \mathbb{R}^I \longrightarrow \mathbb{R}^I, \qquad (\theta_t \omega)(s) = \omega(t+s), \qquad s \geq 0, \ \omega \in \mathbb{R}^I.$$

Intuitively, the translation operator θ_t "cuts and removes" the part of the trajectory ω up to time t. Given a random variable Y, we denote by $Y \circ \theta_t$ the *translated random variable* defined by
$$(Y \circ \theta_t)(\omega) := Y(\theta_t(\omega)), \qquad \omega \in \mathbb{R}^I.$$

Note that $(X_s \circ \theta_t)(\omega) = \omega(t+s) = X_{t+s}(\omega)$ or, more simply,
$$X_s \circ \theta_t = X_{t+s}.$$

In the following statement, we denote by
$$E_x[Y] := E[Y \mid X_0 = x]$$

a version of the conditional expectation function of Y given X_0 (cf. Definition 4.2.16 in [113]) and $\mathscr{F}^X_{0,\infty} = \sigma(X_s, s \geq 0)$ (cf. (2.2.6)).

7.3 The Homogeneous Case

Theorem 7.3.1 (Strong Markov Property in the Homogeneous Case [!]) *Let X be the canonical version of a strong Markov process with time-homogeneous transition law. For every a.s. finite stopping time τ and for every $Y \in b\mathscr{F}_{0,\infty}^X$, we have*

$$E_{X_\tau}[Y] = E[Y \circ \theta_\tau \mid \mathscr{F}_\tau]. \tag{7.3.1}$$

Proof For clarity, we explicitly observe that the left-hand side of (7.3.1) indicates the function $E_x[Y]$ evaluated at $x = X_\tau$. If X satisfies the strong Markov property (7.1.1), we have

$$E[\varphi(X_h) \circ \theta_\tau \mid \mathscr{F}_\tau] = E[\varphi(X_{\tau+h}) \mid \mathscr{F}_\tau]$$
$$= \int_{\mathbb{R}} p(\tau, X_\tau; \tau+h, dy)\varphi(y) =$$

(by the homogeneity assumption)

$$= \int_{\mathbb{R}} p(0, X_\tau; h, dy)\varphi(y) = E_{X_\tau}[\varphi(X_h)]$$

which proves (7.3.1) for $Y = \varphi(X_h)$ with $h \geq 0$ and $\varphi \in b\mathscr{B}$. The general case is proved as in Theorem 2.2.4, first extending (7.3.1) to the case

$$Y = \prod_{i=1}^{n} \varphi_i(X_{h_i})$$

with $0 \leq h_1 < \cdots < h_n$ and $\varphi_1, \ldots, \varphi_n \in b\mathscr{B}$, and finally using the second Dynkin's theorem. □

All the results on Markov processes encountered thus far seamlessly extend to the multidimensional case, where processes take values in $\mathbb{R}^d$, without encountering any significant difficulty. The following Theorem 7.3.2 is preliminary to the study of the relationship between Markov processes and harmonic functions: we recall that a harmonic function is a solution of the Laplace operator or more generally of a partial differential equation of elliptic type. We assume the following general hypotheses:

- D is an open set in $\mathbb{R}^d$;
- X is the canonical version of a strong Markov process with values in $\mathbb{R}^d$;
- X is continuous and has a time-homogeneous transition law p;
- $X_0 \in D$ a.s.;
- $\tau_D < \infty$ a.s. where τ_D is the exit time of X from D (cf. Example 6.2.3).

We denote by ∂D the boundary of D and observe that, based on the assumptions made, $X_{\tau_D} \in \partial D$ a.s. In the following statement, $E_x[\cdot] \equiv E[\cdot \mid X_0 = x]$ indicates the conditional expectation function given X_0.

Theorem 7.3.2 *Let $\varphi \in b\mathscr{B}(\partial D)$. If*[4]

$$u(x) = E_x\left[\varphi(X_{\tau_D})\right] \tag{7.3.2}$$

then we have:

(i) *the process $(u(X_{t \wedge \tau_D}))_{t \geq 0}$ is a martingale with respect to the filtration $(\mathscr{F}^X_{t \wedge \tau_D})_{t \geq 0}$;*

(ii) *for every $y \in D$ and $\epsilon > 0$ such that $D(y, \epsilon) := \{z \in \mathbb{R}^d \mid |z - y| < \epsilon\} \subseteq D$ we have*

$$u(x) = E_x\left[u\left(X_{\tau_{D(y,\epsilon)}}\right)\right] \tag{7.3.3}$$

where $\tau_{D(y,\epsilon)}$ indicates the exit time of X from $D(y, \epsilon)$.

Proof The proof is based on the crucial remark that if τ is a stopping time and $\tau \leq \tau_D$, then we have

$$X_{\tau_D} \circ \theta_\tau = X_{\tau_D}. \tag{7.3.4}$$

More explicitly, for every $\omega \in \mathbb{R}^I$ we have

$$(X_{\tau_D} \circ \theta_\tau)(\omega) = X_{\tau_D}(\theta_\tau(\omega)) = X_{\tau_D}(\omega)$$

since the trajectory ω and the trajectory $\theta_\tau(\omega)$, obtained by cutting and removing the part of ω up to the instant $\tau(\omega)$, exit D for the first time at the same point $X_{\tau_D}(\omega)$.

Let us prove (i): for $0 \leq s \leq t$ we have

$$E\left[u(X_{t \wedge \tau_D}) \mid \mathscr{F}_{s \wedge \tau_D}\right] = E\left[E_{X_{t \wedge \tau_D}}\left[\varphi(X_{\tau_D})\right] \mid \mathscr{F}_{s \wedge \tau_D}\right] =$$

(by the strong Markov property (7.3.1), since $\varphi(X_{\tau_D}) \in b\mathscr{F}^X_{0,\infty}$)

$$= E\left[E\left[\varphi(X_{\tau_D}) \circ \theta_{t \wedge \tau_D} \mid \mathscr{F}_{t \wedge \tau_D}\right] \mid \mathscr{F}_{s \wedge \tau_D}\right] =$$

(by (7.3.4) with $\tau = t \wedge \tau_D$)

$$= E\left[E\left[\varphi(X_{\tau_D}) \mid \mathscr{F}_{t \wedge \tau_D}\right] \mid \mathscr{F}_{s \wedge \tau_D}\right] =$$

---

[4] Formula (7.3.2) means that u is a version of the conditional expectation function of $\varphi(X_{\tau_D})$ given X_0.

7.3 The Homogeneous Case

(since $\mathscr{F}_{s\wedge\tau_D} \subseteq \mathscr{F}_{t\wedge\tau_D}$)

$$= E\left[\varphi(X_{\tau_D}) \mid \mathscr{F}_{s\wedge\tau_D}\right] =$$

(reapplying the strong Markov property (7.3.1))

$$= E_{X_{s\wedge\tau_D}}\left[\varphi(X_{\tau_D})\right] = u(X_{s\wedge\tau_D}).$$

Now let us prove (ii). If $x \notin D(y, \epsilon)$, $\tau_{D(y,\epsilon)} = 0$ and the thesis is an obvious consequence of Example 4.2.18 in [113]. If $x \in D(y, \epsilon)$, we observe that $\tau_{D(y,\epsilon)} \leq \tau_D < \infty$ a.s. since X is continuous and applying the optional sampling theorem, in the form of Theorem 8.5.4, to the martingale $M_t := u(X_{t\wedge\tau_D})$ we have

$$M_0 = E\left[M_{\tau_{D(y,\epsilon)}} \mid \mathscr{F}_0^X\right]$$

that is

$$u(X_0) = E\left[u(X_{\tau_{D(y,\epsilon)}}) \mid X_0\right]$$

which proves (7.3.3). □

Chapter 8
Continuous Martingales

> *Il non poter essere soddisfatto da alcuna cosa terrena, nè, per dir così, dalla terra intera; considerare l'ampiezza inestimabile dello spazio, il numero e la mole maravigliosa dei mondi, e trovare che tutto è poco e piccino alla capacità dell'animo proprio; immaginarsi il numero dei mondi infinito, e l'universo infinito, e sentire che l'animo e il desiderio nostro sarebbe ancora più grande che sii fatto universo; e sempre accusare le cose d'insufficienza e di nullità, e patire mancamento e vòto, e però noia, pare a me il maggior segno di grandezza e di nobiltà, che si vegga della natura umana.*[1]
>
> Giacomo Leopardi

In this chapter, we extend some important results from the discrete to the continuous case, such as the optional sampling theorem and Doob's maximal inequalities for martingales. The general strategy consists of three steps:

- the results are first extended from the discrete case, in which the number of time instants is *finite*, to the case in which the time instants are the so-called *dyadic rationals* defined by

$$\mathscr{D} := \bigcup_{n \geq 1} \mathscr{D}_n, \qquad \mathscr{D}_n := \left\{ \tfrac{k}{2^n} \mid k \in \mathbb{N}_0 \right\} = \left\{ 0, \tfrac{1}{2^n}, \tfrac{2}{2^n}, \tfrac{3}{2^n}, \dots \right\}.$$

---

[1] *The inability to be satisfied by any earthly thing, nor, so to speak, by the entire earth; to consider the immeasurable vastness of space, the wondrous number and magnitude of the worlds, and to find that everything is small and insufficient for the capacity of one's own soul; to imagine the number of worlds as infinite, and the universe as infinite, and to feel that our soul and desire would still be greater than this vast universe; and to always accuse things of inadequacy and nullity, and to suffer from lack and emptiness, and therefore boredom - this, to me, seems the greatest sign of greatness and nobility that one can perceive in human nature.* Translation by J. Galassi

We observe that $\mathscr{D}_n \subseteq \mathscr{D}_{n+1}$ for every $n \in \mathbb{N}$ and $\mathscr{D}$ is a countable set dense in $\mathbb{R}_{\geq 0}$;
- under the assumption of right-continuity of the trajectories, it is almost immediate to extend the validity of the results from the dyadic to the continuous case;
- finally, the assumption of continuity of the trajectories is not restrictive since *every martingale admits a modification with càdlàg trajectories*: the proof is based on Doob's maximal inequalities (which allow to prove that the trajectories *do not diverge* almost surely) and on the upcrossing lemma (which allows to prove that the trajectories *do not oscillate* almost surely). The third fundamental ingredient is Vitali's convergence theorem (Theorem C.0.2 in [113]) which guarantees the preservation of the martingale property when taking the limit.

In the second part of the chapter, we introduce some remarkable martingale spaces that will play a central role in the theory of stochastic integration. We also give the definition of *local martingale*, a notion that generalizes that of martingale by weakening the integrability assumptions.

8.1 Optional Sampling and Maximal Inequalities

Consider a filtered probability space $(\Omega, \mathscr{F}, P, \mathscr{F}_t)$. In this section, we do not assume the usual conditions on the filtration. Hereafter, fixed $T > 0$, we use the notation

$$\mathscr{D}(T) := \bigcup_{n \geq 1} \mathscr{D}_{T,n}, \qquad \mathscr{D}_{T,n} := \left\{ \tfrac{Tk}{2^n} \mid k = 0, 1, \ldots, 2^n \right\}, \qquad n \in \mathbb{N}.$$
(8.1.1)

Lemma 8.1.1 (Doob's Maximal Inequalities on Dyadics) *Let* $X = (X_t)_{t \geq 0}$ *be a martingale or a non-negative sub-martingale. For every* $T, \lambda > 0$ *and* $p > 1$, *we have*

$$P\left(\sup_{t \in \mathscr{D}(T)} |X_t| \geq \lambda \right) \leq \frac{E[|X_T|]}{\lambda},$$
(8.1.2)

$$E\left[\sup_{t \in \mathscr{D}(T)} |X_t|^p \right] \leq \left(\frac{p}{p-1} \right)^p E\left[|X_T|^p \right].$$
(8.1.3)

Proof If X is a martingale, then $|X|$ is a non-negative sub-martingale by Proposition 1.4.12. Therefore, it suffices to prove the thesis for a non-negative sub-martingale X. Fixed $T > 0$, for each $n \in \mathbb{N}$ we consider the process $(X_t)_{t \in \mathscr{D}_{T,n}}$

8.1 Optional Sampling and Maximal Inequalities

which is a non-negative *discrete* sub-martingale with respect to the filtration $(\mathscr{F}_t)_{t \in \mathscr{D}_{T,n}}$ and set

$$M_n := \sup_{t \in \mathscr{D}_{T,n}} X_t, \qquad M := \sup_{t \in \mathscr{D}(T)} X_t.$$

Fix $\varepsilon > 0$. Recalling that $\mathscr{D}_{T,n} \subseteq \mathscr{D}_{T,n+1}$, by Beppo Levi's theorem we have[2]

$$P(M > \lambda - \varepsilon) = \lim_{n \to \infty} P(M_n > \lambda - \varepsilon) \leq$$

(by Doob's maximal inequality for discrete sub-martingales, Theorem 6.1.11)

$$\leq \frac{E[X_T]}{\lambda - \varepsilon}.$$

Formula (8.1.2) follows from the arbitrariness of ε.

Now let $p > 1$. Since $\mathscr{D}_{T,n} \subseteq \mathscr{D}_{T,n+1}$ and $M_n^p = \sup_{t \in \mathscr{D}_{T,n}} X_t^p$, we have $0 \leq M_n^p \nearrow M = \sup_{t \in \mathscr{D}(T)} X_t^p$ as $n \to \infty$. Then, by Beppo Levi's theorem, we have

$$E[M^p] = \lim_{n \to \infty} E[M_n^p] \leq$$

(by Doob's maximal inequality for discrete sub-martingales, Theorem 6.1.11)

$$\leq \left(\frac{p}{p-1}\right)^p E[X_T^p].$$

□

In the following statements, we will always assume the hypothesis of right-continuity of the processes: we will see in Sect. 8.2 that, if the filtration satisfies the usual conditions, every martingale admits a càdlàg modification.

Theorem 8.1.2 (Doob's Maximal Inequalities [!]) *Let $X = (X_t)_{t \geq 0}$ be a right-continuous martingale (or a non-negative sub-martingale). For every $T, \lambda > 0$ and*

---

[2] Note that

$$P(M > \lambda - \varepsilon) = E\left[\mathbb{1}_{(M > \lambda - \varepsilon)}\right] = \lim_{n \to \infty} E\left[\mathbb{1}_{(M_n > \lambda - \varepsilon)}\right] = \lim_{n \to \infty} P(M_n > \lambda - \varepsilon),$$

since the sequence $\mathbb{1}_{(M_n > \lambda - \varepsilon)}$ is monotonically increasing.

$p > 1$ we have

$$P\left(\sup_{t\in[0,T]} |X_t| \geq \lambda\right) \leq \frac{E\left[|X_T|\right]}{\lambda}, \qquad (8.1.4)$$

$$E\left[\sup_{t\in[0,T]} |X_t|^p\right] \leq \left(\frac{p}{p-1}\right)^p E\left[|X_T|^p\right]. \qquad (8.1.5)$$

Proof The thesis is an immediate consequence of Lemma 8.1.1 since if X has right-continuous trajectories then $\sup_{t\in[0,T]} |X_t| = \sup_{t\in\mathscr{D}(T)} |X_t|$. □

In analogy with the discrete case, we have the following simple

Corollary 8.1.3 (Doob's Maximal Inequalities [!]) *Let $X = (X_t)_{t\geq 0}$ be a right-continuous martingale (or a non-negative sub-martingale). For every $\lambda > 0$, $p > 1$ and τ stopping time such that $\tau \leq T$ a.s. for some T, we have*

$$P\left(\sup_{t\in[0,\tau]} |X_t| \geq \lambda\right) \leq \frac{E\left[|X_\tau|\right]}{\lambda},$$

$$E\left[\sup_{t\in[0,\tau]} |X_t|^p\right] \leq \left(\frac{p}{p-1}\right)^p E\left[|X_\tau|^p\right].$$

Proof We will see later (cf. Corollary 8.4.1) that stopping a right-continuous martingale results in a martingale. Then the thesis follows from Theorem 8.1.2 applied to $(X_{t\wedge\tau})_{t\geq 0}$. □

To extend some results on stopping times and martingales from the discrete case to the continuous one, the following technical approximation result is useful.

Lemma 8.1.4 *Let $\tau : \Omega \longrightarrow [0,+\infty]$ be a stopping time. There exists a sequence $(\tau_n)_{n\in\mathbb{N}}$ of discrete stopping times (cf. Definition 6.1.1)*

$$\tau_n : \Omega \longrightarrow \{\tfrac{k}{2^n} \mid k = 1, 2, \ldots, n2^n\}$$

such that:

(i) $\tau_n \longrightarrow \tau$ as $n \to \infty$;
(ii) $\tau_{n+1}(\omega) \leq \tau_n(\omega)$ if $n > \tau(\omega)$.

Proof For each $n \in \mathbb{N}$ we set

$$\tau_n(\omega) = \begin{cases} \frac{k}{2^n} & \text{if } \frac{k-1}{2^n} \leq \tau(\omega) < \frac{k}{2^n} \text{ for } k \in \{1, 2, \ldots, n2^n\}, \\ n & \text{if } \tau(\omega) \geq n. \end{cases}$$

8.1 Optional Sampling and Maximal Inequalities

For every $\omega \in \Omega$ and $n \in \mathbb{N}$ such that $\tau(\omega) < n$ we have

$$\tau_n(\omega) - \tfrac{1}{2^n} \leq \tau(\omega) \leq \tau_n(\omega)$$

which proves (i) and (ii). Finally, for every fixed $n \in \mathbb{N}$, τ_n is a discrete stopping time with respect to the filtration defined by $\mathscr{F}_{\frac{k}{2^n}}$ for $k = 0, 1, \ldots, n2^n$, since we have

$$\left(\tau_n = \tfrac{k}{2^n}\right) = \left(\tfrac{k-1}{2^n} \leq \tau < \tfrac{k}{2^n}\right) \in \mathscr{F}_{\frac{k}{2^n}}, \qquad k = 0, 1, \ldots, n2^n - 1,$$

$$(\tau_n = n) = \left(\tau \geq n - \tfrac{1}{2^n}\right) = \left(\tau < n - \tfrac{1}{2^n}\right)^c \in \mathscr{F}_{n-\frac{1}{2^n}} \subseteq \mathscr{F}_n.$$

□

Remark 8.1.5 Based on (ii) of Lemma 8.1.4, if $\tau(\omega) < \infty$, the approximating sequence $(\tau_n(\omega))_{n \in \mathbb{N}}$ has the property of being *monotonically decreasing* at least for large n. On the other hand, if $\tau(\omega) = \infty$ then $\tau_n(\omega) = n$.

We give a first version of the optional sampling theorem: we will see a second one, with weaker assumptions on stopping times, in Theorem 8.5.4.

Theorem 8.1.6 (Optional Sampling Theorem [!!!]) *Let $X = (X_t)_{t \geq 0}$ be a right-continuous sub-martingale. If τ_1 and τ_2 are stopping times such that $\tau_1 \leq \tau_2 \leq T$ for some $T > 0$, then we have*

$$X_{\tau_1} \leq E\left[X_{\tau_2} \mid \mathscr{F}_{\tau_1}\right].$$

Proof Suppose that X is a right-continuous martingale. Consider the sequences $(\tau_{i,n})_{n \in \mathbb{N}}$, $i = 1, 2$, constructed as in Lemma 8.1.4, of discrete stopping times such that $\tau_{i,n} \xrightarrow[n \to \infty]{} \tau_i$: by construction we also have $\tau_{1,n} \leq \tau_{2,n}$ for every $n \in \mathbb{N}$. Let $\bar{\tau}_{i,n} = \tau_{i,n} \wedge T$. Due to the monotonicity property of $\bar{\tau}_{i,n}$ (cf. Lemma 8.1.4-(ii)) and the right-continuity of X, we have $X_{\bar{\tau}_{i,n}} \xrightarrow[n \to \infty]{} X_{\tau_i}$. On the other hand, by the discrete version of the optional sampling theorem (cf. Theorem 6.1.10) we have

$$X_{\bar{\tau}_{i,n}} = E\left[X_T \mid \mathscr{F}_{\bar{\tau}_{i,n}}\right] \tag{8.1.6}$$

and therefore, by Proposition C.0.7 in [113] (and Remark C.0.8 in [113]), the sequences $(X_{\bar{\tau}_{i,n}})_{n \in \mathbb{N}}$ are uniformly integrable. Then, by Vitali's convergence theorem C.0.2 in [113], we also have convergence in $L^1(\Omega, P)$:

$$X_{\bar{\tau}_{i,n}} \xrightarrow[n \to \infty]{L^1} X_{\tau_i}, \qquad i = 1, 2. \tag{8.1.7}$$

Again by the optional sampling Theorem 6.1.10, we have

$$X_{\bar{\tau}_{1,n}} = E\left[X_{\bar{\tau}_{2,n}} \mid \mathscr{F}_{\bar{\tau}_{1,n}}\right]$$

so that, conditioning on $\mathscr{F}_{\bar{\tau}_1}$ and using the tower property, we get

$$E\left[X_{\bar{\tau}_{1,n}} \mid \mathscr{F}_{\bar{\tau}_1}\right] = E\left[X_{\bar{\tau}_{2,n}} \mid \mathscr{F}_{\bar{\tau}_1}\right].$$

The thesis follows by taking the limit as $n \to \infty$, thanks to (8.1.7) and remembering that the convergence in $L^1(\Omega, P)$ of $X_{\bar{\tau}_{i,n}}$ implies the convergence of the conditional expectations $E\left[X_{\bar{\tau}_{i,n}} \mid \mathscr{F}_{\bar{\tau}_1}\right]$ (cf. Theorem 4.2.10 in [113]).

If X is a sub-martingale, the proof is completely analogous except for the fact that uniform integrability cannot be deduced directly from (8.1.6) but requires using a slightly more subtle argument: for details, we refer to [6], Theorem 5.13. □

The following useful result shows that the martingale property is *equivalent* to the property of having constant expectation over time, at least if we also consider random times (more precisely, bounded stopping times).

Theorem 8.1.7 ([!]) *Let $X = (X_t)_{t \geq 0}$ be an adapted, right-continuous and absolutely integrable (i.e., such that $X_t \in L^1(\Omega, P)$ for every $t \geq 0$) process. Then X is a martingale if and only if $E[X_\tau] = E[X_0]$ for every bounded[3] stopping time τ.*

Proof If X is a right-continuous martingale[4] then it is constant on average on bounded stopping times by the optional sampling Theorem 8.1.6. Conversely, since X is adapted by hypothesis, it remains only to verify that

$$E[X_t \mathbb{1}_A] = E[X_s \mathbb{1}_A], \qquad s \leq t, \ A \in \mathscr{F}_s.$$

To this end, we consider

$$\tau := s \mathbb{1}_A + t \mathbb{1}_{A^c}$$

which is easily verified to be a bounded stopping time. Then by hypothesis, we have

$$E[X_0] = E[X_\tau] = E[X_s \mathbb{1}_A] + E[X_t \mathbb{1}_{A^c}],$$
$$E[X_0] = E[X_t] = E[X_t \mathbb{1}_A] + E[X_t \mathbb{1}_{A^c}],$$

and subtracting one equation from the other yields the thesis. □

---

[3] There exists $T > 0$ such that $\tau \leq T$.

[4] Under the usual conditions on the filtration, this assumption is not restrictive since we will see in Sect. 8.2 that every martingale admits a càdlàg modification.

8.2 Càdlàg Martingales

In this section, we prove that, *under the usual conditions on the filtration, every martingale admits a càdlàg modification* and thus the right-continuity assumption made in the statements of the previous section can be removed. We first prove that a martingale can only have jump discontinuities (with jumps of finite size) on the dyadic rationals of $\mathbb{R}_{\geq 0}$.

Lemma 8.2.1 *Let $X = (X_t)_{t \in \mathscr{D}}$ be a martingale or a non-negative sub-martingale. There exists a negligible event N such that, for every $t \geq 0$, the limits*

$$\lim_{\substack{s \to t^- \\ s \in \mathscr{D}}} X_s(\omega), \qquad \lim_{\substack{s \to t^+ \\ s \in \mathscr{D}}} X_s(\omega) \tag{8.2.1}$$

exist and are finite for every $\omega \in \Omega \setminus N$. Moreover, if $\sup_{t \in \mathscr{D}} E[|X_t|] < \infty$ then also the limit

$$\lim_{\substack{t \to +\infty \\ t \in \mathscr{D}}} X_t(\omega) \tag{8.2.2}$$

exists and is finite, for $\omega \in \Omega \setminus N$.

Proof The idea of the proof is as follows. The fact that the limits in (8.2.1) diverge or do not exist is possible only in two cases: if $\sup_{t \in \mathscr{D}} |X_t(\omega)| = \infty$ or if there exists a non-trivial interval $[a, b]$ that is "crossed" by X an infinite number of times. Doob's maximal inequality and the upcrossing lemma exclude these two possibilities or, more precisely, imply that they occur only for ω belonging to a negligible event.

Consider first the case where $\kappa := \sup_{t \in \mathscr{D}} E[|X_t|] < \infty$. Fixed $n \in \mathbb{N}$, we apply the maximal inequality (6.1.7) and the upcrossing Lemma 6.1.13 to the non-negative discrete sub-martingale $(|X_t|)_{t \in \mathscr{D}_n \cap [0,n]}$: for every $\lambda > 0$ and $0 \leq a < b$, we have

$$P\left(\max_{t \in \mathscr{D}_n \cap [0,n]} |X_t| \geq \lambda\right) \leq \frac{E[|X_n|]}{\lambda} \leq \frac{\kappa}{\lambda},$$

$$E[\nu_{n,a,b}] \leq \frac{E[(|X_n| - a)^+]}{b - a} \leq \frac{\kappa}{b - a},$$

where $\nu_{n,a,b}$ is the number of upcrossings of $(|X_t|)_{t \in \mathscr{D}_n \cap [0,n]}$ on $[a, b]$. Taking the limit as $n \to \infty$ and using Beppo Levi's theorem, we have

$$P\left(\sup_{t \in \mathscr{D}} |X_t| \geq \lambda\right) \leq \frac{\kappa}{\lambda}, \qquad E[\nu_{a,b}] \leq \frac{\kappa}{b - a},$$

where $\nu_{a,b}$ is the number of upcrossings of $(|X_t|)_{t\in\mathscr{D}}$ on $[a,b]$. This implies the existence of two negligible events N_0 and $N_{a,b}$ for which

$$\sup_{t\in\mathscr{D}} |X_t| < \infty \text{ on } \Omega \setminus N_0, \qquad \nu_{a,b} < \infty \text{ on } \Omega \setminus N_{a,b}.$$

Also the event

$$N := \bigcup_{\substack{a,b\in\mathbb{Q} \\ 0\leq a<b}} N_{a,b} \cup N_0$$

is negligible: for every $\omega \in \Omega \setminus N$ we have that $\sup_{t\in\mathscr{D}} |X_t(\omega)| < \infty$ and, on every interval with non-negative rational endpoints, there are only a finite number of upcrossings of $|X(\omega)|$; consequently the limits in (8.2.1) and (8.2.2) exist and are finite on $\Omega \setminus N$.

Now consider the case where X is a generic martingale. For every $n \in \mathbb{N}$, we can apply what has just been proven to the stopped process $(X_{t\wedge n})_{t\in\mathscr{D}}$. Indeed it is immediate to verify that $(X_{t\wedge n})_{t\in\mathscr{D}}$ is a martingale and

$$\sup_{t\in\mathscr{D}} E\left[|X_{t\wedge n}|\right] \leq E\left[|X_n|\right]$$

since, by Proposition 1.4.12, $(|X_{t\wedge n}|)_{t\in\mathscr{D}}$ is a sub-martingale. Hence the limits in (8.2.1) exist and are finite almost surely for $t \leq n$. The thesis follows from the arbitrariness of $n \in \mathbb{N}$. □

The argument used in the second part of the proof of Lemma 8.2.1 is easily adapted to prove the following

Theorem 8.2.2 ([!]) *Let* $X = (X_n)_{n\in\mathbb{N}}$ *be a discrete martingale such that* $\sup_{n\in\mathbb{N}} E\left[|X_n|\right] < \infty$. *Then, there exists and is a.s. finite the pointwise limit*

$$X_\infty := \lim_{n\to\infty} X_n.$$

The usual conditions, in particular the right-continuity of the filtration, play a crucial role in the proof of the next result.

Theorem 8.2.3 ([!]) *Assume that the filtered probability space* $(\Omega, \mathscr{F}, P, \mathscr{F}_t)$ *satisfies the usual conditions. Then every martingale (or non-negative sub-martingale)* $X = (X_t)_{t\geq 0}$ *on* $(\Omega, \mathscr{F}, P, \mathscr{F}_t)$ *admits a modification that is still a martingale (respectively, non-negative sub-martingale) with càdlàg trajectories.*

8.3 The Space $\mathcal{M}^{c,2}$ of Square-Integrable Continuous Martingales

Proof We only prove the case where X is a martingale. By Lemma 8.2.1 the trajectories of $(X_t)_{t \in \mathcal{D}}$ have finite right and left limits almost surely. Then the process

$$\widetilde{X}_t := \lim_{\substack{s \to t^+ \\ s \in \mathcal{D}}} X_s, \qquad t \geq 0,$$

is well defined and has càdlàg trajectories. Let us prove that

$$\widetilde{X}_t = E[X_T \mid \mathscr{F}_t], \qquad 0 \leq t \leq T; \tag{8.2.3}$$

this implies that $\widetilde{X}_t = X_t$ almost surely, i.e., $\widetilde{X}$ is a modification of X, and consequently also that $\widetilde{X}$ is a martingale.

Let us prove (8.2.3) by verifying the two properties of conditional expectation. First of all, by definition $\widetilde{X}_t \in m\mathscr{F}_{t+} = m\mathscr{F}_t$ thanks to the usual conditions. Secondly, since X is a martingale, for every $A \in \mathscr{F}_t$ we have

$$E[X_s \mathbb{1}_A] = E[X_T \mathbb{1}_A], \qquad s \in [t, T]. \tag{8.2.4}$$

Taking the limit in (8.2.4) as $s \to t^+$, with $s \in \mathcal{D} \cap (t, T]$, we get $E[\widetilde{X}_t \mathbb{1}_A] = E[X_T \mathbb{1}_A]$ which proves (8.2.3). Convergence is justified by Vitali's Theorem C.0.2 in [113] since $X_s = E[X_T \mid \mathscr{F}_s]$, with $s \in \mathcal{D} \cap (t, T]$, is uniformly integrable by Proposition C.0.7 in [113]. □

Example 8.2.4 ([!]) Let $X \in L^1(\Omega, P)$. Under the assumptions of Theorem 8.2.3, the martingale $M_t := E[X \mid \mathscr{F}_t]$ admits a càdlàg version.

> In light of Theorem 8.2.3 from now on, given a martingale with respect to a filtration that verifies usual conditions, *we implicitly assume to always consider a càdlàg version of it.*

8.3 The Space $\mathcal{M}^{c,2}$ of Square-Integrable Continuous Martingales

In this section we introduce the space of processes on which we will build the stochastic integral and prove that it is a Banach space.

Definition 8.3.1 For $T > 0$, we denote by $\mathcal{M}_T^{c,2}$ the space of continuous square-integrable martingales $X = (X_t)_{t \in [0,T]}$ and set

$$\|X\|_T := \|X_T\|_{L^2(\Omega, P)} = \sqrt{E[X_T^2]}.$$

Moreover, we denote by $\mathcal{M}^{c,2}$ the space of continuous martingales $X = (X_t)_{t \geq 0}$ such that $X_t \in L^2(\Omega, P)$ for every $t \geq 0$.

Remark 8.3.2 Note that $\|\cdot\|_T$ *is a semi-norm in* $\mathcal{M}_T^{c,2}$, in the sense that $\|X\|_T = 0$ if and only if X is *indistinguishable* from the null process. This fact is a consequence of the continuity assumption of X and Doob's maximal inequality according to which we have

$$E\left[\sup_{t \in [0,T]} X_t^2\right] \leq 4E\left[X_T^2\right] = 4\|X\|_T^2.$$

By identifying indistinguishable processes in $\mathcal{M}_T^{c,2}$ and thus considering $\mathcal{M}_T^{c,2}$ as the space of *equivalence classes of processes* (in the sense of indistinguishability), we obtain a complete normed space.

Proposition 8.3.3 ($\mathcal{M}_T^{c,2}, \|\cdot\|_T$) *is a Banach space.*

Proof Let $(X_n)_{n \in \mathbb{N}}$ be a Cauchy sequence in $\mathcal{M}_T^{c,2}$ with respect to $\|\cdot\|_T$. It is enough to show that $(X_n)_{n \in \mathbb{N}}$ admits a convergent subsequence in $\mathcal{M}_T^{c,2}$.

By Doob's maximal inequality (8.1.4), for every $\varepsilon > 0$ and $n, m \in \mathbb{N}$ we have

$$P\left(\sup_{t \in [0,T]} |X_{n,t} - X_{m,t}| \geq \varepsilon\right) \leq \frac{E\left[|X_{n,T} - X_{m,T}|\right]}{\varepsilon} \leq$$

(by Hölder's inequality)

$$\leq \frac{E\left[|X_{n,T} - X_{m,T}|^2\right]^{\frac{1}{2}}}{\varepsilon} = \frac{\|X_n - X_m\|_T}{\varepsilon}.$$

Consequently, for every $k \in \mathbb{N}$ there exists $n_k \in \mathbb{N}$ such that

$$P\left(\sup_{t \in [0,T]} |X_{n,t} - X_{m,t}| \geq \frac{1}{k}\right) \leq \frac{1}{2^k}, \qquad n, m \geq n_k,$$

and, by Borel-Cantelli's Lemma 1.3.28 in [113], $X_{n_k,\cdot}$ converges uniformly on $[0, T]$ almost surely: the limit value, which we denote by X, is a continuous process (we can set to zero the discontinuous trajectories).

Fix $t \in [0, T]$: by Doob's inequality (8.1.5), also $(X_{n_k,t})_{k \in \mathbb{N}}$ is a Cauchy sequence in $L^2(\Omega, P)$ which is a complete space and, by the uniqueness of the limit, converges to X_t in the sense that

$$\lim_{k \to \infty} E\left[|X_t - X_{n_k,t}|^2\right] = 0. \tag{8.3.1}$$

In particular, if $t = T$, we have

$$\lim_{k \to \infty} \|X - X_{n_k}\|_T = 0.$$

Finally, we prove that X is a martingale. For $0 \leq s \leq t \leq T$ and $G \in \mathscr{F}_s$ we have

$$E\left[X_{n_k,t} \mathbb{1}_G\right] = E\left[X_{n_k,s} \mathbb{1}_G\right]$$

since $X_{n_k} \in \mathscr{M}_T^{c,2}$. Taking the limit as $n \to \infty$ thanks to (8.3.1) we have $E[X_t \mathbb{1}_G] = E[X_s \mathbb{1}_G]$ which proves the thesis. $\square$

8.4 The Space $\mathscr{M}^{c,\mathrm{loc}}$ of Continuous Local Martingales

One of the main motivations for the introduction of stopping times is the use of so-called "localization" techniques, which allow for relaxation of the integrability assumptions. In this section, we analyze the specific case of martingales.

Consider a filtered space $(\Omega, \mathscr{F}, P, \mathscr{F}_t)$ satisfying the usual conditions. The concept of local martingale extends that of martingale by removing the integrability condition of the process. This allows to include important classes of processes (for example, stochastic integrals) that are martingales only if stopped (or "localized"). We first observe that, as in the discrete case (cf. Proposition 6.1.7), the martingale property is preserved by stopping the process.

Corollary 8.4.1 (Stopped Martingale) *Let $X = (X_t)_{t \geq 0}$ be a (càdlàg) martingale and τ_0 a stopping time. Then also the stopped process $(X_{t \wedge \tau_0})_{t \geq 0}$ is a martingale.*

Proof Since X is càdlàg and adapted by hypothesis, by Proposition 6.2.29 we have $X_{t \wedge \tau_0} \in m\mathscr{F}_{t \wedge \tau_0} \subseteq m\mathscr{F}_t$. Moreover, by Theorem 8.1.6 $X_{t \wedge \tau_0} = E\left[X_t \mid \mathscr{F}_{t \wedge \tau_0}\right] \in L^1(\Omega, P)$ for every $t \geq 0$. Again by Theorem 8.1.6, for every bounded stopping time τ we have $E\left[X_{\tau \wedge \tau_0}\right] = E[X_0]$ and therefore the thesis follows from Theorem 8.1.7. $\square$

Definition 8.4.2 (Local Martingale) We say that $X = (X_t)_{t \geq 0}$ is a local martingale if $X_0 \in m\mathscr{F}_0$ and there exists a non-decreasing sequence $(\tau_n)_{n \in \mathbb{N}}$ of stopping times, called *localizing sequence* for X, such that:

(i) $\tau_n \nearrow \infty$ as $n \to \infty$;
(ii) for every $n \in \mathbb{N}$, the stopped and translated process $(X_{t \wedge \tau_n} - X_0)_{t \geq 0}$ is a martingale.

We denote by $\mathscr{M}^{c,\mathrm{loc}}$ the *space of continuous local martingales*.

By Corollary 8.4.1, every (càdlàg) martingale is a local martingale with localizing sequence $\tau_n \equiv \infty$.

Example 8.4.3 Consider the constant process $X = (X_t)_{t \geq 0}$ with $X_t \equiv X_0 \in m\mathscr{F}_0$ for every $t \geq 0$. If $X_0 \in L^1(\Omega, P)$ then X is a martingale. If $X_0 \notin L^1(\Omega, P)$, the process X is not a martingale due to the lack of integrability but is obviously a local martingale: in fact, setting $\tau_n \equiv \infty$, we have $X_{t \wedge \tau_n} - X_0 \equiv 0$.

Example 8.4.4 Let W be a Brownian motion on $(\Omega, \mathscr{F}, P, \mathscr{F}_t)$ and $Y \in m\mathscr{F}_0$. Then the process

$$X_t := YW_t$$

is adapted. Moreover, if $Y \in L^1(\Omega, P)$, being $W_t = W_t - W_0$ and Y independent, we also have $X_t \in L^1(\Omega, P)$ for every $t \geq 0$ and

$$E[YW_t \mid \mathscr{F}_s] = YE[W_t \mid \mathscr{F}_s] = YW_s, \qquad s \leq t,$$

so that X is a martingale.

Without further assumptions on Y apart from the $\mathscr{F}_0$-measurability, the process X may not be a martingale due to the lack of integrability but is still a local martingale: the idea is to remove the trajectories where Y is "too large" by setting

$$\tau_n := \begin{cases} 0 & \text{if } |Y| > n, \\ \infty & \text{if } |Y| \leq n, \end{cases}$$

which defines an increasing sequence of stopping times (note that $(\tau_n \leq t) = (|Y| > n) \in \mathscr{F}_0 \subseteq \mathscr{F}_t$). Then, for every $n \in \mathbb{N}$, the process

$$t \mapsto X_{t \wedge \tau_n} = X_t \mathbb{1}_{(\tau_n = \infty)} = W_t Y \mathbb{1}_{(|Y| \leq n)}$$

is a martingale since it is of the type $W_t \bar{Y}$ where $\bar{Y} = Y \mathbb{1}_{(|Y| \leq n)}$ is a bounded $\mathscr{F}_0$-measurable random variable.

Exercise 8.4.5 (Brownian Motion with Random Initial Value) Let $W = (W_t)_{t \geq 0}$ be a Brownian motion on $(\Omega, \mathscr{F}, P, \mathscr{F}_t)$. Given $t_0 \geq 0$ and $Z \in m\mathscr{F}_{t_0}$, let

$$W_t^{t_0, Z} := W_t - W_{t_0} + Z, \qquad t \geq t_0.$$

The process $W^{t_0, Z}$ has an initial value (at time t_0) equal to Z, is continuous, adapted and has independent and stationary increments, equal to the increments of a standard Brownian motion. If $Z \in L^1(\Omega, P)$ then $(W_t^{t_0, Z})_{t \geq t_0}$ is a martingale; in general, $W^{t_0, Z}$ is a local martingale with localizing sequence $\tau_n \equiv \infty$.

We also notice that, given any distribution μ, it is not difficult to construct a Brownian motion W^μ with initial distribution $W_0^\mu \sim \mu$ on the space $(\Omega \times \mathbb{R}, \mathscr{F} \otimes \mathscr{B}, P \otimes \mu)$.

8.4 The Space $\mathscr{M}^{c,\mathrm{loc}}$ of Continuous Local Martingales

Remark 8.4.6 ([!]) If X is a local martingale with localizing sequence $(\tau_n)_{n\in\mathbb{N}}$ then:

(i) X has a modification with càdlàg trajectories that is constructed from the existence of a càdlàg modification of each martingale $X_{t\wedge\tau_n}$.

> Hereafter, *the fact that a local martingale is càdlàg will be always implicitly assumed by convention;*

(ii) X is adapted since $X_0 \in m\mathscr{F}_0$ by definition and $X_t - X_0$ is the pointwise limit of $X_{t\wedge\tau_n} - X_0$ which is $m\mathscr{F}_t$-measurable by definition of martingale;
(iii) a priori X_t does not have any integrability property;
(iv) if X has càdlàg trajectories then there exists a localizing sequence $(\bar\tau_n)_{n\in\mathbb{N}}$ such that

$$|\bar\tau_n| \leq n, \qquad |X_{t\wedge\bar\tau_n}| \leq n, \qquad t \geq 0,\ n \in \mathbb{N}.$$

Indeed, by Proposition 6.2.7, the exit time σ_n of $|X|$ from the interval $[-n, n]$ is a stopping time; moreover, since X is càdlàg (and therefore every trajectory of X is bounded on every compact time interval) we have $\sigma_n \nearrow \infty$. Then

$$\bar\tau_n := \tau_n \wedge \sigma_n \wedge n$$

is a localizing sequence for X: in particular, since $X_{t\wedge\tau_n} - X_0$ is a martingale, by Corollary 8.4.1, $X_{t\wedge\bar\tau_n} - X_0 = X_{(t\wedge\bar\tau_n)\wedge(\sigma_n\wedge n)} - X_0$ also is a martingale;

(v) if there exists $Y \in L^1(\Omega, P)$ such that $|X_t| \leq Y$ for every $t \geq 0$, then X is a martingale: in fact for $s \leq t$ we have $X_{s\wedge\tau_n} - X_0 = E\left[X_{t\wedge\tau_n} - X_0 \mid \mathscr{F}_s\right]$ which, thanks to the integrability hypothesis, is equivalent to

$$X_{s\wedge\tau_n} = E\left[X_{t\wedge\tau_n} \mid \mathscr{F}_s\right]. \tag{8.4.1}$$

The thesis follows by taking the limit as $n \to \infty$ and using the dominated convergence theorem for the conditional expectation. Notice that, in particular, every bounded local martingale is a true martingale. Convergence in (8.4.1) is a very delicate issue: for example, there exist uniformly integrable local martingales that are not martingales;[5]

---

[5] See, for example, Chapter 2 in [37].

(vi) if $X \geq 0$ then X is a super-martingale because, arguing as in the previous point and using Fatou's lemma instead of the dominated convergence theorem, we obtain

$$X_s \geq E[X_t \mid \mathscr{F}_s], \qquad 0 \leq s \leq t \leq T. \qquad (8.4.2)$$

Moreover, if $E[X_T] = E[X_0]$ then $(X_t)_{t \in [0,T]}$ is a true martingale. In fact, from (8.4.2) it is easy to deduce

$$E[X_0] \geq E[X_t] \geq E[X_T], \qquad 0 \leq t \leq T,$$

and therefore from the assumption we get $E[X_t] = E[X_0]$ for every $t \in [0, T]$. If it were $X_s > E[X_t \mid \mathscr{F}_s]$ on a non-negligible event, we would have a contradiction from (8.4.2).

8.5 Uniformly Square-Integrable Martingales

In this section we prove a further version of the optional sampling theorem. Let $(\Omega, \mathscr{F}, P, \mathscr{F}_t)$ be a filtered space satisfying the usual conditions. To deal with the case where the time index varies in $\mathbb{R}_{\geq 0}$ we introduce a integrability condition that will allow to easily reduce to the case $[0, T]$ by using stopping times.

Definition 8.5.1 Let $p \geq 1$. We say that a process $X = (X_t)_{t \geq 0}$ is *uniformly in L^p* if

$$\sup_{t \geq 0} E\left[|X_t|^p\right] < \infty.$$

Proposition 8.5.2 *Let $X = (X_t)_{t \geq 0}$ be a martingale. The following statements are equivalent:*

(i) X is uniformly in L^2;
(ii) there exists a $\mathscr{F}_\infty$-measurable[6] random variable $X_\infty \in L^2(\Omega, P)$ such that

$$X_t = E[X_\infty \mid \mathscr{F}_t], \qquad t \geq 0.$$

In this case, we also have

$$E\left[\sup_{t \geq 0} X_t^2\right] \leq 4 E\left[X_\infty^2\right]. \qquad (8.5.1)$$

---

[6] Recall the definition of $\mathscr{F}_\infty$ in (6.2.5).

8.5 Uniformly Square-Integrable Martingales

Proof [(ii) $\Rightarrow$ (i)] By Jensen's inequality, we have

$$E\left[X_t^2\right] = E\left[E\left[X_\infty \mid \mathscr{F}_t\right]^2\right] \leq E\left[E\left[X_\infty^2 \mid \mathscr{F}_t\right]\right] = E\left[X_\infty^2\right] < \infty. \quad (8.5.2)$$

[(i) $\Rightarrow$ (ii)] Consider the discrete martingale $(X_n)_{n\in\mathbb{N}}$. By Theorem 8.2.2, for almost every $\omega \in \Omega$, there exists and is finite the limit

$$X_\infty(\omega) := \lim_{n\to\infty} X_n(\omega);$$

we also set $X_\infty(\omega) = 0$ for the ω for which such limit does not exist or is not finite. Clearly, $X_\infty \in m\mathscr{F}_\infty$ and also $X_\infty \in L^2(\Omega, P)$ since by Fatou's lemma, we have

$$E\left[X_\infty^2\right] \leq \lim_{n\to\infty} E\left[X_n^2\right] \leq \sup_{t\geq 0} E\left[X_t^2\right] < \infty$$

by assumption. Thanks to Remark C.0.10 in [113], $(X_n)_{n\in\mathbb{N}}$ is uniformly integrable and thus by Vitali's Theorem C.0.2 in [113], X_n converges to X_∞ in $L^1(\Omega, P)$: from this, it also follows that

$$X_n = E[X_\infty \mid \mathscr{F}_n], \qquad n \in \mathbb{N}; \quad (8.5.3)$$

indeed, using the definition of conditional expectation, it is sufficient to observe that for every $A \in \mathscr{F}_n$, we have

$$0 = \lim_{N\to\infty} E\left[(X_n - X_N)\mathbb{1}_A\right] = E\left[(X_n - X_\infty)\mathbb{1}_A\right].$$

Then, given $t \geq 0$ and taking $n \geq t$, we have

$$X_t = E[X_n \mid \mathscr{F}_t] = E[E[X_\infty \mid \mathscr{F}_n] \mid \mathscr{F}_t] = E[X_\infty \mid \mathscr{F}_t].$$

Finally, for every $n \in \mathbb{N}$, by Doob's maximal inequality, we have

$$E\left[\sup_{t\in[0,n]} X_t^2\right] \leq 4E\left[X_n^2\right] \leq$$

(by (8.5.3) and proceeding as in the proof of (8.5.2))

$$\leq 4E\left[X_\infty^2\right]$$

and (8.5.1) follows by taking the limit as $n \to +\infty$, by Beppo Levi's theorem. $\square$

Example 8.5.3 A real Brownian motion W is not uniformly in L^2 since $E\left[W_t^2\right] = t$. However, for any fixed $T > 0$, the process $X_t := W_{t\wedge T}$ is a martingale that is uniformly in L^2 with $X_\infty = W_T$.

The next result is a version of the optional sampling theorem for martingales that are uniformly in L^2. Such a integrability condition is necessary as is evident from the following example: given a real Brownian motion W and $a > 0$, consider the stopping time $\tau_a = \inf\{t \geq 0 \mid W_t \geq a\}$. We have seen in Remark 7.2.3-(ii) that $\tau_a < \infty$ a.s. but

$$0 = W_0 < E\left[W_{\tau_a}\right] = a.$$

Theorem 8.5.4 (Optional Sampling Theorem [!]) *Let $X = (X_t)_{t \geq 0}$ be a (càdlàg) martingale that is uniformly in L^2. If τ_1 and τ_2 are stopping times such that $\tau_1 \leq \tau_2 < \infty$, then we have*

$$X_{\tau_1} = E\left[X_{\tau_2} \mid \mathscr{F}_{\tau_1}\right].$$

Proof We begin by proving that if $X = (X_t)_{t \geq 0}$ is a (càdlàg) sub-martingale that is uniformly in L^2, then for every stopping time τ such that $P(\tau < \infty) = 1$, we have

$$X_0 \leq E\left[X_\tau \mid \mathscr{F}_0\right]. \tag{8.5.4}$$

First, we observe that by (8.5.1) we have $X_\tau \in L^2(\Omega, P)$. Applying the optional sampling Theorem 8.1.6 with the sequence of bounded stopping times $\tau \wedge n$, we have

$$X_0 \leq E\left[X_{\tau \wedge n} \mid \mathscr{F}_0\right].$$

Taking the limit as $n \to \infty$, we obtain (8.5.4) by the dominated convergence theorem since

$$|X_{\tau \wedge n}| \leq 1 + \sup_{t \geq 0} X_t^2 \in L^1(\Omega, P)$$

thanks to (8.5.1).

To prove the thesis, it is sufficient to verify that for every $A \in \mathscr{F}_{\tau_1}$, we have

$$E\left[X_{\tau_1} \mathbb{1}_A\right] = E\left[X_{\tau_2} \mathbb{1}_A\right]. \tag{8.5.5}$$

Consider

$$\tau := \tau_1 \mathbb{1}_A + \tau_2 \mathbb{1}_{A^c}$$

which is a stopping time since

$$(\tau < t) = (A \cap (\tau_1 < t)) \cup \left(A^c \cap (\tau_2 < t)\right) \in \mathscr{F}_t, \qquad t \geq 0.$$

Then, by (8.5.4), we have

$$E[X_0] = E[X_\tau] = E\left[X_{\tau_1}\mathbb{1}_A\right] + E\left[X_{\tau_2}\mathbb{1}_{A^c}\right],$$
$$E[X_0] = E\left[X_{\tau_1}\right] = E\left[X_{\tau_1}\mathbb{1}_A\right] + E\left[X_{\tau_1}\mathbb{1}_{A^c}\right],$$

and this proves (8.5.5). □

8.6 Key Ideas to Remember

We distill the chapter's key findings and essential concepts for easy comprehension upon initial perusal, setting aside the intricacies of technical or secondary details. As usual, if you have any doubt about what the following succinct statements mean, please review the corresponding section.

- Section 8.1: the optional sampling theorem and Doob's maximal inequalities extend without difficulty from discrete to continuous martingales.
- Section 8.2: under the usual conditions, every martingale admits a càdlàg modification; therefore the continuity assumption of Sect. 8.1 is actually not restrictive.
- Section 8.3: the space $\mathscr{M}^{c,2}$ of continuous square-integrable martingales X on $[0, T]$ is a Banach space, equipped with the L^2 norm of the final value, $\|X_T\|_{L^2(\Omega, P)}$.
- Section 8.4: a local martingale is a process that can be approximated by true martingales through a localizing sequence of stopping times. In the definition of a local martingale, no assumptions are made regarding the integrability of the process or conditions on the initial data. Important classes of processes, including stochastic integrals, fall under the category of local martingales, as they are martingales only when stopped. Every bounded local martingale is a true martingale, and every non-negative local martingale is a supermartingale.
- Section 8.5: we introduce the class of uniformly square-integrable martingales and another version of the optional sampling theorem in which the boundedness assumption on stopping times is removed.

Main notations used or introduced in this chapter:

Symbol	Description	Page
$\mathscr{D}$	Dyadic rationals	133
$\mathscr{D}(T)$	Dyadic rationals of $[0, T]$	134
$\mathscr{M}^{c,2}$	Continuous square-integrable martingales	141
$\mathscr{M}^{c,\text{loc}}$	Continuous local martingales	143

Chapter 9
Theory of Variation

> *The traditional professor writes a, says b, and means c; but it should be d.*
>
> *George Pólya*

In this chapter, we review some basic concepts of deterministic integration theory in the sense of Riemann-Stieltjes and Lebesgue-Stieltjes. We shall see that, unfortunately, the trajectories of a Brownian motion (and, in general, of a martingale) do not have sufficient regularity to use such theories to define the Brownian integral in a deterministic sense, path by path. To understand this fact, it is necessary to introduce the concepts of first and second (or quadratic) variation of a function, which are crucial in the construction of the stochastic integral. In the second part of the chapter, we introduce an important class of stochastic processes called *semimartingales*. A semimartingale is the sum of a local martingale with a process whose trajectories are of bounded variation: under appropriate assumptions, such decomposition is unique. We prove a particular version of the fundamental Doob-Meyer decomposition theorem: if X is a martingale, then X^2 is a semimartingale, i.e., it can be decomposed into the sum of a martingale and a process of bounded variation: the latter is the so-called *quadratic variation process of X*. The results of this chapter provide the background for the construction of the stochastic integral that we will present in the next chapter.

9.1 Riemann-Stieltjes Integral

In this section, we recall some classical results on integration in a deterministic framework. Given $T > 0$, a partition of the interval $[0, T]$ is a set of the form $\pi = \{t_0, t_1, \ldots, t_N\}$ with $0 = t_0 < t_1 < \cdots < t_N = T$. We denote by $\mathscr{P}_T$ the set of partitions of $[0, T]$. Given a function

$$g : [0, T] \longrightarrow \mathbb{R}^d$$

the *first variation* of g with respect to the partition $\pi \in \mathscr{P}_T$ is defined as

$$V(g;\pi) := \sum_{k=1}^{N} |g(t_k) - g(t_{k-1})|.$$

Definition 9.1.1 (BV Function) We say that g is of bounded variation on $[0, T]$, and we write $g \in \mathrm{BV}_T$, if

$$V_T(g) := \sup_{\pi \in \mathscr{P}_T} V(g;\pi) < \infty.$$

We say that

$$g : \mathbb{R}_{\geq 0} \longrightarrow \mathbb{R}^d$$

is locally of bounded variation, and we write $g \in \mathrm{BV}$, if $g|_{[0,T]} \in \mathrm{BV}_T$ for every $T > 0$.

Note that the function $t \mapsto V_t(g)$ is increasing and non-negative.

Example 9.1.2 ([!])

(i) Let $d = 1$. If g is a monotone function on $[0, T]$ then $g \in \mathrm{BV}_T$. In fact, if, for example, g is increasing then

$$V(g;\pi) = \sum_{k=1}^{N} |g(t_k) - g(t_{k-1})| = \sum_{k=1}^{N} (g(t_k) - g(t_{k-1})) = g(T) - g(0)$$

for every $\pi \in \mathscr{P}_T$. In the case $d = 1$, monotonicity is almost a characterization: it is known that $g \in \mathrm{BV}_T$ if and only if g is the difference of increasing monotone functions, $g = g_+ - g_-$. Moreover, if g is continuous then g_+ and g_- are continuous as well.

(ii) It is not difficult to show that, if g is continuous then

$$V_T(g) = \lim_{|\pi| \to 0} V(g;\pi) \qquad (9.1.1)$$

where

$$|\pi| := \max_{1 \leq k \leq N} |t_k - t_{k-1}|$$

is called the *mesh* of π (i.e. the length of the longest subinterval). Interpreting $t \mapsto g(t)$ as a trajectory (or parametrized curve) in $\mathbb{R}^d$, the fact that $g \in \mathrm{BV}_T$ means that g is *rectifiable*, in the sense that the length of g can be computed

9.1 Riemann-Stieltjes Integral

as the supremum of the lengths of polygonal approximations:[1] by definition, $V_T(g)$ is the length of g. Equation (9.1.1) does not hold if g is discontinuous: for example, fixed $s \in \,]0, T[$, the function

$$g(t) = \begin{cases} 1 & \text{if } t = s, \\ 0 & \text{if } t \in [0, s[\cup]s, T], \end{cases}$$

is such that $V(g; \pi) = 2$ for every $\pi \in \mathscr{P}_T$ such that $s \in \pi$ and $V(g; \pi) = 0$ for every $\pi \in \mathscr{P}_T$ such that $s \notin \pi$.

(iii) If $g \in \text{Lip}([0, T]; \mathbb{R}^d)$, that is, there exists a constant c such that $|g(t) - g(s)| \leq c|t - s|$ for every $t, s \in [0, T]$, then $g \in \text{BV}_T$ since

$$V(g; \pi) = \sum_{k=1}^{N} |g(t_k) - g(t_{k-1})| \leq c \sum_{k=1}^{N} (t_k - t_{k-1}) = cT$$

for every $\pi \in \mathscr{P}_T$.

(iv) If g is an integral function of the type

$$g(t) = \int_0^t u(s)\,ds, \qquad t \in [0, T],$$

with $u \in L^1([0, T]; \mathbb{R}^d)$ then $g \in \text{BV}_T$ since

$$V(g; \pi) = \sum_{k=1}^{N} \left| \int_{t_{k-1}}^{t_k} u(s)\,ds \right| \leq \sum_{k=1}^{N} \int_{t_{k-1}}^{t_k} |u(s)|\,ds = \|u\|_{L^1},$$

for every $\pi \in \mathscr{P}_T$.

(v) It is not difficult to prove that the function

$$g(t) = \begin{cases} 0 & \text{if } t = 0, \\ t \sin \frac{1}{t} & \text{if } 0 < t \leq T, \end{cases}$$

is continuous but not of bounded variation.

We now introduce the Riemann-Stieltjes integral. Given $\pi = \{t_0, \ldots, t_N\} \in \mathscr{P}_T$, we denote by $\mathscr{T}_\pi$ the family of *point choices relative to* π: an element of $\mathscr{T}_\pi$ is of the form

$$\tau = \{\tau_1, \ldots, \tau_N\}, \qquad \tau_k \in [t_{k-1}, t_k], \qquad k = 1, \ldots, N.$$

---

[1] A polygonal approximation is obtained by connecting a finite number of line segments along the curve.

Given two functions $f, g : [0, T] \longrightarrow \mathbb{R}$, $\pi \in \mathscr{P}_T$ and $\tau \in \mathscr{T}_\pi$, we say that

$$S(f, g; \pi, \tau) := \sum_{k=1}^{N} f(\tau_k)(g(t_k) - g(t_{k-1}))$$

is the *Riemann-Stieltjes sum of f with respect to g*, relative to the partition π and the choice of points τ.

Proposition 9.1.3 (Riemann-Stieltjes Integral) *For every $f \in C[0, T]$ and $g \in BV_T$ there exists and is finite the limit*

$$\lim_{|\pi| \to 0} S(f, g; \pi, \tau). \tag{9.1.2}$$

Such limit is called Riemann-Stieltjes integral of f with respect to g on $[0, T]$ and denoted by

$$\int_0^T f \, dg \quad \text{or} \quad \int_0^T f(t) dg(t).$$

More precisely, for every $\varepsilon > 0$ there exists $\delta_\varepsilon > 0$ such that

$$\left| S(f, g; \pi, \tau) - \int_0^T f \, dg \right| < \varepsilon$$

for every $\pi \in \mathscr{P}_T$, with $|\pi| < \delta_\varepsilon$, and $\tau \in \mathscr{T}_\pi$.

Proof We use the Cauchy criterion and show that for every $\epsilon > 0$ there exists $\delta_\epsilon > 0$ such that

$$\left| S(f, g; \pi', \tau') - S(f, g; \pi'', \tau'') \right| < \epsilon$$

for every $\pi', \pi'' \in \mathscr{P}_T$ such that $|\pi'|, |\pi''| < \delta_\epsilon$ and for every $\tau' \in \mathscr{T}_{\pi'}$ and $\tau'' \in \mathscr{T}_{\pi''}$.

Let $\pi = \pi' \cup \pi'' = \{t_0, \ldots, t_N\}$. Since f is uniformly continuous on the compact interval $[0, T]$, given $\epsilon > 0$ there exists $\delta_\epsilon > 0$ such that, for $|\pi'|, |\pi''| < \delta_\epsilon$, we have

$$\left| S(f, g; \pi', \tau') - S(f, g; \pi'', \tau'') \right| \leq \epsilon \sum_{k=1}^{N} |g(t_k) - g(t_{k-1})| \leq \epsilon V(g; \pi)$$

which proves the thesis. □

Let us see some particular cases in which it is possible to calculate a Riemann-Stieltjes integral starting from the general definition (9.1.2).

9.1 Riemann-Stieltjes Integral

Example 9.1.4 Fixed $\bar{t} \in \,]0, T[$, let

$$g(t) = \begin{cases} 0 & \text{if } t \in [0, \bar{t}[, \\ 1 & \text{if } t \in [\bar{t}, T]. \end{cases}$$

For every $f \in C[0, T]$, $\pi = \{t_0, \ldots, t_N\} \in \mathscr{P}_T$ and $\tau \in \mathscr{T}_\pi$, let $\bar{k}$ be the index for which $\bar{t} \in \,]t_{\bar{k}-1}, t_{\bar{k}}]$. Then we have

$$S(f, g; \pi, \tau) = f(\tau_{\bar{k}}) \big(g(t_{\bar{k}}) - g(t_{\bar{k}-1})\big) = f(\tau_{\bar{k}}) \xrightarrow[|\pi| \to 0]{} f(\bar{t}).$$

Hence

$$\int_0^T f\, dg = f(\bar{t}).$$

Note that

$$\int_0^T f(t) dg(t) = \int_{[0,T]} f(t) \delta_{\bar{t}}(dt)$$

where the right-hand side is the integral with respect to the Dirac delta measure centered at $\bar{t}$.

Example 9.1.5 Let

$$g(t) = \int_0^t u(s)\, ds, \qquad t \in [0, T],$$

the integral function of Example 9.1.2-(iv), with $u \in L^1([0, T]; \mathbb{R})$. By considering separately the positive and negative parts of u, it is not restrictive to assume $u \geq 0$. Given $\pi \in \mathscr{P}_T$ and $f \in C[0, T]$, we consider the particular choice of points

$$\tau_k \in \arg\min_{[t_{k-1}, t_k]} f, \qquad k = 1, \ldots, N.$$

Then we have

$$S(f, g; \pi, \tau) = \sum_{k=1}^N f(\tau_k)(g(t_k) - g(t_{k-1}))$$

$$= \sum_{k=1}^N f(\tau_k) \int_{t_{k-1}}^{t_k} u(s)\, ds$$

$$\leq \sum_{k=1}^N \int_{t_{k-1}}^{t_k} f(s) u(s)\, ds = \int_0^T f(s) u(s)\, ds.$$

We prove a similar inequality with the choice

$$\tau_k \in \arg\max_{[t_{k-1},t_k]} f, \qquad k = 1, \ldots, N.$$

and, taking the limit as $|\pi| \to 0$, we conclude that

$$\int_0^T f(t)dg(t) = \int_0^T f(t)u(t)dt \equiv \int_0^T f(t)g'(t)dt.$$

The general result that provides the rules for Riemann-Stieltjes integration is the following important Itô's formula.

Theorem 9.1.6 (Deterministic Itô's Formula) *For every $F = F(t,x) \in C^1([0,T] \times \mathbb{R})$ and $g \in BV_T \cap C[0,T]$ we have*

$$F(T, g(T)) - F(0, g(0)) = \int_0^T (\partial_t F)(t, g(t))dt + \int_0^T (\partial_x F)(t, g(t))dg(t)$$

Proof For every $\pi = \{t_0, \ldots, t_N\} \in \mathscr{P}_T$, we have

$$(T, g(T)) - F(0, g(0)) = \sum_{k=1}^N (F(t_k, g(t_k)) - F(t_{k-1}, g(t_{k-1}))) =$$

(by the mean value theorem and the continuity of g, with $\tau', \tau'' \in \mathscr{T}_\pi$)

$$= \sum_{k=1}^N \left((\partial_t F)(\tau'_k, g(\tau''_k))(t_k - t_{k-1}) + (\partial_x F)(\tau'_k, g(\tau''_k))(g(t_k) - g(t_{k-1})) \right)$$

which proves the thesis, taking the limit as $|\pi| \to 0$. $\square$

Remark 9.1.7 When F depends only on x, the Itô's formula becomes

$$F(g(T)) - F(g(0)) = \int_0^T F'(g(t))dg(t)$$

which is sometimes written, especially in the context of stochastic calculus (cf. Notation 10.4.2), in terms of the so-called "differential notation":

$$dF(g(t)) = F'(g(t))dg(t). \tag{9.1.3}$$

The latter formally reminds the usual chain rule for the derivation of composite functions.

9.1 Riemann-Stieltjes Integral

In the multidimensional case where $g = (g_1, \ldots, g_d)$ takes values in $\mathbb{R}^d$, setting $\nabla_x = (\partial_{x_1}, \ldots, \partial_{x_d})$, the Itô's formula becomes

$$F(T, g(T)) - F(0, g(0)) = \int_0^T (\partial_t F)(t, g(t))dt + \int_0^T (\nabla_x F)(t, g(t))dg(t)$$

$$= \int_0^T (\partial_t F)(t, g(t))dt + \sum_{i=1}^d \int_0^T (\partial_{x_i} F)(t, g(t))dg_i(t)$$

or in differential notation

$$dF(t, g(t)) = (\partial_t F)(t, g(t))dt + (\nabla_x F)(t, g(t))dg(t).$$

Example 9.1.8 Let us consider some examples of the application of the deterministic Itô's formula:

(i) for $F(t, x) = x$ we have

$$g(T) - g(0) = \int_0^T dg$$

which generalizes the fundamental theorem of integral calculus;

(ii) for $F(t, x) = f(t)x$, with $f \in C^1[0, T]$, we have

$$f(T)g(T) - f(0)g(0) = \int_0^T f'(t)g(t)dt + \int_0^T f(t)dg(t)$$

which generalizes the integration by parts formula. In differential form we have

$$d(f(t)g(t)) = f'(t)g(t)dt + f(t)dg(t) \qquad (9.1.4)$$

which formally resembles the formula for the derivative of a product;

(iii) for $F(t, x) = x^2$ we have

$$\int_0^T g(t)dg(t) = \frac{g^2(T) - g^2(0)}{2}$$

or

$$dg^2(t) = 2g(t)dg(t).$$

9.2 Lebesgue-Stieltjes Integral

Any real-valued a function $g \in BV \cap C(\mathbb{R}_{\geq 0})$ decomposes into the difference $g = g_+ - g_-$ where g_+, g_- are increasing and continuous functions. By Theorem 1.4.33 in [113], g_+ and g_- are associated with two measures on[2] $(\mathbb{R}_{\geq 0}, \mathscr{B})$ which we denote by μ_g^+ and μ_g^-, respectively: we have

$$\mu_g^{\pm}([a,b]) = \mu_g^{\pm}(]a,b]) = g_{\pm}(b) - g_{\pm}(a), \qquad a \leq b.$$

In order to apply Theorem 1.4.33 in [113], it would be sufficient to assume that g is right-continuous (as in Example 9.1.4 where $\mu_g = \delta_{\bar{t}}$). However, to simplify the treatment, here we only consider a continuous function g because we will later study the stochastic integral only with respect to continuous integrators. We denote by

$$|\mu_g| := \mu_g^+ + \mu_g^-$$

the measure defined as the sum of μ_g^+ and μ_g^-. Moreover, for each $H \in \mathscr{B}$ such that at least one of $\mu_g^+(H)$ and $\mu_g^-(H)$ is finite, we set

$$\mu_g(H) = \mu_g^+(H) - \mu_g^-(H). \tag{9.2.1}$$

We say that μ_g is a *signed measure* since it can also take negative values, including $-\infty$.

Definition 9.2.1 (Lebesgue-Stieltjes Measure) Given $g \in BV \cap C(\mathbb{R}_{\geq 0})$, we say that μ_g in (9.2.1) is the Lebesgue-Stieltjes measure associated with g. For each $H \in \mathscr{B}$ and $f \in L^1(H, |\mu_g|)$, we define the *Lebesgue-Stieltjes integral of f with respect to g on H* as

$$\int_H f d\mu_g := \int_H f d\mu_g^+ - \int_H f d\mu_g^-.$$

The Lebesgue-Stieltjes integral generalizes the Riemann-Stieltjes integral, extending the class of integrable functions.

Proposition 9.2.2 (Riemann-Stieltjes vs Lebesgue-Stieltjes) *For every* $f \in C(\mathbb{R}_{\geq 0})$, $g \in BV \cap C(\mathbb{R}_{\geq 0})$ *and* $T > 0$, *we have*

$$\int_0^T f dg = \int_{[0,T]} f d\mu_g.$$

---

[2] We define the measures on $\mathbb{R}_{\geq 0}$ since the space of non-negative real numbers will be the set of time indices for stochastic processes. To apply Theorem 1.4.33 in [113], we can extend the functions g_+, g_- so that they are continuous and constant for $t \leq 0$. All the results of the section obviously hold on $(\mathbb{R}, \mathscr{B})$.

9.2 Lebesgue-Stieltjes Integral

Proof Given $\pi = \{t_0, \ldots, t_N\} \in \mathscr{P}_T$, let us consider the simple functions

$$f_\pi^\pm(t) = \sum_{k=1}^N f(\tau_k^\pm) \mathbb{1}_{]t_{k-1}, t_k]}(t)$$

with

$$\tau_k^+ \in \underset{[t_{k-1}, t_k]}{\arg\max} f, \qquad \tau_k^- \in \underset{[t_{k-1}, t_k]}{\arg\min} f, \qquad k = 1, \ldots, N.$$

Then we have

$$\sum_{k=1}^N f(\tau_k^-)(g_+(t_k) - g_+(t_{k-1})) = \int_{[0,T]} f_\pi^- d\mu_g^+ \leq \int_{[0,T]} f d\mu_g^+ \leq \int_{[0,T]} f_\pi^+ d\mu_g^+$$

$$= \sum_{k=1}^N f(\tau_k^+)(g_+(t_k) - g_+(t_{k-1})).$$

Taking the limit as $|\pi| \to 0$, we obtain

$$\int_0^T f dg_+ = \int_{[0,T]} f d\mu_g^+.$$

Proceeding in a similar manner with g_-, we conclude the proof. $\square$

We prove a technical result that will be used later (see, for example, Theorem 11.2.1).

Proposition 9.2.3 *In a filtered probability space* $(\Omega, \mathscr{F}, P, \mathscr{F}_t)$ *satisfying the usual conditions, let:*

- τ *be a finite (i.e. $\tau < \infty$ a.s.) stopping time;*
- A *be a continuous, increasing, and adapted process with $A_0 = 0$;*
- X *be a non-negative integrable random variable.*

Then we have

$$E\left[\int_0^\tau X dA_t\right] = E\left[\int_0^\tau E[X \mid \mathscr{F}_t] dA_t\right]$$

and, more precisely,

$$E\left[\int_0^\tau X dA_t\right] = E\left[\int_0^\tau M_t dA_t\right]$$

for every càdlàg version M of the martingale $E[X \mid \mathscr{F}_t]$.

Proof First, assume that A and X are bounded a.s. by some $N \in \mathbb{N}$. Fixed $n \in \mathbb{N}$, let $\tau_k = \frac{k\tau}{n}$ for $k = 0, \ldots, n$. We have

$$E\left[\int_0^\tau X dA_t\right] = E\left[\sum_{k=1}^n X\left(A_{\tau_k} - A_{\tau_{k-1}}\right)\right]$$

$$= E\left[\sum_{k=1}^n E\left[X \mid \mathscr{F}_{\tau_k}\right]\left(A_{\tau_k} - A_{\tau_{k-1}}\right)\right]$$

$$= E\left[\sum_{k=1}^n M_{\tau_k}\left(A_{\tau_k} - A_{\tau_{k-1}}\right)\right]$$

$$= E\left[\int_0^\tau M_t^{(n)} dA_t\right]$$

where

$$M_t^{(n)} = M_0 + \sum_{k=1}^n M_{\tau_k} \mathbb{1}_{]\tau_{k-1}, \tau_k]}(t).$$

Due to the right-continuity of M, we have

$$\lim_{n \to \infty} M_t^{(n)}(\omega) = M_t(\omega)$$

for almost every ω such that $t \leq \tau(\omega)$. Given the boundedness of X and therefore of M, the thesis follows from the dominated convergence theorem.

Moving on to the general case, it is sufficient to apply what we have just proved to $X \wedge N$ and $A \wedge N$, using Beppo Levi's theorem to take the limit as $N \to \infty$. □

9.3 Semimartingales

Definition 9.3.1 We say that a process $X = (X_t)_{t \geq 0}$ is

- *increasing* if the trajectories $t \mapsto X_t(\omega)$ are increasing functions[3] for almost every $\omega \in \Omega$;
- *locally of bounded variation* if $X(\omega) \in BV$ for almost every $\omega \in \Omega$ (cf. Definition 9.1.1). For brevity, we often omit the adjective "locally" and simply speak of processes of bounded variation (or *BV processes*), still using the notation BV to indicate the family of such processes;

---

[3] That is, $X_s(\omega) \leq X_t(\omega)$ if $s \leq t$.

9.3 Semimartingales

- *a semimartingale* if it is of the form $X = M + A$ where M is a local martingale and A is an adapted process, of bounded variation and such that $A_0 = 0$.

The interest in semimartingales is due to the fact that we will use such processes as integrators in the Itô stochastic integral. We will restrict our attention to continuous semimartingales, i.e., processes of the form $X = M + A$ with $M \in \mathcal{M}^{c,\text{loc}}$ (cf. Definition 8.4.2) and A continuous, adapted and of bounded variation.

Example 9.3.2 Let $x, \mu, \sigma \in \mathbb{R}$ and W be a standard Brownian motion. The *Brownian motion with drift*

$$X_t := x + \mu t + \sigma W_t, \qquad t \geq 0,$$

is a continuous semimartingale with decomposition $X = M + A$ where $M_t = x + \sigma W_t$ and $A_t = \mu t$. We will prove in Corollary 9.3.7 that the decomposition of a continuous semimartingale is unique.

Remark 9.3.3 A deep result, the Doob-Meyer decomposition theorem, states that every càdlàg sub-martingale is a semimartingale: unlike the discrete case (cf. Theorem 1.4.15), the proof of this fact is far from elementary.

In [121], Cap. IV Theorem 71, it is shown that if X is a continuous local martingale, $X \in \mathcal{M}^{c,\text{loc}}$, with $X_0 = 0$ and $0 < \alpha < \frac{1}{2}$ then the process $|X|^\alpha$ is not a semimartingale unless X is identically zero.

9.3.1 Brownian Motion as a Semimartingale

A Brownian motion W is a continuous martingale and therefore also a semimartingale. To show that its BV part is null (and almost all trajectories of W are not BV), we introduce the concept of *second (or quadratic) variation* of a function g relative to the partition $\pi = \{t_0, t_1, \ldots, t_N\} \in \mathscr{P}_T$:

$$V_T^{(2)}(g; \pi) := \sum_{k=1}^{N} |g(t_k) - g(t_{k-1})|^2. \qquad (9.3.1)$$

Proposition 9.3.4 *If $g \in BV_T \cap C[0, T]$ then*

$$\lim_{|\pi| \to 0} V_T^{(2)}(g; \pi) = 0.$$

Proof Since g is uniformly continuous on the compact interval $[0, T]$, for every $\varepsilon > 0$ there exists $\delta_\varepsilon > 0$ such that

$$\max_{1 \le k \le N} |g(t_k) - g(t_{k-1})| < \epsilon$$

for every $\pi \in \mathscr{P}_T$ such that $|\pi| < \delta_\epsilon$. Consequently,

$$V_T^{(2)}(g; \pi) \le \epsilon \sum_{k=1}^N |g(t_k) - g(t_{k-1})| \le \epsilon V_T(g).$$

□

Example 9.3.5 ([!]) If W is a real Brownian motion, then

$$\lim_{|\pi| \to 0} V_T^{(2)}(W; \pi) = T \qquad \text{in } L^2(\Omega, P), \tag{9.3.2}$$

and consequently, the trajectories of W are not of bounded variation almost surely.
To prove (9.3.2), given a partition $\pi = \{t_0, t_1, \ldots, t_N\} \in \mathscr{P}_T$, we set

$$\delta_k = t_k - t_{k-1}, \qquad \Delta_k = W_{t_k} - W_{t_{k-1}}, \qquad k = 1, \ldots, N,$$

and observe that $E\left[\Delta_k^4\right] = 3\delta_k^2$ and

$$E\left[\Delta_k^2 - \delta_k\right] = 0,$$

$$E\left[\left(\Delta_h^2 - \delta_h\right)\left(\Delta_k^2 - \delta_k\right)\right] = E\left[\left(\Delta_h^2 - \delta_h\right) E\left[\Delta_k^2 - \delta_k \mid \mathscr{F}_{t_h}\right]\right] = 0 \tag{9.3.3}$$

if $h < k$. Then we have

$$E\left[\left(V_T^{(2)}(W; \pi) - T\right)^2\right] = E\left[\left(\sum_{k=1}^N \left(\Delta_k^2 - \delta_k\right)\right)^2\right]$$

$$= \sum_{k=1}^N E\left[\left(\Delta_k^2 - \delta_k\right)^2\right]$$

$$+ 2 \sum_{h<k} E\left[\left(\Delta_h^2 - \delta_h\right)\left(\Delta_k^2 - \delta_k\right)\right] =$$

(since the terms of the second sum are null by (9.3.3))

$$= \sum_{k=1}^N E\left[\Delta_k^4 - 2\Delta_k^2 \delta_k + \delta_k^2\right] =$$

9.3 Semimartingales

(again by (9.3.3))

$$= \sum_{k=1}^{N} 2\delta_k^2 \leq 2|\pi| \sum_{k=1}^{N} \delta_k = 2|\pi|T$$

which proves the thesis.

9.3.2 Semimartingales of Bounded Variation

In Example 9.3.5 we have repeatedly used the martingale property to prove that W has positive quadratic variation and therefore is not of bounded variation. In fact, this result extends to the entire class of continuous local martingales whose trajectories are not of bounded variation unless they are identically zero.

Theorem 9.3.6 ([!]) *Let $X = (X_t)_{t \geq 0}$ be a continuous local martingale, $X \in \mathcal{M}^{c,loc}$. If $X \in BV$ then X is indistinguishable from the process identically equal to X_0.*

Proof Without loss of generality, we can consider $X_0 = 0$. First, we prove the thesis in the case where $X \in BV$ is a bounded continuous martingale: precisely, suppose there exists a constant K such that

$$\sup_{t \geq 0} (|X_t| + V_t(X)) \leq K.$$

Fixed $T > 0$ and $\pi \in \mathcal{P}_T$, we set

$$\Delta_k = X_{t_k} - X_{t_{k-1}}, \qquad \Delta_\pi = \max_{1 \leq k \leq N} |X_{t_k} - X_{t_{k-1}}|.$$

We observe that by identity (1.4.3) we have

$$E\left[(X_{t_k} - X_{t_{k-1}})^2\right] = E\left[X_{t_k}^2 - X_{t_{k-1}}^2\right]$$

and, by the uniform continuity of the trajectories,

$$\lim_{|\pi| \to 0} \Delta_\pi(\omega) = 0, \qquad 0 \leq \Delta_\pi(\omega) \leq 2K, \qquad \omega \in \Omega. \tag{9.3.4}$$

Then we have

$$E\left[X_T^2\right] = E\left[\sum_{k=1}^N \left(X_{t_k}^2 - X_{t_{k-1}}^2\right)\right] = E\left[\sum_{k=1}^N \left(X_{t_k} - X_{t_{k-1}}\right)^2\right]$$

$$\leq E\left[\Delta_\pi V_T(X;\pi)\right] \leq K E\left[\Delta_\pi\right] \qquad (9.3.5)$$

which, as $|\pi| \to 0$, tends to zero by (9.3.4) and the dominated convergence theorem. Hence $E\left[X_T^2\right] = 0$ and by Doob's maximal inequality

$$E\left[\sup_{0 \leq t \leq T} X_t^2\right] \leq 4 E\left[X_T^2\right] = 0.$$

Consequently, by continuity, almost surely the trajectories of X are identically zero on $[0, T]$. Given the arbitrariness of T, we conclude that X is indistinguishable from the null process.

In the general case, we consider a localizing sequence $\bar{\tau}_n$ for which $Y_{n,t} := X_{t \wedge \bar{\tau}_n} \in BV$. We refine this sequence by defining the stopping times

$$\sigma_n = \inf\{t \geq 0 \mid |Y_{n,t}| + V_t(Y_{n,\cdot}) \geq n\}.$$

Also $\tau_n := \bar{\tau}_n \wedge \sigma_n \wedge n$ is a localizing sequence for X: moreover, $X_{t \wedge \tau_n}$ is a bounded continuous martingale, that is constant for $t \geq n$ and whose first variation is bounded by n. As proven above, $X_{t \wedge \tau_n}$ is indistinguishable from the null process and the thesis follows by taking the limit as $n \to \infty$. □

Corollary 9.3.7 ([!]) *Let X be a continuous semimartingale. The decomposition $X = M + A$, with $M \in \mathcal{M}^{c,loc}$ and $A \in BV$ continuous, adapted process such that $A_0 = 0$, is unique.*

Proof If $X = M' + A'$ is another decomposition, then $M - M' = A' - A$ is a continuous local martingale that is locally of bounded variation. By Theorem 9.3.6, M is indistinguishable from M' and A is indistinguishable from A'. □

Remark 9.3.8 Without the continuity assumption, the decomposition of a semimartingale is generally not unique: discontinuities in the paths of a semimartingale can lead to different decompositions. For example, the Poisson process N is increasing and therefore of bounded variation: then $N = M + A$ with $A := N$ and $M := 0$. However, we have also the decomposition given by $A_t := \lambda t$ and $M_t := N_t - \lambda t$, where M is the compensated Poisson process (cf. Proposition 5.3.1).

9.4 Doob's Decomposition and Quadratic Variation Process

In this section, we introduce a fundamental result that underpins the theory of stochastic integration: for every continuous local martingale X there exists an increasing process, called the *quadratic variation process* and denoted by $\langle X \rangle$, which "compensates" the local sub-martingale X^2 in the sense that $X^2 - \langle X \rangle$ is a continuous local martingale. The process $\langle X \rangle$ can be constructed path by path as the limit of the quadratic variation (9.3.1) as $|\pi| \to 0$: this is consistent with what was seen in Example 9.3.5 related to the Brownian motion W for which $\langle W \rangle_t = t$ and the process $W_t^2 - t$ is a continuous martingale.

Recall that $\mathcal{M}^{c,2}$ denotes the space of continuous martingales X such that $X_t \in L^2(\Omega, P)$ for every $t \geq 0$ (cf. Definition 8.3.1) and $\mathcal{M}^{c,\mathrm{loc}}$ denotes the space of continuous local martingales (cf. Definition 8.4.2).

Theorem 9.4.1 (Doob's Decomposition Theorem [[!!]]) *For every $X \in \mathcal{M}^{c,2}$ there exist and are unique (up to indistinguishability) two processes M and $\langle X \rangle$ such that:*

(i) M is a continuous martingale;
(ii) $\langle X \rangle$ is an adapted, continuous and increasing process,[4] such that $\langle X \rangle_0 = 0$;
(iii)

$$X_t^2 = M_t + \langle X \rangle_t, \qquad t \geq 0;$$

(iv)

$$E\left[(X_t - X_s)^2 \mid \mathscr{F}_s\right] = E\left[\langle X \rangle_t - \langle X \rangle_s \mid \mathscr{F}_s\right], \qquad t \geq s \geq 0. \tag{9.4.1}$$

Formula (9.4.1) is the first version of an important identity called Itô's isometry (see Sect. 10.2.1).

More generally, if $X \in \mathcal{M}^{c,\mathrm{loc}}$ then (ii) and (iii) still hold, while (i) is replaced by

(i') $M \in \mathcal{M}^{c,\mathrm{loc}}$.

The process $\langle X \rangle$ is called the quadratic variation process of X and we have

$$\langle X \rangle_t = \lim_{n \to \infty} \sum_{k=1}^{2^n} \left(X_{\frac{tk}{2^n}} - X_{\frac{t(k-1)}{2^n}}\right)^2, \qquad t > 0, \tag{9.4.2}$$

with convergence in probability. More generally, given a continuous semimartingale of the form $S = X + A$, with $X \in \mathcal{M}^{c,\mathrm{loc}}$ and $A \in BV$ adapted, for every $t > 0$ we

---

[4] Clearly $\langle X \rangle$ is also absolutely integrable since $\langle X \rangle_t = X_t^2 - M_t$ with $X_t \in L^2(\Omega, P)$ by hypothesis and $M_t \in L^1(\Omega, P)$ by definition of martingale.

have

$$\langle S \rangle_t := \lim_{n \to \infty} \sum_{k=1}^{2^n} \left(S_{\frac{tk}{2^n}} - S_{\frac{t(k-1)}{2^n}} \right)^2 = \langle X \rangle_t \qquad (9.4.3)$$

in probability and therefore we say that $\langle S \rangle$ is the quadratic variation process of S.

The proof of Theorem 9.4.1 is postponed to Sect. 9.6.

Example 9.4.2 Let $X_t = t + W_t$, where W is a Brownian motion, then by definition $\langle X \rangle_t = \langle W \rangle_t = t$. Note that $E\left[X_t^2 - t\right] = t^2$ and $X_t^2 - t$ is not a martingale.

Remark 9.4.3 Theorem 9.4.1 is a special case of a deep and more general result, known as Doob-Meyer decomposition theorem, which states that *every càdlàg submartingale X of class D (i.e., such that the family of random variables X_τ, with τ stopping time, is uniformly integrable) can be uniquely written in the form $X = M + A$ where M is a continuous martingale and A is an increasing process such that $A_0 = 0$.*

This result was first proved by Meyer in the 1960s of the last century and since then many other proofs have been provided. A particularly concise proof has been recently proposed in [14]: the very intuitive idea is to discretize the process X on the dyadics, use the discrete version of the Doob's decomposition theorem (cf. Theorem 1.4.15) and finally prove that the sequence of discrete decompositions converges to the desired decomposition, using Komlós' Lemma 9.6.1.

Remark 9.4.4 By the optional sampling Theorem 8.1.6, the important identity (9.4.1) is generalized to the case where instead of t, s there are two bounded stopping times τ, σ such that $\sigma \leq \tau \leq T$ a.s. for some $T > 0$.

9.5 Covariation Matrix

We extend the concept of quadratic variation process to the multidimensional case.

Proposition 9.5.1 (Covariation Process) *Let $X, Y \in \mathscr{M}^{c,loc}$ be real-valued processes. The covariation process of X and Y, defined by*

$$\langle X, Y \rangle := \frac{\langle X + Y \rangle - \langle X - Y \rangle}{4}, \qquad (9.5.1)$$

is the unique (up to indistinguishability) process such that

(i) $\langle X, Y \rangle \in BV$ *is adapted, continuous, and such that* $\langle X, Y \rangle_0 = 0$;
(ii) $XY - \langle X, Y \rangle \in \mathscr{M}^{c,loc}$ *and is a true martingale if* $X, Y \in \mathscr{M}^{c,2}$.

9.5 Covariation Matrix

If $X, Y \in \mathscr{M}^{c,2}$, we have

$$E\left[(X_t - X_s)(Y_t - Y_s) \mid \mathscr{F}_s\right] = E\left[\langle X, Y\rangle_t - \langle X, Y\rangle_s \mid \mathscr{F}_s\right], \qquad t \geq s \geq 0, \tag{9.5.2}$$

and

$$\langle X, Y\rangle_t = \lim_{n \to \infty} \sum_{k=1}^{2^n} \left(X_{\frac{tk}{2^n}} - X_{\frac{t(k-1)}{2^n}}\right)\left(Y_{\frac{tk}{2^n}} - Y_{\frac{t(k-1)}{2^n}}\right), \qquad t \geq 0, \tag{9.5.3}$$

in probability.

Proof Given the elementary equality

$$XY = \frac{(X+Y)^2 - (X-Y)^2}{4}$$

it is easy to verify that the process $\langle X, Y\rangle$ defined as in (9.5.1) satisfies properties (i) and (ii). Uniqueness follows directly from Theorem 9.3.6. Formula (9.5.2) follows from the identity

$$E\left[(X_t - X_s)(Y_t - Y_s) \mid \mathscr{F}_s\right] = E\left[X_t Y_t - X_s Y_s \mid \mathscr{F}_s\right]$$

and from the martingale property of $XY - \langle X, Y\rangle$. Formula (9.5.3) is a simple consequence of (9.5.1), applied to $X + Y$ and $X - Y$, and of Proposition 11.2.4 whose proof is given in Chap. 11. □

Remark 9.5.2 By uniqueness, we have $\langle X, X\rangle = \langle X\rangle$. The following properties are direct consequences of definition (9.5.1) of covariation and of (9.5.3):

(i) symmetry: $\langle X, Y\rangle = \langle Y, X\rangle$;
(ii) bi-linearity: $\langle \alpha X + \beta Y, Z\rangle = \alpha \langle X, Z\rangle + \beta \langle Y, Z\rangle$, for $\alpha, \beta \in \mathbb{R}$;
(iii) Cauchy-Schwarz: $|\langle X, Y\rangle| \leq \sqrt{\langle X\rangle \langle Y\rangle}$.

Since the quadratic variation of a continuous BV function is zero (cf. Proposition 9.3.4), the definition of quadratic variation extends to continuous semimartingales in a natural way: recall that in Theorem 9.4.1 we defined the *quadratic variation process* of a continuous semimartingale $S = X + A$, with $X \in \mathscr{M}^{c,\text{loc}}$ and $A \in \text{BV}$ adapted, as $\langle S\rangle := \langle X\rangle$.

Definition 9.5.3 (Covariation Matrix of a Semimartingale) If $S = (S^1, \ldots, S^d)$ is a continuous d-dimensional semimartingale with decomposition $S = X + A$, the *covariation matrix* of S is the $d \times d$ symmetric matrix defined by

$$\langle S\rangle := (\langle X^i, X^j\rangle)_{i,j=1,\ldots,d}.$$

9.6 Proof of Doob's Decomposition Theorem

To prove Theorem 9.4.1 we adapt an argument proposed in [14], based on an interesting and useful result of functional analysis. The classic Bolzano-Weierstrass theorem ensures that from any bounded sequence in the Euclidean space it is possible to extract a convergent subsequence. Although this result does not extend to the infinite-dimensional case, the following lemma shows that it is always possible to construct a convergent sequence of *convex combinations* (subsequences are particular convex combinations) of the elements of the starting sequence. More precisely, given a sequence $(f_n)_{n \in \mathbb{N}}$ in a Hilbert space, we denote by

$$\mathscr{C}_n = \{\lambda_n f_n + \cdots + \lambda_N f_N \mid N \geq n, \; \lambda_n, \ldots, \lambda_N \geq 0, \; \lambda_n + \cdots + \lambda_N = 1\}$$

the family of convex combinations of a finite number of elements of $(f_k)_{k \geq n}$.

Lemma 9.6.1 (Komlós' Lemma [72]) *Let $(f_n)_{n \in \mathbb{N}}$ be a bounded sequence in a Hilbert space. Then there exists a convergent sequence $(g_n)_{n \in \mathbb{N}}$, with $g_n \in \mathscr{C}_n$.*

Proof If $\|f_n\| \leq K$ for each $n \in \mathbb{N}$ then, by the triangle inequality, $\|g\| \leq K$ for each $g \in \mathscr{C}_n$. Therefore, setting

$$a_n := \inf_{g \in \mathscr{C}_n} \|g\|, \qquad n \in \mathbb{N},$$

we have $a_n \leq a_{n+1}$ and $a := \sup_{n \in \mathbb{N}} a_n \leq K$. Then for each $n \in \mathbb{N}$ there exists $g_n \in \mathscr{C}_n$ such that $\|g_n\| \leq a + \frac{1}{n}$. On the other hand, for each $\varepsilon > 0$ there exists $n_\varepsilon \in \mathbb{N}$ such that $\left\|\frac{g_n + g_m}{2}\right\| \geq a - \varepsilon$ for each $n \geq m \geq n_\varepsilon$, simply because $\frac{g_n + g_m}{2} \in \mathscr{C}_n$ and by definition of a. Then, for each $n, m \geq n_\varepsilon$, we have

$$\|g_n - g_m\|^2 = 2\|g_n\|^2 + 2\|g_m\|^2 - \|g_n + g_m\|^2 \leq 4\left(a + \frac{1}{n}\right)^2 - 4(a - \varepsilon)^2$$

which proves that $(g_n)_{n \in \mathbb{N}}$ is a Cauchy sequence and therefore convergent. □

Proof of Theorem 9.4.1 Uniqueness follows directly from Theorem 9.3.6 since if M' and A' satisfy (i), (ii) and (iii) then $M - M'$ is a continuous martingale of bounded variation starting from 0. We prove existence assuming first that $X = (X_t)_{t \in [0,1]}$ is a continuous and bounded martingale:

$$\sup_{t \in [0,1]} |X_t| \leq K \tag{9.6.1}$$

for some positive constant K. This is the difficult part of the proof, in which the main ideas emerge. We proceed step by step.

9.6 Proof of Doob's Decomposition Theorem

Step 1 Fixing $n \in \mathbb{N}$, we introduce the following notation to simplify the calculations on dyadics of $[0, 1]$:

$$X_{n,k} = X_{\frac{k}{2^n}}, \quad A_{n,k} = \sum_{i=1}^{k} \left(X_{n,i} - X_{n,i-1}\right)^2, \quad \mathscr{F}_{n,k} := \mathscr{F}_{\frac{k}{2^n}}, \quad k = 0, 1, \ldots, 2^n.$$

Clearly $k \mapsto X_{n,k}$ and $k \mapsto A_{n,k}$ are processes adapted to the discrete filtration $(\mathscr{F}_{n,k})_{k=0,1,\ldots,2^n}$ and $k \mapsto A_{n,k}$ is increasing. Moreover, the process

$$M_{n,k} := X_{n,k}^2 - A_{n,k}, \quad k = 0, 1, \ldots, 2^n$$

is a discrete martingale. In fact, we have

$$E\left[A_{n,k} - A_{n,k-1} \mid \mathscr{F}_{n,k-1}\right] = E\left[\left(X_{n,k} - X_{n,k-1}\right)^2 \mid \mathscr{F}_{n,k-1}\right] =$$

(by (1.4.3))

$$= E\left[X_{n,k}^2 - X_{n,k-1}^2 \mid \mathscr{F}_{n,k-1}\right] \tag{9.6.2}$$

which proves the martingale property of $M_{n,k}$.

Step 2 This is the crucial point of the proof: we show that

$$\sup_{n \in \mathbb{N}} E\left[A_{n,2^n}^2\right] \leq 36K^4. \tag{9.6.3}$$

Note that, for each fixed $n \in \mathbb{N}$, the final value $A_{n,2^n}$ of the process $A_{n,\cdot}$ is clearly in $L^2(\Omega, P)$, being a finite sum of terms that are bounded by hypothesis: however, the number of such terms increases exponentially in n and this explains the difficulty in proving (9.6.3) which is *a uniform estimate in* $n \in \mathbb{N}$. Here we essentially use the martingale property and the boundedness of X (note that in the general hypotheses X is square-integrable but in (9.6.3) powers of X of order four appear). We have

$$A_{n,2^n}^2 = \sum_{k=1}^{2^n} \left(X_{n,k} - X_{n,k-1}\right)^4 + 2 \sum_{k=1}^{2^n} \sum_{h=k+1}^{2^n} \left(X_{n,k} - X_{n,k-1}\right)^2 \left(X_{n,h} - X_{n,h-1}\right)^2$$

$$= \sum_{k=1}^{2^n} \left(X_{n,k} - X_{n,k-1}\right)^4 + 2 \sum_{k=1}^{2^n} \left(X_{n,k} - X_{n,k-1}\right)^2 \left(A_{n,2^n} - A_{n,k}\right). \tag{9.6.4}$$

By taking the expectation, we estimate the first sum of Eq. (9.6.4) pointwise using Eq. (9.6.1). Then, we apply the tower property in the second sum:

$$E\left[A_{n,2^n}^2\right] \leq 2K^2 \sum_{k=1}^{2^n} E\left[(X_{n,k} - X_{n,k-1})^2\right]$$

$$+ 2 \sum_{k=1}^{2^n} E\left[(X_{n,k} - X_{n,k-1})^2 E\left[A_{n,2^n} - A_{n,k} \mid \mathscr{F}_{n,k}\right]\right] =$$

(by the martingale property (9.6.2) of $M_{n,k} = X_{n,k}^2 - A_{n,k}$)

$$= 2K^2 E\left[A_{n,2^n}\right] + 2 \sum_{k=1}^{2^n} E\left[(X_{n,k} - X_{n,k-1})^2 E\left[X_{n,2^n}^2 - X_{n,k}^2 \mid \mathscr{F}_{n,k}\right]\right] \leq$$

(since $\left|X_{n,2^n}^2 - X_{n,k}^2\right| \leq 2K^2$)

$$\leq 6K^2 E\left[A_{n,2^n}\right] \leq 6K^2 E\left[A_{n,2^n}^2\right]^{\frac{1}{2}}$$

having applied Hölder's inequality in the last step. This concludes the proof of (9.6.3).

Step 3 We extend the discrete martingale $M_{n,\cdot}$ to the whole $[0, 1]$ by setting

$$M_t^{(n)} := E\left[M_{n,2^n} \mid \mathscr{F}_t\right], \qquad t \in [0, 1].$$

For every $t \in \left[\frac{k-1}{2^n}, \frac{k}{2^n}\right]$ we have, by the tower property,

$$M_t^{(n)} = E\left[E\left[M_{n,2^n} \mid \mathscr{F}_{n,k}\right] \mid \mathscr{F}_t\right]$$
$$= E\left[M_{n,k} \mid \mathscr{F}_t\right]$$
$$= E\left[X_{n,k}^2 - A_{n,k} \mid \mathscr{F}_t\right]$$
$$= E\left[X_{n,k}^2 - (X_{n,k} - X_{n,k-1})^2 \mid \mathscr{F}_t\right] - A_{n,k-1}$$
$$= E\left[2X_{n,k}X_{n,k-1} \mid \mathscr{F}_t\right] - X_{n,k-1}^2 - A_{n,k-1}$$
$$= 2X_t X_{n,k-1} - X_{n,k-1}^2 - A_{n,k-1}.$$

9.6 Proof of Doob's Decomposition Theorem

Then, from the continuity of X, it follows that $M^{(n)}$ also is a continuous process. Moreover, by Step 2 the sequence

$$M_1^{(n)} = X_1^2 - A_{n,2^n}$$

is bounded in $L^2(\Omega, P)$. One could prove that $(M_1^{(n)})_{n \in \mathbb{N}}$ is a Cauchy sequence, converging in L^2 norm (and therefore in probability) but the direct proof of this fact is a bit technical and laborious. Therefore, here we prefer to take a shortcut relying on Komlós' Lemma 9.6.1: for each $n \in \mathbb{N}$ there exist non-negative weights $\lambda_n^{(n)}, \ldots, \lambda_{N_n}^{(n)}$ whose sum is equal to one, such that setting

$$\widetilde{M}_{n,t} = \lambda_n^{(n)} M_t^{(n)} + \cdots + \lambda_{N_n}^{(n)} M_t^{(N_n)}, \qquad t \in [0, 1],$$

we have that $\widetilde{M}_{n,1}$ converges in $L^2(\Omega, P)$ to a random variable Z. Let M be a càdlàg version of the martingale defined by

$$M_t := E[Z \mid \mathscr{F}_t], \qquad t \in [0, 1].$$

Since $t \mapsto \widetilde{M}_{n,t}$ is a continuous martingale for each $n \in \mathbb{N}$, by Doob's maximal inequality we have

$$E\left[\sup_{t \in [0,1]} |\widetilde{M}_{n,t} - M_t|^2\right] \leq 4E\left[|\widetilde{M}_{n,1} - M_1|^2\right] = 4E\left[|\widetilde{M}_{n,1} - Z|^2\right].$$

Hence, after taking a subsequence, we have

$$\lim_{n \to \infty} \sup_{t \in [0,1]} |\widetilde{M}_{n,t}(\omega) - M_t(\omega)|^2 = 0, \qquad \omega \in \Omega \setminus F,$$

with F negligible, from which we deduce the existence of a continuous version of M. Consequently, also the process

$$A_t := X_t^2 - M_t$$

is continuous.

To show that A is increasing, we first fix two dyadic numbers $s, t \in [0, 1]$ with $s \leq t$: then there exists $\bar{n}$ such that $s, t \in \mathscr{D}_n$ for every $n \geq \bar{n}$, that is, $s = \frac{k_n}{2^n}$ and $t = \frac{h_n}{2^n}$ for certain $k_n, h_n \in \{0, 1, \ldots, 2^n\}$. Now by construction

$$X_{n,k_n}^2 - M_{n,k_n} = A_{n,k_n} \leq A_{n,h_n} = X_{n,h_n}^2 - M_{n,h_n}$$

and a similar inequality also holds for every convex combination, so in the limit we have $A_s(\omega) \leq A_t(\omega)$ for every $\omega \in \Omega \setminus F$. From the density of dyadic numbers in [0, 1] and the continuity of A, it follows that A is increasing a.s. Finally, we prove (9.4.1): by (1.4.3) we have

$$\begin{aligned} E\left[(X_t - X_s)^2 \mid \mathscr{F}_s\right] &= E\left[X_t^2 - X_s^2 \mid \mathscr{F}_s\right] \\ &= E\left[M_t - M_s \mid \mathscr{F}_s\right] + E\left[A_t - A_s \mid \mathscr{F}_s\right] \\ &= E\left[A_t - A_s \mid \mathscr{F}_s\right]. \end{aligned}$$

Step 4 Now suppose that $X = (X_t)_{t \geq 0}$ is a continuous, not necessarily bounded, martingale but such that $X_t \in L^2(\Omega, P)$ for every $t \geq 0$. We use a localization procedure and define the sequence of stopping times

$$\tau_n = \inf\{t \mid |X_t| \geq n\} \wedge n, \qquad n \in \mathbb{N}.$$

By the continuity of X, we have $\tau_n \nearrow \infty$ as $n \to \infty$. By Corollary 8.4.1, $X_{t \wedge \tau_n}$ is a continuous, bounded martingale that is constant for $t \geq n$: then we can use the previous arguments to show that there exist a continuous square-integrable martingale $M^{(n)}$ and a continuous and increasing process $A^{(n)}$ such that

$$X_{t \wedge \tau_n}^2 = M_t^{(n)} + A_t^{(n)}, \qquad t \geq 0.$$

By uniqueness, for every $m > n$ we have $M_t^{(n)} = M_t^{(m)}$ and $A_t^{(n)} = A_t^{(m)}$ for $t \in [0, \tau_n]$: thus the definition $M_t := M_t^{(n)}$ and $A_t := A_t^{(n)}$ is well posed for every n such that $\tau_n \geq t$. Clearly, M, A are continuous processes, A is increasing and M is a martingale: indeed, if $0 \leq s \leq t$, for every n such that $\tau_n \geq t$ we have

$$M_{s \wedge \tau_n} = E\left[M_{t \wedge \tau_n} \mid \mathscr{F}_s\right].$$

Hence, we conclude by employing the same reasoning as in the proof of Theorem 8.1.6, given that the family $\{M_{t \wedge \tau_n} \mid n \in \mathbb{N}\}$ is uniformly integrable, as guaranteed by Doob's inequality

$$E\left[\sup_{s \in [0,t]} |M_s|^2\right] \leq 4E\left[M_t^2\right]$$

and Remark C.0.10 in [113].

The same localizing sequence can be used to deal with the case where $X \in \mathscr{M}^{c,\text{loc}}$ and in this case it is obvious that $M \in \mathscr{M}^{c,\text{loc}}$.

Step 5 Given the current available tools, proving formulas (9.4.2) and (9.4.3) would require lengthy and laborious calculations. However, since we do not intend to utilize these formulas soon, we opt to defer their proof to a later stage when we will have the Itô's formula at our disposal: this will simplify the proof significantly (cf. Proposition 11.2.4).

□

9.7 Key Ideas to Remember

We highlight the major takeaways from this chapter and the key concepts you should remember after your first read-through, skipping over the technical jargon and less important details. As usual, if you have any doubt about what the following succinct statements mean, please review the corresponding section.

- Section 9.1: to facilitate the understanding of the stochastic integration theory, we recall the definition of the Riemann-Stieltjes integral. It is the natural generalization of the Riemann integral, defined under the assumption that the integrand function is continuous and the integrator is of bounded variation. The main rules of integral calculus are provided by Itô's formula which, in a deterministic version, anticipates the analogous result for the stochastic integral.
- Section 9.2: the Lebesgue integral can be generalized as well. In fact, by Carathéodory's theorem, to each BV function is associated a (signed) measure, called the Lebesgue-Stieltjes measure. The related integral, called the Lebesgue-Stieltjes integral, admits a class of integrable functions much larger than the Riemann-Stieltjes integral.
- Section 9.3: a semimartingale is an adapted process that decomposes into the sum of a local martingale with a BV process. For a continuous semimartingale, this decomposition is unique: in fact, if a process is simultaneously a continuous local martingale and of bounded variation then it is indistinguishable from a constant process. This is due to the fact that a continuous and BV process X has zero quadratic variation and this, in combination with the martingale property, implies (see (9.3.5)) that X is constant. A direct and instructive calculation shows that the quadratic variation process of a Brownian motion W is equal to $\langle W \rangle_T = T$: consequently, almost all trajectories of W are not of bounded variation.
- Section 9.4: the Doob's decomposition theorem states that for every continuous local martingale X there exists an increasing (and therefore BV) process, called the *quadratic variation process* and denoted by $\langle X \rangle$, which "compensates" the local sub-martingale X^2 in the sense that $X^2 - \langle X \rangle$ is a continuous local martingale. In practice, this result states that X^2 is a semimartingale and provides its Doob's decomposition into BV and martingale parts.
- Section 9.6: the general idea of the proof of the Doob's decomposition theorem is simple: the process $\langle X \rangle$ can be constructed path by path as the limit of the quadratic variation process. However, considering the significance of the technical details involved, it is advisable to skip this section during the initial reading.

Main notations used or introduced in this chapter:

Symbol	Description	Page
$\mathscr{P}_T$	Family of partitions π of $[0, T]$	152
$V(g; \pi)$	First variation of the function g relative to π	152
BV_T	Family of functions of bounded variation on $[0, T]$	152
$V_T(g)$	First variation of the function g on $[0, T]$	152
BV	Family of functions locally of bounded variation	152
$\int_0^T f\, dg$	Riemann-Stieltjes integral of f with respect to g on $[0, T]$	154
$dF(g(t)) = F'(g(t))dg(t)$	Deterministic Itô's formula in differential notation	156
μ_g	Lebesgue-Stieltjes measure of $g \in \mathrm{BV} \cap C$	158
$\mathscr{M}^{c,2}$	Continuous square-integrable martingales	141
$\mathscr{M}^{c,\mathrm{loc}}$	Continuous local martingales	143
$V_T^{(2)}(g; \pi)$	Quadratic variation of function g relative to π	161
$\langle X \rangle$	Quadratic variation process	165
$\langle X, Y \rangle$	Covariation process	166

Chapter 10
Stochastic Integral

> *One needs for stochastic integration a six months course to cover only the definitions. What is there to do?*
>
> Paul-André Meyer

In this chapter, we introduce the stochastic integral

$$X_t := \int_0^t u_s dB_s, \qquad t \geq 0,$$

interpreted as a stochastic process with varying integration endpoint.[1] We will assume appropriate hypotheses on the *integrand* process u and the *integrator* process B. The prototype for the integrator is the Brownian motion: since the Brownian trajectories are not of bounded variation, we cannot adopt the deterministic theory of Lebesgue-Stieltjes integration to define the integral path by path. Instead, we will follow the construction due to Kiyosi Itô (1915–2008) which is based on the theory of variation presented in Chap. 9: a crucial ingredient is the assumption that the integrand process u is *progressively measurable*.

The construction of the stochastic integral is in some ways analogous to that of the Lebesgue integral but is decidedly longer and more laborious: it begins with the "simple" processes (i.e., piecewise constant in time) and advances to progressively measurable processes whose trajectories satisfy a weak integrability property with respect to the time variable. An important intermediate step is when u is a "square-integrable process" (cf. Definition 10.1.1); in this case, the stochastic integral has some remarkable properties: it is a continuous square-integrable martingale, i.e., it

---

[1] So we want to define X_t not only as a random variable for fixed t, but as a stochastic process indexed by $t \geq 0$: we will see that this entails some additional difficulty due to the fact that t varies in an uncountable set.

belongs to the space $\mathcal{M}^{c,2}$, the so-called *Itô isometry* holds, and finally, the quadratic variation process is given explicitly by

$$\langle X \rangle_t = \int_0^t u_s^2 d\langle B \rangle_s, \qquad t \geq 0.$$

The last part of the chapter is dedicated to the definition of the stochastic integral in the case where B is a *continuous semimartingale*. We will also introduce the important class of *Itô processes* which are continuous semimartingales that can be uniquely decomposed into the sum of a Lebesgue integral (of a progressively measurable and absolutely integrable process) with a Brownian stochastic integral.

> As stated by Meyer in the quote at the beginning of the chapter, an entire semester course would be needed just to give the definition of the stochastic integral in full details. For those approaching the theory of stochastic integration for the first time, it is advisable to follow the reading scheme indicated in Sect. 10.5, in particular focusing on studying Sects. 10.1 and 10.4, initially skipping Sects. 10.2 and 10.3.

10.1 Integral with Respect to a Brownian Motion

For an introductory purpose, we examine the particular case where B is a real Brownian motion defined on a filtered space $(\Omega, \mathscr{F}, P, \mathscr{F}_t)$. To overcome the problem of irregularity of Brownian trajectories, the idea is to selectively choose the class of integrand processes in order to exploit some probabilistic properties.

Definition 10.1.1 We denote by $\mathbb{L}^2$ the class of processes $u = (u_t)_{t \geq 0}$ such that:

(i) u is progressively measurable with respect to $(\mathscr{F}_t)$ (cf. Definition 6.2.27);
(ii) for every $T \geq 0$ we have

$$E\left[\int_0^T u_t^2 dt\right] < \infty. \qquad (10.1.1)$$

Remark 10.1.2 Property (i) is more than a simple condition of joint measurability in (t, ω) (which would be natural since we are defining an integral): it also incorporates the critical assumption that the information structure of the considered filtration is upheld. Let us remember that, if u is continuous, then (i) is equivalent to the fact that u is adapted to $(\mathscr{F}_t)$.

10.1 Integral with Respect to a Brownian Motion

Remark 10.1.3 As previously mentioned, we restrict our attention to *continuous integrators*. However, it is possible to define the stochastic integral also with respect to càdlàg processes such as the Poisson process. In such cases, it is necessary to impose a more stringent condition on the integrand, essentially requiring it to be approximable by left-continuous processes.[2]

As for the Lebesgue integral, the construction of the stochastic integral takes place in steps, initially considering "simple" processes.

Definition 10.1.4 We say that $u \in \mathbb{L}^2$ is simple if

$$u_t = \sum_{k=1}^{N} \alpha_k \mathbb{1}_{[t_{k-1}, t_k[}(t), \qquad t \geq 0, \qquad (10.1.2)$$

where $0 \leq t_0 < t_1 < \cdots < t_N$ and $\alpha_1, \ldots, \alpha_N$ are random variables such that $P(\alpha_k \neq \alpha_{k+1}) > 0$ for $k = 1, \ldots, N-1$. For every $T \geq t_N$ we set

$$\int_0^T u_t dB_t := \sum_{k=1}^{N} \alpha_k \left(B_{t_k} - B_{t_{k-1}} \right)$$

and define the stochastic integral for two generic integration endpoints a and b, with $0 \leq a \leq b$, as

$$\int_a^b u_t dB_t := \int_0^{t_N} u_t \mathbb{1}_{[a,b[}(t) dB_t. \qquad (10.1.3)$$

In this introductory part, we do not worry about clarifying all the details of the definition of integral, such as the fact that (10.1.3) is well posed because it is independent, up to indistinguishable processes, of the representation (10.1.2) of the process u.

Remark 10.1.5 A simple process is piecewise constant as a function of time and has trajectories that depend on the coefficients $\alpha_1, \ldots, \alpha_N$ which are random. From the fact that $u \in \mathbb{L}^2$ some properties of the variables $\alpha_1, \ldots, \alpha_N$ follow:

(i) since u is progressively measurable and $\alpha_k = u_t \in m\mathscr{F}_t$ for every $t \in [t_{t-k}, t_k[$, then

$$\alpha_k \in m\mathscr{F}_{t_{k-1}}, \qquad k = 1, \ldots, N; \qquad (10.1.4)$$

---

[2] The Poisson process is a BV process and therefore we can define the related stochastic integral in the Lebesgue-Stieltjes sense: however, if the integrand is not continuous from the left, the integral loses the fundamental property of being a (local) martingale: for an intuitive explanation of this fact, see Section 2.1 in [37].

(ii) by the integrability assumption (10.1.1) we have

$$E\left[\int_0^{t_N} u_t^2 dt\right] = \sum_{k=1}^N E\left[\int_0^{t_N} \alpha_k^2 \mathbb{1}_{[t_{k-1},t_k[}(t)dt\right]$$

$$= \sum_{k=1}^N E\left[\alpha_k^2\right](t_k - t_{k-1}) < +\infty$$

and therefore $\alpha_1, \ldots, \alpha_N \in L^2(\Omega, P)$.

We now prove some fundamental properties of the stochastic integral.

Theorem 10.1.6 ([!]) *Given two simple processes* $u, v \in \mathbb{L}^2$, *consider*

$$X_t := \int_0^t u_s dB_s, \qquad Y_t := \int_0^t v_s dB_s, \qquad t \geq 0.$$

For $0 \leq s \leq t \leq T$ *the following properties hold:*

(i) X *is a continuous square-integrable martingale*, $X \in \mathcal{M}^{c,2}$, *and*

$$E\left[\int_s^t u_r dB_r \mid \mathcal{F}_s\right] = 0; \tag{10.1.5}$$

(ii) *the Itô isometry holds*

$$E\left[\left(\int_s^t u_r dB_r\right)^2 \mid \mathcal{F}_s\right] = E\left[\int_s^t u_r^2 dr \mid \mathcal{F}_s\right] \tag{10.1.6}$$

and more generally

$$E\left[\int_s^t u_r dB_r \int_s^t v_r dB_r \mid \mathcal{F}_s\right] = E\left[\int_s^t u_r v_r dr \mid \mathcal{F}_s\right], \tag{10.1.7}$$

$$E\left[\int_s^t u_r dB_r \int_t^T v_r dB_r \mid \mathcal{F}_s\right] = 0; \tag{10.1.8}$$

(iii) *the covariation process of* X *and* Y *(cf. Proposition 9.5.1) is given by*

$$\langle X, Y\rangle_t = \int_0^t u_s v_s ds, \qquad t \geq 0. \tag{10.1.9}$$

Finally, the unconditional versions of formulas (10.1.5), (10.1.6), (10.1.7) *and* (10.1.8) *also hold.*

10.1 Integral with Respect to a Brownian Motion

Proof First, let us observe that formulas (10.1.5), (10.1.6), (10.1.7) and (10.1.8) are equivalent to

$$E[X_t - X_s \mid \mathscr{F}_s] = 0, \tag{10.1.10}$$

$$E\left[(X_t - X_s)^2 \mid \mathscr{F}_s\right] = E[\langle X \rangle_t - \langle X \rangle_s \mid \mathscr{F}_s],$$

$$E[(X_t - X_s)(Y_t - Y_s) \mid \mathscr{F}_s] = E[\langle X, Y \rangle_t - \langle X, Y \rangle_s \mid \mathscr{F}_s],$$

$$E[(X_t - X_s)(Y_T - Y_t) \mid \mathscr{F}_s] = 0.$$

Let us prove (10.1.5), which is equivalent to the martingale property $E[X_t \mid \mathscr{F}_s] = X_s$: referring to (10.1.2) and remembering the notation (10.1.3), it is not restrictive to assume $s = t_k$ and $t = t_h$ for some k, h with $k < h \leq N$. We have

$$E\left[X_{t_h} \mid \mathscr{F}_{t_k}\right] = X_{t_k} + E\left[\int_{t_k}^{t_h} u_r d B_r \mid \mathscr{F}_{t_k}\right]$$

$$= X_{t_k} + \sum_{i=k+1}^{h} E\left[\alpha_i \left(B_{t_i} - B_{i-1}\right) \mid \mathscr{F}_{t_k}\right] =$$

(by (10.1.4) and the tower property)

$$= X_{t_k} + \sum_{i=k+1}^{h} E\left[\alpha_i E\left[B_{t_i} - B_{t_{i-1}} \mid \mathscr{F}_{t_{i-1}}\right] \mid \mathscr{F}_{t_k}\right] = X_{t_k}$$

where the last equality follows from the independence and stationarity of Brownian increments for which we have

$$E\left[B_{t_i} - B_{t_{i-1}} \mid \mathscr{F}_{t_{i-1}}\right] = E\left[B_{t_i} - B_{t_{i-1}}\right] = 0$$

for every $i = 1, \ldots, N$.

Regarding Itô's isometry, still assuming that $s = t_k$ and $t = t_h$, we have

$$E\left[\left(\int_s^t u_r d B_r\right)^2 \mid \mathscr{F}_s\right]$$

$$= E\left[\left(X_{t_h} - X_{t_k}\right)^2 \mid \mathscr{F}_{t_k}\right]$$

$$= E\left[\left(\sum_{i=k+1}^{h} \alpha_i \left(B_{t_i} - B_{t_{i-1}}\right)\right)^2 \mid \mathscr{F}_{t_k}\right]$$

$$= \sum_{i=k+1}^{h} E\left[\alpha_i^2 \left(B_{t_i} - B_{t_{i-1}}\right)^2 \mid \mathscr{F}_{t_k}\right]$$

$$+ \frac{1}{2} \sum_{k+1 \leq i < j \leq h} E\left[\alpha_i \left(B_{t_i} - B_{t_{i-1}}\right) \alpha_j \left(B_{t_j} - B_{t_{j-1}}\right) \mid \mathscr{F}_{t_k}\right] =$$

(by (10.1.4) and the tower property)

$$= \sum_{i=k+1}^{h} E\left[\alpha_i^2 E\left[\left(B_{t_i} - B_{t_{i-1}}\right)^2 \mid \mathscr{F}_{t_{i-1}}\right] \mid \mathscr{F}_{t_k}\right]$$

$$+ \frac{1}{2} \sum_{k+1 \leq i < j \leq h} E\left[\alpha_i \left(B_{t_i} - B_{t_{i-1}}\right) \alpha_j E\left[B_{t_j} - B_{t_{j-1}} \mid \mathscr{F}_{t_{j-1}}\right] \mid \mathscr{F}_{t_k}\right] =$$

(since $B_{t_j} - B_{t_{j-1}}$ is independent of $\mathscr{F}_{t_{j-1}}$)

$$= \sum_{i=k+1}^{h} E\left[\alpha_i^2 (t_i - t_{i-1}) \mid \mathscr{F}_{t_k}\right]$$

$$= \sum_{i=k+1}^{h} E\left[\int_s^t \alpha_i^2 \mathbb{1}_{[t_{i-1}, t_i[}(r) dr \mid \mathscr{F}_s\right]$$

$$= E\left[\int_s^t u_r^2 dr \mid \mathscr{F}_s\right].$$

Formula (10.1.7) is proven in a similar way. Regarding (10.1.8), it is enough to observe that

$$E\left[\int_s^t u_r dB_r \int_t^T v_r dB_r \mid \mathscr{F}_s\right]$$

$$= E\left[\int_s^T u_r \mathbb{1}_{[s,t[}(r) dB_r \int_s^T v_r \mathbb{1}_{[t,T[}(r) dB_r \mid \mathscr{F}_s\right] =$$

(by (10.1.7))

$$= E\left[\int_s^T u_r v_r \mathbb{1}_{[s,t[}(r) \mathbb{1}_{[t,T[}(r) dr\right] = 0.$$

Finally, $\langle X, Y \rangle$ in (10.1.9) is a BV process that is adapted, continuous, and such that $\langle X, Y \rangle_0 = 0$. Recalling Proposition 9.5.1, to prove that $\langle X, Y \rangle$ is the covariation

10.1 Integral with Respect to a Brownian Motion

process of X and Y, it is enough to verify that $XY - \langle X, Y \rangle$ is a martingale. For $0 \leq s \leq t$, we have

$$E[X_t Y_t \mid \mathscr{F}_s] = X_s Y_s + E[(X_t - X_s)(Y_t - Y_s) \mid \mathscr{F}_s] + 2X_s E[Y_t - Y_s \mid \mathscr{F}_s] =$$

(by (10.1.7) and since $E[Y_t - Y_s \mid \mathscr{F}_s] = 0$ by (10.1.10))

$$= X_s Y_s + E\left[\int_s^t u_r v_r dr \mid \mathscr{F}_s\right]$$

$$= X_s Y_s + E[\langle X, Y \rangle_t - \langle X, Y \rangle_s \mid \mathscr{F}_s]$$

which proves the thesis. □

Thanks to Itô's isometry (10.1.6), the stochastic integral extends to the case of integrands in $\mathbb{L}^2$ with an approximation procedure using simple processes. The following density result holds, whose proof is postponed to Sect. 10.1.1.

Lemma 10.1.7 *Let $u \in \mathbb{L}^2$. For every $T > 0$ there exists a sequence $(u_n)_{n \in \mathbb{N}}$ of simple processes in $\mathbb{L}^2$ that converges to u in the $L^2(\Omega \times [0, T])$-norm:*

$$\lim_{n \to \infty} E\left[\int_0^T (u_s - u_{n,s})^2 ds\right] = 0. \qquad (10.1.11)$$

Given $u \in \mathbb{L}^2$, we consider an approximating sequence $(u_n)_{n \in \mathbb{N}}$ of simple processes as in Lemma 10.1.7 for a fixed $T > 0$. Then, $(u_n)_{n \in \mathbb{N}}$ is a Cauchy sequence in $L^2([0, T] \times \Omega)$ and by Itô's isometry we have

$$\lim_{n,m \to \infty} E\left[\left(\int_0^T u_{n,s} dB_s - \int_0^T u_{m,s} dB_s\right)^2\right]$$

$$= \lim_{n,m \to \infty} E\left[\int_0^T (u_{n,s} - u_{m,s})^2 ds\right] = 0.$$

Hence, also the sequence of stochastic integrals is a Cauchy sequence in $L^2(\Omega, P)$, thereby ensuring the existence of

$$\int_0^T u_s dB_s := \lim_{n \to \infty} \int_0^T u_{n,s} dB_s.$$

With this procedure, the stochastic integral is defined for a fixed T as a limit in $L^2(\Omega, P)$-norm, i.e., only up to a negligible event. We will see in Sect. 10.2.3 that, thanks to Doob's maximal inequality, it is possible to construct the integral as a stochastic process (varying the integration endpoint) by defining it as a limit in the space of martingales $\mathscr{M}^{c,2}$. By approximation, the properties of Theorem 10.1.6 remain valid under the assumption that $u \in \mathbb{L}^2$.

In Sect. 10.2.4 we will further extend the integral to the case of integrands in $u \in \mathbb{L}^2_{\text{loc}}$, that is, u is progressively measurable and satisfies the mild integrability condition

$$\int_0^T u_t^2 dt < \infty \qquad T > 0, \text{ a.s.} \qquad (10.1.12)$$

which is considerably weaker than (10.1.1): for example, every adapted continuous process u belongs to $\mathbb{L}^2_{\text{loc}}$ since the integral in (10.1.12), on the compact interval $[0, T]$, is finite by the continuity of the trajectories of u. On the other hand, $u_t = \exp(B_t^4)$ is in $\mathbb{L}^2_{\text{loc}}$ but not[3] in $\mathbb{L}^2$. Theorem 10.1.6 does not extend to the case of $u \in \mathbb{L}^2_{\text{loc}}$, however, we will prove that in this case the integral process is a local martingale.

10.1.1 Proof of Lemma 10.1.7

To prove the density of the class of simple processes in the space $\mathbb{L}^2$, we use the following consequence of Proposition B.3.3 in [113], namely the so-called "continuity in mean" of absolutely integrable functions.

Corollary 10.1.8 (Continuity in Mean) *If $f \in L^1(\mathbb{R})$ then for almost every $x \in \mathbb{R}$ we have*

$$\lim_{h \to 0} \frac{1}{h} \int_x^{x+h} |f(x) - f(y)| dy = 0.$$

We prove Lemma 10.1.7 initially assuming that u is continuous. Fixed $T > 0$, for $n \in \mathbb{N}$, we denote by

$$t_{n,k} = \frac{Tk}{2^n}, \qquad k = 0, \ldots, 2^n, \qquad (10.1.13)$$

the dyadic numbers of $[0, T]$ and define the simple process

$$u_{n,t} = \sum_{k=1}^{2^n} \alpha_{n,k} \mathbb{1}_{[t_{n,k-1}, t_{n,k}[}, \qquad \alpha_{n,k} = u_{t_{n,k-1}} \mathbb{1}_{\{|u_{t_{n,k-1}}| \le n\}}, \qquad t \in [0, T].$$

Then (10.1.11) follows from the dominated convergence theorem.

---

[3] Since

$$E\left[\int_0^T e^{2B_t^4} dt\right] = \int_\mathbb{R} \int_0^T e^{2x^4} \frac{1}{\sqrt{2\pi t}} e^{-\frac{x^2}{2t}} dt dx = +\infty.$$

10.2 Integral with Respect to Continuous Square-Integrable Martingales

To conclude, it is enough to prove that every $u \in \mathbb{L}^2$ can be approximated in the $L^2([0, T] \times \Omega)$-norm by a sequence $(u_n)_{n \in \mathbb{N}}$ of continuous processes in $\mathbb{L}^2$. To this end, we define[4]

$$u_{n,t} := \fint_{\left(t-\frac{1}{n}\right) \vee 0}^{t} u_s \, ds, \qquad 0 < t \leq T, \; n \in \mathbb{N}.$$

Note that u_n is continuous and adapted (and therefore progressively measurable). Moreover, we have

$$E\left[\int_0^T (u_t - u_{n,t})^2 \, dt\right] = E\left[\int_0^T \left(\fint_{\left(t-\frac{1}{n}\right) \vee 0}^{t} (u_t - u_s) \, ds\right)^2 dt\right] \leq$$

(by Jensen's inequality)

$$\leq E\left[\int_0^T \fint_{\left(t-\frac{1}{n}\right) \vee 0}^{t} (u_t - u_s)^2 \, ds \, dt\right]$$

$$= \int_0^T \fint_{\left(t-\frac{1}{n}\right) \vee 0}^{t} E\left[(u_t - u_s)^2\right] ds \, dt. \qquad (10.1.14)$$

Now, by Corollary 10.1.8 we have

$$\lim_{n \to \infty} \fint_{\left(t-\frac{1}{n}\right) \vee 0}^{t} E\left[(u_t - u_s)^2\right] ds = 0 \qquad \text{a.e.}$$

and therefore we can take the limit in (10.1.14) as $n \to \infty$ and conclude using the Lebesgue dominated convergence theorem.

10.2 Integral with Respect to Continuous Square-Integrable Martingales

We assume that the integrator process B belongs to the class $\mathscr{M}^{c,2}$, i.e., B is a continuous martingale such that $B_t \in L^2(\Omega, P)$ for every $t \geq 0$. The construction of the stochastic integral is similar to the case of a Brownian motion with some additional technicalities.

We denote by $\langle B \rangle$ the quadratic variation process defined in Theorem 9.4.1: $\langle B \rangle$ is a continuous and increasing process associated with the Lebesgue-Stieltjes

---

[4] Here $\fint_a^b u_s \, ds = \frac{1}{b-a} \int_a^b u_s \, ds$ for $a < b$.

measure $\mu_{\langle B \rangle}$ (cf. Sect. 9.2). We let

$$\int_{[a,b]} f \, d\mu_{\langle B \rangle} \quad \text{or} \quad \int_a^b f(t) d\langle B \rangle_t, \qquad 0 \leq a \leq b,$$

indicate the integral with respect to $\mu_{\langle B \rangle}$. For example, if B is a Brownian motion then $\langle B \rangle_t = t$ and the corresponding Lebesgue-Stieltjes measure is simply the Lebesgue measure, as seen in Sect. 10.1.

Definition 10.2.1 We denote by $\mathbb{L}^2_B$ the class of processes $u = (u_t)_{t \geq 0}$ such that:

(i) u is progressively measurable;
(ii) for every $T \geq 0$ we have

$$E\left[\int_0^T u_t^2 d\langle B \rangle_t\right] < \infty. \tag{10.2.1}$$

Generally, the process B will be fixed once and for all and therefore, if there is no risk of confusion, we will simply write $\mathbb{L}^2$ instead of $\mathbb{L}^2_B$.

At a later stage, we will weaken the integrability condition (ii) by requiring that u belongs to the following class.

Definition 10.2.2 We denote by $\mathbb{L}^2_{B,\text{loc}}$ (or, more simply, $\mathbb{L}^2_{\text{loc}}$) the class of processes u such that

(i) u is progressively measurable;
(ii') for every $T \geq 0$ we have

$$\int_0^T u_t^2 d\langle B \rangle_t < \infty \qquad \text{a.s.} \tag{10.2.2}$$

Property (ii') is a very weak integrability condition that is automatically verified if, for example, u has continuous trajectories or, more generally, locally bounded ones (note that the integration domain in (10.2.2) is compact). Formula (10.2.2) is equivalent to $P(u \in L^2([0, T], \mu_{\langle B \rangle})) = 1$.

10.2.1 Integral of Indicator Processes

Consider a very particular class of integrands that, with respect to the temporal variable, are indicator functions of an interval: precisely, an *indicator process* is a stochastic process of the form

$$u_t = \alpha \mathbb{1}_{[t_0, t_1[}(t), \qquad t \geq 0, \tag{10.2.3}$$

10.2 Integral with Respect to Continuous Square-Integrable Martingales

where α is a $\mathscr{F}_{t_0}$-*measurable and bounded* random variable (i.e., such that $|\alpha| \leq c$ a.s. for some positive constant c) and $t_1 > t_0 \geq 0$.

Remark 10.2.3 Every indicator process u belongs to $\mathbb{L}^2$: in fact, u is càdlàg and adapted, therefore progressively measurable; moreover, u satisfies (10.2.1) since

$$E\left[\int_0^T u_t^2 d\langle B\rangle_t\right] = E\left[\alpha^2 \left(\langle B\rangle_{T\wedge t_1} - \langle B\rangle_{T\wedge t_0}\right)\right] \leq c^2 E\left[\langle B\rangle_{T\wedge t_1} - \langle B\rangle_{T\wedge t_0}\right] < \infty$$

for every $T \geq 0$.

The definition of the stochastic integral of an indicator process is elementary and completely explicit: it is defined, path by path, by multiplying α by an increment of B.

Definition 10.2.4 (Stochastic Integral of Indicator Processes) Let u be the indicator process in (10.2.3) and $B \in \mathscr{M}^{c,2}$. For every $T \geq t_1$ we set

$$\int_0^T u_t dB_t := \alpha \left(B_{t_1} - B_{t_0}\right) \tag{10.2.4}$$

and we define the stochastic integral for two generic integration endpoints a and b, with $0 \leq a \leq b$, as

$$\int_a^b u_t dB_t := \int_0^{t_1} u_t \mathbb{1}_{[a,b[}(t) dB_t. \tag{10.2.5}$$

Remark 10.2.5 If $[t_0, t_1[\cap[a, b[\neq \emptyset$, the integral in the right-hand side of (10.2.5) is defined by (10.2.4) interpreting $u_t \mathbb{1}_{[a,b[}(t)$ as the simple process $\alpha \mathbb{1}_{[t_0 \vee a, t_1 \wedge b[}(t)$ and choosing $T = t_1$. Otherwise, it is understood that the integral is null by definition.

Remark 10.2.6 ([!]) Being defined in terms of increments of B, the stochastic integral does not depend on the initial value B_0. Moreover, X is an adapted and continuous process.

In the next result, we establish some fundamental properties of the stochastic integral. The second part of the proof is based on the remarkable identity (9.4.1), valid for every $B \in \mathscr{M}^{c,2}$, which we recall here:

$$E\left[(B_t - B_s)^2 \mid \mathscr{F}_s\right] = E\left[\langle B\rangle_t - \langle B\rangle_s \mid \mathscr{F}_s\right], \qquad 0 \leq s \leq t. \tag{10.2.6}$$

Throughout the chapter, we insist on providing the explicit expression of the quadratic variation of the stochastic integral or the covariation of two integrals: the reason is that they appear in the most important tool for calculating stochastic integrals, Itô's formula, which we will present in Chap. 11.

Theorem 10.1.6 has the following natural extension.

Theorem 10.2.7 ([!]) *Let*

$$X_t := \int_0^t u_s dB_s, \qquad Y_t := \int_0^t v_s dB_s, \qquad t \geq 0,$$

where u, v are indicator processes and $B \in \mathcal{M}^{c,2}$. For $0 \leq s \leq t \leq T$, the following properties hold:

(i) *X is a continuous square-integrable martingale, $X \in \mathcal{M}^{c,2}$, and we have*

$$E\left[\int_s^t u_r dB_r \mid \mathcal{F}_s\right] = 0; \tag{10.2.7}$$

(ii) *the Itô isometry holds*

$$E\left[\left(\int_s^t u_r dB_r\right)^2 \mid \mathcal{F}_s\right] = E\left[\int_s^t u_r^2 d\langle B\rangle_r \mid \mathcal{F}_s\right] \tag{10.2.8}$$

and more generally

$$E\left[\int_s^t u_r dB_r \int_s^t v_r dB_r \mid \mathcal{F}_s\right] = E\left[\int_s^t u_r v_r d\langle B\rangle_r \mid \mathcal{F}_s\right], \tag{10.2.9}$$

$$E\left[\int_s^t u_r dB_r \int_t^T v_r dB_r \mid \mathcal{F}_s\right] = 0; \tag{10.2.10}$$

(iii) *the covariation process of X and Y is given by*

$$\langle X, Y\rangle_t = \int_0^t u_s v_s d\langle B\rangle_s, \qquad t \geq 0. \tag{10.2.11}$$

Proof By Remark 10.2.5 it is not restrictive to assume $u = \alpha \mathbb{1}_{[s,t[}$ and $v = \beta \mathbb{1}_{[s,t[}$ with bounded $\alpha, \beta \in m\mathcal{F}_s$.

(i) We have

$$E\left[\int_s^t u_r dB_r \mid \mathcal{F}_s\right] = E\left[\alpha(B_t - B_s) \mid \mathcal{F}_s\right] = \alpha E\left[B_t - B_s \mid \mathcal{F}_s\right] = 0$$

where we have exploited the fact that $\alpha \in m\mathcal{F}_s$ and the martingale property of B. This proves (10.2.7) which is equivalent to the martingale property of X. Clearly $X_T \in L^2(\Omega, P)$ for every $T \geq 0$ since X_T is the product of the bounded random variable α, times an increment of B which is square-integrable.

10.2 Integral with Respect to Continuous Square-Integrable Martingales

(ii) We directly prove (10.2.9): we have

$$E\left[\int_s^t u_r dB_r \int_s^t v_r dB_r \mid \mathcal{F}_s\right] = E\left[\alpha\beta(B_t - B_s)^2 \mid \mathcal{F}_s\right]$$
$$= \alpha\beta E\left[(B_t - B_s)^2 \mid \mathcal{F}_s\right] =$$

(by the crucial formula (10.2.6))

$$= \alpha\beta E\left[\langle B\rangle_t - \langle B\rangle_s \mid \mathcal{F}_s\right]$$
$$= E\left[\alpha\beta(\langle B\rangle_t - \langle B\rangle_s) \mid \mathcal{F}_s\right]$$
$$= E\left[\int_s^t u_r v_r d\langle B\rangle_r \mid \mathcal{F}_s\right].$$

The proof of (10.2.9) is analogous.

(iii) The process $\langle X, Y\rangle$ in (10.2.11) is adapted, continuous and locally of bounded variation since it is the difference of increasing processes

$$\langle X, Y\rangle_t = \int_0^t (u_s v_s)^+ d\langle B\rangle_s - \int_0^t (u_s v_s)^- d\langle B\rangle_s.$$

Moreover, $\langle X, Y\rangle_0 = 0$. To conclude, it is enough to prove that $XY - \langle X, Y\rangle$ is a martingale: we have

$$X_t Y_t = \left(X_s + \int_s^t u_r dB_r\right)\left(Y_s + \int_s^t v_r dB_r\right)$$
$$= X_s Y_s + \int_s^t u_r dB_r \int_s^t v_r dB_r + X_s \int_s^t v_r dB_r + Y_s \int_s^t u_r dB_r$$

and therefore

$$E[X_t Y_t \mid \mathcal{F}_s] = X_s Y_s + E\left[\int_s^t u_r dB_r \int_s^t v_r dB_r \mid \mathcal{F}_s\right]$$
$$+ X_s E\left[\int_s^t v_r dB_r \mid \mathcal{F}_s\right] + Y_s E\left[\int_s^t u_r dB_r \mid \mathcal{F}_s\right] =$$

(by (10.2.9) and (10.2.7))

$$= X_s Y_s + E\left[\int_s^t u_r v_r d\langle B\rangle_r \mid \mathcal{F}_s\right]$$

so

$$E[X_t Y_t - \langle X, Y \rangle_t \mid \mathscr{F}_s] = X_s Y_s - \langle X, Y \rangle_s.$$

□

Remark 10.2.8 Formulas (10.2.7), (10.2.8), (10.2.9), (10.2.10), and (10.2.11) can be rewritten in the form

$$E[X_t - X_s \mid \mathscr{F}_s] = 0,$$

$$E\left[(X_t - X_s)^2 \mid \mathscr{F}_s\right] = E[\langle X \rangle_t - \langle X \rangle_s \mid \mathscr{F}_s],$$

$$E[(X_t - X_s)(Y_t - Y_s) \mid \mathscr{F}_s] = E[\langle X, Y \rangle_t - \langle X, Y \rangle_s \mid \mathscr{F}_s],$$

$$E[(X_t - X_s)(Y_T - Y_t) \mid \mathscr{F}_s] = 0.$$

By taking the expected value, we also obtain the unconditional versions of Itô's isometry:

$$E\left[\left(\int_s^t u_r dB_r\right)^2\right] = E\left[\int_s^t u_r^2 d\langle B \rangle_r\right], \tag{10.2.12}$$

$$E\left[\int_s^t u_r dB_r \int_s^t v_r dB_r\right] = E\left[\int_s^t u_r v_r d\langle B \rangle_r\right],$$

$$E\left[\int_s^t u_r dB_r \int_t^T v_r dB_r\right] = 0, \tag{10.2.13}$$

and (10.2.11) with $u = v$ becomes

$$\langle X \rangle_t = \int_0^t u_s^2 d\langle B \rangle_s, \qquad t \geq 0.$$

10.2.2 Integral of Simple Processes

In this section, we extend the class of integrable processes to simple processes: they are sums of indicator processes like those considered in the previous section. Due to linearity, the definition of stochastic integral extends, path by path, in an elementary and explicit way. The fundamental properties of the integral remain valid: the martingale property and Itô's isometry.

10.2 Integral with Respect to Continuous Square-Integrable Martingales

Definition 10.2.9 (Simple Process) A simple process u is a process of the form

$$u_t = \sum_{k=1}^{N} u_{k,t}, \qquad u_{k,t} := \alpha_k \mathbb{1}_{[t_{k-1}, t_k[}(t), \tag{10.2.14}$$

where:

(i) $0 \le t_0 < t_1 < \cdots < t_N$;
(ii) α_k is a bounded $\mathscr{F}_{t_{k-1}}$-measurable random variable for each $k = 1, \ldots, N$.

One can also require that $P(\alpha_k \ne \alpha_{k+1}) > 0$, for $k = 1, \ldots, N-1$, so that the representation (10.2.14) of u is unique.

Definition 10.2.10 (Stochastic Integral of Simple Processes) Let u be a simple process of the form (10.2.14) and let $B \in \mathscr{M}^{c,2}$. The stochastic integral of u with respect to B is the stochastic process

$$\int_0^t u_s \, dB_s := \sum_{k=1}^{N} \int_0^t u_{k,s} \, dB_s = \sum_{k=1}^{N} \alpha_k \left(B_{t \wedge t_k} - B_{t \wedge t_{k-1}} \right).$$

Theorem 10.2.11 *Theorem 10.2.7 remains valid under the assumption that u, v are simple processes.*

Proof The continuity and the martingale property (10.2.7) are immediate due to linearity. As by Itô's isometry (10.2.9), first we can write v in the form (10.2.14) with respect to the same choice of $t_0, \ldots, t_N$, for certain $v_{k,t} = \beta_k \mathbb{1}_{[t_{k-1}, t_k[}(t)$: note that

$$u_t v_t = \sum_{k=1}^{N} u_{k,t} \sum_{h=1}^{N} v_{h,t} = \sum_{k=1}^{N} \alpha_k \beta_k \mathbb{1}_{[t_{k-1}, t_k[}(t). \tag{10.2.15}$$

Then we have

$$E\left[\int_s^t u_r \, dB_r \int_s^t v_r \, dB_r \mid \mathscr{F}_s \right]$$

$$= E\left[\sum_{k=1}^{N} \int_s^t u_{k,r} \, dB_r \sum_{h=1}^{N} \int_s^t v_{h,r} \, dB_r \mid \mathscr{F}_s \right]$$

$$= \sum_{k=1}^{N} E\left[\int_s^t u_{k,r} \, dB_r \int_s^t v_{k,r} \, dB_r \mid \mathscr{F}_s \right]$$

$$\times 2 \sum_{h<k} E\left[\int_{t_{h-1}}^{t_h} u_{h,r} \mathbb{1}_{[s,t[}(r) \, dB_r \int_{t_{k-1}}^{t_k} v_{k,r} \mathbb{1}_{[s,t[}(r) \, dB_r \mid \mathscr{F}_s \right] =$$

(by (10.2.8) and (10.2.10))

$$= \sum_{k=1}^{N} E\left[\int_s^t u_{k,r} v_{k,r} d\langle B\rangle_r \mid \mathscr{F}_s\right] =$$

(by (10.2.15))

$$= E\left[\int_s^t u_r v_r d\langle B\rangle_r \mid \mathscr{F}_s\right].$$

Finally, the fact that $\langle X, Y\rangle$ in (10.2.11) is the covariation process of X and Y is proven as in the proof of Theorem 10.2.7-(iii). □

10.2.3 Integral in $\mathbb{L}^2$

In this section, we extend the class of integrands by exploiting the density of simple processes in $\mathbb{L}_B^2$ (cf. Definition 10.2.1). The stochastic integral is now defined as a limit in $\mathscr{M}^{c,2}$ and therefore, recalling Remark 8.3.2, as an *equivalence class* and no longer path by path. However, the fundamental properties of the integral remain valid: the martingale property and Itô's isometry. As usual, since B is fixed, we simply write $\mathbb{L}^2$ instead of $\mathbb{L}_B^2$.

Lemma 10.1.7 has the following generalization, which is proven with a technical trick: the idea is to make a change of time variable to "realign" the continuous and increasing process $\langle B\rangle_t$ to the Brownian case in which $\langle B\rangle_t \equiv t$; for details, we refer to Lemma 2.2.7 in [67].

Lemma 10.2.12 *Let $u \in \mathbb{L}^2$. For every $T > 0$ there exists a sequence $(u_n)_{n\in\mathbb{N}}$ of simple processes such that*

$$\lim_{n\to\infty} E\left[\int_0^T (u_s - u_{n,s})^2 d\langle B\rangle_s\right] = 0.$$

We recall the convention according to which $\mathscr{M}_T^{c,2}$ is the space of equivalence classes (according to indistinguishability) of continuous square-integrable martingales $X = (X_t)_{t\in[0,T]}$, equipped with the norm

$$\|X\|_T := \sqrt{E\left[X_T^2\right]}.$$

By Proposition 8.3.3, $(\mathscr{M}_T^{c,2}, \|\cdot\|_T)$ is a Banach space.

10.2 Integral with Respect to Continuous Square-Integrable Martingales

We now see how to define the stochastic integral of $u \in \mathbb{L}^2$. Given $T > 0$ and an approximating sequence $(u_n)_{n \in \mathbb{N}}$ of simple processes as in Lemma 10.2.12, we denote by

$$X_{n,t} = \int_0^t u_{n,s} dB_s, \qquad t \in [0, T], \tag{10.2.16}$$

the sequence of their respective stochastic integrals. By Theorem 10.2.11 $X_n \in \mathcal{M}_T^{c,2}$ and by Itô's isometry (10.2.8) we have

$$\|X_n - X_m\|_T^2 = E\left[\left(\int_0^T (u_{n,t} - u_{m,t}) dB_t\right)^2\right] = E\left[\int_0^T (u_{n,t} - u_{m,t})^2 d\langle B\rangle_t\right].$$

It follows that $(X_n)_{n \in \mathbb{N}}$ is a Cauchy sequence in $(\mathcal{M}_T^{c,2}, \|\cdot\|_T)$ and therefore there exists

$$X := \lim_{n \to \infty} X_n \qquad \text{in } \mathcal{M}_T^{c,2}. \tag{10.2.17}$$

Proposition 10.2.13 (Stochastic Integral of $\mathbb{L}^2$ Processes) *The limit process $X = (X_t)_{t \in [0,T]}$ in (10.2.17) is independent of the approximating sequence and is called the stochastic integral process of u with respect to B on $[0, T]$ and denoted by*

$$X_t = \int_0^t u_s dB_s, \qquad t \in [0, T].$$

Proof Let X be the limit in (10.2.17) defined from the approximating sequence $(u_n)_{n \in \mathbb{N}}$. Let $(v_n)_{n \in \mathbb{N}}$ be another approximating sequence for u and

$$Y_{n,t} = \int_0^t v_{n,s} dB_s, \qquad t \in [0, T]. \tag{10.2.18}$$

Then $\|Y_n - X\|_T \leq \|Y_n - X_n\|_T + \|X_n - X\|_T$ and it is enough to observe that, again by Itô's isometry, we have

$$\|Y_n - X_n\|_T^2 = E\left[\left(\int_0^T (v_{n,t} - u_{n,t}) dB_t\right)^2\right]$$
$$= E\left[\int_0^T (v_{n,t} - u_{n,t})^2 d\langle B\rangle_t\right] \xrightarrow[n \to \infty]{} 0.$$

$\square$

Remark 10.2.14 ([!]) By construction, the Itô stochastic integral

$$X_t = \int_0^t u_s dB_s, \qquad (10.2.19)$$

with $u \in \mathbb{L}^2$ and $B \in \mathcal{M}^{c,2}$, *is an equivalence class* in $\mathcal{M}^{c,2}$: each representative of this class is a continuous martingale, uniquely determined up to indistinguishable processes. From this perspective, unless a particular choice of the representative has been made, the single trajectories of the stochastic integral process are not defined and it does not make sense to consider $X_t(\omega)$ for a specific $\omega \in \Omega$.

Theorem 10.2.15 *Theorem 10.2.7 remains valid under the assumption that $u, v \in \mathbb{L}^2$.*

Proof Let $(u_n)_{n \in \mathbb{N}}$ and $(v_n)_{n \in \mathbb{N}}$ be sequences of simple processes, approximating u and v in $(\mathcal{M}_T^{c,2}, \|\cdot\|_T)$, respectively. We denote by $(X_n)_{n \in \mathbb{N}}$ and $(Y_n)_{n \in \mathbb{N}}$ the corresponding stochastic integrals in (10.2.16) and (10.2.18). Equations (10.2.7) and (10.2.8) are a direct consequence of the fact that $X_{n,t} \to X_t$ in $L^2(\Omega, P)$ (and therefore also in $L^1(\Omega, P)$) and $X_{n,t} Y_{n,t} \to X_t Y_t$ in $L^1(\Omega, P)$, together with the general fact that[5] if $Z_n \to Z$ in $L^1(\Omega, P)$ then $E[Z_n \mid \mathcal{G}] \to E[Z \mid \mathcal{G}]$ in $L^1(\Omega, P)$. The proof of (10.2.11) is identical to that of Theorem 10.2.7-(iii). □

Remark 10.2.16 ([!]) Let $B \in \mathcal{M}^{c,2}$ and $u \in \mathbb{L}_B^2$. By Theorem 10.2.15, the integral X in (10.2.19) belongs to $\mathcal{M}^{c,2}$ and therefore, in turn, *can be used as an integrator*. Since

$$\langle X \rangle_t = \int_0^t u_s^2 d\langle B \rangle_s,$$

we have that $v \in \mathbb{L}_X^2$ if v is progressively measurable and satisfies

$$E\left[\int_0^t v_s^2 d\langle X \rangle_s\right] = E\left[\int_0^t v_s^2 u_s^2 d\langle B \rangle_s\right] < \infty$$

for every $t \geq 0$. In this case, we have

$$\int_0^t v_s dX_s = \int_0^t v_s u_s dB_s$$

which can be verified directly for simple u, v and, in general, by approximation.

---

[5] By Jensen's inequality, we have

$$E\left[|E[Z_n \mid \mathcal{G}] - E[Z \mid \mathcal{G}]|\right] \leq E\left[E[|Z_n - Z| \mid \mathcal{G}]\right] = E[|Z_n - Z|].$$

10.2 Integral with Respect to Continuous Square-Integrable Martingales

In particular, if B is a Brownian motion, then the Lebesgue-Stieltjes measure associated with $\langle X \rangle$ is absolutely continuous with respect to the Lebesgue measure, with density u^2.

We now give two propositions whose statements seem almost obvious but actually, in light of Remark 10.2.14, require a rigorous proof. Both results are proven using an approximation procedure, technical and somewhat tedious.

Proposition 10.2.17 ([!]) *Suppose that $u, v \in \mathbb{L}^2$ are modifications on an event F in the sense that, for every $t \in [0, T]$, $u_t(\omega) = v_t(\omega)$ for every $\omega \in F \setminus N$ where N is a negligible event. Then the corresponding integral processes*

$$X_t = \int_0^t u_s \, dB_s, \qquad Y_t = \int_0^t v_s \, dB_s,$$

are indistinguishable on F, that is, $\sup_{t \subset [0,T]} |X_t(\omega) - Y_t(\omega)| = 0$ *for $\omega \in F \setminus N$.*

Proof Let us consider the approximations u_n and v_n defined as in Lemma 10.2.12. By construction, for every $n \in \mathbb{N}$ and $t \in [0, T]$, $u_{n,t} = v_{n,t}$ almost surely on F. It follows that the relative integrals $(X_{n,t})_{t \in [0,T]}$ in (10.2.16) and $(Y_{n,t})_{t \in [0,T]}$ in (10.2.18) are modifications on F. Taking the limit in n, we deduce that $(X_t)_{t \in [0,T]}$ and $(Y_t)_{t \in [0,T]}$ are modifications on F: the thesis follows from the continuity of X and Y. □

Remark 10.2.18 Suppose that, for some $T > 0$, we have

$$\int_0^T u_t \, dB_t = \int_0^T v_t \, dB_t$$

where $u, v \in \mathbb{L}^2$ and B is a Brownian motion. Then $P(u = v \text{ a.e. on } [0, T]) = 1$, that is, almost surely the trajectories of u and v are a.e. equal on $[0, T]$. Indeed, by Itô's isometry, we have

$$E\left[\int_0^T (u_t - v_t)^2 dt\right] = E\left[\left(\int_0^T (u_t - v_t) dB_t\right)^2\right] = 0$$

which proves the thesis.

Proposition 10.2.19 (Integral with Random Integration Endpoint [!]) *Let X in (10.2.19) be the stochastic integral process of $u \in \mathbb{L}^2$ with respect to $B \in \mathcal{M}^{c,2}$. Let τ be a stopping time such that $0 \leq \tau \leq T$ for some $T > 0$. Then $(u_t \mathbb{1}_{(t \leq \tau)})_{t \geq 0} \in \mathbb{L}^2$ and*

$$X_\tau = \int_0^\tau u_s \, dB_s = \int_0^T u_s \mathbb{1}_{(s \leq \tau)} dB_s \qquad a.s.$$

Proof First, we observe that, by Proposition 10.2.17, if $F \in \mathscr{F}_t$ then

$$\mathbb{1}_F \int_t^T u_s dB_s = \int_t^T \mathbb{1}_F u_s dB_s \quad \text{a.s.} \tag{10.2.20}$$

The measurability condition on F is essential because it ensures that the integral on the right-hand side of (10.2.20) is well-defined, being the integrand progressively measurable on $[t, T]$.

Now we recall the notation (10.1.13), $t_{n,k} := \frac{Tk}{2^n}$, for the dyadic numbers of $[0, T]$ and we use the usual discretization of τ:

$$\tau_n = \sum_{k=1}^{2^n} t_{n,k} \mathbb{1}_{F_{n,k}}$$

with

$$F_{n,1} = \left(0 \leq \tau \leq \tfrac{T}{2^n}\right), \qquad F_{n,k} = \left(t_{n,k-1} < \tau \leq t_{n,k}\right), \qquad k = 2, \ldots, 2^n.$$

We note that $(F_{n,k})_{k=1,\ldots,2^n}$ forms a partition of Ω with $F_{n,k} \in \mathscr{F}_{t_{n,k}}$ and $(\tau_n)_{n \in \mathbb{N}}$ is a decreasing sequence of stopping times that converges to τ. By continuity, we have $X_{\tau_n} \to X_\tau$. Moreover, setting

$$Y = \int_0^T u_s \mathbb{1}_{(s \leq \tau)} dB_s, \qquad Y_n = \int_0^T u_s \mathbb{1}_{(s \leq \tau_n)} dB_s,$$

using Itô's isometry, it is easy to prove that $Y_n \to Y$ in $L^2(\Omega, P)$ and therefore also almost surely.

To prove the thesis, i.e., the fact that $X_\tau = Y$ a.s., it is sufficient to verify that $X_{\tau_n} = Y_n$ a.s. for each $n \in \mathbb{N}$. Now, on $F_{n,k}$ we have

$$X_{\tau_n} = X_{t_{n,k}} = \int_0^T u_s dB_s - \int_{t_{n,k}}^T u_s dB_s,$$

and therefore

$$X_{\tau_n} = \int_0^T u_s dB_s - \sum_{k=1}^{2^n} \mathbb{1}_{F_{n,k}} \int_{t_{n,k}}^T u_s dB_s. \tag{10.2.21}$$

10.2 Integral with Respect to Continuous Square-Integrable Martingales

On the other hand,

$$Y_n = \int_0^T u_s \left(1 - \mathbb{1}_{(s > \tau_n)}\right) dB_s$$

$$= \int_0^T u_s dB_s - \sum_{k=1}^{2^n} \int_{t_{n,k}}^T u_s \mathbb{1}_{F_{n,k}} dB_s =$$

(by (10.2.20), with probability one)

$$= \int_0^T u_s dB_s - \sum_{k=1}^{2^n} \mathbb{1}_{F_{n,k}} \int_{t_{n,k}}^T u_s dB_s$$

which, combined with (10.2.21), proves the thesis. □

10.2.4 Integral in $\mathbb{L}^2_{loc}$

Weakening the integrability condition on the integrand from $\mathbb{L}^2$ to $\mathbb{L}^2_{loc}$, some of the fundamental properties of the integral are lost, including the martingale property and Itô's isometry. However, we will prove that the integral is a *local martingale* and provide a "surrogate" for Itô's isometry, Lemma 10.2.25.

We recall that $u \in \mathbb{L}^2_{loc}$ if it is progressively measurable and, for every $t > 0$,

$$A_t := \int_0^t u_s^2 d\langle B \rangle_s < \infty \qquad \text{a.s.} \qquad (10.2.22)$$

The process A is continuous, adapted, and increasing; moreover, A is non-negative since $A_0 = 0$ (see Fig. 10.1).

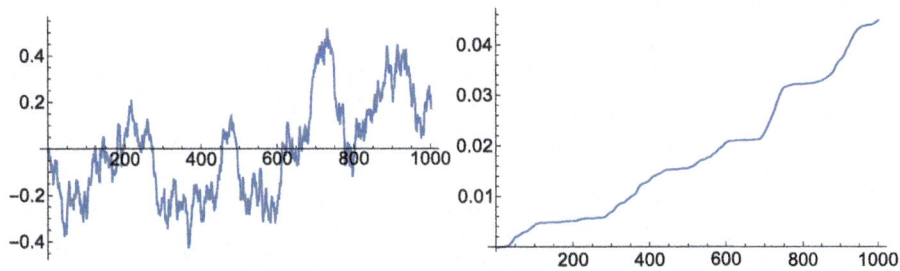

Fig. 10.1 On the left: plot of a trajectory of a Brownian motion W. **On the right:** plot of the related trajectory of $A_t = \int_0^t W_s^2 ds$, corresponding to the process in (10.2.22) with $u = W$ and B Brownian motion

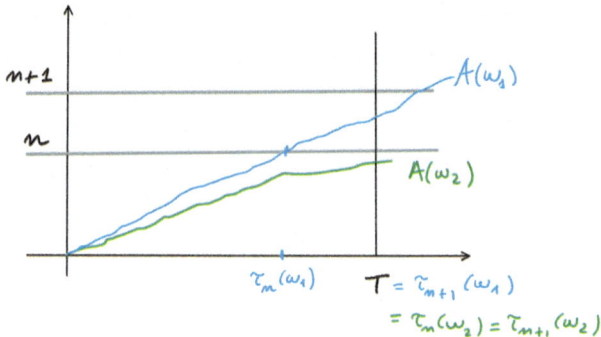

Fig. 10.2 Plot of two trajectories of the process A in (10.2.22) and the corresponding stopping times τ_n and τ_{n+1} in (10.2.23)

Remark 10.2.20 ([!]) Note that the class $\mathbb{L}^2$ depends on the fixed probability measure, as opposed to $\mathbb{L}^2_{\text{loc}}$ that is invariant with respect to equivalent[6] probability measures.

Let us fix $T > 0$ and consider the sequence of stopping times defined by

$$\tau_n = T \wedge \inf\{t \geq 0 \mid A_t \geq n\}, \qquad n \in \mathbb{N}, \tag{10.2.23}$$

and represented in Fig. 10.2. Due to the continuity of A, we have $\tau_n \nearrow T$ almost surely, and thus the sequence of events $F_n := (\tau_n = T)$ is such that $F_n \nearrow \Omega \setminus N$ with $P(N) = 0$. Truncating u at time τ_n, we define the process

$$u_{n,t} := u_t \mathbb{1}_{(t \leq \tau_n)}, \qquad t \in [0, T],$$

which is progressively measurable and such that

$$E\left[\int_0^t u_{n,s}^2 d\langle B\rangle_s\right] = E\left[\int_0^{t \wedge \tau_n} u_s^2 d\langle B\rangle_s\right] \leq n, \qquad t \in [0, T].$$

Thus, $u_n \in \mathbb{L}^2$ and the corresponding integral

$$X_{n,t} := \int_0^t u_{n,s} dB_s = \int_0^{t \wedge \tau_n} u_s dB_s, \qquad t \in [0, T], \tag{10.2.24}$$

belongs to $\mathcal{M}^{c,2}$ according to Theorem 10.2.15. Moreover, for every $n, h \in \mathbb{N}$, almost surely for every $t \in [0, T]$ we have

$$u_{n,t} = u_{n+h,t} = u_t \quad \text{on } F_n,$$

---

[6] Equivalent measures have the same certain (and, therefore, also negligible) events.

10.2 Integral with Respect to Continuous Square-Integrable Martingales

and therefore the processes $(X_{n,t})_{t\in[0,n]}$ and $(X_{n+h,t})_{t\in[0,n]}$ are indistinguishable on F_n thanks to Proposition 10.2.17. Hence, the following definition is well-posed:

Definition 10.2.21 (Stochastic Integral of Processes in $\mathbb{L}^2_{loc}$) The stochastic integral of $u \in \mathbb{L}^2_{loc}$ with respect to $B \in \mathcal{M}^{c,2}$ on $[0, T]$ is the continuous and adapted process $X = (X_t)_{t\in[0,T]}$ that on F_n is indistinguishable from X_n in (10.2.24) for every $n \in \mathbb{N}$. As usual, we write

$$X_t = \int_0^t u_s dB_s, \qquad t \in [0, T]. \tag{10.2.25}$$

We will see later, in Proposition 10.2.26, that

$$\int_0^t u_s dB_s = \lim_{n\to\infty} \int_0^t u_{n,s} dB_s$$

with convergence *in probability*.

Remark 10.2.22 As already observed earlier, the stochastic integral is defined as an equivalence class of indistinguishable processes. The previous definition and in particular the notation (10.2.25) are well-posed in the sense that if X and $\bar{X}$ denote respectively the stochastic integral processes of u with respect to B on the intervals $[0, T]$ and $[0, \bar{T}]$ with $T \le \bar{T}$ then, by an approximation procedure starting from simple processes, we get that X and $\bar{X}|_{[0,T]}$ are indistinguishable processes. Consequently, the *Itô stochastic integral process of u with respect to B* denoted by

$$X_t = \int_0^t u_s dB_s, \qquad t \ge 0.$$

is well-defined.

Proposition 10.2.19 has the following simple generalization.

Proposition 10.2.23 (Integral with Random Integration Endpoint) *Let X be the stochastic integral process of $u \in \mathbb{L}^2_{loc}$ with respect to $B \in \mathcal{M}^{c,2}$. Let τ be a stopping time such that $0 \le \tau \le T$ for some $T > 0$. Then $(u_t \mathbb{1}_{(t\le\tau)})_{t\ge 0} \in \mathbb{L}^2_{loc}$ and*

$$X_\tau = \int_0^\tau u_s dB_s = \int_0^T u_s \mathbb{1}_{(s\le\tau)} dB_s \qquad a.s.$$

Proof It is clear that $(u_t \mathbb{1}_{(t\le\tau)})_{t\ge 0} \in \mathbb{L}^2_{loc}$. Let $(\tau_n)_{n\in\mathbb{N}}$ be the sequence of stopping times in (10.2.23). By definition on the event $F_n = (\tau_n = T)$, we have

$$X_\tau = \int_0^\tau u_s \mathbb{1}_{(s\le\tau_n)} dB_s =$$

(by Proposition 10.2.19, since $u_s \mathbb{1}_{(s \leq \tau_n)} \in \mathbb{L}^2$)

$$= \int_0^T u_s \mathbb{1}_{(s \leq \tau_n)} \mathbb{1}_{(s \leq \tau)} dB_s =$$

(since $\tau_n = T \geq \tau$ on F_n)

$$= \int_0^T u_s \mathbb{1}_{(s \leq \tau)} dB_s.$$

The thesis follows from the arbitrariness of n. □

Extending the class of integrands from $\mathbb{L}^2$ to $\mathbb{L}^2_{loc}$, we lose the martingale property, however, we have the following

Theorem 10.2.24 ([!]) Let

$$X_t = \int_0^t u_s dB_s, \qquad Y_t = \int_0^t v_s dB_s$$

with $u, v \in \mathbb{L}^2_{loc}$ and $B \in \mathcal{M}^{c,2}$. Then:

(i) X is a continuous local martingale, i.e., $X \in \mathcal{M}^{c,loc}$, and

$$\tau_n := n \wedge \inf\{t \geq 0 \mid A_t \geq n\}, \qquad n \in \mathbb{N},$$

with A in (10.2.22), is a localizing sequence for X (cf. Definition 8.4.2);
(ii) the covariation process of X and Y is

$$\langle X, Y \rangle_t = \int_0^t u_s v_s d\langle B \rangle_s, \qquad t \geq 0.$$

Proof By Proposition 10.2.23 (with the choice $\tau = t \wedge \tau_n$ and $T = t$), for every $t \geq 0$ we have

$$X_{t \wedge \tau_n} = \int_0^t u_s \mathbb{1}_{(s \leq \tau_n)} dB_s \qquad \text{a.s.}$$

and therefore, by continuity, $X_{t \wedge \tau_n}$ is a version of the stochastic integral of the process $u_s \mathbb{1}_{(s \leq \tau_n)}$ which belongs to $\mathbb{L}^2$. It follows that $X_{t \wedge \tau_n}$ is a continuous martingale and therefore X is a local martingale with localizing sequence $(\tau_n)_{n \in \mathbb{N}}$.

Now let $A_t = \int_0^t u_s v_s d\langle B \rangle_s$ and

$$\tau_n = n \wedge \inf\{t \geq 0 \mid \langle X \rangle_t + \langle Y \rangle_t \geq n\}, \qquad n \in \mathbb{N}.$$

10.2 Integral with Respect to Continuous Square-Integrable Martingales

By Theorem 10.2.15 (cf. (10.2.11)) and the Cauchy-Schwarz inequality of Remark 9.5.2-(iii), the process

$$(XY - A)_{t \wedge \tau_n} = X_{t \wedge \tau_n} Y_{t \wedge \tau_n} - A_{t \wedge \tau_n} = X_{t \wedge \tau_n} Y_{t \wedge \tau_n} - \int_0^t u_s v_s \mathbb{1}_{(s \leq \tau_n)} d\langle B \rangle_s$$

is a martingale: it follows that $XY - A \in \mathcal{M}^{c, \text{loc}}$ with localizing sequence $(\tau_n)_{n \in \mathbb{N}}$ and therefore $A = \langle X, Y \rangle$. □

For the stochastic integral of $u \in \mathbb{L}^2_{\text{loc}}$, we no longer have a fundamental tool such as Itô's isometry: in many situations it can be conveniently replaced by the following lemma.

Lemma 10.2.25 ([!]) *Let*

$$X_t = \int_0^t u_s dB_s, \qquad \langle X \rangle_t = \int_0^t u_s^2 d\langle B \rangle_s,$$

with $u \in \mathbb{L}^2_{\text{loc}}$ and $B \in \mathcal{M}^{c,2}$. For every $t, \varepsilon, \delta > 0$ we have

$$P(|X_t| \geq \varepsilon) \leq P(\langle X \rangle_t \geq \delta) + \frac{\delta}{\varepsilon^2}.$$

Proof Let

$$\tau_\delta = \inf\{s > 0 \mid \langle X \rangle_s \geq \delta\}, \qquad \delta > 0.$$

Given $t, \varepsilon > 0$, we have

$$P(|X_t| \geq \varepsilon) = P((|X_t| \geq \varepsilon) \cap (\tau_\delta \leq t)) + P((|X_t| \geq \varepsilon) \cap (\tau_\delta > t)) \leq$$

(since $(\tau_\delta \leq t) = (\langle X \rangle_t \geq \delta)$)

$$\leq P(\langle X \rangle_t \geq \delta) + P((|X_t| \geq \varepsilon) \cap (\tau_\delta > t))$$

and therefore it remains to prove that

$$P((|X_t| \geq \varepsilon) \cap (\tau_\delta > t)) \leq \frac{\delta}{\varepsilon^2}.$$

Now we have

$$P\left(\left(\left|\int_0^t u_s dB_s\right| \geq \varepsilon\right) \cap (t < \tau_\delta)\right) = P\left(\left(\left|\int_0^t u_s \mathbb{1}_{(s < \tau_\delta)} dB_s\right| \geq \varepsilon\right) \cap (t < \tau_\delta)\right)$$

$$\leq P\left(\left|\int_0^t u_s \mathbb{1}_{(s < \tau_\delta)} dB_s\right| \geq \varepsilon\right) \leq$$

(by Chebyshev's inequality (3.1.3) in [113])

$$\leq \frac{1}{\varepsilon^2} E\left[\left|\int_0^t u_s \mathbb{1}_{(s<\tau_\delta)} dB_s\right|^2\right] =$$

(by Itô's isometry, since $u_s \mathbb{1}_{(s<\tau_\delta)} \in \mathbb{L}^2$)

$$= \frac{1}{\varepsilon^2} E\left[\int_0^t u_s^2 \mathbb{1}_{(s<\tau_\delta)} d\langle B\rangle_s\right] \leq \frac{\delta}{\varepsilon^2}.$$

□

10.2.5 Stochastic Integral as a Riemann-Stieltjes Integral

The following result shows that the stochastic integral of $u \in \mathbb{L}_{loc}^2$ can also be defined by approximation, as we did for $u \in \mathbb{L}^2$, provided that we use convergence in probability instead of in $L^2(\Omega, P)$-norm.

Proposition 10.2.26 *Let $u, u_n \in \mathbb{L}_{loc}^2$, $n \in \mathbb{N}$, such that*

$$\int_0^t |u_{n,s} - u_s|^2 d\langle B\rangle_s \xrightarrow[n\to\infty]{P} 0. \tag{10.2.26}$$

Then

$$\int_0^t u_{n,s} dB_s \xrightarrow[n\to\infty]{P} \int_0^t u_s dB_s.$$

Proof The thesis is an immediate consequence of Itô's isometry in the form of Lemma 10.2.25: fixed $\varepsilon > 0$ and setting $\delta = \varepsilon^3$, we have

$$\lim_{n\to\infty} P\left(\left|\int_0^t (u_{n,s} - u_s) dB_s\right| \geq \varepsilon\right)$$

$$\leq \lim_{n\to\infty} P\left(\int_0^t |u_{n,s} - u_s|^2 d\langle B\rangle_s \geq \delta\right) + \varepsilon = \varepsilon$$

thanks to assumption (10.2.26). □

As a simple application of Proposition 10.2.26, we prove that, in the case where the integrand is a continuous process, the stochastic integral is indeed the limit in probability of the Riemann-Stieltjes sums in which the integrand is evaluated at the *left endpoint* of each interval of the partition: this is consistent with the construction of the Itô integral, which crucially exploits the hypothesis

10.2 Integral with Respect to Continuous Square-Integrable Martingales

of progressive measurability of the integrand. The following result is also the basis of the *numerical approximation methods for the stochastic integral*.

Corollary 10.2.27 ([!]) *Let u be a continuous and adapted process, $B \in \mathcal{M}^{c,2}$, and $(\pi_n)_{n \in \mathbb{N}}$ be a sequence of partitions of $[0, t]$, with $\pi_n = (t_{n,k})_{k=0,\ldots,m_n}$, such that $\lim_{n \to \infty} |\pi_n| = 0$. Then*

$$\sum_{k=1}^{m_n} u_{t_{n,k-1}} \left(B_{t_{n,k}} - B_{t_{n,k-1}} \right) \xrightarrow[n \to \infty]{P} \int_0^t u_s dB_s.$$

Proof Setting

$$u_{n,s} = \sum_{k=1}^{m_n} u_{t_{n,k-1}} \mathbb{1}_{[t_{n,k-1}, t_{n,k}[}(s)$$

we have that $u_n \in \mathbb{L}^2_{\text{loc}}$ and

$$\sum_{k=1}^{m_n} u_{t_{n,k-1}} \left(B_{t_{n,k}} - B_{t_{n,k-1}} \right) = \int_0^t u_{n,s} dB_s.$$

Moreover, by the continuity of u and the dominated convergence theorem, we have

$$\lim_{n \to \infty} \int_0^t |u_{n,s} - u_s|^2 d\langle B \rangle_s = 0 \quad \text{a.s.}$$

The thesis follows from Proposition 10.2.26. □

A useful consequence of Corollary 10.2.27 is the following

Corollary 10.2.28 ([!]) *Assume that, for $i = 1, 2$, the process $B^i \in \mathcal{M}^{c,2}$ and the continuous adapted process u^i are defined on $(\Omega^i, \mathscr{F}^i, P^i)$. Moreover, let*

$$X_t^i = \int_0^t u_s^i dB_s^i.$$

If $(u^1, B^1) \stackrel{d}{=} (u^2, B^2)$ (i.e. (u^1, B^1) and (u^2, B^2) are equal in law) then we also $(u^1, B^1, X^1) \stackrel{d}{=} (u^2, B^2, X^2)$.

A similar result holds under much more general assumptions: in this regard, see, for example, Exercise IV.5.16 in [123].

10.3 Integral with Respect to Continuous Semimartingales

In the previous sections, we assumed that the integrating process B is a continuous square-integrable martingale. Now we extend the definition of the stochastic integral to the case where the integrator, here denoted by S, is a *continuous semimartingale*: precisely, by Definition 9.3.1, S is an adapted and continuous process of the form

$$S = A + B$$

where $A \in \mathrm{BV}$ is such that $A_0 = 0$ and $B \in \mathscr{M}^{c,\mathrm{loc}}$. We use the notation

$$\int_0^t u_r\, dS_r$$

to indicate the stochastic integral of the process u with respect to S: it is defined as the sum

$$\int_0^t u_r\, dS_r := \int_0^t u_r\, dA_r + \int_0^t u_r\, dB_r$$

where the two integrals on the right-hand side have the meaning that we now explain.

Let μ_A be the Lebesgue-Stieltjes measure[7] associated with A and defined path by path: we denote by

$$\int_0^t u_r\, dA_r := \int_{[0,t]} u_r\, \mu_A(dr)$$

the corresponding Lebesgue-Stieltjes integral. In order for this integral to be well-defined, we require that $u \in \mathbb{L}^2_{S,\mathrm{loc}}$ according to the following

Definition 10.3.1 $\mathbb{L}^2_{S,\mathrm{loc}}$ is the class of progressively measurable processes u such that

$$\int_{[0,t]} |u_r|\, |\mu_A|(dr) + \int_0^t u_r^2\, d\langle B\rangle_r < \infty \qquad \text{a.s.}$$

for every $t \geq 0$.

---

[7] According to Definition 9.2.1, μ_A is a signed measure.

As for the integral with respect to $B \in \mathcal{M}^{c,\text{loc}}$, one can use a localization procedure entirely analogous[8] to that of Sect. 10.2.4. In conclusion, recalling Definition 9.5.3 of quadratic variation of a semimartingale, we have the following

Proposition 10.3.2 *Let $S = A + B$ be a continuous semimartingale and $u \in \mathbb{L}^2_{S,\text{loc}}$. The stochastic integral process*

$$X_t := \int_0^t u_r dS_r = \int_0^t u_r dA_r + \int_0^t u_r dB_r, \qquad t \geq 0,$$

is a continuous semimartingale with quadratic variation process

$$\langle X \rangle_t = \int_0^t u_r^2 d\langle B \rangle_r, \qquad t \geq 0. \tag{10.3.1}$$

---

[8] Let $(\tau_n)_{n \in \mathbb{N}}$ be a localizing sequence for B: as in Remark 8.4.6-(iv) we can assume $|B_{t \wedge \tau_n}| \leq n$ so that $B_n := (B_{t \wedge \tau_n})_{t \geq 0} \in \mathcal{M}^{c,2}$. If $u \in \mathbb{L}^2_{S,\text{loc}}$ then

$$\int_0^t u_r^2 d\langle B_n \rangle_r \leq \int_0^t u_r^2 d\langle B \rangle_r < \infty \qquad \text{a.s.}$$

and therefore $u \in \mathbb{L}^2_{B_n,\text{loc}}$ and the integral

$$Y_{n,t} := \int_0^t u_r dB_{n,r}$$

is well-defined. On the event $F_{n,T} := (T \leq \tau_n)$ we have a.s.

$$\sup_{0 \leq t \leq T} |Y_{n,t} - Y_{m,t}| = 0, \qquad m \geq n.$$

This is true if u is simple and in general it can be proved by approximation, as Proposition 10.2.17. Since $F_{n,T} \nearrow F_T$ with $P(F_T) = 1$, we define the integral

$$Y_t = \int_0^t u_r dB_r, \qquad 0 \leq t \leq T,$$

as the equivalence class of continuous and adapted processes that, for each $n \in \mathbb{N}$, are indistinguishable from $(Y_{n,t})_{t \in [0,T]}$ on $F_{n,T}$. If Y and $\bar{Y}$ indicate respectively the stochastic integral processes of u on the intervals $[0, T]$ and $[0, \bar{T}]$ with $T \leq \bar{T}$, then Y and $\bar{Y}|_{[0,T]}$ are indistinguishable on $[0, T]$. Therefore, the Itô stochastic integral process of $u \in \mathbb{L}^2_{S,\text{loc}}$ with respect to $B \in \mathcal{M}^{c,\text{loc}}$ is well defined:

$$Y_t = \int_0^t u_r dB_r, \qquad t \geq 0.$$

We have $Y \in \mathcal{M}^{c,\text{loc}}$ with quadratic variation process

$$\langle Y \rangle_t = \int_0^t u_r^2 d\langle B \rangle_r, \qquad t \geq 0,$$

and a localizing sequence for Y is given by $\bar{\tau}_n = \tau_n \wedge \tau_n'$ where $\tau_n' = \inf\{t \geq 0 \mid \langle I \rangle_t \geq n\}$.

In the next section, we deal with the particular case where $A_t = t$ and B is a Brownian motion.

10.4 Scalar Itô Processes

An Itô process is specific type of continuous semimartingale, which can be expressed as the sum of a Lebesgue integral and a stochastic integral. In this section, W denotes a real Brownian motion.

Definition 10.4.1 (Itô Process [!]) An *Itô process* is a process of the form

$$X_t = X_0 + \int_0^t u_s ds + \int_0^t v_s dW_s, \tag{10.4.1}$$

where:

(i) $X_0 \in m\mathscr{F}_0$;
(ii) $u \in \mathbb{L}^1_{\text{loc}}$, that is, u is progressively measurable and such that

$$\int_0^t |u_s| ds < \infty, \qquad \text{a.s.}$$

for any $t \geq 0$;
(iii) $v \in \mathbb{L}^2_{\text{loc}}$, that is, v is progressively measurable and such that[9]

$$\int_0^t |v_s|^2 ds < \infty \qquad \text{a.s.}$$

for any $t \geq 0$.

Notation 10.4.2 (Differential Notation [[!])] To indicate the Itô process in (10.4.1), the so-called "differential notation" is often used:

$$dX_t = u_t dt + v_t dW_t. \tag{10.4.2}$$

This notation, in addition to being more compact, has the merit of evoking the expressions of classical differential calculus. In rigorous terms, dX_t is neither a "derivative" nor a "differential of the process X". These terms have not been defined; rather, it is a symbol that holds significance solely within the context of expression (10.4.2): such expression, in turn, is a writing whose precise meaning is given by the integral equation (10.4.1). When we talk about *stochastic differential calculus*, we refer to this type of symbolic calculation whose true meaning is

---

[9] Remember that $\langle W \rangle_s = s$.

10.4 Scalar Itô Processes

given by the corresponding integral expressions: therefore, it is actually a *stochastic integral* calculus.

The process in (10.4.1) is a continuous semimartingale and can therefore act as an integrator itself, in fact, we have $X = A + M$ where:

- the process

$$A_t := \int_0^t u_s ds$$

is continuous, adapted, and of bounded variation according to Example 9.1.2-(iv), and is called the *drift* of X;
- the stochastic integral process

$$M_t := X_0 + \int_0^t v_s dW_s$$

is a continuous local martingale and is called the *diffusive part or diffusion* of X.

By formula (10.3.1), the quadratic variation process of X is

$$\langle X \rangle_t = \int_0^t v_s^2 ds,$$

or, in differential notation,

$$d\langle X \rangle_t = v_t^2 dt.$$

Remark 10.4.3 ([!]) *The representation of an Itô process is unique* in the following sense: if X is the process in (10.4.2) and we also have

$$dX_t = u'_t dt + v'_t dW_t,$$

with $u' \in \mathbb{L}^1_{\text{loc}}$ and $v' \in \mathbb{L}^2_{\text{loc}}$, then

$$P\left(v = v' \text{ a.e.}\right) = P\left(u = u' \text{ a.e.}\right) = 1.$$

In particular, if u, u', v, v' are continuous, then u is indistinguishable from u' and v is indistinguishable from v'.

Indeed, the process

$$M_t := \int_0^t v_s dW_s - \int_0^t v'_s dW_s = \int_0^t u'_s ds - \int_0^t u_s ds$$

is a continuous local martingale, of bounded variation, which, by Theorem 9.3.6, is indistinguishable from the null process. Consider

$$\tau_n := n \wedge \inf\{t \geq 0 \mid A_t \geq n\}, \qquad A_t := \int_0^t (v_s - v'_s)^2 ds, \qquad n \in \mathbb{N},$$

the usual localizing sequence for M. Then we have

$$0 = E\left[\left(\int_0^{\tau_n} (v_s - v'_s) dW_s\right)^2\right] = E\left[\left(\int_0^n (v_s - v'_s) \mathbb{1}_{[0,\tau_n]}(s) dW_s\right)^2\right]$$

$$= E\left[\int_0^n (v_s - v'_s)^2 \mathbb{1}_{[0,\tau_n]}(s) ds\right]$$

where the second and third equalities are due respectively to Proposition 10.2.23 and Itô's isometry. Taking the limit as $n \to \infty$, by Beppo Levi's theorem, we have

$$E\left[\int_0^\infty (v_s - v'_s)^2 ds\right] = 0$$

and therefore $P\left(v = v' \text{ a.e.}\right) = 1$. On the other hand, by Proposition B.3.2 in [113], we also have that

$$P\left(u = u' \text{ a.e.}\right) = 1.$$

10.5 Key Ideas to Remember

We summarize the contents of the chapter and provide a roadmap for reading, glossing over technical and secondary aspects. As usual, if you have any doubt about what the following succinct statements mean, please review the corresponding section.

- Section 10.1: when approaching these topics for the first time, it is preferable to select some content and postpone the general treatment and in-depth studies to a later time. In particular, it is best to first consider only the case where the integrator is a Brownian motion. As for the integrand, the crucial assumption is that it is a progressively measurable process; the construction of the Brownian integral takes place in three steps, gradually widening the class of integrands:

  (1) the definition of the integral of *simple* processes is explicit: it is a Riemann sum of Brownian increments. In this case, three fundamental properties of the integral are directly proven:

10.5 Key Ideas to Remember

   (i) it is a continuous martingale;
   (ii) Ito's isometry;
   (iii) there is an explicit expression for the quadratic variation process;

   (2) the stochastic integral extends by density to integrands in $\mathbb{L}^2$. The three fundamental properties remain valid;
   (3) with a localization procedure using stopping times (which stop the quadratic variation process when it exceeds some level), the stochastic integral extends to integrands in the much wider class $\mathbb{L}^2_{\text{loc}}$. In this case, the first two fundamental properties are lost, or rather, they remain valid in a weakened form.

- Section 10.2: the construction of the stochastic integral extends to the case where the integrator process is in $\mathcal{M}^{c,2}$ and essentially analogous properties to those of the Brownian integral hold. The integration endpoint can also be random, provided it is a stopping time (see Proposition 10.2.23).
- Section 10.3: we further extend the definition of the stochastic integral to the case where the integrator is a continuous semimartingale.
- Section 10.4: an Itô process is a particular continuous semimartingale that is the sum of a Lebesgue integral with an integrand in $\mathbb{L}^1_{\text{loc}}$ (drift term) and a Brownian integral with an integrand in $\mathbb{L}^2_{\text{loc}}$ (diffusive part): in differential notation, it is written as $dX_t = u_t dt + v_t dW_t$. The decomposition of an Itô process into drift and diffusive parts is unique and the quadratic variation process is $d\langle X \rangle_t = v_t^2 dt$.

Main notations used or introduced in this chapter:

Symbol	Description	Page
$\int_0^t u_s dB_s$	Stochastic integral with integrand u and integrator B	175
$\mathbb{L}^2$	Progressively measurable processes in $L^2(\Omega \times [0, T])$	176
$\mathbb{L}^2_{\text{loc}}$	Progressively measurable processes in $L^2([0, T])$ a.s.	182
$\mathcal{M}^{c,2}$	Space of continuous martingales X, with $X_t \in L^2(\Omega, P)$ for any t	183
$\mu_{\langle B \rangle}$	Lebesgue-Stieltjes measure of the increasing process $\langle B \rangle$	184
$\int_a^b f(t) d\langle B \rangle_t$	Lebesgue-Stieltjes integral with respect to the increasing process $\langle B \rangle$	183
$\mathbb{L}^2_B$	Progressively measurable processes in $L^2(\Omega \times [0, T], P \otimes \mu_{\langle B \rangle})$	184
$\mathbb{L}^2_{B,\text{loc}}$	Progressively measurable processes in $L^2([0, T], \mu_{\langle B \rangle})$ a.s.	184
$\mathcal{M}^{c,2}_T$	Continuous square-integrable martingales	190
$\|X\|_T = \sqrt{E[X_T^2]}$	Norm in $\mathcal{M}^{c,2}_T$	190
$\mathbb{L}^1_{\text{loc}}$	Progressively measurable processes in $L^1([0, T])$ a.s.	204

Chapter 11
Itô's Formula

> *To put meaning in one's life may end in madness, But life without meaning is the torture Of restlessness and vague desire- It is a boat longing for the sea and yet afraid.*
>
> *Edgar Lee Master*

Ito's formula is the most important tool in stochastic differential calculus. In this chapter, we present several versions that provide the general rules of stochastic calculus and generalize the analogous deterministic formula of Theorem 9.1.6 for the Lebesgue-Stieltjes integral.

11.1 Itô's Formula for Continuous Semimartingales

Although the case of semimartingales is very general, we immediately give this version of Itô's formula because it has the advantage of having a compact expression and an intuitive proof. Recall that a continuous semimartingale is an adapted and continuous process of the form $X = A+M$ with $A \in \mathrm{BV}$ such that $A_0 = 0$ and $M \in \mathscr{M}^{c,\mathrm{loc}}$, that is, M is a continuous local martingale according to Definition 8.4.2.

We denote by $\langle X \rangle$ the *quadratic variation process* of X: by Theorem 9.4.1, we have $\langle X \rangle \equiv \langle M \rangle$ where $\langle M \rangle$ is the unique continuous increasing process such that $\langle M \rangle_0 = 0$ and $M^2 - \langle M \rangle$ is a local martingale. For example, if X is a Brownian motion then $A \equiv 0$ and the quadratic variation process is deterministic: $\langle X \rangle_t = t$ for $t \geq 0$. More generally, if X is an Itô process of the form $dX_t = u_t dt + v_t dW_t$ (cf. Definition 10.4.1) then $d\langle X \rangle_t = v_t^2 dt$.

Theorem 11.1.1 (Itô's Formula [!!!]) *Let X be a continuous real semimartingale and $F \in C^2(\mathbb{R})$. Then almost surely, for every $t \geq 0$ we have*

$$F(X_t) = F(X_0) + \int_0^t F'(X_s)dX_s + \frac{1}{2}\int_0^t F''(X_s)d\langle X \rangle_s \qquad (11.1.1)$$

or, in differential notation,

$$dF(X_t) = F'(X_t)dX_t + \frac{1}{2}F''(X_t)d\langle X\rangle_t. \tag{11.1.2}$$

Idea of the Proof Given a partition $\pi = \{t_0, \ldots, t_N\}$ of $[0, t]$, we write the difference $F(X_t) - F(X_0)$ as a telescoping sum and then expand it in a Taylor series up to the second order: we obtain

$$F(X_t) - F(X_0) = \sum_{k=1}^{N} \left(F(X_{t_k}) - F(X_{t_k})\right)$$

$$= \sum_{k=1}^{N} F'(X_{t_{k-1}})\left(X_{t_k} - X_{t_{k-1}}\right) + \frac{1}{2}\sum_{k=1}^{N} F''(X_{t_{k-1}})\left(X_{t_k} - X_{t_{k-1}}\right)^2$$

$$+ \text{"remainder"}.$$

Finally, we prove that, in an appropriate sense, the limits exist

$$\sum_{k=1}^{N} F'(X_{t_{k-1}})\left(X_{t_k} - X_{t_{k-1}}\right) \longrightarrow \int_0^t F'(X_s)dX_s,$$

$$\sum_{k=1}^{N} F''(X_{t_{n,k-1}})\left(X_{t_k} - X_{t_{k-1}}\right)^2 \longrightarrow \int_0^t F''(X_s)d\langle X\rangle_s$$

as $|\pi| \to 0$ and the remainder term is negligible. The detailed proof, which involves more technical intricacies, is presented in Sect. 11.3.

Remark 11.1.2 Compared to the deterministic version (9.1.3), in Itô's formula (11.1.2) an additional second-order term appears, which comes from the quadratic variation of X: the factor $\frac{1}{2}$ appearing in front of it is the coefficient of the Taylor series expansion of F.

Likewise, we establish a more comprehensive version of Itô's formula.

Theorem 11.1.3 (Itô's Formula) *Let X be a continuous real semimartingale and $F = F(t, x) \in C^{1,2}(\mathbb{R}_{\geq 0} \times \mathbb{R})$. Then almost surely, for every $t \geq 0$ we have*

$$F(t, X_t) = F(0, X_0) + \int_0^t (\partial_t F)(s, X_s)ds + \int_0^t (\partial_x F)(s, X_s)dX_s$$

$$+ \frac{1}{2}\int_0^t (\partial_{xx} F)(s, X_s)d\langle X\rangle_s$$

11.1 Itô's Formula for Continuous Semimartingales

or, in differential notation,

$$dF(t, X_t) = \partial_t F(t, X_t)dt + (\partial_x F)(t, X_t)dX_t + \frac{1}{2}(\partial_{xx}F)(t, X_t)d\langle X\rangle_t.$$

11.1.1 Itô's Formula for Brownian Motion

We consider Itô's formula for a real Brownian motion W and delve into several illustrative examples. Recall that the quadratic variation process of W is simply $\langle W\rangle_t = t$.

Corollary 11.1.4 (Itô's Formula for Brownian Motion) *For every $F = F(t, x) \in C^{1,2}(\mathbb{R}_{\geq 0} \times \mathbb{R})$ we have*

$$F(t, W_t) = F(0, W_0) + \int_0^t (\partial_t F)(s, W_s)ds + \int_0^t (\partial_x F)(s, W_s)dW_s$$
$$+ \frac{1}{2}\int_0^t (\partial_{xx}F)(s, W_s)ds$$

or, in differential notation,

$$dF(t, W_t) = \left(\partial_t F + \frac{1}{2}\partial_{xx}F\right)(t, W_t)dt + (\partial_x F)(t, W_t)dW_t.$$

Example 11.1.5

(i) if $F(t, x) = f(t)x$, with $f \in C^1(\mathbb{R})$, we have

$$\partial_t F(t, x) = f'(t)x, \qquad \partial_x F(t, x) = f(t), \qquad \partial_{xx} F(t, x) = 0.$$

Then we have

$$f(t)W_t = \int_0^t f'(s)W_s ds + \int_0^t f(s)dW_s$$

which corresponds to the deterministic integration by parts formula of Example 9.1.8-(ii). In differential form, we equivalently have

$$d(f(t)W_t) = f'(t)W_t dt + f(t)dW_t$$

which resembles the usual formula for the derivation of a product;

(ii) if $F(t, x) = x^2$ we have

$$\partial_t F(t, x) = 0, \qquad \partial_x F(t, x) = 2x, \qquad \partial_{xx} F(t, x) = 2,$$

and therefore

$$W_t^2 = 2 \int_0^t W_s dW_s + t$$

or, in differential form,

$$dW_t^2 = 2W_t dW_t + dt;$$

(iii) if $F(t, x) = e^{at+\sigma x}$, with $a, \sigma \in \mathbb{R}$, we have

$$\partial_t F(t, x) = aF(t, x), \qquad \partial_x F(t, x) = \sigma F(t, x), \qquad \partial_{xx} F(t, x) = \sigma^2 F(t, x),$$

and therefore, setting $X_t = e^{at+\sigma W_t}$, we obtain

$$X_t = 1 + a \int_0^t X_s ds + \sigma \int_0^t X_s dW_s + \frac{\sigma^2}{2} \int_0^t X_s ds$$

or, in

$$dX_t = \left(a + \frac{\sigma^2}{2}\right) X_t dt + \sigma X_t dW_t.$$

With the choice $a = -\frac{\sigma^2}{2}$, the drift of the process vanishes, and we obtain

$$X_t = 1 + \int_0^t \sigma X_s dW_s$$

which is a continuous martingale: specifically, $X_t = e^{\sigma W_t - \frac{\sigma^2}{2}t}$ is the *exponential martingale* introduced in Proposition 4.4.1.

Remark 11.1.6 ([!]) Itô's formula shows that every stochastic process of the form $X_t = F(t, W_t)$, with F sufficiently regular, is an Itô process according to Definition 10.4.1: in particular, X is a semimartingale, and Itô's formula provides the explicit expression of the decomposition (unique up to indistinguishable processes) of X into the sum $X = A + M$ where the process of bounded variation

$$A_t := \int_0^t \left(\partial_t F + \frac{1}{2} \partial_{xx} F\right)(s, W_s) ds$$

11.1 Itô's Formula for Continuous Semimartingales

is the *drift* of X and the local martingale[1]

$$M_t := X_0 + \int_0^t (\partial_x F)(s, W_s) dW_s$$

is the *diffusive part* of X.

Note that if F solves the heat equation

$$\partial_t F(t, x) + \frac{1}{2} \partial_{xx} F(t, x) = 0, \qquad t > 0, \ x \in \mathbb{R}, \tag{11.1.3}$$

then the drift of X vanishes and therefore X is a local martingale. Conversely, if X is a local martingale then by Remark 10.4.3 we have that

$$(\partial_t F + \frac{1}{2} \partial_{xx} F)(t, W_t) = 0 \tag{11.1.4}$$

in the sense of indistinguishability and this implies[2] that F solves the heat equation (11.1.3).

11.1.2 Itô's Formula for Itô Processes

Let X be an Itô process of the form

$$dX_t = \mu_t dt + \sigma_t dW_t \tag{11.1.5}$$

with $\mu \in \mathbb{L}^1_{\text{loc}}$ and $\sigma \in \mathbb{L}^2_{\text{loc}}$. In Sect. 10.4 we saw that X is a continuous semimartingale with quadratic variation process

$$\langle X \rangle_t = \int_0^t \sigma_s^2 ds$$

that is, $d\langle X \rangle_t = \sigma_t^2 dt$. Hence we have the following further version of Itô's formula.

---

[1] We find here the result of Theorem 4.4.3, proven in the context of Markov process theory!

[2] The *stochastic* equation (11.1.4) is equivalent to the *deterministic* equation (11.1.3): just observe that if f is a continuous function such that $f(W_t) = 0$ a.s. for a $t > 0$ then $f \equiv 0$: in fact, if it were $f(\bar{x}) > 0$ for a $\bar{x} \in \mathbb{R}$ then we would also have $f(x) > 0$ for $|x - \bar{x}| < r$ for some $r > 0$ sufficiently small; this leads to a contradiction since, the Gaussian density being strictly positive, we would have

$$0 < E\left[f(W_t) \mathbb{1}_{(|W_t - \bar{x}| < r)}\right] = 0.$$

Corollary 11.1.7 (Itô's Formula for Itô Processes) *Let X be the Itô process in* (11.1.5). *For each $F = F(t, x) \in C^{1,2}(\mathbb{R}_{\geq 0} \times \mathbb{R})$ we have*

$$F(t, X_t) = F(0, X_0) + \int_0^t (\partial_t F)(s, X_s)ds + \int_0^t (\partial_x F)(s, X_s)dX_s$$
$$+ \frac{1}{2} \int_0^t (\partial_{xx} F)(s, X_s)\sigma_s^2 ds \qquad (11.1.6)$$

or equivalently

$$dF(t, X_t) = \left(\partial_t F + \mu_t \partial_x F + \frac{\sigma_t^2}{2} \partial_{xx} F \right)(t, X_t)dt + \sigma_t \partial_x F(t, X_t)dW_t.$$

Example 11.1.8 ([!!]) Let us calculate the stochastic differential of the process

$$Y_t = e^{t \int_0^t W_s dW_s}.$$

First of all, we notice that we cannot use Itô's formula for Brownian motion from Corollary 11.1.4 *because Y_t is not a function of W_t but depends on $(W_s)_{s \in [0,t]}$, that is, on the entire trajectory of W in the interval $[0, t]$*. The general criterion to correctly apply Itô's formula is to first analyze how Y_t depends on the variable t, distinguishing the "deterministic" from the "stochastic" dependence: in this example, we highlight in bold the deterministic dependence

$$\mathbf{t} \mapsto \exp\left(\mathbf{t} \int_0^t W_s dW_s \right)$$

and the stochastic dependence

$$\mathbf{t} \mapsto \exp\left(t \int_0^{\mathbf{t}} W_s dW_s \right)$$

to establish that

$$Y_t = F(t, X_t), \qquad F(t, x) = e^{tx}, \qquad X_t = \int_0^t W_s dW_s,$$

and therefore $dX_t = W_t dW_t$ and $d\langle X \rangle_t = W_t^2 dt$. Then we can apply Itô's formula (11.1.6): since

$$\partial_t F(t, x) = x F(t, x), \qquad \partial_x F(t, x) = t F(t, x), \qquad \partial_{xx} F(t, x) = t^2 F(t, x),$$

11.1 Itô's Formula for Continuous Semimartingales

we get

$$dY_t = \left(X_t + \frac{(tW_t)^2}{2}\right) Y_t dt + tW_t Y_t dW_t.$$

Example 11.1.9 ([!]) Consider an Itô process with *deterministic* coefficients

$$X_t = x + \int_0^t \mu(s)ds + \int_0^t \sigma(s)dW_s$$

with $x \in \mathbb{R}$, $\mu \in L^1_{\text{loc}}(\mathbb{R}_{\geq 0})$ and $\sigma \in L^2_{\text{loc}}(\mathbb{R}_{\geq 0})$. As an application of Itô's formula (11.1.6), we prove that

$$X_t \sim \mathcal{N}_{m(t),\mathscr{C}(t)}, \qquad m(t) := x + \int_0^t \mu(s)ds, \qquad \mathscr{C}(t) := \int_0^t \sigma^2(s)ds,$$

for every $t \geq 0$. In fact, we can easily calculate the characteristic function of X: first, for every $\eta \in \mathbb{R}$ we have

$$de^{i\eta X_t} = e^{i\eta X_t}\left(i\eta dX_t - \frac{\eta^2}{2}d\langle X \rangle_t\right)$$

$$= e^{i\eta X_t}\left(a(t,\eta)dt + i\eta\sigma(t)dW_t\right), \qquad a(t,\eta) := i\eta\mu(t) - \frac{\eta^2\sigma^2(t)}{2}.$$

Applying the expected value and being null the expectation of the stochastic integral, we have

$$\varphi_{X_t}(\eta) = e^{i\eta x} + E\left[\int_0^t a(s,\eta)e^{i\eta X_s}ds\right]$$

$$= e^{i\eta x} + \int_0^t a(s,\eta)\varphi_{X_s}(\eta)ds;$$

equivalently, $t \mapsto \varphi_{X_t}(\eta)$ solves the Cauchy problem

$$\begin{cases} \frac{d}{dt}\varphi_{X_t}(\eta) = a(t,\eta)\varphi_{X_t}(\eta), \\ \varphi_{X_0}(\eta) = e^{i\eta x}, \end{cases}$$

so that

$$\varphi_{X_t}(\eta) = e^{i\eta m(t) - \frac{\eta^2}{2}\mathscr{C}(t)}$$

and this proves the thesis.

Example 11.1.10 ([!]) Given

$$X_t := \int_0^t W_s ds \tag{11.1.7}$$

we have $X_t \sim \mathcal{N}_{0,\frac{t^3}{3}}$. In fact, by Itô's formula, we have

$$d(tW_t) = t dW_t + W_t dt$$

that is

$$X_t = tW_t - \int_0^t s dW_s = \int_0^t (t-s) dW_s.$$

We note that the expression of X in (11.1.7) is that of an Itô process, while

$$\int_0^t (t-s) dW_s$$

is not written in the form of an Itô process: to circumvent this problem, we define the Itô process

$$Y_t^{(a)} := \int_0^t (a-s) dW_s$$

dependent on the parameter $a \in \mathbb{R}$. We know that

$$Y_t^{(a)} \sim \mathcal{N}_{0,\frac{t^3}{3}+at(a-t)}$$

and the thesis follows from the fact that $X_t = Y_t^{(t)}$.

11.2 Some Consequences of Itô's Formula

11.2.1 Burkholder-Davis-Gundy Inequalities

We prove some classical inequalities that are a basic tool in the study of martingales and stochastic differential equations.

Theorem 11.2.1 (Burkholder-Davis-Gundy [!]) *Let X be a continuous local martingale such that $X_0 = 0$ a.s. and τ an a.s. finite stopping time (i.e., such that*

11.2 Some Consequences of Itô's Formula

$\tau < \infty$ a.s.). *For every $p > 0$ there exist two positive constants c_p, C_p such that*

$$c_p E\left[\langle X\rangle_\tau^{p/2}\right] \le E\left[\sup_{t\in[0,\tau]} |X_t|^p\right] \le C_p E\left[\langle X\rangle_\tau^{p/2}\right]. \tag{11.2.1}$$

In (11.2.1), $\langle X\rangle$ denotes the quadratic variation process of X.

Proof We only prove the case $p \ge 2$ in which it is possible to give an elementary proof based on Itô's formula. For the general case, see, for example, Proposition 3.26 in [67]. The case $p = 2$ follows from Itô's isometry (9.4.1) and therefore it is sufficient to consider $p > 2$.

We begin by proving the second inequality. It is not restrictive to assume $E\left[\langle X\rangle_\tau^{p/2}\right] > 0$ otherwise there is nothing to prove. Let

$$\bar{X}_\tau = \sup_{t\in[0,\tau]} |X_t|$$

and assume for the moment that $\bar{X}_\tau \le n$ a.s. for some $n \in \mathbb{N}$. Then, by Doob's maximal inequality, Corollary 8.1.3, we have

$$E\left[\bar{X}_\tau^p\right] \le c_p E\left[|X_\tau|^p\right] =$$

(by Itô's formula, noting that the function $x \mapsto |x|^p$ is of class C^2 since $p \ge 2$)

$$= c_p E\left[\int_0^\tau p|X_t|^{p-1} dX_t\right] + \frac{c_p}{2} E\left[\int_0^\tau p(p-1)|X_t|^{p-2} d\langle X\rangle_t\right] =$$

(since the first term is null because the stochastic integral is a martingale, given the boundedness assumption of $\bar{X}_\tau$)

$$= c'_p E\left[\int_0^\tau |X_t|^{p-2} d\langle X\rangle_t\right]$$

$$\le c'_p E\left[\int_0^\tau \bar{X}_\tau^{p-2} d\langle X\rangle_t\right]$$

$$= c'_p E\left[\bar{X}_\tau^{p-2}\langle X\rangle_\tau\right] \le$$

(by Hölder's inequality with exponents $\frac{p}{p-2}$ and $\frac{p}{2}$)

$$\le c'_p E\left[\bar{X}_\tau^p\right]^{\frac{p-2}{p}} E\left[\langle X\rangle_\tau^{p/2}\right]^{\frac{2}{p}}$$

and from this inequality, the thesis easily follows. To remove the boundedness assumption, it is sufficient to apply the result just proved to the stopping time

$\tau_n = \inf\{t \geq 0 \mid |X_t| \geq n\} \wedge \tau$ and then take the limit as $n \to \infty$ using Beppo Levi's theorem.

Let us now prove the first inequality: with the usual localization argument based on Beppo Levi's theorem, it is not restrictive to assume that τ, $\bar{X}_\tau$ and $\langle X \rangle_\tau$ are bounded by a positive constant. We also assume $E\left[\bar{X}_\tau^p\right] > 0$ otherwise there is nothing to prove. Let $r = \frac{p}{2} > 1$ and $A = \langle X \rangle$. By the deterministic Itô's formula, Theorem 9.1.6 and formula (9.1.4), we have

$$dA_t^r = rA_t^{r-1}dA_t,$$

$$dA_t^r = d\left(A_t A_t^{r-1}\right) = A_t dA_t^{r-1} + A_t^{r-1}dA_t,$$

and inserting the first into the second equality we get

$$dA_t^r = A_t dA_t^{r-1} + \frac{1}{r}dA_t^r$$

that is

$$(r-1)A_\tau^r = r\int_0^\tau A_t dA_t^{r-1}.$$

Since also

$$A_\tau^r = A_\tau \int_0^\tau dA_t^{r-1} = \int_0^\tau A_\tau dA_t^{r-1},$$

we finally obtain

$$A_\tau^r = r\int_0^\tau (A_\tau - A_t) dA_t^{r-1}.$$

Then we have

$$E\left[A_\tau^r\right] = rE\left[\int_0^\tau (A_\tau - A_t) dA_t^{r-1}\right] =$$

(by Proposition 9.2.3 and since $A_t = E[A_t \mid \mathscr{F}_t]$)

$$= rE\left[\int_0^\tau E[A_\tau - A_t \mid \mathscr{F}_t] dA_t^{r-1}\right] =$$

(by (9.4.1) and (1.4.3) (see also Remark 9.4.4), remembering the notation $A = \langle X \rangle$)

$$= rE\left[\int_0^\tau E\left[X_\tau^2 - X_t^2 \mid \mathscr{F}_t\right] d\langle X \rangle_t^{r-1}\right]$$

11.2 Some Consequences of Itô's Formula

$$\le rE\left[\int_0^\tau E\left[\bar{X}_\tau^2 \mid \mathscr{F}_t\right] d\langle X\rangle_t^{r-1}\right] =$$

(again by Proposition 9.2.3)

$$= rE\left[\int_0^\tau \bar{X}_\tau^2 d\langle X\rangle_t^{r-1}\right] = rE\left[\bar{X}_\tau^2 \langle X\rangle_\tau^{r-1}\right].$$

To conclude, just apply Hölder's inequality with exponents r, $\frac{r}{r-1}$ and finally divide by $E\left[\langle X\rangle_\tau^r\right]^{\frac{r-1}{r}}$. □

We have the following immediate

Corollary 11.2.2 ([!]) *Let $\sigma \in \mathbb{L}^2$ and W be a real Brownian motion. For every $p \ge 2$ and $T > 0$ we have*

$$E\left[\sup_{0\le t\le T}\left|\int_0^t \sigma_s dW_s\right|^p\right] \le c_p T^{\frac{p-2}{2}} E\left[\int_0^T |\sigma_s|^p ds\right] \tag{11.2.2}$$

where c_p is a positive constant that depends only on p.

Proof It is enough[3] to consider $p > 2$. Applying the Burkholder-Davis-Gundy inequality to the continuous martingale

$$X_t = \int_0^t \sigma_s dW_s,$$

we obtain

$$E\left[\sup_{0\le t\le T} |X_t|\right] \le c_p E\left[\langle X\rangle_T^{p/2}\right] = c_p E\left[\left(\int_0^T \sigma_t^2 dt\right)^{p/2}\right].$$

The thesis follows by applying Hölder's inequality with exponents $\frac{p}{2}$ and $\frac{p}{p-2}$. □

Remark 11.2.3 Assume $p > 4$ and

$$X_t := \int_0^t \sigma_s dW_s \quad \text{with} \quad E\left[\int_0^T |\sigma_s|^p ds\right] < \infty.$$

Combining estimate (11.2.2) with Kolmogorov's continuity theorem, we have that the integral process X admits a version with α-Hölder continuous trajectories for every $\alpha \in [0, \frac{1}{2} - \frac{2}{p}[$.

---

[3] The case $p = 2$ corresponds to Itô's isometry.

11.2.2 Quadratic Variation Process

We prove formula (9.4.2) that we left pending.

Proposition 11.2.4 *Let X be a continuous local martingale with quadratic variation process $\langle X \rangle$. We have*

$$\langle X \rangle_t = \lim_{n \to \infty} \sum_{k=1}^{2^n} \left(X_{\frac{tk}{2^n}} - X_{\frac{t(k-1)}{2^n}} \right)^2, \qquad t \geq 0,$$

in probability. Moreover, if $S = A + X$ is a continuous semimartingale, with $A \in BV$ and $X \in \mathcal{M}^{c,loc}$, we have

$$\lim_{n \to \infty} \sum_{k=1}^{2^n} \left(S_{\frac{tk}{2^n}} - S_{\frac{t(k-1)}{2^n}} \right)^2 = \langle X \rangle_t, \qquad t \geq 0, \qquad (11.2.3)$$

in probability.

Proof As usual, we denote by $t_{n,k} = \frac{tk}{2^n}$, $k = 0, \ldots, 2^n$, the dyadic rationals of the interval $[0, t]$. We first assume that X is a bounded continuous local martingale, $|X| \leq K$ with K positive constant. Given $n \in \mathbb{N}$ and $k \in \{1, \ldots, 2^n\}$, we consider the process

$$Y_s := X_s - X_{t_{n,k-1}}, \qquad s \geq t_{n,k-1},$$

and observe that $\langle Y \rangle_s = \langle X \rangle_s - \langle X \rangle_{t_{n,k-1}}$: indeed, it is enough to observe that

$$Y_s^2 - \left(\langle X \rangle_s - \langle X \rangle_{t_{n,k-1}} \right) = X_s^2 - \langle X \rangle_s + M_s, \qquad M_s := -2 X_s X_{t_{n,k-1}} + X_{t_{n,k-1}}^2 + \langle X \rangle_{t_{n,k-1}},$$

and it is easily verified that $(M_s)_{s \geq t_{n,k-1}}$ is a martingale. Applying Itô's formula, we have

$$dY_s^2 = 2 Y_s dY_s + d\langle Y \rangle_s$$

and in integral form over $[t_{n,k}, t_{n,k-1}]$

$$\left(X_{t_{n,k}} - X_{t_{n,k-1}} \right)^2 = 2 \int_{t_{n,k-1}}^{t_{n,k}} \left(X_s - X_{t_{n,k-1}} \right) dY_s + \langle X \rangle_{t_{n,k}} - \langle X \rangle_{t_{n,k-1}}$$

that is

$$\left(X_{t_{n,k}} - X_{t_{n,k-1}} \right)^2 - \left(\langle X \rangle_{t_{n,k}} - \langle X \rangle_{t_{n,k-1}} \right) = 2 \int_{t_{n,k-1}}^{t_{n,k}} \left(X_s - X_{t_{n,k-1}} \right) dY_s.$$

11.3 Proof of Itô's Formula

Summing over k, we obtain

$$R_n := \sum_{k=1}^{2^n} \left(X_{t_{n,k}} - X_{t_{n,k-1}}\right)^2 - \langle X \rangle_t = 2 \sum_{k=1}^{2^n} \int_{t_{n,k-1}}^{t_{n,k}} \left(X_s - X_{t_{n,k-1}}\right) dY_s.$$

Thanks to the Itô isometry in the form (10.2.12) and (10.2.13) (also remember the Theorem 10.2.15), we have

$$E\left[R_n^2\right] = 4 \sum_{k=1}^{2^n} E\left[\int_{t_{n,k-1}}^{t_{n,k}} \left(X_s - X_{t_{n,k-1}}\right)^2 d\langle Y \rangle_s\right]$$

$$= 4E\left[\int_0^t \sum_{k=1}^{2^n} \left(X_s - X_{t_{n,k-1}}\right)^2 \mathbb{1}_{[t_{n,k-1}, t_{n,k}]}(s) d\langle Y \rangle_s\right]$$

and taking the limit, by the dominated convergence theorem, we have $\lim_{n \to \infty} E\left[R_n^2\right] = 0$. Therefore, in this particular case, we prove the convergence in L^2 norm which obviously implies convergence in probability.

To remove the boundedness assumption on X, it is sufficient to use a localization argument proving the thesis for the bounded martingale $X_{t \wedge \tau_n}$, with

$$\tau_n = t \wedge \inf\{s \geq 0 \mid |X_s| \geq n, \ \langle X \rangle_s \geq n, \ V_s(A) \geq n\}, \qquad n \in \mathbb{N},$$

to then let n tend to infinity: with this procedure, we can prove convergence in probability. The proof of (11.2.3) is similar and is omitted. □

11.3 Proof of Itô's Formula

We prove Theorem 11.1.1. Let $X = A + M$ be a continuous real-valued semimartingale where A is an adapted, continuous, and locally of bounded variation process and $M \in \mathscr{M}^{c,\text{loc}}$. In Theorem 9.4.1 we defined the quadratic variation process $\langle M \rangle$ as the unique (up to indistinguishability) adapted, continuous, increasing process such that $\langle M \rangle_0 = 0$ and $M^2 - \langle M \rangle \in \mathscr{M}^{c,\text{loc}}$. Moreover, if M is square-integrable, i.e., $M \in \mathscr{M}^{c,2}$, then we have the important identities

$$E\left[(M_t - M_s)^2 \mid \mathscr{F}_s\right] = E\left[M_t^2 - M_s^2 \mid \mathscr{F}_s\right] \qquad (11.3.1)$$

$$= E\left[\langle M \rangle_t - \langle M \rangle_s \mid \mathscr{F}_s\right], \qquad 0 \leq s \leq t. \qquad (11.3.2)$$

Even though it is a calculation we have already done, it is useful to remember that (11.3.1) simply comes from

$$E\left[(M_t - M_s)^2 \mid \mathscr{F}_s\right] = E\left[M_t^2 - 2M_t M_s + M_s^2 \mid \mathscr{F}_s\right]$$
$$= E\left[M_t^2 \mid \mathscr{F}_s\right] - 2M_s E\left[M_t \mid \mathscr{F}_s\right] + M_s^2 =$$

(by the martingale property of M)

$$= E\left[M_t^2 \mid \mathscr{F}_s\right] - M_s^2.$$

Instead, (11.3.2) is equivalent to the martingale property of $M^2 - \langle M \rangle$. The proof of Itô's formula is essentially based on these two identities. Another ingredient is the uniform estimate (9.6.3) of the L^2 norm of the quadratic variation of M on the dyadics.

We divide the proof of Theorem 11.1.1 into four steps.

First Step Consider the continuous semimartingale $X = A + M$. Since (11.1.1) is an equality of continuous processes, it is sufficient to prove that they are modifications: in other words, we can prove the thesis for a fixed $t > 0$. We set

$$\tau_n = t \wedge \inf\{s \geq 0 \mid |X_s| \geq n, \ \langle X \rangle_s \geq n, \ V_s(A) \geq n\}, \qquad n \in \mathbb{N},$$

where $V_s(A)$ denotes the first variation process of A on $[0, s]$ (cf. Definition 9.1.1). By continuity, $\tau_n \nearrow \infty$ a.s. and therefore it is enough to prove Itô's formula for $X_{t \wedge \tau_n}$ for each $n \in \mathbb{N}$: equivalently, it is enough to prove for each fixed $\bar{N} \in \mathbb{N}$ that (11.1.1) holds in the case where the processes $|X|$, $|M|$, A, $\langle X \rangle$ and $V(A)$ are bounded by $\bar{N}$. In this case, it is not restrictive to assume that the function F has compact support, possibly modifying it outside $[-\bar{N}, \bar{N}]$. At first, we also assume that $F \in C^3(\mathbb{R})$.

We use the notation (8.1.1) for the dyadics

$$\mathscr{D}(t) = \{t_{n,k} = \tfrac{tk}{2^n} \mid k = 0, \ldots, 2^n, \ n \in \mathbb{N}\}$$

of $[0, t]$ and indicate with $\Delta_{n,k} Y = Y_{t_{n,k}} - Y_{t_{n,k-1}}$ the increment of a generic process Y. Moreover, let $\mathscr{F}_{n,k} := \mathscr{F}_{t_{n,k}}$ and

$$\delta_n(Y) = \sup_{\substack{s, r \in \mathscr{D}(t) \\ |s-r| < \frac{1}{2^n}}} |Y_s - Y_r|, \qquad n \in \mathbb{N}.$$

11.3 Proof of Itô's Formula

Expanding in Taylor series up to the second order with Lagrange remainder, we obtain

$$F(X_t) - F(X_0) = \sum_{k=1}^{2^n} \left(F(X_{t_{n,k}}) - F(X_{t_{n,k-1}}) \right)$$

$$= \sum_{k=1}^{2^n} F'(X_{t_{n,k-1}}) \Delta_{n,k} X + \frac{1}{2} \sum_{k=1}^{2^n} F''(X_{t_{n,k-1}}) \left(\Delta_{n,k} X \right)^2 + R_n \quad (11.3.3)$$

with

$$|R_n| \leq \|F'''\|_\infty \sum_{k=1}^{2^n} \left(\Delta_{n,k} X \right)^3. \quad (11.3.4)$$

In the next two steps, we estimate the individual terms in (11.3.3) to show that they converge to the corresponding terms in (11.1.1) and $R_n \longrightarrow 0$ as $n \to \infty$.

Second Step Regarding the first sum in (11.3.3), we have

$$\sum_{k=1}^{2^n} F'(X_{t_{n,k-1}}) \Delta_{n,k} X = I_n^{1,A} + I_n^{1,M}$$

where, by Proposition 9.1.3,

$$I_n^{1,A} := \sum_{k=1}^{2^n} F'(X_{t_{n,k-1}}) \Delta_{n,k} A \xrightarrow[n \to \infty]{} \int_0^t F'(X_s) dA_s \quad (11.3.5)$$

with the integral understood in the Riemann-Stieltjes sense (or Lebesgue-Stieltjes, by Proposition 9.2.2) and

$$I_n^{1,M} := \sum_{k=1}^{2^n} F'(X_{t_{n,k-1}}) \Delta_{n,k} M \xrightarrow[n \to \infty]{} \int_0^t F'(X_s) dM_s$$

in probability, by Corollary 10.2.27.

Third Step Regarding the second sum in (11.3.3), we have

$$\sum_{k=1}^{2^n} F''(X_{t_{n,k-1}}) (\Delta_{n,k} X)^2 = I_n^{2,A} + 2 I_n^{2,AM} + I_n^{2,M}$$

where

$$I_n^{2,A} := \sum_{k=1}^{2^n} F''(X_{t_{n,k-1}})(\Delta_{n,k}A)^2, \quad I_n^{2,AM} := \sum_{k=1}^{2^n} F''(X_{t_{n,k-1}})(\Delta_{n,k}A)(\Delta_{n,k}M),$$

$$I_n^{2,M} := \sum_{k=1}^{2^n} F''(X_{t_{n,k-1}})(\Delta_{n,k}M)^2.$$

Now we have

$$|I_n^{2,A}| \le \|F''\|_\infty \delta_n(A) V_t(A) \le \bar{N}\|F''\|_\infty \delta_n(A) \xrightarrow[n \to \infty]{} 0 \quad \text{a.s.}$$

by the uniform continuity of the trajectories of A on $[0, t]$. A similar result holds for $I_n^{2,AM}$. Recalling that by definition $\langle X \rangle = \langle M \rangle$, it remains to prove that

$$I_n^{2,M} \xrightarrow[n \to \infty]{} \int_0^t F''(X_s) d\langle M \rangle_s.$$

Since, analogously to (11.3.5), we almost surely have

$$\sum_{k=1}^{2^n} F''(X_{t_{n,k-1}}) \Delta_{n,k}\langle M \rangle \xrightarrow[n \to \infty]{} \int_0^t F''(X_s) d\langle M \rangle_s,$$

we prove that

$$\sum_{k=1}^{2^n} F''(X_{t_{n,k-1}}) \left((\Delta_{n,k}M)^2 - \Delta_{n,k}\langle M \rangle\right) \xrightarrow[n \to \infty]{} 0$$

in $L^2(\Omega, P)$ norm. Setting $G_{n,k} = F''(X_{t_{n,k-1}})\left((\Delta_{n,k}M)^2 - \Delta_{n,k}\langle M \rangle\right)$, expanding the square of the sum, we have

$$E\left[\left(\sum_{k=1}^{2^n} G_{n,k}\right)^2\right] = E\left[\sum_{k=1}^{2^n} G_{n,k}^2\right]$$

since the double products cancel out: in fact, if $h < k$, we have

$$E\left[G_{n,h} G_{n,k}\right] = E\left[G_{n,h} F''(X_{t_{n,k-1}}) E\left[(\Delta_{n,k}M)^2 - \Delta_{n,k}\langle M \rangle \mid \mathscr{F}_{n,k-1}\right]\right] = 0$$

11.3 Proof of Itô's Formula

due to (11.3.2). Now, by the elementary inequality $(x+y)^2 \leq 2x^2 + 2y^2$, we have

$$E\left[\sum_{k=1}^{2^n} G_{n,k}^2\right] \leq 2\|F''\|_\infty E\left[\sum_{k=1}^{2^n}\left((\Delta_{n,k}M)^4 + (\Delta_{n,k}\langle M\rangle)^2\right)\right]$$

$$\leq 2\|F''\|_\infty E\left[\delta_n^2(M)\sum_{k=1}^{2^n}(\Delta_{n,k}M)^2 + \delta_n(M)V_t(\langle M\rangle)\right] \leq$$

(applying Hölder's inequality to the first term)

$$\leq 2\|F''\|_\infty \left(E\left[\delta_n^4(M)\right]^{\frac{1}{2}} E\left[\left(\sum_{k=1}^{2^n}(\Delta_{n,k}M)^2\right)^2\right]^{\frac{1}{2}} + \bar{N}E\left[\delta_n(\langle M\rangle)\right]\right) \xrightarrow[n\to\infty]{} 0$$

since:

- $\delta_n(M) \leq 2\bar{N}$ and $\delta_n(M) \xrightarrow[n\to\infty]{} 0$ almost everywhere by the uniform continuity of M on $[0, t]$: consequently, $E\left[\delta_n^4(M)\right] \to 0$ by the dominated convergence theorem. Similarly, $E[\delta_n(\langle M\rangle)] \xrightarrow[n\to\infty]{} 0$;

- $\sup_{n\in\mathbb{N}} E\left[\left(\sum_{k=1}^{2^n}(\Delta_{n,k}M)^2\right)^2\right] \leq 16\bar{N}^4$ by estimate (9.6.3).

Based on (11.3.4), the proof that

$$\lim_{n\to\infty} E\left[|R_n|^2\right] = 0$$

is entirely analogous.

Fourth Step We conclude the proof by removing the additional regularity assumption on F. Given $F \in C^2(\mathbb{R})$ with compact support, consider a sequence $(F_n)_{n\in\mathbb{N}}$ of C^3 functions that converge uniformly to F along with their first and second derivatives. We apply Itô's formula to F_n and let n tend to infinity: we have $F_n(X_s) \xrightarrow[n\to\infty]{} F(X_s)$ for every $s \in [0, t]$. By the dominated convergence theorem, we have a.s.

$$\lim_{n\to\infty} \int_0^t \left(F_n'(X_s) - F'(X_s)\right) dA_s = \lim_{n\to\infty} \int_0^t \left(F_n''(X_s) - F''(X_s)\right) d\langle X\rangle_s = 0$$

and by Itô's isometry

$$\lim_{n\to\infty} E\left[\left(\int_0^t \left(F_n'(X_s) - F'(X_s)\right) dM_s\right)^2\right]$$
$$= \lim_{n\to\infty} E\left[\int_0^t \left(F_n'(X_s) - F'(X_s)^2\right) d\langle M\rangle_s\right] = 0.$$

11.4 Key Ideas to Remember

We outline the chapter's main findings and essential concepts, omitting technical details. As usual, if you have any doubt about what the following succinct statements mean, please review the corresponding section.

- Section 11: the significance of the quadratic variation process becomes apparent in the outline of Itô's formula proof: in particular, it introduces an additional term that modifies the usual rules of deterministic integral calculus. The Itô's formula provides the Doob's decomposition of a process that is a sufficiently regular function of a continuous semimartingale, providing the expression of the drift and the diffusive part.
- Sections 11.1.1 and 11.1.2: the heat operator appears in the drift term of Itô's formula for the Brownian motion: a process of the form $X_t = F(t, W_t)$ is a (local) martingale if and only if the function F is a solution of the heat equation. An application of Itô's formula shows that Itô processes with deterministic coefficients have normal distribution.
- Section 11.2: the Burkholder-Davis-Gundy inequality generalizes the Itô isometry and provides a comparison between the L^p norm of a continuous local martingale X and the $L^{p/2}$ norm of the related quadratic variation process $\langle X \rangle$.

Main notations used or introduced in this chapter:

Symbol	Description	Page
$\mathcal{M}^{c,2}$	Continuous square-integrable martingales	141
$\mathcal{M}^{c,\text{loc}}$	Continuous local martingales	143
$\langle X \rangle$	Quadratic variation process	165

Chapter 12
Multidimensional Stochastic Calculus

> *Tu, tu non mi basti mai davvero non mi basti mai tu, tu dolce terra mia dove non sono stato mai.*[1]
>
> *Lucio Dalla*

In this chapter, we extend the definitions and results of the previous chapters to the multidimensional case. We do not introduce any really new concepts; however, some results, such as Itô's formula, become technically more complicated and for this reason, some formal rules introduced in Sect. 12.3 can be useful for practical calculations.

12.1 Multidimensional Brownian Motion

Definition 12.1.1 (d-Dimensional Brownian Motion) Let $W=(W_t^1,\ldots,W_t^d)_{t\geq 0}$ be a stochastic process with values in $\mathbb{R}^d$ defined on a filtered probability space $(\Omega, \mathscr{F}, P, \mathscr{F}_t)$. We say that W is a d-dimensional Brownian motion if it satisfies the following properties:

 (i) $W_0 = 0$ a.s.;
 (ii) W is a.s. continuous;
 (iii) W is adapted;
 (iv) $W_t - W_s$ is independent of $\mathscr{F}_s$ for every $t \geq s \geq 0$;
 (v) $W_t - W_s \sim \mathcal{N}_{0,(t-s)I}$ for every $t \geq s \geq 0$, where I denotes the $d \times d$ identity matrix.

---

[1] *You, you're never enough for me
truly, you're never enough for me
you, you sweet land of mine
where I have never been before.*

A multidimensional Brownian motion is a vector of independent real Brownian motions: indeed, we have

Proposition 12.1.2 *If $W = (W^1, \ldots, W^d)$ is a d-dimensional Brownian motion on $(\Omega, \mathscr{F}, P, \mathscr{F}_t)$ then:*

(i) any component W^i, for $i = 1, \ldots, d$, is a real Brownian motion on $(\Omega, \mathscr{F}, P, \mathscr{F}_t)$;

(ii) $W_t^i - W_s^i$ and $W_t^j - W_s^j$ are independent random variables for every $i \neq j$ and $t \geq s \geq 0$;

(iii) the covariation matrix of W is $\langle W \rangle_t = tI$ or, in differential notation,

$$d\langle W^i, W^j \rangle_t = \delta_{ij} dt \tag{12.1.1}$$

where δ_{ij} is the Kronecker delta

$$\delta_{ij} = \begin{cases} 1 & \text{if } i = j, \\ 0 & \text{if } i \neq j; \end{cases}$$

(iv) if A is an orthogonal $d \times d$ matrix then the process defined by $B_t := AW_t$ is still a d-dimensional Brownian motion. If instead A is a generic $N \times d$ matrix then B satisfies properties (i), (ii), (iii) and (iv) of Definition 12.1.1 and $B_t - B_s \sim \mathcal{N}_{0,(t-s)\mathscr{C}}$ for every $0 \leq s \leq t$, where $\mathscr{C} = AA^*$. The covariation matrix of B coincides with its covariance matrix $\langle B \rangle_t = \text{cov}(B_t) = t\mathscr{C}$. We say that B is an N-dimensional correlated Brownian motion.

Proof Properties (i) and (ii) follow from the fact that, for $t > s \geq 0$, the increment $W_t - W_s$ has Gaussian density

$$\frac{1}{(2\pi(t-s))^{\frac{d}{2}}} e^{-\frac{|x|^2}{2(t-s)}} = \prod_{i=1}^{d} \frac{1}{\sqrt{2\pi(t-s)}} e^{-\frac{x_i^2}{2(t-s)}}, \quad x \in \mathbb{R}^d,$$

which is the product of standard one-dimensional Gaussians: in particular, independence follows from Theorem 2.3.23 in [113].

As for (iii), by point (i) we have $\langle W^i \rangle_t = \langle W^i, W^i \rangle_t = t$ for each $i = 1, \ldots, d$. For $i \neq j$ it is a simple exercise[2] to prove that $W^i W^j$ is a martingale and therefore $\langle W^i, W^j \rangle_t = 0$.

---

[2] For $t \geq s \geq 0$, we have

$$E\left[W_t^i W_t^j \mid \mathscr{F}_s\right] = E\left[\left(W_t^i - W_s^i\right) W_t^j \mid \mathscr{F}_s\right] + W_s^i E\left[W_t^j \mid \mathscr{F}_s\right] = W_s^i W_s^j$$

since

$$E\left[\left(W_t^i - W_s^i\right) W_t^j \mid \mathscr{F}_s\right] = E\left[\left(W_t^i - W_s^i\right)\left(W_t^j - W_s^j\right) \mid \mathscr{F}_s\right] + W_s^j E\left[W_t^i - W_s^i \mid \mathscr{F}_s\right]$$

12.1 Multidimensional Brownian Motion

Point (iv) is a simple check based on Proposition 2.5.15 in [113]. □

Example 12.1.3 ([!]) Let W be a two-dimensional Brownian motion. Setting

$$A = \begin{pmatrix} 1 & 0 \\ \rho & \sqrt{1-\rho^2} \end{pmatrix}$$

with $\rho \in [-1, 1]$, we have

$$\mathscr{C} = AA^* = \begin{pmatrix} 1 & \rho \\ \rho & 1 \end{pmatrix}.$$

The two-dimensional correlated Brownian motion $B := AW$ is such that

$$B_t^1 = W_t^1, \qquad B_t^2 = \rho W_t^1 + \sqrt{1-\rho^2} W_t^2,$$

are scalar Brownian motions and

$$\text{cov}(B_t^1, B_t^2) = \langle B^1, B^2 \rangle_t = \rho t.$$

In this section, we briefly show how to define the stochastic integral of multidimensional processes, focusing in particular on Brownian motion and Itô processes. For simplicity, we only deal with the case where the integrator is in $\mathscr{M}^{c,2}$ even though all the results extend directly to integrators that are continuous semimartingales. Hereafter, d and N denote two natural numbers.

Definition 12.1.4 Let $B = (B^1, \ldots, B^d) \in \mathscr{M}^{c,2}$ be a d-dimensional process. Consider a process $u = (u^{ij})$ with values in the space of matrices of dimension $N \times d$. We write $u \in \mathbb{L}_B^2$ (or simply $u \in \mathbb{L}^2$) if $u^{ij} \in \mathbb{L}_{B^j}^2$ for each $i = 1, \ldots, N$ and $j = 1, \ldots, d$. The class $\mathbb{L}_{\text{loc}}^2 \equiv \mathbb{L}_{B,\text{loc}}^2$ is defined in an analogous way. The stochastic integral of u with respect to B is the N-dimensional process, defined component by component as

$$\int_0^t u_s dB_s := \left(\sum_{j=1}^d \int_0^t u_s^{ij} dB_s^j \right)_{i=1,\ldots,N}$$

for $t \geq 0$.

Theorem 12.1.5 ([!]) *Let*

$$X_t = \int_0^t u_s dB_s^1, \qquad Y_t = \int_0^t v_s dB_s^2,$$

$$= E\left[\left(W_t^i - W_s^i \right) \left(W_t^j - W_s^j \right) \right] = 0$$

by the independence of increments.

with B^1, B^2 one-dimensional processes in $\mathcal{M}^{c,2}$ and u, v one-dimensional processes respectively in $\mathbb{L}^2_{B^1,loc}$ and $\mathbb{L}^2_{B^2,loc}$. Then:

(i)
$$\langle X, Y \rangle_t = \int_0^t u_s v_s d\langle B^1, B^2 \rangle_s; \qquad (12.1.2)$$

(ii) if $u \in \mathbb{L}^2_{B^1}$ and $v \in \mathbb{L}^2_{B^2}$ then the following version of Itô's isometry holds

$$E\left[\int_t^T u_s dB^1_s \int_t^T v_s dB^2_s \mid \mathscr{F}_t\right] = E\left[\int_t^T u_s v_s d\langle B^1, B^2 \rangle_s \mid \mathscr{F}_t\right], \quad 0 \le t \le T. \qquad (12.1.3)$$

Proof When u and v are indicator processes, (12.1.3) is proven by repeating the proof of Theorem 10.2.7-(ii) with the only difference that, instead of (10.2.6), we use (9.5.2) in the form

$$E\left[(B^1_T - B^1_t)(B^2_T - B^2_t) \mid \mathscr{F}_t\right] = E\left[\langle B^1, B^2 \rangle_T - \langle B^1, B^2 \rangle_t \mid \mathscr{F}_t\right], \quad 0 \le t \le T.$$

The proof of (12.1.2) is completely analogous to the case where $B^1 = B^2$. □

Corollary 12.1.6 *If $W = (W^1, \ldots, W^d)$ is a d-dimensional Brownian motion (cf. Definition 12.1.1) on $(\Omega, \mathscr{F}, P, \mathscr{F}_t)$ then for each $u, v \in \mathbb{L}^2_W$ we have*

$$E\left[\int_t^T u_s dW^i_s \int_t^T v_s dW^j_s \mid \mathscr{F}_t\right]$$
$$= \delta_{ij} E\left[\int_t^T u_s v_s ds \mid \mathscr{F}_t\right], \qquad 0 \le t \le T, \, i, j = 1, \ldots, d. \qquad (12.1.4)$$

Proof Equation (12.1.4) follows directly from (12.1.3) and point (iii) of Proposition 12.1.2. □

Remark 12.1.7 The components of the covariation matrix (cf. Definition 9.5.3) of the integral process

$$X_t = \int_0^t u_s dB_s$$

are

$$\langle X \rangle^{ij}_t = \langle \sum_{h=1}^d \int_0^t u^{ih}_s dB^h_s, \sum_{k=1}^d \int_0^t u^{jk}_s dB^k_s \rangle =$$

(by (12.1.2))

$$= \sum_{h,k=1}^{d} \int_0^t u_s^{ih} u_s^{jk} d\langle B^h, B^k \rangle_s \qquad (12.1.5)$$

for $i, j = 1, \ldots, N$.

12.2 Multidimensional Itô Processes

Definition 12.2.1 (Itô Process [!]) Let W be a d-dimensional Brownian motion. An N-dimensional *Itô process* is a process of the form

$$X_t = X_0 + \int_0^t u_s ds + \int_0^t v_s dW_s \qquad (12.2.1)$$

where:

(i) $X_0 \in m\mathscr{F}_0$ is an N-dimensional random variable;
(ii) u is an N-dimensional process in $\mathbb{L}_{\text{loc}}^1$, i.e., u is progressively measurable and such that, for every $t \geq 0$,

$$\int_0^t |u_s| ds < \infty, \qquad \text{a.s.}$$

(iii) v is a process in $\mathbb{L}_{\text{loc}}^2$ with values in the space of $N \times d$ matrices, i.e., v is progressively measurable and such that, for every $t \geq 0$,

$$\int_0^t |v_s|^2 ds < \infty \qquad \text{a.s.}$$

where $|v|$ denotes the Hilbert-Schmidt norm of the matrix v, i.e., the Euclidean norm in $\mathbb{R}^{N \times d}$, defined by

$$|v|^2 = \sum_{i=1}^{N} \sum_{j=1}^{d} (v^{ij})^2.$$

In differential notation, we write

$$dX_t = u_t dt + v_t dW_t.$$

Combining (12.1.5) with the fact that $\langle w \rangle_t = tI$, we obtain the following

Proposition 12.2.2 *Let X be the Itô process in (12.2.1). The covariation matrix of X is*

$$\langle X \rangle_t = \int_0^t v_s v_s^* ds, \qquad t \geq 0,$$

or, in differential notation,

$$d\langle X^i, X^j \rangle_t = \mathscr{C}_t^{ij} dt, \qquad \mathscr{C}^{ij} := (vv^*)^{ij} = \sum_{k=1}^d v^{ik} v^{jk}. \tag{12.2.2}$$

Proposition 12.2.3 (Itô Isometry) *For every $N \times d$ matrix $v \in \mathbb{L}^2$ and d-dimensional Brownian motion W, we have*

$$E\left[\left|\int_0^t v_s dW_s\right|^2\right] = E\left[\int_0^t |v|^2 ds\right].$$

Proof We have

$$E\left[\left|\int_0^t v_s dW_s\right|^2\right] = \sum_{i=1}^N E\left[\left(\sum_{j=1}^d \int_0^t v_s^{ij} dW_s^j\right)^2\right] =$$

(by (12.1.4))

$$= \sum_{i=1}^N \sum_{j=1}^d E\left[\left(\int_0^t v_s^{ij} dW_s^j\right)^2\right] =$$

(by the scalar Itô isometry)

$$= \sum_{i=1}^N \sum_{j=1}^d E\left[\int_0^t (v_s^{ij})^2 ds\right].$$

□

Example 12.2.4 In the simplest case where u, v are constants, we have

$$X_t = X_0 + ut + vW_t,$$

that is, X is a correlated Brownian motion with drift.

12.3 Multidimensional Itô's Formula

Theorem 12.3.1 (Itô's Formula for Continuous Semimartingales) *Let $X = (X^1, \ldots, X^d)$ be a continuous d-dimensional semimartingale and $F = F(t, x) \in C^{1,2}(\mathbb{R}_{\geq 0} \times \mathbb{R}^d)$. Then almost surely, for every $t \geq 0$ we have*

$$F(t, X_t) = F(0, X_0) + \int_0^t (\partial_t F)(s, X_s) ds + \sum_{j=1}^d \int_0^t (\partial_{x_j} F)(s, X_s) dX_s^j$$

$$+ \frac{1}{2} \sum_{i,j=1}^d \int_0^t (\partial_{x_i x_j} F)(s, X_s) d\langle X^i, X^j \rangle_s$$

or, in the differential notation,

$$dF(t, X_t) = \partial_t F(t, X_t) dt + \sum_{j=1}^d (\partial_{x_j} F)(t, X_t) dX_t^j + \frac{1}{2} \sum_{i,j=1}^d (\partial_{x_i x_j} F)(t, X_t) d\langle X^i, X^j \rangle_t.$$

Below we examine two particularly important cases in which we use the expressions (12.1.1) and (12.2.2) of the covariations $\langle X^i, X^j \rangle$:

(i) if W is a d-dimensional Brownian motion (cf. Definition 12.1.1) we have

$$d\langle W^i, W^j \rangle_t = \delta_{ij} dt \qquad (12.3.1)$$

where δ_{ij} is the Kronecker delta;

(ii) if X is an Itô process of the form

$$dX_t = \mu_t dt + \sigma_t dW_t \qquad (12.3.2)$$

where μ is an N-dimensional process in $\mathbb{L}^1_{\text{loc}}$ and σ is an $N \times d$ matrix in $\mathbb{L}^2_{\text{loc}}$, then

$$d\langle X^i, X^j \rangle_t = \mathscr{C}_t^{ij} dt, \qquad \mathscr{C}^{ij} = (\sigma \sigma^*)^{ij}, \qquad (12.3.3)$$

that is, recalling the notation $\langle X \rangle$ for the covariation matrix of X (cf. Definition 9.5.3),

$$d\langle X \rangle_t = \mathscr{C}_t dt.$$

Corollary 12.3.2 (Itô's Formula for Brownian Motion) *Let W be a d-dimensional Brownian motion. For every $F = F(t, x) \in C^{1,2}(\mathbb{R}_{\geq 0} \times \mathbb{R}^d)$ we* have

$$F(t, W_t) = F(0,0) + \int_0^t (\partial_t F)(s, W_s)ds + \sum_{j=1}^d \int_0^t (\partial_{x_j} F)(s, W_s)dW_s^j$$
$$+ \frac{1}{2}\int_0^t (\Delta F)(s, W_s)ds$$

where Δ is the Laplace operator in $\mathbb{R}^d$:

$$\Delta = \sum_{j=1}^d \partial_{x_j x_j}.$$

In differential notation, we have

$$dF(t, W_t) = \left(\partial_t F + \frac{1}{2}\Delta F\right)(t, W_t)dt + (\nabla_x F)(t, W_t)dW_t,$$

where $\nabla_x = (\partial_{x_1}, \ldots, \partial_{x_d})$ denotes the spatial gradient.

Example 12.3.3 (Quadratic Martingale) Let us compute the stochastic differential of $|W_t|^2$ where W is an N-dimensional Brownian motion. In this case

$$F(x) = |x|^2 = x_1^2 + \cdots + x_N^2, \qquad \partial_{x_i} F(x) = 2x_i, \qquad \partial_{x_i x_j} F(x) = 2\delta_{ij},$$

where δ_{ij} is the Kronecker delta. Therefore, we have

$$d|W_t|^2 = Ndt + 2W_t dW_t = Ndt + 2\sum_{i=1}^N W_t^i dW_t^i.$$

It follows that the process $X_t = |W_t|^2 - Nt$ is a martingale.

Corollary 12.3.4 (Itô's Formula for Itô Processes [!]) *Let X be an Itô process in $\mathbb{R}^N$ of the form (12.3.2). For every $F = F(t, x) \in C^{1,2}(\mathbb{R}_{\geq 0} \times \mathbb{R}^N)$ we have*

$$F(t, X_t) = F(0, X_0) + \int_0^t (\partial_t F)(s, X_s)ds + \sum_{j=1}^N \int_0^t (\partial_{x_j} F)(s, X_s)dX_s^j$$
$$+ \frac{1}{2}\sum_{i,j=1}^N \int_0^t (\partial_{x_i x_j} F)(s, X_s)\mathscr{C}_s^{ij} ds$$

12.3 Multidimensional Itô's Formula

where $\mathscr{C} = \sigma\sigma^*$. In differential notation, we hav

$$dF(t, X_t) = \left(\partial_t F + \frac{1}{2}\sum_{i,j=1}^{N} \mathscr{C}_s^{ij}\partial_{x_i x_j} F + \sum_{j=1}^{N} \mu_t^j \partial_{x_j} F\right)(t, X_t)dt$$

$$+ \sum_{j=1}^{N}\sum_{k=1}^{d} \sigma_t^{jk} \partial_{x_j} F(t, X_t) dW_t^k.$$

Example 12.3.5 (Exponential Martingale) Let

$$dY_t = \sigma_t dW_t$$

with σ of dimension $N \times d$ and W a d-dimensional Brownian motion. Recall that the covariation matrix of Y is $d\langle Y\rangle_t = \sigma_t \sigma_t^* dt$. Given $\eta \in \mathbb{R}^N$, let

$$M_t^\eta = \exp\left(\langle \eta, Y_t\rangle - \frac{1}{2}\langle\langle Y\rangle_t \eta, \eta\rangle\right) = \exp\left(\langle \eta, Y_t\rangle - \frac{1}{2}\int_0^t |\sigma_s^*\eta|^2 ds\right).$$

We apply Itô's formula with $F(x) = e^{\langle x, \eta\rangle}$ and

$$dX_t = dY_t - \frac{1}{2}\sigma_t\sigma_t^*\eta dt.$$

We have $M_t^\eta = F(X_t)$ and

$$\partial_{x_i} F(x) = \eta_i F(x), \qquad \partial_{x_i x_j} F(x) = \eta_i \eta_j F(x),$$

so that

$$dM_t^\eta = X_t\left(\eta dX_t + \frac{1}{2}\langle \sigma_t\sigma_t^*\eta, \eta\rangle dt\right) = X_t \eta dY_t = X_t \sum_{i=1}^{N}\sum_{j=1}^{d} \eta_i \sigma_t^{ij} dW_t^j.$$

In particular, it follows that M^η is a positive local martingale (and therefore a supermartingale by Remark 8.4.6-(vi)).

Proposition 4.4.2 has the following multidimensional generalization: we consider the exponential martingale

$$M_t^\eta := e^{i\langle \eta, W_t\rangle + \frac{|\eta|^2}{2}t}, \qquad t \geq 0, \ \eta \in \mathbb{R}^d, \tag{12.3.4}$$

where i is the imaginary unit and W is a d-dimensional Brownian motion.

Proposition 12.3.6 *Let W be a d-dimensional, continuous, and adapted process on the space $(\Omega, \mathscr{F}, P, \mathscr{F}_t)$ and such that $W_0 = 0$ a.s. If for every $\eta \in \mathbb{R}^d$ the process M^η in (12.3.4) is a martingale, then W is a Brownian motion.*

Remark 12.3.7 (Formal Rules for Covariations [!]) Let X be the Itô process in (12.3.2) with components

$$dX_t^i = \mu_t^i dt + \sum_{k=1}^d \sigma_t^{ik} dW_t^k, \quad i = 1, \ldots, N. \tag{12.3.5}$$

To determine the coefficients of the second derivatives in Itô's formula, we need to calculate the covariation matrix $\langle X \rangle = (\langle X^i, X^j \rangle)$ which we know to be given by $d\langle X \rangle_t = \sigma_t \sigma_t^* dt$ by (12.3.3). From a practical standpoint, the calculation of $\sigma \sigma^*$ can be cumbersome and it is therefore preferable to use the following rule of thumb: we write

$$d\langle X^i, X^j \rangle = dX^i * dX^j$$

and calculate the product "$*$" on the right-hand side as a product of the "polynomials" dX^i in (12.3.5) according to the following calculation rules

$$dt * dt = dt * dW_t^i = dW_t^i * dt = 0, \quad dW_t^i * dW_t^j = \delta_{ij} dt, \tag{12.3.6}$$

where δ_{ij} is the Kronecker delta.

Example 12.3.8 Suppose $N = d = 2$ in (12.3.5) and calculate the stochastic differential of the product of $Z_t = X_t^1 X_t^2$. We have $Z_t = F(X_t)$ where $F(x_1, x_2) = x_1 x_2$ and

$$\partial_{x_1} F(x) = x_2, \quad \partial_{x_2} F(x) = x_1, \quad \partial_{x_1 x_1} F(x) = \partial_{x_2 x_2} F(x) = 0,$$
$$\partial_{x_1 x_2} F(x) = \partial_{x_2 x_1} F(x) = 1.$$

Consequently,

$$d(X_t^1 X_t^2) = X_t^1 dX_t^2 + X_t^2 dX_t^1 + d\langle X^1, X^2 \rangle_t$$
$$= X_t^1 dX_t^2 + X_t^2 dX_t^1 + \left(\sigma_t^{11} \sigma_t^{21} + \sigma_t^{12} \sigma_t^{22} \right) dt.$$

Moreover, regarding the quadratic variation of X^1, we have

$$d\langle X^1 \rangle_t = \left((\sigma_t^{11})^2 + (\sigma_t^{12})^2 \right) dt.$$

12.3 Multidimensional Itô's Formula

Example 12.3.9 Let us calculate the stochastic differential of the process

$$Y_t = e^{tW_t^1} \int_0^t W_s^2 dW_s^1$$

where (W^1, W^2) is a standard two-dimensional Brownian motion. Proceeding as in Example 11.1.8, we identify the function $F = F(t, x_1, x_2) = e^{tx_1} x_2$ and the Itô process

$$dX_t^1 = dW_t^1, \qquad dX_t^2 = W_t^2 dW_t^1$$

in order to apply Itô's formula. We have

$$\partial_t F = x_1 F, \quad \partial_{x_1} F = tF, \quad \partial_{x_2} F = e^{tx_1}, \quad \partial_{x_1 x_1} F = t^2 F, \quad \partial_{x_1 x_2} F = te^{tx_1},$$

$$\partial_{x_2 x_2} F = 0,$$

and by the formal rules (12.3.6) for the calculation of covariation processes

$$d\langle X^1 \rangle_t = dt, \qquad d\langle X^1, X^2 \rangle_t = W_t^2 dt.$$

Consequently

$$dY_t = W_t^1 Y_t dt + tY_t dW_t^1 + e^{tW_t^1} dW_t^2 + \frac{1}{2}\left(t^2 Y_t + 2te^{tW_t^1} W_t^2\right) dt.$$

Finally, we give the multidimensional version of Corollary 11.2.2 on the L^p estimates for the stochastic integral. We omit the proof which is similar to the scalar case.

Corollary 12.3.10 ([!]) *Let $\sigma \in \mathbb{L}^2$, an $N \times d$-dimensional matrix, and W a d-dimensional Brownian motion. For every $p \geq 2$ and $T > 0$ we have*

$$E\left[\sup_{0 \leq t \leq T} \left|\int_0^t \sigma_s dW_s\right|^p\right] \leq cT^{\frac{p-2}{2}} E\left[\int_0^T |\sigma_s|^p ds\right] \tag{12.3.7}$$

where $|\sigma|$ indicates the Hilbert-Schmidt norm[3] of σ and c is a positive constant that depends only on p, N, and d.

---

[3] That is, the Euclidean norm in $\mathbb{R}^{N \times d}$.

12.4 Lévy's Characterization and Correlated Brownian Motion

We recall expression (12.3.1) of the covariations of a standard Brownian motion W.

Theorem 12.4.1 (Lévy's Characterization of a Brownian Motion) *Let X be a d-dimensional process defined on the space $(\Omega, \mathscr{F}, P, (\mathscr{F}_t))$ and such that $X_0 = 0$ a.s. Then X is a Brownian motion if and only if X is a continuous local martingale such that*

$$\langle X^i, X^j \rangle_t = \delta_{ij} t, \qquad t \geq 0. \tag{12.4.1}$$

Proof We use Proposition 12.3.6 and verify that, for every $\eta \in \mathbb{R}^d$, the exponential process

$$M_t^\eta := e^{i\eta X_t + \frac{|\eta|^2}{2} t}$$

is a martingale. By Itô's formula we have

$$dM_t^\eta = M_t^\eta \left(\frac{|\eta|^2}{2} dt + i\eta dX_t - \frac{1}{2} \sum_{i,j=1}^d \eta_i \eta_j d\langle X^i, X^j \rangle_t \right) =$$

(by assumption (12.4.1))

$$= M_t^\eta i\eta dX_t$$

and therefore, by Theorem 10.2.24, M^η is a continuous local martingale. On the other hand M^η is also a true martingale being a bounded process, hence the thesis. □

Corollary 12.4.2 *Let $\alpha = (\alpha^1, \ldots, \alpha^d)$ be a d-dimensional progressively measurable process such that $|\alpha_t| = 1$ for $t \geq 0$ almost surely. For every d-dimensional Brownian motion W, the process*

$$B_t := \int_0^t \alpha_s dW_s$$

is a real Brownian motion.

Proof By Theorem 10.2.15 B is a continuous martingale and by assumption

$$\langle B \rangle_t = \int_0^t |\alpha_s|^2 ds = t.$$

The thesis follows from Theorem 12.4.1. □

12.4 Lévy's Characterization and Correlated Brownian Motion

Definition 12.4.3 (Correlated Brownian Motion) Let α be a progressively measurable process with values in the space of matrices of dimension $N \times d$, whose rows α^i are such that $|\alpha_t^i| = 1$ for $t \geq 0$ almost surely. Given a standard d-dimensional Brownian motion W, the process

$$B_t := \int_0^t \alpha_s dW_s$$

is called *correlated Brownian motion*.

By Corollary 12.4.2, each component of B is a real Brownian motion and by (12.3.3) we have

$$\langle B^i, B^j \rangle_t = \int_0^t \rho_s^{ij} ds$$

where $\rho_t = \alpha_t \alpha_t^*$ is called *correlation matrix of B*. Moreover, we have

$$\text{cov}(B_t) = \int_0^t E[\rho_s] ds,$$

since

$$\text{cov}(B_t^i, B_t^j) = E\left[B_t^i B_t^j\right] = E\left[\sum_{k=1}^d \int_0^t \alpha_s^{ik} dW_s^k \sum_{h=1}^d \int_0^t \alpha_s^{jh} dW_s^h\right] =$$

(by the Itô isometry, Proposition 12.2.3)

$$= E\left[\int_0^t \sum_{k=1}^d \alpha_s^{ik} \alpha_s^{jk} ds\right] = \int_0^t E\left[\rho_s^{ij}\right] ds.$$

When σ is orthogonal, we have $N = d$, $\alpha^* = \alpha^{-1}$ and therefore $\alpha^i \cdot \alpha^j = \delta_{ij}$ for each pair of rows: in this particular case, B is also a standard d-dimensional Brownian motion according to Definition 12.1.1.

Example 12.4.4 (Itô's Formula for Correlated Brownian Motion [!]) In some applications, it is natural to use Itô processes defined in terms of a correlated Brownian motion $dB_t = \alpha_t dW_t$ as in Definition 12.4.3. For example, in the Black&Scholes financial model [19], the stochastic dynamics governing the prices of N risky assets can be described by the following equations

$$dS_t^i = \mu_t^i S_t^i dt + \sigma_t^i S_t^i dB_t^i, \qquad i = 1, \ldots, N, \tag{12.4.2}$$

or alternatively by

$$dS_t^i = \mu_t^i S_t^i dt + \sum_{j=1}^d v_t^{ij} S_t^i dW_t^j, \qquad i = 1, \ldots, N, \qquad (12.4.3)$$

where W is a standard d-dimensional Brownian motion. In (12.4.3), the dynamics of the i-th asset explicitly involves all Brownian motions $W^1, \ldots, W^d$ and the diffusion coefficients v^{ij} incorporate the correlations between the different assets. The dynamics described in Eq. (12.4.2) may offer greater convenience, as the i-th asset depends only on the real Brownian motion B^i: the coefficient σ^i, usually called *volatility*, is an indicator of the "riskiness" of the i-th asset; the dependence between the different assets is implicit in B through the correlation matrix $\rho = \alpha\alpha^*$, for which $d\langle B\rangle_t = \rho_t dt$. In this context, it is often preferred to assign the dynamics (12.4.2) instead of (12.4.3), to keep separate the volatility structures of individual securities from that of correlation.

In the case of correlated Brownian motion, the formal calculation rules of Remark 12.3.7 change to

$$dt * dt = dt * dB_t^i = dB_t^i * dt = 0, \qquad dB_t^i * dB_t^j = \varrho_t^{ij} dt. \qquad (12.4.4)$$

For example, let us assume the dynamics (12.4.2) with $N = 2$ and let B be two-dimensional Brownian motion defined as in Example 12.1.3, with correlation matrix

$$\begin{pmatrix} 1 & \varrho \\ \varrho & 1 \end{pmatrix}, \qquad \varrho \in [-1, 1].$$

Then we have

$$d\frac{S_t^1}{S_t^2} = \frac{dS_t^1}{S_t^2} - \frac{S_t^1}{(S_t^2)^2} dS_t^2 + \frac{1}{2}\left(-\frac{2}{(S_t^2)^2} d\langle S^1, S^2\rangle_t + \frac{2S_t^1}{(S_t^2)^3} d\langle S^2\rangle_t\right)$$

$$= \frac{S_t^1}{S_t^2}\left(\mu_t^1 - \mu_t^2 - \varrho_t \sigma_t^1 \sigma_t^2 + (\sigma_t^2)^2\right) dt + \frac{S_t^1}{S_t^2}(\sigma_t^1 dB_t^1 - \sigma_t^2 dB_t^2).$$

12.5 Key Ideas to Remember

We summarize the most significant findings of the chapter and the fundamental concepts to be retained from an initial reading, while disregarding the more technical or secondary matters. As usual, if you have any doubt about what the following succinct statements mean, please review the corresponding section.

12.5 Key Ideas to Remember

- Sections 12.1, 12.2, and 12.3: these sections contain the multidimensional extension of the main concepts of stochastic integration. Since several technical and non-substantial complications arise, the rules of thumb of Remark 12.3.7 come in handy when applying Itô's formula.
- Section 12.4: a classical result by Lévy provides a characterization of a Brownian motion in terms of the martingale property and the expression of the covariation matrix. In certain applications, such as in finance (see Example 12.4.4), it's common to employ correlated Brownian motion and the associated Itô's formula.

Symbols introduced in this chapter:

Symbol	Description	Page
$\int_0^t u_s dB_s := \left(\sum_{j=1}^d \int_0^t u_s^{ij} dB_s^j \right)_{i=1,\ldots,N}$	Multidimensional stochastic integral	229
$\Delta = \sum_{j=1}^d \partial_{x_j x_j}$	Laplace operator in $\mathbb{R}^d$	234

Chapter 13
Changes of Measure and Martingale Representation

> *It has been suggested that an army of monkeys might be trained to pound typewriters at random in the hope that ultimately great works of literature would be produced. Using a coin for the same purpose may save feeding and training expenses and free the monkeys for other monkey business.*
>
> *William Feller*

In this chapter, we present two classic results:

- Girsanov Theorem 13.3.3, which states that the process obtained by adding a drift to a Brownian motion is still a Brownian motion under a new probability measure;
- the martingale representation Theorem 13.5.1, according to which every local martingale with respect to the Brownian filtration admits a representation in terms of stochastic integral and consequently has a continuous version.

These results can be combined to examine the effect of a change of probability measure on the expression of the drift of an Itô process. In the treatment of these problems, a central role is played by exponential martingales.

13.1 Change of Measure and Itô Processes

Consider a d-dimensional Brownian motion W on a filtered space $(\Omega, \mathscr{F}, P, \mathscr{F}_t)$ and a d-dimensional process $\lambda \in \mathbb{L}^2_{\text{loc}}$. Applying Itô's formula to the exponential process

$$M_t^\lambda := \exp\left(-\int_0^t \lambda_s dW_s - \frac{1}{2}\int_0^t |\lambda_s|^2 ds\right), \qquad t \in [0,T], \tag{13.1.1}$$

we obtain

$$dM_t^\lambda = -M_t^\lambda \lambda_t dW_t. \tag{13.1.2}$$

Thus, M^λ is a local martingale, sometimes called *exponential martingale*. Being positive, M^λ is a super-martingale (cf. Remark (8.4.6)-(vi)) and in particular

$$E\left[M_t^\lambda\right] \leq M_0^\lambda = 1, \qquad t \in [0, T].$$

Furthermore, M^λ is a true martingale on $[0, T]$ if and only if $E\left[M_T^\lambda\right] = 1$.

Exponential martingales have an interesting connection with changes of probability measure. Recall that two probabilities P, Q on a measurable space $(\Omega, \mathscr{F})$ are said to be equivalent if they have the same certain and negligible events: in this case we write $Q \sim P$. By the Radon-Nikodym Theorem B.1.3 in [113], for each probability Q, equivalent to P, there is a random variable Z that is a.s. strictly positive and such that

$$Q(A) = \int_A Z dP, \qquad A \in \mathscr{F};$$

in particular, we have $E^P[Z] = 1$. Z is called the Radon-Nikodym derivative of Q with respect to P and is denoted by the symbol $Z = \frac{dQ}{dP}$. Note that it is equivalent to assign $Q \sim P$ or a strictly positive r.v. Z such that $E^P[Z] = 1$.

The following theorem states that there is a one-to-one correspondence between the measures Q, equivalent to P, and the processes $\lambda \in \mathbb{L}_{loc}^2$ such that M^λ is a martingale. Moreover, a change of probability measure corresponds to a change of drift of the Brownian motion (and the related Itô processes).

Theorem 13.1.1 (Changes of Measure and Drift [!!]) *Let $W = (W_t)_{t \in [0,T]}$ be a d-dimensional Brownian motion on the space $(\Omega, \mathscr{F}, P)$ equipped with the standard Brownian filtration[1] $\mathscr{F}^W$. We have:*

(i) *if Q is a probability measure equivalent to P then there exists $\lambda \in \mathbb{L}_{loc}^2$ such that*

$$\frac{dQ}{dP} = M_T^\lambda \tag{13.1.3}$$

where M^λ is the exponential martingale in (13.1.1);

(ii) *conversely, if $\lambda \in \mathbb{L}_{loc}^2$ is such that M^λ is a true martingale then (13.1.3) defines a probability measure $Q \sim P$.*

---

[1] The filtration obtained by completing the filtration generated by W so that it satisfies the usual conditions.

13.1 Change of Measure and Itô Processes

Moreover, if $Q \sim P$:

(a) almost surely we have

$$M_t^\lambda = E^P \left[\frac{dQ}{dP} \mid \mathscr{F}_t^W \right], \qquad t \in [0, T]; \tag{13.1.4}$$

(b) the process

$$W_t^\lambda := W_t + \int_0^t \lambda_s ds \tag{13.1.5}$$

is a Brownian motion on $(\Omega, \mathscr{F}, Q, \mathscr{F}_t^W)$;
(c) if X is an Itô process of the form

$$dX_t = b_t dt + \sigma_t dW_t \tag{13.1.6}$$

with $b \in \mathbb{L}_{loc}^1$ and $\sigma \in \mathbb{L}_{loc}^2$, then

$$dX_t = (b_t - \sigma_t \lambda_t)dt + \sigma_t dW_t^\lambda. \tag{13.1.7}$$

We will prove Theorem 13.1.1 in Sect. 13.5.1, as a corollary of the two main results of this chapter, Girsanov theorem and the Brownian martingale representation theorem.

13.1.1 An Application: Risk-Neutral Valuation of Financial Derivatives

In some applications, we are interested in replacing the drift b_t of an Itô process of the form (13.1.6) with a suitable drift $r_t \in \mathbb{L}_{loc}^1$. Theorem 13.1.1, states that this is possible by changing the probability measure provided that there exists a process $\lambda \in \mathbb{L}_{loc}^2$ such that $r_t = b_t - \sigma_t \lambda_t$ and M^λ in (13.1.1) is a martingale. In this section, we present a specific application in the field of mathematical finance.

In the one-dimensional Black&Scholes model [19] of Example 12.4.4, the price S of a risky asset has the following stochastic dynamics

$$dS_t = \mu S_t dt + \sigma S_t dW_t, \tag{13.1.8}$$

where W is a real Brownian motion on $(\Omega, \mathscr{F}, P, \mathscr{F}_t)$ and μ, σ are two real parameters called *expected return rate* and *volatility*, respectively. We assume $\sigma > 0$ in order not to cancel the random effect of the Brownian motion that describes the riskiness[2] of the asset. Moreover, it is reasonable to assume $\mu > r$ where r denotes the risk-free interest rate:[3] this is economically motivated by the fact that investors, to take on the risk of investing in the asset S, expect a return rate $\mu > r$, more remunerative than the bank account. In financial jargon, P is called the "real-world measure" because the dynamics (13.1.8) under the measure P intends to describe the real evolution of the risky asset: precisely, the parameters μ, σ of the model are those that could be estimated by means of econometric methods applied to real data, such as a historical series of stock prices. This statistical estimation is typically conducted with the intention of *predicting* the future price trend based on past data.

In mathematical finance, starting from model (13.1.8), another probability measure Q is introduced as in Theorem 13.1.1 with λ equal to the constant process

$$\lambda = \frac{\mu - r}{\sigma} \in \mathbb{R}_+. \tag{13.1.9}$$

The choice of λ is such that the dynamics of S becomes

$$dS_t = rS_t dt + \sigma S_t dW_t^\lambda,$$

thus formally analogous[4] to (13.1.8) but with the expected return rate equal to the risk-free rate. The measure Q does not intend to describe the real dynamics of the stock: Q is called "risk-neutral measure" or also "martingale measure" because the process $\widetilde{S}_t := e^{-rt} S_t$ of the *discounted asset price*[5] is a Q-martingale[6] and, in particular, we have

$$S_0 = e^{-rT} E^Q [S_t]. \tag{13.1.10}$$

Formula (13.1.10) is a *risk-neutral valuation formula*, according to which the current price S_0 is fair in the sense that it is equal to the expected value of the discounted future price.

The measure Q is employed to assess specific financial instruments known as *derivatives*, whose value is determined at a future time T based on S_T: precisely, a

---

[2] If $\sigma = 0$, (13.1.8) reduces to an ordinary differential equation

$$dS_t = \mu S_t dt$$

with deterministic solution $S_t = S_0 e^{\mu t}$: the latter is called a *compound capitalization formula* with interest rate μ.

[3] The interest rate paid by the bank account which is assumed to be the risk-free investment of reference.

[4] $W_t^\lambda = W_t + \lambda t$ is a real Brownian motion under the measure Q.

[5] The discount factor e^{-rt} eliminates the "time value" of prices.

[6] As opposed to the real measure P under which, being $\mu > r$, the discounted price is a sub-martingale: this describes the expectation of a higher return compared to a bank account, considering the riskiness of the asset.

"payoff function" φ is given and the random variable $\varphi(S_T)$ represents the value of the derivative at time T. For consistency with formula (13.1.10), the (discounted) expected value in the risk-neutral measure

$$e^{-rT} E^Q [\varphi(S_T)] \tag{13.1.11}$$

is called "risk-neutral price", at the initial time, of the derivative with payoff φ. The expected value in (13.1.11) can be calculated explicitly using the fact that S_T has log-normal distribution, returning the famous *Black&Scholes formula*.

The parameter λ in (13.1.9) is called "market price of risk" because it is defined as the ratio between the return differential $\mu - r$ required to assume the risk of investing in S and the volatility σ that measures the riskiness of S.

Unlike P, the measure Q does not have a "statistical" purpose and does not reflect the actual probabilities of events; rather, it is an artificial measure under which all market prices (of the bank account, of the stock S and of the derivative $\varphi(S_T)$) are deemed fair: the purposes of Q are mainly the valuation of derivatives and the study of some fundamental properties of financial models, such as *absence of arbitrage* and *completeness*. For a full treatment of these topics, we refer, for example, to [111, 112] and [115].

13.2 Integrability of Exponential Martingales

In this section, we give some conditions on the process λ that guarantee that the exponential martingale (13.1.1) is a true martingale.

Proposition 13.2.1 *Assume that*

$$\int_0^T |\lambda_t|^2 dt \leq \kappa \quad a.s. \tag{13.2.1}$$

for a certain constant κ. Then the exponential martingale M^λ in (13.1.1) is a true martingale and

$$E\left[\sup_{0 \leq t \leq T} \left(M_t^\lambda \right)^p \right] < \infty, \qquad p \geq 1.$$

We prove Proposition 13.2.1 at the end of the section.

Notation 13.2.2 For every process X, we set

$$\bar{X}_T := \sup_{0 \leq t \leq T} |X_t|.$$

Consider the integral process

$$Y_t := \int_0^t \lambda_s dW_s, \qquad t \in [0, T], \qquad (13.2.2)$$

where the Brownian motion W and $\lambda \in \mathbb{L}^2_{loc}$ are both d-dimensional processes.[7] Under condition (13.2.1), the Burkholder-Davis-Gundy inequality provides the following summability estimate for Y: for every $p > 0$, we have

$$E\left[\bar{Y}_T^p\right] \leq cE\left[\langle Y \rangle_T^{p/2}\right] \leq c\kappa^{p/2}.$$

In fact, a stronger, exponential-type integrability estimate holds; to prove it we need the following

Lemma 13.2.3 *For every continuous non-negative super-martingale $Z = (Z_t)_{t \in [0,T]}$, we have*

$$P\left(\sup_{0 \leq t \leq T} Z_t \geq \varepsilon\right) \leq \frac{E[Z_0]}{\varepsilon}, \qquad \varepsilon > 0.$$

Proof Fix $\varepsilon > 0$, and let

$$\tau := \inf\{t \geq 0 \mid Z_t \geq \varepsilon\} \wedge T.$$

Then τ is a bounded stopping time and by the optional sampling Theorem 8.1.6, we have

$$E[Z_0] \geq E[Z_\tau] \geq E\left[Z_\tau \mathbb{1}_{(\bar{Z}_T \geq \varepsilon)}\right] \geq \varepsilon P(\bar{Z}_T \geq \varepsilon).$$

$\square$

Proposition 13.2.4 (Exponential Integrability) *Let Y be the stochastic integral in (13.2.2) with $\lambda \in \mathbb{L}^2$ satisfying the condition (13.2.1). Then we have*

$$P\left(\bar{Y}_T \geq \epsilon\right) \leq 2e^{-\frac{\epsilon^2}{2\kappa}}, \qquad \epsilon > 0, \qquad (13.2.3)$$

---

[7] Then, more explicitly,

$$Y_t = \sum_{j=1}^d \int_0^t \lambda_s^j dW_s^j.$$

We note that $M_t^\lambda = \exp\left(-Y_t - \tfrac{1}{2}\langle Y \rangle_t\right)$.

13.2 Integrability of Exponential Martingales

and consequently there exists $\alpha = \alpha(\kappa) > 0$ *such that*

$$E\left[e^{\alpha \bar{Y}_T^2}\right] < \infty. \tag{13.2.4}$$

Proof For every $\alpha > 0$, the process

$$Z_t^\alpha = e^{\alpha Y_t - \frac{\alpha^2}{2}\langle Y \rangle_t},$$

is a continuous, positive supermartingale. Furthermore, under the condition (13.2.1), for every $\epsilon > 0$ and $t \in [0, T]$, we have

$$(Y_t \geq \epsilon) = \left(e^{\alpha Y_t} \geq e^{\alpha \epsilon}\right) \subseteq \left(Z_t^\alpha \geq e^{\alpha \epsilon - \frac{\alpha^2 \kappa}{2}}\right).$$

Hence

$$P\left(\sup_{0 \leq t \leq T} Y_t \geq \epsilon\right) \leq P\left(\sup_{0 \leq t \leq T} Z_t^\alpha \geq e^{\alpha \epsilon - \frac{\alpha^2 \kappa}{2}}\right) \leq e^{-\alpha \epsilon + \frac{\alpha^2 \kappa}{2}}$$

by Lemma 13.2.3, since $E[Z_0^\alpha] = 1$. Choosing $\alpha = \frac{\epsilon}{\kappa}$ in order to minimize the last term, we get

$$P\left(\sup_{0 \leq t \leq T} Y_t \geq \epsilon\right) \leq e^{-\frac{\epsilon^2}{2\kappa}}.$$

An analogous estimate holds for $-Y$ and this proves (13.2.3). Finally, (13.2.4) is an immediate consequence of (13.2.3), Proposition 3.1.6 in [113] and Example 3.1.7 in [113]. □

Remark 13.2.5 Proposition 13.2.4 extends to the case where σ is a $N \times d$-dimensional process: in this case we have

$$P\left(\bar{Y}_T \geq \epsilon\right) \leq 2N e^{-\frac{\epsilon^2}{2\kappa N}}, \qquad \epsilon > 0, \tag{13.2.5}$$

and there exists $\alpha = \alpha(\kappa, N) > 0$ such that

$$E\left[e^{\alpha \bar{Y}_T^2}\right] < \infty.$$

Indeed, it is enough to note that

$$(\bar{Y}_T \geq \epsilon) \subseteq \left(\bar{Y}_T^j \geq \frac{\epsilon}{\sqrt{N}}\right)$$

for at least one component Y^j, with $j \in \{1, \ldots, N\}$, of Y. Therefore, we have

$$P\left(\bar{Y}_t \geq \epsilon\right) \leq \sum_{j=1}^{N} P\left(\bar{Y}_T^j \geq \frac{\epsilon}{\sqrt{N}}\right)$$

and the thesis follows.

Proof of Proposition 13.2.1 For every $\varepsilon > 0$, by (13.2.3) we have

$$P\left(\sup_{0 \leq t \leq T} M_t^\lambda \geq \varepsilon\right) \leq P\left(\sup_{0 \leq t \leq T} e^{|Y_t|} \geq \varepsilon\right) = P\left(\bar{Y}_T \geq \log \varepsilon\right) \leq 2e^{-\frac{(\log \varepsilon)^2}{2\kappa}}.$$

and consequently, by Proposition 3.1.6 in [113], we have

$$E\left[\sup_{0 \leq t \leq T}(M_t^\lambda)^p\right] = p \int_0^\infty \varepsilon^{p-1} P\left(\sup_{0 \leq t \leq T} M_t^\lambda \geq \varepsilon\right) d\varepsilon < \infty. \qquad (13.2.6)$$

In particular for $p = 2$ we have

$$E\left[\int_0^T \lambda_t^2 (M_t^\lambda)^2 dt\right] \leq E\left[\sup_{0 \leq t \leq T}(M_t^\lambda)^2 \int_0^T \lambda_t^2 dt\right] \leq$$

(by assumption (13.2.1))

$$\leq \kappa E\left[\sup_{0 \leq t \leq T}(M_t^\lambda)^2\right] < \infty$$

by (13.2.6). Therefore $\lambda M^\lambda \in \mathbb{L}^2$ and from (13.1.2) it follows that M^λ is a martingale. □

A more general condition that guarantees the martingale property for the exponential process M^λ is given by the following classical result by Novikov [100].

Theorem 13.2.6 (Novikov's Condition) *If $\lambda \in \mathbb{L}_{loc}^2$ is such that*

$$E\left[\exp\left(\frac{1}{2}\int_0^T |\lambda_s|^2 ds\right)\right] < \infty \qquad (13.2.7)$$

then the process M^λ in (13.1.1) is a martingale.

Remark 13.2.7 Condition (13.2.7) is sharp in the sense that, for every $0 < \alpha < \frac{1}{2}$, there exists a process $\lambda \in \mathbb{L}_{loc}^2$ that satisfies

$$E\left[\exp\left(\alpha \int_0^T |\lambda_s|^2 ds\right)\right] < \infty$$

and is such that M^λ in (13.1.1) is not a martingale: for details see Chapter 6 in [90].

13.3 Girsanov Theorem

Let W be a d-dimensional Brownian motion on the space $(\Omega, \mathscr{F}, P, \mathscr{F}_t)$. In Sect. 13.2 we have provided sufficient conditions on $\lambda \in \mathbb{L}_{loc}^2$ for the exponential process

$$M_t^\lambda := \exp\left(-\int_0^t \lambda_s dW_s - \frac{1}{2}\int_0^t |\lambda_s|^2 ds\right), \qquad t \in [0, T]. \tag{13.3.1}$$

to be a true martingale and thus in particular $E\left[M_T^\lambda\right] = 1$: in this case

$$Q(A) := \int_A M_T^\lambda dP, \qquad A \in \mathscr{F},$$

is a probability measure on $(\Omega, \mathscr{F})$ with Radon-Nikodym derivative

$$\frac{dQ}{dP} = M_T^\lambda. \tag{13.3.2}$$

The proof of the following lemma is based on the Bayes' formula of Theorem 4.2.14 in [113]: for every $X \in L^1(\Omega, Q)$ we have

$$E^Q[X \mid \mathscr{F}_t] = \frac{E^P\left[XM_T^\lambda \mid \mathscr{F}_t\right]}{E^P\left[M_T^\lambda \mid \mathscr{F}_t\right]} \qquad t \in [0, T]. \tag{13.3.3}$$

Lemma 13.3.1 *Suppose that M^λ in (13.3.1) is a P-martingale and let Q be the probability measure in (13.3.2). A process $X = (X_t)_{t \in [0,T]}$ is a Q-martingale if and only if $(X_t M_t^\lambda)_{t \in [0,T]}$ is a P-martingale.*

Proof Since M^λ is adapted and strictly positive, it is clear that X is adapted if and only if XM^λ is. Moreover, we have

$$E^Q[|X_t|] = E^P\left[|X_t|M_T^\lambda\right] = E^P\left[E^P\left[|X_t|M_T^\lambda \mid \mathscr{F}_t\right]\right] =$$

(since X is adapted and M^λ is a P-martingale)

$$= E^P\left[|X_t|E^P\left[M_T^\lambda \mid \mathscr{F}_t\right]\right] = E^P\left[|X_t|M_t^\lambda\right],$$

and thus $X_t \in L^1(\Omega, Q)$ if and only if $X_t M_t^\lambda \in L^1(\Omega, P)$. Similarly, for $s \leq t$ we have

$$E^P\left[X_t M_T^\lambda \mid \mathscr{F}_s\right] = E^P\left[E^P\left[X_t M_T^\lambda \mid \mathscr{F}_t\right] \mid \mathscr{F}_s\right] = E^P\left[X_t M_t^\lambda \mid \mathscr{F}_s\right].$$

Then from (13.3.3) with $X = X_t$ we have

$$E^Q\left[X_t \mid \mathscr{F}_s\right] = \frac{E^P\left[X_t M_T^\lambda \mid \mathscr{F}_s\right]}{E^P\left[M_T^\lambda \mid \mathscr{F}_s\right]} = \frac{E^P\left[X_t M_t^\lambda \mid \mathscr{F}_s\right]}{M_s^\lambda},$$

which proves the thesis. □

Remark 13.3.2 Under the assumptions of Lemma 13.3.1, the process

$$\left(M_t^\lambda\right)^{-1} = \exp\left(\int_0^t \lambda_s dW_s + \frac{1}{2}\int_0^t |\lambda_s|^2 ds\right).$$

is a Q-martingale since $M^\lambda \left(M^\lambda\right)^{-1}$ is obviously a P-martingale. Moreover, for every absolutely integrable random variable X, we have

$$E^P[X] = E^P\left[X\left(M_T^\lambda\right)^{-1} M_T^\lambda\right] = E^Q\left[X\left(M_T^\lambda\right)^{-1}\right]$$

and therefore

$$\frac{dP}{dQ} = \left(M_T^\lambda\right)^{-1}.$$

In particular, *P, Q are equivalent measures*, in the sense that they have the same certain and negligible events, since they have mutually strictly positive densities.

A Brownian motion is a martingale and therefore is a "driftless process": Girsanov theorem states that if a drift is added to a Brownian motion, this process is still a Brownian motion with respect to a new probability measure. To understand this result, which at first glance seems a bit strange, it is helpful to keep in mind the elementary Example 1.4.8 at the end of which we observed that the martingale property *is not a property of the paths of the process, but rather depends on the probability measure under consideration.*

13.3 Girsanov Theorem

Theorem 13.3.3 (Girsanov [!!]) *If W is a Brownian motion and M^λ in (13.3.1) is a martingale on the space $(\Omega, \mathscr{F}, P, \mathscr{F}_t)$, then the process*

$$W_t^\lambda := W_t + \int_0^t \lambda_s ds, \qquad t \in [0, T],$$

is a Brownian motion on $(\Omega, \mathscr{F}, Q, \mathscr{F}_t)$ with $\frac{dQ}{dP} = M_T^\lambda$.

Proof By Proposition 12.3.6 on the characterization of a Brownian motion, it is sufficient to show that, for every $\eta \in \mathbb{R}^d$, the process

$$X_t^\eta := e^{i\eta W_t^\lambda + \frac{|\eta|^2}{2} t}, \qquad t \in [0, T],$$

is a Q-martingale (i.e., a martingale under the measure Q): equivalently, by Lemma 13.3.1, we prove that the process

$$X_t^\eta M_t^\lambda = \exp\left(i\eta W_t + i \int_0^t \eta \lambda_s ds + \frac{|\eta|^2 t}{2} - \int_0^t \lambda_s dW_s - \frac{1}{2} \int_0^t |\lambda_s|^2 ds\right)$$

$$= \exp\left(-\int_0^t (\lambda_s - i\eta) dW_s - \frac{1}{2} \sum_{j=1}^d \int_0^t \left(\lambda_s^j - i\eta^j\right)^2 ds\right)$$

is a P-martingale. Under the boundedness condition (13.2.1), the thesis follows from Lemma 13.2.1, which also holds for complex-valued processes and in particular for $\lambda - i\eta$.

For the general case we use a localization argument: we consider the sequence of stopping times

$$\tau_n = \inf\left\{t \geq 0 \mid \int_0^t |\lambda_s|^2 ds \geq n\right\} \wedge T, \qquad n \in \mathbb{N}.$$

By Lemma 13.2.1, the process $(X_{t \wedge \tau_n}^\eta M_{t \wedge \tau_n}^\lambda)$ is a P-martingale and

$$\mathbb{E}^P\left[X_{t \wedge \tau_n}^\eta M_{t \wedge \tau_n}^\lambda \mid \mathscr{F}_s\right] = X_{s \wedge \tau_n}^\eta M_{s \wedge \tau_n}^\lambda, \qquad s \leq t,\ n \in \mathbb{N}.$$

Hence, to prove that $X^\eta Z$ is a martingale, it is sufficient to show that $(X_{t \wedge \tau_n}^\eta M_{t \wedge \tau_n}^\lambda)$ converges to $(X_t^\eta M_t^\lambda)$ in L^1-norm as n tends to infinity. Since

$$\lim_{n \to \infty} X_{t \wedge \tau_n}^\eta = X_t^\eta \qquad \text{a.s.}$$

and $0 \leq X_{t \wedge \tau_n}^\eta \leq e^{\frac{|\eta|^2 T}{2}}$, it is enough to prove that

$$\lim_{n \to \infty} M_{t \wedge \tau_n}^\lambda = M_t^\lambda \qquad \text{in } L^1(\Omega, P).$$

Let

$$M_{n,t} = \min\{M^\lambda_{t\wedge\tau_n}, M^\lambda_t\};$$

we have $0 \leq M_{n,t} \leq M^\lambda_t$ and by the dominated convergence theorem

$$\lim_{n\to\infty} \mathbb{E}\left[M_{n,t}\right] = \mathbb{E}\left[M^\lambda_t\right].$$

On the other hand

$$\mathbb{E}\left[\left|M^\lambda_t - M^\lambda_{t\wedge\tau_n}\right|\right] = \mathbb{E}\left[M^\lambda_t - M_{n,t}\right] + \mathbb{E}\left[M^\lambda_{t\wedge\tau_n} - M_{n,t}\right] =$$

(since $\mathbb{E}\left[M^\lambda_t\right] = \mathbb{E}\left[M^\lambda_{t\wedge\tau_n}\right] = 1$)

$$= 2\mathbb{E}\left[M^\lambda_t - M_{n,t}\right]$$

which proves the thesis. □

13.4 Approximation by Exponential Martingales

Another reason for interest in exponential martingales is the fact that they are a useful approximation tool. Hereafter, W is a Brownian motion on the space $(\Omega, \mathscr{F}, P)$ equipped with the standard Brownian filtration $\mathscr{F}^W$: the choice of this particular filtration is crucial for the validity of the following results. The next theorem is the main ingredient in the proof of the Brownian martingale representation theorem that we will present in Sect. 13.5.

> The proofs of this section are a bit technical and can be skipped at a first reading.

Theorem 13.4.1 *The space of linear combinations of random variables of the form*

$$M^\lambda_T = \exp\left(-\int_0^T \lambda(t)dW_t - \frac{1}{2}\int_0^T \lambda(t)^2 dt\right),$$

with λ deterministic function in $L^\infty([0, T])$, is dense in $L^2(\Omega, \mathscr{F}^W_T)$.

13.4 Approximation by Exponential Martingales

The proof of Theorem 13.4.1 is based on the following

Lemma 13.4.2 *Let $(t_n)_{n \in \mathbb{N}}$ be a sequence dense in $[0, T]$. The family of random variables of the form*

$$\varphi(W_{t_1}, \ldots, W_{t_n}), \qquad \varphi \in C_0^\infty(\mathbb{R}^n), \quad n \in \mathbb{N},$$

is dense in $L^2(\Omega, \mathscr{F}_T^W)$.

Proof The discrete filtration defined by

$$\mathscr{G}_n := \sigma(W_{t_1}, \ldots, W_{t_n}), \qquad n \in \mathbb{N},$$

is such that $\sigma(\mathscr{G}_n, n \in \mathbb{N}) = \mathscr{G}_T^W$ where $\mathscr{G}^W$ denotes the filtration generated by Brownian motion. Given $X \in L^2(\Omega, \mathscr{F}_T^W)$, later we will prove that

$$\lim_{n \to \infty} E\left[|X - X_n|^2\right] = 0, \qquad X_n := E[X \mid \mathscr{G}_n], \quad n \in \mathbb{N}. \tag{13.4.1}$$

Since $X_n \in m\mathscr{G}_n$, by Doob's Theorem 2.3.3 in [113] we have

$$X_n = \varphi_n(W_{t_1}, \ldots, W_{t_n})$$

for some measurable function φ_n that is square-integrable respect to the law $\mu_{W_{t_1}, \ldots, W_{t_n}}$: by density, φ_n can be approximated in L^2 by a sequence $(\varphi_{n,k})_{k \in \mathbb{N}}$ in $C_0^\infty(\mathbb{R}^n)$ and we also have

$$\lim_{k \to \infty} \varphi_{n,k}(W_{t_1}, \ldots, W_{t_n}) = X_n, \qquad \text{in } L^2(\Omega, P),$$

which proves the thesis.

It remains to prove (13.4.1). By Doob's maximal inequality (8.1.3) we have

$$E\left[\sup_{n \in \mathbb{N}} X_n^2\right] \leq 4E\left[X^2\right] < \infty. \tag{13.4.2}$$

Then, by Theorem 8.2.2 on the convergence of discrete martingales, there exists the a.s. pointwise limit

$$M := \lim_{n \to \infty} X_n.$$

Moreover, since

$$(X_n - M)^2 \leq 2(X_n^2 + M^2) \leq 2 \sup_{n \in \mathbb{N}} X_n^2,$$

by (13.4.2) and the dominated convergence theorem, we also have

$$\lim_{n \to \infty} X_n = M \quad \text{in } L^2(\Omega, P).$$

Setting $M_n = E[M \mid \mathcal{G}_n]$, we have

$$E\left[(X_n - M_n)^2\right] = E\left[(X_n - E[M \mid \mathcal{G}_n])^2\right] = E\left[(E[X_n - M \mid \mathcal{G}_n])^2\right] \leq$$

(by Jensen's inequality)

$$\leq E\left[(X_n - M)^2\right] \xrightarrow[n \to \infty]{} 0.$$

To conclude, let us prove that $M = E\left[X \mid \mathcal{F}_T^W\right] = X$ so that $M = X$ almost surely. First, $M \in m\mathcal{G}_T^W \subseteq m\mathcal{F}_T^W$; then, fixed $\bar{n} \in \mathbb{N}$, for $Z \in b\mathcal{G}_{\bar{n}}$ and $n \geq \bar{n}$ we have

$$E[Z(M - X)] = E[ZE[M - X \mid \mathcal{G}_n]] = E[Z(M_n - X_n)] \xrightarrow[\bar{n} \leq n \to \infty]{} 0 \tag{13.4.3}$$

due to (13.4.3). Since the elements of $\mathcal{F}_T^W$ and $\mathcal{G}_T^W$ differ only for negligible events, it follows that $M = E\left[X \mid \mathcal{F}_T^W\right]$. □

Proof of Theorem 13.4.1 It is sufficient to prove that if $X \in L^2(\Omega, \mathcal{F}_T^W)$ and, for every $\lambda \in L^\infty([0, T])$,

$$\langle X, M_T^\lambda \rangle_{L^2(\Omega)} = E\left[XM_T^\lambda\right] = 0 \tag{13.4.4}$$

then $X = 0$ almost surely.

From (13.4.4), choosing a piecewise constant λ, we have

$$F(\eta) := E\left[X e^{\eta_1 W_{t_1} + \cdots + \eta_n W_{t_n}}\right] = 0, \quad \eta \in \mathbb{R}^n, \ t_1, \ldots, t_n \in [0, T],$$

and the analytic extension of F to $\mathbb{C}^n$, by the theorem of analytic continuation, is identically zero. Then, for every $\varphi \in C_0^\infty(\mathbb{R}^n)$, by Theorem 2.5.6 in [113] on Fourier inversion, we have

$$E\left[X\varphi(W_{t_1}, \ldots, W_{t_n})\right] = E\left[\frac{X}{(2\pi)^n} \int_{\mathbb{R}^n} e^{-i(\eta_1 W_{t_1} + \cdots + \eta_n W_{t_n})} \hat{\varphi}(\eta) d\eta\right]$$

$$= \frac{1}{(2\pi)^n} \int_{\mathbb{R}^n} \hat{\varphi}(\eta) E\left[e^{-i(\eta_1 W_{t_1} + \cdots + \eta_n W_{t_n})} X\right] d\eta = 0,$$

and the thesis follows from Lemma 13.4.2. □

13.5 Representation of Brownian Martingales

The Brownian stochastic integral constructed in Chap. 10 is a continuous local martingale. The following result shows that, conversely, every local martingale with respect to the standard Brownian filtration $\mathscr{F}^W$ admits a representation as a stochastic integral.

Theorem 13.5.1 (Representation of Brownian Martingales [!!!]) *Let W be a Brownian motion on the space $(\Omega, \mathscr{F}, P)$ equipped with the standard Brownian filtration $\mathscr{F}^W$. If $X = (X_t)_{t \in [0,T]}$ is a càdlàg version of a local martingale on $(\Omega, \mathscr{F}, P, \mathscr{F}^W)$ then there exists a unique $u \in \mathbb{L}^2_{loc}$ such that*

$$X_t = X_0 + \int_0^t u_s dW_s, \qquad t \in [0, T]. \tag{13.5.1}$$

In particular, X is an a.s. continuous process.

Remark 13.5.2 Theorem 13.5.1 strengthens the result proven in Sect. 8.2 as it states that every Brownian local martingale admits a continuous modification, not just a càdlàg one.

Before presenting the proof of Theorem 13.5.1, we preface it with the following proposition, grounded on the approximation results elaborated in Sect. 13.4.

Proposition 13.5.3 ([!]) *For every random variable $X \in L^2(\Omega, \mathscr{F}^W_T)$ there exists a unique $u \in \mathbb{L}^2$ such that*

$$X = E[X] + \int_0^T u_t dW_t. \tag{13.5.2}$$

Proof We restrict our attention to the one-dimensional case for simplicity. As for uniqueness, if $u, v \in \mathbb{L}^2$ satisfy (13.5.2), then

$$\int_0^T (u_t - v_t) dW_t = 0$$

and from Itô's isometry it follows that $P(u = v \text{ a.e. on } [0, T]) = 1$ (cf. Remark 10.2.18).

As for the existence part, the proof is straightforward if X is of the form

$$X = M_T^\lambda := \exp\left(-\int_0^T \lambda(t) dW_t - \frac{1}{2} \int_0^T \lambda(t)^2 dt\right) \tag{13.5.3}$$

with $\lambda \in L^\infty([0, T])$ deterministic function. Indeed, by Itô's formula we have

$$X = 1 - \int_0^T \lambda(t) M_t^\lambda dW_t$$

with $\lambda M^\lambda \in \mathbb{L}^2$ by Proposition 13.2.1 and therefore, in particular, $E[X] = E[M_T^\lambda] = 1$ by the martingale property.

In general, according to Theorem 13.4.1 every $X \in L^2(\Omega, \mathscr{F}_T^W)$ is approximated in L^2 by a sequence $(X_n)_{n \in \mathbb{N}}$ of linear combinations of variables of the form (13.5.3) for which

$$X_n = E[X_n] + \int_0^T u_{n,t} dW_t \tag{13.5.4}$$

with $u_n \in \mathbb{L}^2$. By Itô's isometry we have

$$E\left[(X_n - X_m)^2\right] = (E[X_n - X_m])^2 + E\left[\int_0^T (u_{n,t} - u_{m,t})^2 dt\right],$$

and thus $(u_n)_{n \in \mathbb{N}}$ is a Cauchy sequence in $\mathbb{L}^2$. The thesis follows by taking the limit in (13.5.4). □

Proof of Theorem 13.5.1 The uniqueness of u follows from the uniqueness of the representation of an Itô process (cf. Remark 10.4.3).

As for the existence, let us first consider the case where X is a martingale such that $X_T \in L^2(\Omega, P)$. By Theorem 13.5.3, there exists $u \in \mathbb{L}^2$ such that

$$X_T = E[X_T] + \int_0^T u_t dW_t,$$

from which (13.5.1) follows, simply by applying the conditional expectation to $\mathscr{F}_t^W$ for every $t \in [0, T]$. In particular, we have demonstrated that X possesses a continuous modification.

Now we remove the assumption $X_T \in L^2(\Omega, P)$ and prove that every $\mathscr{F}^W$-martingale X admits a continuous modification. Since $X_T \in L^1(\Omega, P)$ and $L^2(\Omega, P)$ is dense in $L^1(\Omega, P)$, there exists a sequence $(Y_n)_{n \in \mathbb{N}}$ of random variables in $L^2(\Omega, P)$ such that

$$E[|Y_n - X_T|] \leq \frac{1}{2^n}, \qquad n \in \mathbb{N}.$$

For the previous point, the sequence of martingales

$$X_{n,t} := E\left[Y_n \mid \mathscr{F}_t^W\right], \qquad t \in [0, T],$$

13.5 Representation of Brownian Martingales

admits a continuous modification and by Doob's maximal inequality, Theorem 8.1.2, we have

$$P\left(\sup_{t\in[0,T]} |X_{n,t} - X_t| \geq \frac{1}{k}\right) \leq kE\left[|X_{n,T} - X_T|\right] \leq \frac{k}{2^n}, \qquad k, n \in \mathbb{N}.$$

From Borel-Cantelli's Lemma 1.3.28 in [113], it follows that, almost surely, $(X_n)_{n\in\mathbb{N}}$ converges uniformly on $[0, T]$ to the martingale X, which is therefore a.s. continuous.

If X is a local martingale, consider a localizing sequence $(\tau_n)_{n\in\mathbb{N}}$: the process $X_{t\wedge\tau_n} - X_0$ is a martingale and, as we have just proved, admits a continuous modification. Since

$$X_t 1\!\!1_{(\tau_n \geq T)} = X_{t\wedge\tau_n} 1\!\!1_{(\tau_n \geq T)}, \qquad t \in [0, T],\ n \in \mathbb{N}, \tag{13.5.5}$$

we deduce that also X admits a continuous modification.

Finally, we prove (13.5.1) under the assumption that X is a continuous local martingale. By Remark 8.4.6, there exists a localizing sequence $(\tau_n)_{n\in\mathbb{N}}$ such that $X_{t\wedge\tau_n} - X_0$ is a continuous and bounded martingale for every $n \in \mathbb{N}$. Then there exists a sequence $(u_n)_{n\in\mathbb{N}}$ in $\mathbb{L}^2$ such that

$$X_{t\wedge\tau_n} = X_0 + \int_0^t u_{n,s} dW_s, \qquad t \in [0, T]. \tag{13.5.6}$$

By (13.5.5) and Proposition 10.2.26, we can take the limit in (13.5.6) to conclude the proof. □

13.5.1 Proof of Theorem 13.1.1

By the Brownian martingale representation Theorem 13.5.1, there exists $u \in \mathbb{L}^2_{\text{loc}}$ such that the process M in (13.1.4) admits the representation

$$M_t = 1 + \int_0^t u_s dW_s, \qquad t \in [0, T].$$

Note that $\lambda_t := -\frac{u_t}{M_t}$ belongs to $\mathbb{L}^2_{\text{loc}}$ since M is an adapted, continuous, and strictly positive process. Consequently, we have

$$M_t = 1 - \int_0^t M_s \lambda_s dW_s, \qquad t \in [0, T],$$

that is, M solves a linear stochastic differential equation of which the exponential martingale M^λ in (13.1.1) is the unique[8] solution. Hence $M = M^\lambda$ in the sense of indistinguishability.

By construction, M is a martingale and therefore, by Girsanov Theorem 13.3.3, W^λ in (13.1.5) is a Brownian motion on $(\Omega, \mathscr{F}, Q, \mathscr{F}_t^W)$. Finally, we have

$$dX_t = b_t dt + \sigma_t dW_t =$$

(by (13.1.5))

$$= b_t dt + \sigma_t(dW_t^\lambda - \lambda_t dt)$$

from which (13.1.7) follows.

13.6 Key Ideas to Remember

We summarize the main findings and key concepts of the chapter, omitting technical details. As usual, if you have any doubt about what the following succinct statements mean, please review the corresponding section.

- Section 13.1: the exponential martingale M^λ in (13.1.1), with $\lambda \in \mathbb{L}^2_{\text{loc}}$, is the main tool used throughout the chapter. If M^λ is a true martingale, then it can be used as a density (or Radon-Nikodym derivative) to define a measure Q equivalent to the initially considered measure P. The process W^λ in (13.1.5), obtained by adding a drift λ to a Brownian motion, is a Brownian motion under the new measure Q. The idea is that there is a correspondence between changes in drift of a Brownian motion (and related Itô processes) and changes in probability measure: the drift coefficient λ acts as the exponent of the martingale M^λ, which is the Radon-Nikodym derivative of the change of measure.
- Section 13.1.1: the results on changes in drift and measure (often referred to as "Girsanov's change of measure" in financial jargon) are pivotal in modern financial derivatives valuation theory. It is worth noting that a Girsanov's change of measure alters the drift term of an Itô process while *leaving the diffusion coefficient unchanged.*
- Section 13.2: we provide sufficient conditions on the process λ for M^λ to be a true martingale. Novikov's condition is a classic condition that is often used in probability theory and mathematical finance.
- Section 13.3: the proof of Girsanov theorem is a relatively direct consequence of Proposition 4.4.2 that characterizes Brownian motion in terms of exponential martingales.

---

[8] The fact that M^λ is a solution is a simple check with Itô's formula. For uniqueness, it is not difficult to adapt the proof of Theorem 17.1.1 that we will prove later.

13.6 Key Ideas to Remember   261

- Sections 13.4 and 13.5: the proof of the Brownian martingale representation theorem is quite challenging and is based on a density result of exponential martingales in the space $L^2(\Omega, \mathscr{F}_T^W)$ where $\mathscr{F}^W$ indicates the standard Brownian filtration (which satisfies the usual conditions). A significant corollary is the fact that every local Brownian martingale admits a continuous modification.

Main notations used or introduced in this chapter:

Symbol	Description	Page
M^λ	Exponential martingale solution of $dM_t^\lambda = -M_t^\lambda \lambda_t dW_t$	244
$Q \sim P$	Equivalence between measures P and Q	244
$\frac{dQ}{dP}$	Radon-Nikodym derivative of Q with respect to P	244
$\tilde{X}_T = \sup_{0 \leq t \leq T} |X_t|$	Maximum process	247
W^λ	Brownian motion with drift λ	253
$\mathscr{G}^W$	Filtration generated by Brownian motion	255
$\mathscr{F}^W$	Standard Brownian filtration	257

Chapter 14
Stochastic Differential Equations

> *It seems fair to say that all differential equations are better models of the world when a stochastic term is added and that their classical analysis is useful only if it is stable in an appropriate sense to such perturbations.*
>
> *David Mumford*

Starting from this chapter, we begin the study of Stochastic Differential Equations, hereafter abbreviated as SDEs. As anticipated in Sect. 2.6, such equations were originally introduced for the construction of continuous Markov processes or diffusions. Over time, SDEs have become increasingly important in stochastic modeling across a wide range of fields. SDEs generalize deterministic differential equations by incorporating a random perturbation factor, which allows them to model systems that are subject to uncertainty. In addition, SDEs can be used to construct explicit examples of continuous semimartingales.

In this chapter, we introduce the notion of solution to an SDE and the related problems of existence and uniqueness. These problems have a dual formulation, in a weak and strong sense. We give a very particular existence and uniqueness result from which some peculiarities of SDEs compared to the usual deterministic equations can be deduced, including the so-called "regularization by noise" effect. We see that it is possible to transfer the study of an SDE to a canonical setting and analyze the relationship between weak and strong solvability. Finally, we prove some preliminary estimates of continuous dependence and integrability of solutions.

14.1 Solving SDEs: Concepts of Existence and Uniqueness

Hereafter, $N, d \in \mathbb{N}$ and $0 \leq t_0 < T$ are fixed constants. An SDE is an expression of the form

$$dX_t = b(t, X_t)dt + \sigma(t, X_t)dW_t \qquad (14.1.1)$$

where W is a d-dimensional Brownian motion and

$$b = b(t, x) :]t_0, T[\times \mathbb{R}^N \longrightarrow \mathbb{R}^N, \qquad \sigma = \sigma(t, x) :]t_0, T[\times \mathbb{R}^N \longrightarrow \mathbb{R}^{N \times d}, \qquad (14.1.2)$$

are measurable functions:[1] b is called the *drift* coefficient and σ the *diffusion* coefficient of the SDE. In (14.1.2) $\mathbb{R}^{N \times d}$ indicates the space of matrices of dimension $N \times d$. To simplify the presentation, we will always assume the following

Assumption 14.1.1 The functions b, σ are measurable and locally bounded in x uniformly in t (in short, we write $b, \sigma \in L^{\infty}_{\text{loc}}(]t_0, T[\times \mathbb{R}^N)$): precisely, for each $n \in \mathbb{N}$ there is a constant κ_n such that

$$|b(t, x)| + |\sigma(t, x)| \leq \kappa_n, \qquad t \in]t_0, T[, \ |x| \leq n.$$

Remark 14.1.2 Though introducing slightly denser notation, we opt to generalize the initial time as t_0 rather than strictly setting it to zero. We anticipate that this approach will enhance comprehension of the theory of "strong solutions" discussed in Chap. 17, along with pivotal results like the flow property of solutions and parameter dependence estimates. Beginning from Chap. 18, for the sake of simplicity, we will revert to setting t_0 as zero.

Before giving the definition of solution to an SDE, it is necessary to properly set the problem through the following

Definition 14.1.3 (Set-up) A set-up $(W, \mathscr{F}_t)$ on $[t_0, T]$ consists of:

- a filtered probability space $(\Omega, \mathscr{F}, P, (\mathscr{F}_t)_{t \in [t_0, T]})$;

---

[1] More generally, it is possible to study equations whose coefficients depend stochastically on the time variable. This type of equation intervenes, for example, in the study of optimal control problems and stochastic filtering. We will restrict our attention to deterministic coefficients. We refer, for example, to [77] and [66] for a general treatment.

14.1 Solving SDEs: Concepts of Existence and Uniqueness

- a d-dimensional Brownian motion[2] $W = (W_t)_{t \in [t_0,T]}$ on $(\Omega, \mathscr{F}, P, \mathscr{F}_t)$, starting at time t_0.

Remark 14.1.4 We explicitly note that $\mathscr{F}_{t_0}$ is independent of W_t for $t \geq t_0$ and therefore also from the standard Brownian filtration $(\mathscr{F}_t^W)_{t \in [t_0, T]}$ that verifies the usual conditions.

Definition 14.1.5 (Solution of an SDE) A *solution of the SDE with coefficients b, σ on the set-up* $(W, \mathscr{F}_t)$ is an N-dimensional process $X = (X_t)_{t \in [t_0, T]}$ defined on the same space as W and such that:

(i) X is continuous and adapted, i.e., $X_t \in m\mathscr{F}_t$ for every $t \in [t_0, T]$;
(ii) almost surely we have[3]

$$X_t = X_{t_0} + \int_{t_0}^t b(s, X_s)ds + \int_{t_0}^t \sigma(s, X_s)dW_s, \qquad t \in [t_0, T]. \qquad (14.1.4)$$

---

[2] On the probability space $(\Omega, \mathscr{F}, P, (\mathscr{F}_t)_{t \in [t_0,T]})$, we say that $W = (W_t)_{t \in [t_0,T]}$ is a Brownian motion starting at time t_0 if:

(i) $W_{t_0} = 0$ a.s.;
(ii) W is a.s. continuous;
(iii) W is adapted to $(\mathscr{F}_t)_{t \in [t_0,T]}$;
(iv) $W_t - W_s$ is independent of $\mathscr{F}_s$ for every $t_0 \leq s \leq t \leq T$;
(v) $W_t - W_s \sim \mathcal{N}_{0,(t-s)I}$ for every $t_0 \leq s \leq t \leq T$, where I denotes the $d \times d$ identity matrix.

For example, let $B = (B_t)_{t \geq 0}$ be a standard Brownian motion on $(\Omega, \mathscr{F}, P, (\mathscr{F}_t)_{t \geq 0})$; then $W_t := B_t - B_{t_0}$ is a Brownian motion starting at time t_0 on $(\Omega, \mathscr{F}, P)$ with respect to the filtration $(\mathscr{F}_t)_{t \geq t_0}$ or even with respect to the standard filtration defined by

$$\mathscr{F}_t^W := \sigma(\mathscr{G}_t^W \cup \mathcal{N}), \qquad \mathscr{G}_t^W := \sigma(W_s, t_0 \leq s \leq t), \qquad t_0 \leq t \leq T.$$

Note that there is a strict inclusion $\mathscr{F}_t^W \subset \mathscr{F}_t^B$ in the case $t_0 > 0$. Moreover, since the stochastic integral depends only on the Brownian increments (cf. Corollary 10.2.27), we have a.s.

$$\int_{t_0}^t u_s dB_s = \int_{t_0}^t u_s dW_s, \qquad t \geq t_0.$$

[3] That is, there exists a version of the stochastic integral

$$t \longmapsto \int_{t_0}^t \sigma(s, X_s)dW_s$$

such that (14.1.4) holds for every $t \in [t_0, T]$ almost surely. We explicitly note that, under the local boundedness Assumption 14.1.1, we have

$$\int_{t_0}^T |b(t, X_t)|dt + \int_{t_0}^T |\sigma(t, X_t)|^2 dt < \infty \qquad \text{a.s.} \qquad (14.1.3)$$

and therefore the integrals in (14.1.4) are well defined.

To indicate that X is a solution of the SDE with coefficients b, σ on the set-up $(W, \mathscr{F}_t)$ we write

$$X \in \text{SDE}(b, \sigma, W, \mathscr{F}_t).$$

It is customary to associate an SDE with an "initial condition" that can be assigned *pointwise* through a random variable $Z \in m\mathscr{F}_{t_0}$ if the set-up $(W, \mathscr{F}_t)$ has been previously fixed or, as we will see later, *in law* through a distribution μ_0 on $\mathbb{R}^N$.

Definition 14.1.6 (Strong Solution of an SDE) Given a set-up $(W, \mathscr{F}_t)$ and an initial datum $Z \in m\mathscr{F}_{t_0}$, we denote by

$$\mathscr{F}^{Z,W} = (\mathscr{F}_t^{Z,W})_{t \in [t_0, T]}$$

the filtration generated by W and Z, completed so that it satisfies the usual conditions.[4] We say that a solution $X \in \text{SDE}(b, \sigma, W, \mathscr{F}_t)$, such that $X_{t_0} = Z$, is a *strong solution* if it is adapted to the filtration $\mathscr{F}^{Z,W}$.

Remark 14.1.7 ([!]) Strong solutions are characterized by the property of being adapted to the filtration $\mathscr{F}^{Z,W}$: since $\mathscr{F}^{Z,W}$ is the smallest[5] filtration with respect to which a solution of the SDE can be defined, this measurability condition is the most restrictive possible.

If the initial datum is deterministic, i.e., $Z \in \mathbb{R}^N$, then a strong solution is adapted to the standard Brownian filtration $\mathscr{F}^W$. This means that, through the SDE, a process (the solution) X is associated with W and X is a "functional" of W, meaning that X_t can be expressed as a function of the process $(W_s)_{s \in [t_0, t]}$. This remark is relevant since in various applications, such as in signal theory, W represents a set of observed data that are used as "input" for a dynamical system (formalized by the SDE) that produces the solution X as "output": in this case, it is crucial that the output can be represented as a function of the input data. In other fields, such as mathematical finance, it may be sufficient to consider a weaker notion of solution, in particular if one is only interested in applications where what is relevant is the law of the solution.

Example 14.1.8 When the coefficients $b = b(t)$ and $\sigma = \sigma(t)$ of the SDE (14.1.1) are deterministic L^∞ functions of the time variable only, the solution of the corresponding SDE is the Itô process

$$X_t = Z + \int_{t_0}^t b(s)ds + \int_{t_0}^t \sigma(s)dW_s.$$

---

[4] By Theorem 6.2.22 and the independence of Z from $\mathscr{F}^W$ (cf. Remark 14.1.4), W is a Brownian motion also with respect to $\mathscr{F}^{Z,W}$.

[5] The smallest filtration verifying the usual conditions.

14.1 Solving SDEs: Concepts of Existence and Uniqueness

We recall from Example 11.1.9 that if also the initial datum is deterministic, then X_t is a Gaussian process.

Next, we present two formulations of the problem concerning the existence of solutions to an SDE.

Definition 14.1.9 (Solvability of an SDE) We say that the SDE with coefficients b, σ is solvable

- **in the weak sense,** if for every distribution μ_0 on $\mathbb{R}^N$ there exist a set-up $(W, \mathscr{F}_t)$ and a solution $X \in \text{SDE}(b, \sigma, W, \mathscr{F}_t)$ such that $X_{t_0} \sim \mu_0$;
- **in the strong sense,** if for every set-up $(W, \mathscr{F}_t)$ and $Z \in m\mathscr{F}_{t_0}$ there exists a strong solution $X \in \text{SDE}(b, \sigma, W, \mathscr{F}_t^{Z,W})$ such that $X_{t_0} = Z$ a.s.

Although it may seem counter-intuitive, it is possible for a process to satisfy an equation of the type

$$X_t = x + \int_0^t b(s, X_s)ds + \int_0^t \sigma(s, X_s)dW_s$$

with deterministic initial datum $x \in \mathbb{R}^N$, *and not be adapted to* $\mathscr{F}^W$: in other words, in some instances, a solution X needs to possess additional randomness beyond that induced by the Brownian motion with respect to which the SDE is formulated. A famous example is due to Tanaka [139] (see also [154]): here we describe the general idea and refer to Section 9.2.1 in [112] or Example 3.5, Chapter 5 in [67] for details.

Example 14.1.10 (Tanaka [!]) Consider the scalar (i.e., with $N = d = 1$) SDE

$$dX_t = \sigma(X_t)dW_t \qquad (14.1.5)$$

with null drift and initial datum, $b = Z = 0$, and diffusion coefficient

$$\sigma(x) = \text{sgn}(x) := \begin{cases} 1 & \text{if } x \geq 0, \\ -1 & \text{if } x < 0. \end{cases}$$

To prove that the SDE (14.1.5) is solvable in the weak sense, consider a Brownian motion X defined on the space $(\Omega, \mathscr{F}, P, \mathscr{F}^X)$. The process

$$W_t := \int_0^t \sigma(X_s)dX_s \qquad (14.1.6)$$

is a continuous martingale with quadratic variation $\langle W \rangle_t = t$ and consequently, by Theorem 12.4.1, it is also a Brownian motion on $(\Omega, \mathscr{F}, P, \mathscr{F}^X)$. Since $\sigma^2 \equiv 1$, from the definition $dW_t = \sigma(X_t)dX_t$ we obtain

$$dX_t = \sigma^2(X_t)dX_t = \sigma(X_t)dW_t$$

which means that X is a solution of the SDE (14.1.5) with respect to W, i.e., $X \in \text{SDE}(0, \sigma, W, \mathscr{F}^X)$, with null initial datum. The crucial point is that it can be proved[6] that W, defined by (14.1.6), is adapted to the standard filtration $\mathscr{F}^{|X|}$ of the absolute value process $|X|$: if X were adapted to $\mathscr{F}^W$ then it should also be adapted to $\mathscr{F}^{|X|}$ and this is absurd. This example may seem a bit pathological because the coefficient σ is a discontinuous function: more recently Barlow [7] has shown that for every $\alpha < \frac{1}{2}$ there exists an α-Hölder continuous function σ that is bounded above and below by positive constants, and such that the SDE (14.1.5) is solvable in the weak sense but not in the strong sense.

In conclusion, an SDE can be solvable in the weak sense without being solvable in the strong sense: weak solvability is less restrictive because it gives the freedom to choose the space, the Brownian motion, and the filtration with respect to which to write the SDE. On the contrary, strong solutions are constrained to be adapted to the standard filtration $\mathscr{F}^{Z,W}$ of the initial datum Z and the Brownian motion W.

Just like for existence, there exist different notions of uniqueness for the solution of an SDE.

Definition 14.1.11 (Uniqueness for an SDE) We say that for the SDE with coefficients b, σ there is uniqueness

- **in the strong sense,** if the fact that $X \in \text{SDE}(b, \sigma, W, \mathscr{F}_t)$ and $Y \in \text{SDE}(b, \sigma, W, \mathscr{G}_t)$ with $X_{t_0} = Y_{t_0}$ a.s. implies that X and Y are indistinguishable processes;
- **in the weak sense (or in law),** if the fact that $X \in \text{SDE}(b, \sigma, W, \mathscr{F}_t)$ and $Y \in \text{SDE}(b, \sigma, B, \mathscr{G}_t)$, with $X_{t_0} \stackrel{d}{=} Y_{t_0}$, implies that $(X, W) \stackrel{d}{=} (Y, B)$ or, equivalently, (X, W) and (Y, B) have the same finite-dimensional distributions.

In the definition of strong uniqueness, the two processes X and Y are defined on the *same* probability space $(\Omega, \mathscr{F}, P)$ and are solutions of the SDE on the setups $(W, \mathscr{F}_t)$ and $(W, \mathscr{G}_t)$, respectively: here W is a Brownian motion with respect to both filtrations $(\mathscr{F}_t)$ and $(\mathscr{G}_t)$ which can be different. Strong uniqueness is also known as "pathwise uniqueness". In the definition of uniqueness in law, the processes X and Y can be solutions on different set-ups $(W, \mathscr{F}_t)$ and $(B, \mathscr{G}_t)$, even defined on different probability spaces.

Example 14.1.12 ([!]) For the SDE in Example 14.1.10, there is weak but not strong uniqueness. In fact, every solution X of the SDE (14.1.5) is a local martingale with $\langle X \rangle_t = t$ and therefore, by Lévy's characterization Theorem 12.4.1, X is a Brownian motion: hence there is uniqueness in law.

On the other hand, if X is the weak solution constructed in Example 14.1.10, we can verify that also $-X$ is a solution of the SDE and therefore there is no strong

---

[6] Here the Meyer-Tanaka formula is used: see, for example, Section 5.3.2 in [112] or Section 2.11 in [37].

14.2 Weak Existence and Uniqueness via Girsanov Theorem

uniqueness: in fact, since $\sigma(-x) = -\sigma(x)$ if $x \neq 0$, we have

$$\int_0^t \sigma(-X_s)dW_s = -\int_0^t \sigma(X_s)dW_s + 2\int_0^t \mathbb{1}_{(X_s=0)}dW_s$$
$$= -\int_0^t \sigma(X_s)dW_s \qquad \text{a.s.}$$

since, by Itô's isometry,

$$E\left[\left(\int_0^t \mathbb{1}_{(X_s=0)}dW_s\right)^2\right] = \int_0^t E\left[\mathbb{1}_{(X_s=0)}\right]ds = 0.$$

Here we used the fact that $P(X_s = 0) = 0$ for every $s \geq 0$ since X is a Brownian motion.

Remark 14.1.13 ([!]) Theorem 14.3.6, by Yamada and Watanabe, states that *if an SDE is solvable in the strong sense then it is also solvable in the weak sense.* Furthermore, *strong uniqueness implies uniqueness in law*: while this result may seem intuitive, its proof is not straightforward; indeed, strong uniqueness pertains to solutions defined on the same space, whereas proving weak uniqueness requires dealing with solutions that may be defined on different spaces. Finally, we also have that *if for an SDE there is strong uniqueness then every solution is a strong solution*.

Remark 14.1.14 Recently, a further notion of uniqueness for SDEs, called "path-by-path uniqueness", has also been studied: see in this regard [31, 48] and [130].

14.2 Weak Existence and Uniqueness via Girsanov Theorem

There are many ways to prove weak existence and uniqueness for an SDE. In this section, we examine a very particular technique that exploits the results on changes of measure of Chap. 13. The following remarkable Theorem 14.2.3 is an example of the so-called "regularizing effect of Brownian motion", whereby weak existence and uniqueness for an SDE are obtained under minimal regularity assumptions on the drift coefficient, which is here assumed to be only measurable and bounded. Under such assumptions, the corresponding ordinary differential equation (without the Brownian part) does not generally have a unique solution as shown by the well-known

Example 14.2.1 (Peano's Brush) The SDE (14.1.1) with $b(t, x) = |x|^\alpha$, $\sigma = 0$ and null initial datum reduces to the Volterra integral equation

$$X_t = \int_0^t |X_s|^\alpha ds. \qquad (14.2.1)$$

Formula (14.2.1) has the null function as its unique solution if $\alpha \geq 1$, while if $\alpha \in {]}0,1[$ there are infinite solutions of the form

$$X_t = \begin{cases} 0 & \text{if } 0 \leq t \leq s, \\ \left(\frac{t-s}{\beta}\right)^\beta & \text{if } s \leq t \leq T, \end{cases}$$

where $\beta = \frac{1}{1-\alpha}$ and $s \in [0, T]$.

A similar phenomenon also occurs in the stochastic case.

Example 14.2.2 (Itô and Watanabe [64] [!]) The SDE

$$dX_t = 3X_t^{\frac{1}{3}}dt + 3X_t^{\frac{2}{3}}dW_t, \qquad X_0 = 0,$$

has infinite strong solutions of the form

$$X_t^{(a)} = \begin{cases} 0 & \text{for } 0 \leq t < \tau_a, \\ W_t^3 & \text{for } t \geq \tau_a, \end{cases}$$

where $a \in [0, +\infty]$ and $\tau_a = \inf\{t \geq a \mid W_t = 0\}$. For $a = +\infty$ and $a = 0$, we have the solutions $X_t^{(+\infty)} \equiv 0$ and $X_t^{(0)} = W_t^3$, respectively.

In light of the previous examples, the following result is quite surprising and documents the regularizing effect of Brownian motion.

Theorem 14.2.3 (Zvonkin [154], Veretennikov [144]) *Suppose that the coefficient*

$$b : {]}0, T[\times \mathbb{R}^d \longrightarrow \mathbb{R}^d$$

is a Borel-measurable and bounded function. Then the SDE

$$dX_t = b(t, X_t)dt + dW_t \qquad (14.2.2)$$

is solvable in the weak sense and the solution is unique in law.

Proof

Existence Let μ_0 be a distribution on $\mathbb{R}^d$ and let X be a d-dimensional Brownian motion with initial value $X_0 \sim \mu_0$ (cf. Exercise 8.4.5) defined on the space $(\Omega, \mathscr{F}, P, \mathscr{F}_t)$. By the boundedness of b and Proposition 13.2.1, we have that

$$M_t := \exp\left(\int_0^t b(s, X_s)dX_s - \frac{1}{2}\int_0^t |b(s, X_s)|^2 ds\right), \qquad t \in [0, T],$$
$$(14.2.3)$$

14.2 Weak Existence and Uniqueness via Girsanov Theorem

is a martingale. Then, by Theorem 13.1.1, the process

$$W_t := X_t - X_0 - \int_0^t b(s, X_s)ds \tag{14.2.4}$$

is a standard Brownian motion under the measure Q defined by $\frac{dQ}{dP} = M_T$. Formula (14.2.4) shows that X is a weak solution of the SDE (14.2.2) under the measure Q. Moreover

$$Q(X_0 \in H) = E^P[\mathbb{1}_H(X_0) M_T] = E^P\left[\mathbb{1}_H(X_0) E^P[M_T \mid \mathscr{F}_0]\right] = P(X_0 \in H)$$

by the martingale property of the process M, and therefore $X_0 \sim \mu_0$ under Q.

Uniqueness Let $X^{(i)}$, $i = 1, 2$, be solutions of the SDE (14.2.2) on the setups $(W^{(i)}, \mathscr{F}_t^{(i)})$ defined on the spaces $(\Omega_i, \mathscr{F}^{(i)}, P_i)$, respectively. Assume that $X_0^{(1)}$ and $X_0^{(2)}$ are equal in law. Again, by the boundedness of b and Proposition 13.2.1, the processes

$$M_t^{(i)} := \exp\left(-\int_0^t b(s, X_s^{(i)})dW_s^{(i)} - \frac{1}{2}\int_0^t |b(s, X_s^{(i)})|^2 ds\right), \qquad t \in [0, T], \tag{14.2.5}$$

are martingales. From Theorem 13.1.1 it follows that

$$X_t^{(i)} = X_0^{(i)} + \int_0^t b(s, X_s^{(i)})ds + W_t^{(i)} \tag{14.2.6}$$

are Brownian motions respectively on the spaces $(\Omega_i, \mathscr{F}^{(i)}, Q_i, \mathscr{F}_t^{(i)})$ where $\frac{dQ_i}{dP} = M_T^{(i)}$. Therefore, the law of $X^{(1)}$ in Q_1 is equal to the law of $X^{(2)}$ in Q_2: from (14.2.5), (14.2.6), and Corollary 10.2.28, it follows that the law of $(X^{(1)}, W^{(1)}, M^{(1)})$ in Q_1 is equal to the law of $(X^{(2)}, W^{(2)}, M^{(2)})$ in Q_2. Finally, for every $0 \leq t_1 < \cdots < t_n \leq T$ and $H \in \mathscr{B}_{2nd}$ we have

$$P_1((X_{t_1}^{(1)}, W_{t_1}^{(1)}, \ldots, X_{t_n}^{(1)}, W_{t_n}^{(1)}) \in H) = \int_{\Omega_1} \mathbb{1}_H(X_{t_1}^{(1)}, W_{t_1}^{(1)}, \ldots, X_{t_n}^{(1)}, W_{t_n}^{(1)}) \frac{dQ_1}{M_T^{(1)}}$$

$$= \int_{\Omega_2} \mathbb{1}_H(X_{t_1}^{(2)}, W_{t_1}^{(2)}, \ldots, X_{t_n}^{(2)}, W_{t_n}^{(2)}) \frac{dQ_2}{M_T^{(2)}}$$

$$= P_2((X_{t_1}^{(2)}, W_{t_1}^{(2)}, \ldots, X_{t_n}^{(2)}, W_{t_n}^{(2)}) \in H)$$

which proves the thesis.

$\square$

Remark 14.2.4 Theorem 14.2.3 can be extended in various directions. Using the Novikov condition (Theorem 13.2.6) to prove that the process in (14.2.3) is a martingale, one proves the existence of a weak solution of the SDE (14.2.2) under the more general assumption of linear growth in x (in addition to measurability) of the coefficient b: for more details see, for example, Proposition 5.3.6 in [67].

In Sect. 18.4 we will prove a "strong version" of Theorem 14.2.3, under the more restrictive assumption that $b = b(t, x)$ is a bounded and Hölder continuous function in the variable x, uniformly in t.

14.3 Weak vs Strong Solutions: The Yamada-Watanabe Theorem

We examine the relationship between strong and weak solvability. For simplicity, we assume $t_0 = 0$ and, given $N, d \in \mathbb{N}$ and $T > 0$, we consider an SDE with coefficients

$$b = b(t, x) :]0, T[\times \mathbb{R}^N \longrightarrow \mathbb{R}^N, \qquad \sigma = \sigma(t, x) :]0, T[\times \mathbb{R}^N \longrightarrow \mathbb{R}^{N \times d}.$$

Furthermore, we let μ_0 be a distribution on $\mathbb{R}^N$ that we will use as the initial condition.

> Since the result of this section are rather technical, on a first reading it is recommended to read the statements and skip the proofs.

Definition 14.3.1 (Weak Solution of an SDE) The SDE with coefficients b, σ and initial law μ_0 is *solvable in the weak sense* if there exist a set-up $(W, \mathscr{F}_t)$ and a solution $X \in \mathrm{SDE}(b, \sigma, W, \mathscr{F}_t)$ such that $X_0 \sim \mu_0$. In this case, almost surely

$$X_t = X_0 + \int_0^t b(s, X_s) ds + \int_0^t \sigma(s, X_s) dW_s, \qquad t \in [0, T], \qquad (14.3.1)$$

and we say that *the pair (X, W) is a weak solution of the SDE with coefficients b, σ and initial law μ_0*.

Remark 14.3.2 ([!]) To prove that an SDE is solvable in the weak sense, it is necessary to construct not only the process X but also the set-up $(W, \mathscr{F}_t)$ on which the SDE is written: for this reason, the weak solution is typically referred to as the *pair (X, W)*, not just the process X.

We now see that *it is always possible to transfer the problem of weak solvability of an SDE to a "canonical setting"*.

14.3 Weak vs Strong Solutions: The Yamada-Watanabe Theorem

Notation 14.3.3 Given $n \in \mathbb{N}$, we denote by

$$\mathbf{\Omega}_n = C([0, T]; \mathbb{R}^n)$$

the space of continuous n-dimensional trajectories equipped with the filtration $(\mathscr{G}_t^n)_{t \in [0,T]}$ generated by the identity process

$$\mathbf{X}_t(w) := w(t), \qquad w \in \mathbf{\Omega}_n, \ t \in [0, T],$$

and the Borel σ-algebra[7] $\mathscr{G}_T^n$.

Remark 14.3.4 If the process (X, W), defined on the space $(\Omega, \mathscr{F}, P)$, is a solution of the SDE (14.3.1) then its law $\mu_{X,W}$ is the distribution on $\mathbf{\Omega}_{N+d} = \mathbf{\Omega}_N \times \mathbf{\Omega}_d$ defined by

$$\mu_{X,W}(H) = P((X, W) \in H), \qquad H \in \mathscr{G}_T^{N+d}.$$

Hereafter, we will repeatedly use the fact that $\mathbf{\Omega}_{N+d}$ is a Polish space on which, thanks to Theorem 4.3.2 in [113], it is possible to define a *regular version of the conditional probability*. The following lemma is a crucial ingredient in all subsequent analysis.

Lemma 14.3.5 (Transfer of Solutions [!]) *If (X, W) is a weak solution of the SDE with coefficients b, σ and initial law μ_0 on the space $(\Omega, \mathscr{F}, P)$, then the canonical process $(\mathbf{X}, \mathbf{W})$ defined by*

$$\mathbf{X}_t(x, w) := x(t), \quad \mathbf{W}_t(x, w) := w(t), \quad (x, w) \in \mathbf{\Omega}_{N+d}, \ t \in [0, T],$$

is a weak solution of the SDE with coefficients b, σ and initial law μ_0 on the space $(\mathbf{\Omega}_{N+d}, \mathscr{G}_T^{N+d}, \mu_{X,W})$.

Proof We have the scheme

$$(\Omega, \mathscr{F}, P) \xrightarrow{(X,W)} (\mathbf{\Omega}_{N+d}, \mathscr{G}_T^{N+d}, \mu_{X,W}) \xrightarrow{(\mathbf{X},\mathbf{W})} (\mathbf{\Omega}_{N+d}, \mathscr{G}_T^{N+d})$$

and by construction, $(X, W) \stackrel{d}{=} (\mathbf{X}, \mathbf{W})$. The fact that $\mathbf{W}$ is a Brownian motion is a consequence[8] of the equality in law of (X, W) and $(\mathbf{X}, \mathbf{W})$. Suppose for the moment

---

[7] We saw in Proposition 3.2.1 that, in the space of continuous trajectories, the σ-algebra generated by cylinders (or, equivalently, by the identity process) coincides with the Borel σ-algebra.

[8] In particular, it is sufficient to show the independence of the increments using the characteristic function: for details, see for example Lemma IV.1.2 in [63].

that the initial law is $\mu_0 = \delta_{x_0}$ for some $x_0 \in \mathbb{R}^N$ and therefore $\mathbf{X}_0 = x_0$ almost surely. Letting

$$J_t := \int_0^t b(s, X_s)ds + \int_0^t \sigma(s, X_s)dW_s,$$

$$\mathbf{J}_t := \int_0^t b(s, \mathbf{X}_s)ds + \int_0^t \sigma(s, \mathbf{X}_s)d\mathbf{W}_s,$$

we have that (X, W, J) and $(\mathbf{X}, \mathbf{W}, \mathbf{J})$ are equal in law by Corollary 10.2.28. Therefore, $\mathbf{X} - x_0 - \mathbf{J}$ is indistinguishable from the null process, and this proves the thesis.

The case where the initial datum $\mathbf{X}_0$ is random can be handled by conditioning on $\mathbf{X}_0$. Precisely, to lighten the notation, let $\mathbf{P} := \mu_{X,W}$: by Theorem 4.3.2 in [113], there exists a regular version

$$\mathbf{P}(\cdot \mid \mathbf{X}_0) = \left(\mathbf{P}_{x,w}(\cdot \mid \mathbf{X}_0)\right)_{(x,w) \in \Omega_{d+N}}$$

of the conditional probability of $\mathbf{P}$ given $\mathbf{X}_0$. For $\mathbf{P}$-almost every $(x, w) \in \Omega_{N+d}$, under the measure $\mathbf{P}_{x,w}(\cdot \mid \mathbf{X}_0)$, the process $(\mathbf{X}, \mathbf{W})$ has the same law as $(\hat{X}, W)$ where $(\hat{X}, W)$ is the solution of the SDE with coefficients b, σ and initial datum $\hat{X}_0 = x(0)$. Then, for what has been proven previously, for $\mathbf{P}$-almost every $(x, w) \in \Omega_{N+d}$, under the measure $\mathbf{P}_{x,w}(\cdot \mid \mathbf{X}_0)$, the process $(\mathbf{X}, \mathbf{W})$ is a solution of the SDE with coefficients b, σ and initial datum $x(0)$. To conclude, it is sufficient to observe that, for

$$Z := \sup_{t \in [0,T]} \left| \mathbf{X}_t - \mathbf{X}_0 - \int_0^t b(s, \mathbf{X}_s)ds - \int_0^t \sigma(s, \mathbf{X}_s)d\mathbf{W}_s \right|$$

by the law of total probability, we have $E[Z] = E[E[Z \mid \mathbf{X}_0]] = 0$. $\square$

The following result establishes the relationships between solvability and uniqueness for an SDE in the weak and strong sense, according to Definitions 14.1.9 and 14.1.11.

Theorem 14.3.6 (Yamada and Watanabe [149] [!])

(i) *Strong solvability implies weak solvability;*
(ii) *strong uniqueness implies weak uniqueness;*
(iii) *weak solvability and strong uniqueness together imply strong solvability.*

Proof We provide a detailed outline of the proof, and direct readers to Chapter 8 in [136] for a comprehensive treatment.[9]

(i) In order to infer weak solvability from strong solvability, we only have to construct a set-up. More precisely, given a distribution μ_0 on $\mathbb{R}^N$, we consider the canonical space $\mathbb{R}^N \times \Omega_d$ equipped with the product measure $\mu_0 \otimes \mu_W$,

---

[9] Further reference sources are Theorem 21.14 and Lemma 21.17 in [66] and Section V.17 in [124].

14.3 Weak vs Strong Solutions: The Yamada-Watanabe Theorem

where μ_W is the law of the d-dimensional Brownian motion, and with the filtration $(\mathscr{G}_t)_{t\in[0,T]}$ generated by the identity process

$$(\mathbf{Z}, \mathbf{W}) : \mathbb{R}^N \times \mathbf{\Omega}_d$$
$$\longrightarrow \mathbb{R}^N \times \mathbf{\Omega}_d, \qquad \mathbf{Z}(z, w) = z, \ \mathbf{W}_t(z, w) = w(t), \ t \in [0, T].$$

Then $\mathbf{Z} \sim \mu_0$ is $\mathscr{G}_0$-measurable and $\mathbf{W}$ is a Brownian motion (with respect to $\mathscr{G}_t$). Hence, by the hypothesis of strong solvability, there exists a solution $\mathbf{X}$ related to the set-up $(\mathbf{W}, \mathscr{G}_t)$ and such that $\mathbf{X}_0 = \mathbf{Z} \sim \mu_0$.

(ii) We omit the case where the initial datum is random: this can be treated in a completely analogous way to the second part of the proof of Lemma 14.3.5 (for details, see, for example, Proposition IX.1.4 in [123]).

We thus consider two solutions $X^i \in \text{SDE}(b, \sigma, W^i, \mathscr{F}^i_t)$ such that $X^i_0 = x \in \mathbb{R}^N$ almost surely, for $i = 1, 2$. We prove that the hypothesis of strong uniqueness implies that (X^1, W^1) and (X^2, W^2) are equal in law. The problem is that the solutions X^1 and X^2 are generally defined on different sample spaces: so the idea is to construct versions of X^1 and X^2 that are solutions of the SDE on the same space and with respect to the same Brownian motion. To this end, we construct a canonical space on which *three* processes are defined: a Brownian motion and the versions of X^1 and X^2.

By Theorem 4.3.4 in [113] (and Remark 4.3.5 in [113]), there exists a regular version

$$\mu_{X^i|W^i} = (\mu_{X^i|W^i}(\cdot; w))_{w \in \mathbf{\Omega}_d}$$

of the law of X^i conditioned on W^i: for each $w \in \mathbf{\Omega}_d$, $\mu_{X^i|W^i}(\cdot; w)$ is a distribution on the Borel σ-algebra $\mathscr{G}^N_T$ of $\mathbf{\Omega}_N$ and we have[10]

$$\int_A \mu_{X^i|W^i}(H; w) \mu_W(dw) = E\left[E\left[\mathbb{1}_H(X^i) \mid W^i\right] \mathbb{1}_A(W^i)\right]$$
$$= \mu_{X^i, W^i}(H \times A), \qquad (H, A) \in \mathscr{G}^N_T \times \mathscr{G}^d_T. \tag{14.3.2}$$

Now, on $\mathbf{\Omega}_N \times \mathbf{\Omega}_N \times \mathbf{\Omega}_d$ we define the probability measure[11]

$$\mathbf{P}(H \times K \times A) := \int_A \mu_{X^1|W^1}(H; w) \mu_{X^2|W^2}(K; w) \mu_W(dw),$$
$$(H, K, A) \in \mathscr{G}^N_T \times \mathscr{G}^N_T \times \mathscr{G}^d_T, \tag{14.3.3}$$

---

[10] Here $\mu_W \equiv \mu_{W^i}$, $i = 1, 2$, is the Wiener measure on $\mathbf{\Omega}_d$.
[11] $\mathbf{P}$ extends to the product σ-algebra $\mathscr{G}^N_T \otimes \mathscr{G}^N_T \otimes \mathscr{G}^d_T = \mathscr{G}^{2N+d}_T$.

and denote by $(\mathbf{X}^1, \mathbf{X}^2, \mathbf{W})$ the canonical process on such space. Taking respectively $H = \mathbf{\Omega}_N$ or $K = \mathbf{\Omega}_N$ in (14.3.3), by (14.3.2) we have

$$(\mathbf{X}^i, \mathbf{W}) \stackrel{d}{=} (X^i, W^i), \qquad i = 1, 2; \tag{14.3.4}$$

we deduce in particular that $\mathbf{W}$ is a Brownian motion under the measure $\mathbf{P}$ and, as in the proof of Lemma 14.3.5, $(\mathbf{X}^1, \mathbf{W})$ and $(\mathbf{X}^2, \mathbf{W})$ are both solutions of the SDE with coefficients b, σ and with initial datum x. By strong uniqueness, we have that $\mathbf{X}^1$ and $\mathbf{X}^2$ are indistinguishable under the measure $\mathbf{P}$ and therefore

$$(X^1, W^1) \stackrel{d}{=} (\mathbf{X}^1, \mathbf{W}) = (\mathbf{X}^2, \mathbf{W}) \stackrel{d}{=} (X^2, W^2).$$

(iii) Again, we consider only the case of a deterministic initial datum. Let $X \in \text{SDE}(b, \sigma, W, \mathscr{F}_t)$ be a solution with initial datum $X_0 = x \in \mathbb{R}^N$ a.s. We apply the construction of point (ii) with $X^1 = X^2 = X$, that is, we construct on the space $\mathbf{\Omega}_N \times \mathbf{\Omega}_N \times \mathbf{\Omega}_d$ the measure $\mathbf{P}$ as in (14.3.3) and the canonical process $(\mathbf{X}^1, \mathbf{X}^2, \mathbf{W})$ where $\mathbf{X}^1, \mathbf{X}^2$ are equal in law to X and are solutions of the SDE with respect to the Brownian motion $\mathbf{W}$.

We consider the conditional probability $\mathbf{P}(\cdot \mid \mathbf{W}) = (\mathbf{P}_w(\cdot \mid \mathbf{W}))_{w \in \mathbf{\Omega}_d}$ and the related conditional laws

$$\mu_{\mathbf{X}^i \mid \mathbf{W}}(H) = \mathbf{P}(\mathbf{X}^i \in H \mid \mathbf{W}), \qquad H \in \mathbf{\Omega}_N, \ i = 1, 2,$$

noting that $\mu_{\mathbf{X}^i \mid \mathbf{W}} = \mu_{X \mid W}$ by (14.3.4). We have[12] that the random variables $\mathbf{X}^1$ and $\mathbf{X}^2$ are *simultaneously equal a.s. and independent* in $\mathbf{P}_w(\cdot \mid \mathbf{W})$ for almost every $w \in \mathbf{\Omega}_d$ and therefore[13] $\mathbf{X}^1$ and $\mathbf{X}^2$ have a Dirac delta distribution under $\mathbf{P}_w(\cdot \mid \mathbf{W})$. In other terms, for almost every $w \in \mathbf{\Omega}_d$ we have $\mu_{X \mid W}(H; w) = \mu_{\mathbf{X}^i \mid \mathbf{W}}(H; w) = \delta_{F(w)}$ for some measurable map F from $\mathbf{\Omega}_d$

---

[12] Indeed, by the strong uniqueness hypothesis, we have $\mathbf{P}(\mathbf{X}^1 = \mathbf{X}^2) = 1$ so that

$$E\left[\mathbf{P}(\mathbf{X}^1 = \mathbf{X}^2 \mid \mathbf{W})\right] = E\left[\mathbf{P}(\mathbf{X}^1 = \mathbf{X}^2)\right] = 1$$

and since $\mathbf{P}(\mathbf{X}^1 = \mathbf{X}^2 \mid \mathbf{W}) \leq 1$, we also have $\mathbf{P}_w(\mathbf{X}^1 = \mathbf{X}^2 \mid \mathbf{W}) = 1$ for almost every $w \in \mathbf{\Omega}_d$. Moreover, from definition (14.3.3) of $\mathbf{P}$, it is not difficult to verify that the joint conditional law of $\mathbf{X}^1, \mathbf{X}^2$ is the product of the marginals

$$\mu_{\mathbf{X}^1, \mathbf{X}^2 \mid \mathbf{W}}(H \times K) = \mathbf{P}\left((\mathbf{X}^1, \mathbf{X}^2) \in H \times K \mid \mathbf{W}\right) = \mu_{X \mid W}(H) \mu_{X \mid W}(K)$$

$$= \mu_{\mathbf{X}^1 \mid \mathbf{W}}(H) \mu_{\mathbf{X}^2 \mid \mathbf{W}}(K), \qquad H, K \in \mathbf{\Omega}_N,$$

from which the independence for almost every $w \in \mathbf{\Omega}_d$.

[13] As an exercise, prove that if X, Y are real random variables on a space $(\Omega, \mathscr{F}, P)$, that are equal a.s. and independent, then $X \sim \delta_{x_0}$ for some $x_0 \in \mathbb{R}$. Prove that an analogous result holds for X, Y with values in the space $\mathbf{\Omega}_n$.

to Ω_N and therefore $X = F(W)$ a.s. To conclude, it is necessary to show that X is adapted to the standard Brownian filtration $\mathscr{F}^W$: for the proof of this fact, based on the properties of the regular version of conditional probability, we refer[14] to Problem 3.21 on page 310 in [67]. □

Remark 14.3.7 ([!]) In Remark 14.1.7 we pointed out that strong solutions differ from weak ones by the property of being adapted to the standard Brownian filtration (assuming for simplicity that the initial datum is deterministic). This measurability property is well expressed by the functional dependence $X = F(W)$ shown in the previous proof: in particular, a strong solution (X, W) can be defined on the canonical space Ω_d. On the contrary, Lemma 14.3.5 shows that it is possible to "transport" every weak solution to the canonical space $\Omega_N \times \Omega_d$. This means that weak solutions generally require a richer sample space, in which the trajectories of a solution (that are elements of Ω_N) are not necessarily functionals of the Brownian trajectories (that are elements of Ω_d): this is the case of Tanaka's Example 14.1.10.

14.4 Standard Assumptions and Preliminary Estimates

In this section, we introduce additional assumptions on the coefficients that enable us to obtain useful estimates for the solutions of SDEs.

Definition 14.4.1 (Standard Assumptions) The coefficients b, σ satisfy the standard assumptions on $]t_0, T[$ if there exist two positive constants c_1, c_2 such that

$$|b(t,x)| + |\sigma(t,x)| \leq c_1(1 + |x|), \tag{14.4.1}$$

$$|b(t,x) - b(t,y)| + |\sigma(t,x) - \sigma(t,y)| \leq c_2|x - y|, \tag{14.4.2}$$

for every $t \in]t_0, T[$ and $x, y \in \mathbb{R}^N$.

Formulas (14.4.1) and (14.4.2) are *linear growth* and *global Lipschitz continuity* conditions in x uniform in $t \in]t_0, T[$, respectively. We note that, under Assumption 14.1.1, (14.4.2) implies (14.4.1). In some results, we will weaken (14.4.2) by requiring *local* Lipschitz continuity in x.

Example 14.4.2 (Geometric Brownian Motion) Consider the SDE with linear coefficients

$$dX_t = \mu X_t dt + \sigma X_t dW_t \tag{14.4.3}$$

---

[14] In fact, in [67] more is proved (see also Remark 2 on page 310 in [123]): highlighting the dependence on the initial datum $x \in \mathbb{R}^N$, the function $F = F(x, w)$ is jointly measurable and, for $Z \in m\mathscr{F}_0$, $X = F(Z, W)$ is a strong solution of the SDE with random initial datum $X_0 = Z$.

where μ, σ are real parameters. In this case, $b(t, x) = \mu x$ and $\sigma(t, x) = \sigma x$, so the standard assumptions are obviously satisfied. As in Example 11.1.5-(iii), a direct application of Itô's formula shows that

$$X_t = X_0 e^{\left(\mu - \frac{\sigma^2}{2}\right)t + \sigma W_t}$$

is a solution of (14.4.3). The process X, known as *geometric Brownian motion*, is used to represent the dynamics of a risky financial asset price in the classical Black-Scholes model [19]. The model generalizes to the case of time-dependent coefficients, $\mu = \mu(t), \sigma = \sigma(t) \in L^\infty(\mathbb{R}_{\geq 0})$: also in this case, it is easy to determine the explicit expression of the solution.

Several constants are introduced in the estimates that we prove in this section. Since it is essential to keep track of them, we introduce the following

Convention 14.4.3 To indicate that a constant c depends *solely and exclusively* on the values of the parameters $\alpha_1, \ldots, \alpha_n$, we will write $c = c(\alpha_1, \ldots, \alpha_n)$.

Lemma 14.4.4 ([!]) *Let X, Y be adapted and a.s. continuous processes and $p \geq 2$. Then:*

- *if b, σ satisfy the linear growth condition (14.4.1), there exists a positive constant $\bar{c}_1 = \bar{c}_1(T, d, N, p, c_1)$, such that*

$$E\left[\sup_{t_0 \leq t \leq t_1} \left|\int_{t_0}^t b(s, X_s)ds + \int_{t_0}^t \sigma(s, X_s)dW_s\right|^p\right]$$
$$\leq \bar{c}_1(t_1 - t_0)^{\frac{p-2}{2}} \int_{t_0}^{t_1} \left(1 + E\left[\sup_{t_0 \leq r \leq s} |X_r|^p\right]\right) ds \qquad (14.4.4)$$

for every $t_1 \in]t_0, T[$;

- *if b, σ satisfy the global Lipschitz condition (14.4.2), there exists a positive constant $\bar{c}_2 = \bar{c}_2(T, d, N, p, c_2)$ such that*

$$E\left[\sup_{t_0 \leq t \leq t_1} \left|\int_{t_0}^t (b(s, X_s) - b(s, Y_s))\, ds + \int_{t_0}^t (\sigma(s, X_s) - \sigma(s, Y_s))\, dW_s\right|^p\right]$$
$$\leq \bar{c}_2(t_1 - t_0)^{\frac{p-2}{2}} \int_{t_0}^{t_1} E\left[\sup_{t_0 \leq r \leq s} |X_r - Y_r|^p\right] ds \qquad (14.4.5)$$

for every $t_1 \in]t_0, T[$.

Proof We recall the elementary inequality

$$|x_1 + \cdots + x_n|^p \leq n^{p-1}\left(|x_1|^p + \cdots |x_n|^p\right), \qquad x_1, \ldots, x_n \in \mathbb{R}^N, \, n \in \mathbb{N}. \qquad (14.4.6)$$

14.4 Standard Assumptions and Preliminary Estimates

By Hölder's inequality, we have

$$E\left[\sup_{t_0 \leq t \leq t_1} \left|\int_{t_0}^{t} b(s, X_s)ds\right|^p\right]$$

$$\leq (t_1 - t_0)^{p-1} E\left[\int_{t_0}^{t_1} |b(s, X_s)|^p ds\right] \leq$$

(by (14.4.1))

$$\leq (t_1 - t_0)^{p-1} c_1^p \int_{t_0}^{t_1} E\left[(1 + |X_s|)^p\right] ds \leq$$

(by (14.4.6))

$$\leq 2^{p-1}(t_1 - t_0)^{p-1} c_1^p \int_{t_0}^{t_1} \left(1 + E\left[|X_s|^p\right]\right) ds$$

$$\leq 2^{p-1}(t_1 - t_0)^{p-1} c_1^p \int_{t_0}^{t_1} \left(1 + E\left[\sup_{t_0 \leq r \leq s} |X_r|^p\right]\right) ds.$$

Similarly, by Burkholder-Davis-Gundy's inequality, in the version of Corollary 12.3.10, there exists a constant $c = c(d, N, p)$ such that

$$E\left[\sup_{t_0 \leq t \leq t_1} \left|\int_{t_0}^{t} \sigma(s, X_s)dW_s\right|^p\right] \leq c(t_1 - t_0)^{\frac{p-2}{2}} E\left[\int_{t_0}^{t_1} |\sigma(s, X_s)|^p ds\right] \leq$$

(proceeding as for the previous estimate)

$$\leq c(t_1 - t_0)^{\frac{p-2}{2}} 2^{p-1} c_1^p \int_{t_0}^{t_1} \left(1 + E\left[\sup_{t_0 \leq r \leq s} |X_r|^p\right]\right) ds.$$

This proves (14.4.4).

Again, by Hölder's inequality, we have

$$E\left[\sup_{t_0 \leq t \leq t_1} \left|\int_{t_0}^{t} (b(s, X_s) - b(s, Y_s))\, ds\right|^p\right]$$

$$\leq (t_1 - t_0)^{p-1} E\left[\int_{t_0}^{t_1} |b(s, X_s) - b(s, Y_s)|^p ds\right] \leq$$

(by (14.4.2))

$$\leq (t_1 - t_0)^{p-1} c_2^p \int_{t_0}^{t_1} E\left[|X_s - Y_s|^p\right] ds$$

$$\leq (t_1 - t_0)^{p-1} c_2^p \int_{t_0}^{t_1} E\left[\sup_{t_0 \leq r \leq s} |X_r - Y_r|^p\right] ds.$$

Similarly, by Corollary 12.3.10, we have

$$E\left[\sup_{t_0 \leq t \leq t_1} \left|\int_{t_0}^{t} (\sigma(s, X_s) - \sigma(s, Y_s)) dW_s\right|^p\right]$$

$$\leq c_p (t_1 - t_0)^{\frac{p-2}{2}} E\left[\int_{t_0}^{t_1} |\sigma(s, X_s) - \sigma(s, Y_s)|^p ds\right] \leq$$

(proceeding as for the previous estimate, by (14.4.2))

$$\leq c_p (t_1 - t_0)^{\frac{p-2}{2}} c_2^p \int_{t_0}^{t_1} E\left[\sup_{t_0 \leq r \leq s} |X_r - Y_r|^p\right] ds.$$

This proves (14.4.5). □

14.5 Some A Priori Estimates

In this section, we prove some polynomial and exponential integrability estimates for the solutions of SDEs *whose coefficients satisfy the linear growth assumption* (14.4.1). We use the term "a priori" estimates because *condition (14.4.1) alone is not enough to ensure the existence of a solution:* existence is therefore implicitly assumed as a hypothesis. The following estimates have considerable theoretical importance (for example, for the proof of Feynman-Kac's Theorem 15.4.4) and practical applications (for example, for the results of continuous dependence on parameters of Sect. 17.4 and the study of the convergence of numerical approximation schemes for SDEs). On the other hand, the proofs of this section, technical and not very informative, can be skipped at first reading.

To lighten the notation, in this section we assume $t_0 = 0$ and for each stochastic process X we set

$$\bar{X}_t = \sup_{0 \leq s \leq t} |X_s|.$$

14.5 Some A Priori Estimates

Hereafter, we will repeatedly use the following classic

Lemma 14.5.1 (Grönwall) *Consider $v \in L^1([0, T])$ such that*

$$v(t) \leq a + b \int_0^t v(s)ds, \qquad t \in [0, T],$$

where a and b are non-negative real numbers. Then we have

$$v(t) \leq ae^{bt}, \qquad t \in [0, T].$$

In Grönwall's lemma, the integrability assumption of v is necessary: a counter-example is given by $v(t) = 0$ for $t = 0$ and $v(t) = \frac{1}{t}$ for $t > 0$, with $a = 0$ and $b = 1$. If we add the assumptions $v \geq 0$ and $a = 0$ to the hypotheses of Grönwall's lemma, then we have $v \equiv 0$.

Theorem 14.5.2 (A Priori L^p Estimates) *Let $X = (X_t)_{t \in [0,T]}$ be a solution of the SDE*

$$dX_t = b(t, X_t)dt + \sigma(t, X_t)dW_t,$$

with b, σ satisfying the linear growth assumption (14.4.1). Then for every $T > 0$ and $p \geq 2$ there exists a positive constant $c = c(T, p, d, N, c_1)$ such that

$$E\left[\sup_{0 \leq t \leq T} |X_t|^p\right] \leq c(1 + E\left[|X_0|^p\right]). \qquad (14.5.1)$$

Proof It is not restrictive to assume $E[|X_0|^p] < \infty$ otherwise the thesis is obvious. The general idea of the proof is simple: from estimate (14.4.4) we have

$$v(t) := E\left[\bar{X}_t^p\right]$$

$$\leq 2^{p-1}\left(E\left[|X_0|^p\right] + \bar{c}_1 \int_0^t \left(1 + E\left[\bar{X}_s^p\right]ds\right)\right), \qquad t \in [0, T],$$

or equivalently

$$v(t) \leq c\left(1 + E\left[|X_0|^p\right] + \int_0^t v(s)ds\right), \qquad t \in [0, T],$$

and therefore the thesis would follow directly from Grönwall's lemma.

As a matter of fact, to apply Grönwall's lemma, it is necessary to know a priori[15] that $v \in L^1([0, T])$. For this reason, it is necessary to proceed more carefully using

---

[15] Based on what has been proven so far, we do not even know if v is a continuous function.

a technical localization argument. Let

$$\tau_n = \inf\{t \in [0, T] \mid |X_t| \geq n\}, \qquad n \in \mathbb{N},$$

with the convention $\min \emptyset = T$. Being X a.s. continuous, we have that τ_n is an increasing sequence of stopping times such that $\tau_n \nearrow T$ a.s. With b_n, σ_n as in (17.1.2), we have

$$X_{t \wedge \tau_n} = X_0 + \int_0^{t \wedge \tau_n} b(s, X_s) ds + \int_0^{t \wedge \tau_n} \sigma(s, X_s) dW_s$$

$$= X_0 + \int_0^t b_n(s, X_{s \wedge \tau_n}) ds + \int_0^t \sigma_n(s, X_{s \wedge \tau_n}) dW_s.$$

The coefficients $b_n = b_n(t, x)$ and $\sigma_n = \sigma_n(t, x)$, although stochastic, satisfy the linear growth condition (14.4.1) with the same constant c_1: the proof of estimate (14.4.4) can be repeated in a substantially identical way to the case of deterministic b, σ, to obtain

$$v_n(t_1) := E\left[\sup_{0 \leq t \leq t_1} |X_{t \wedge \tau_n}|^p\right]$$

$$\leq 2^{p-1}\left(E\left[|X_0|^p\right] + \bar{c}_1 \int_0^{t_1} \left(1 + \underbrace{E\left[\sup_{0 \leq r \leq s} |X_{r \wedge \tau_n}|^p\right]}_{= v_n(s)}\right) ds\right), \qquad t_1 \in [0, T],$$

or equivalently

$$v_n(t_1) \leq c\left(1 + E\left[|X_0|^p\right] + \int_0^{t_1} v_n(s) ds\right), \qquad t_1 \in [0, T],$$

with c positive constant that depends only on T, p, d, N, c_1 and not on n. We observe that v_n is a measurable and bounded function since $|X_{t \wedge \tau_n}| \leq |X_0| \mathbb{1}_{(|X_0| \geq n)} + n \mathbb{1}_{(|X_0| < n)}$ and therefore $v_n(t) \leq E[(|X_0| + n)^p] < +\infty$: then by Grönwall's lemma we have

$$E\left[\sup_{0 \leq t \leq T} |X_{t \wedge \tau_n}|^p\right] = v_n(T) \leq c e^{cT}\left(1 + E\left[|X_0|^p\right]\right),$$

and taking the limit as n goes to infinity, we get (14.5.1) by Beppo Levi's theorem. $\square$

If the diffusive coefficient σ is bounded, a stronger integrability estimate than that of Theorem 14.5.2 holds.

14.5 Some A Priori Estimates

Theorem 14.5.3 (A Priori Exponential Estimate) *Let $X = (X_t)_{t \in [0,T]}$ be the solution of the SDE*

$$dX_t = b(t, X_t)dt + \sigma(t, X_t)dW_t,$$

with b satisfying the linear growth assumption (14.4.1) and σ bounded by a constant κ, i.e., $|\sigma(t, x)| \leq \kappa$ for $(t, x) \in [0, T] \times \mathbb{R}^N$. Then there exist two positive constants α and c, depending only on T, κ, c_1 and N, such that

$$E\left[e^{\alpha \bar{X}_T^2}\right] \leq cE\left[e^{c|X_0|^2}\right], \qquad \bar{X}_T := \sup_{0 \leq t \leq T} |X_t|.$$

Proof Let

$$\bar{M}_T = \sup_{0 \leq t \leq T} \left| \int_0^t \sigma(s, X_s) dW_s \right|.$$

Given $\delta > 0$, almost surely on $(\bar{M}_T < \delta)$ we have

$$|X_t| < |X_0| + c_1 \int_0^t (1 + \bar{X}_s) ds + \delta, \qquad t \in [0, T],$$

so that, by Grönwall's lemma,

$$\bar{X}_T < (|X_0| + c_1 T + \delta) e^{c_1 T}.$$

Consequently

$$\left(\bar{X}_T \geq (|X_0| + c_1 T + \delta) e^{c_1 T} \right) \subseteq \left(\bar{M}_T \geq \delta \right)$$

and by Proposition 13.2.4 (and estimate (13.2.5)) there exists a positive constant c, depending only on N, κ and T, such that[16]

$$P\left(\bar{X}_T \geq (|X_0| + c_1 T + \delta) e^{c_1 T} \mid X_0 \right) \leq c e^{-\frac{\delta^2}{c}}. \qquad (14.5.2)$$

Let $\lambda = (|X_0| + c_1 T + \delta) e^{c_1 T}$ and observe that

$$\delta = \lambda e^{-c_1 T} - |X_0| - c_1 T \geq \frac{\lambda}{2} e^{-c_1 T} \qquad \text{if } \lambda \geq \bar{a}|X_0| + \bar{b} \qquad (14.5.3)$$

---

[16] Provided that we switch to the canonical setting by means of Lemma 14.3.5 (this is not restrictive since the thesis depends only on the law of X), a regular version of the conditional probability exists and estimate (14.5.2) holds pointwise as a consequence of Proposition 13.2.4.

with $\bar{a} := 2e^{c_1 T}$ and $\bar{b} := 2c_1 T e^{c_1 T}$. So, combining (14.5.2) and (14.5.3), we have

$$P\left(\bar{X}_T \geq \lambda \mid X_0\right) \leq c e^{-\bar{c}\lambda^2}, \qquad \lambda \geq \bar{a}|X_0| + \bar{b}, \qquad (14.5.4)$$

with $c, \bar{c}$ positive constants depending only on T, κ, c_1 and N. Now we apply the Proposition 3.1.6 in [113] with $f(\lambda) = e^{\alpha \lambda^2}$, where the constant $\alpha > 0$ will be determined later: we have

$$E\left[e^{\alpha \bar{X}_T^2} \mid X_0\right] = 1 + 2\alpha \int_0^\infty \lambda e^{\alpha \lambda^2} P\left(\bar{X}_T \geq \lambda \mid X_0\right) d\lambda \leq$$

(by (14.5.4))

$$\leq 1 + 2\alpha \int_0^{\bar{a}|X_0|+\bar{b}} \lambda e^{\alpha \lambda^2} d\lambda + 2\alpha c \int_{\bar{a}|X_0|+\bar{b}}^{+\infty} \lambda e^{\lambda^2(\alpha - \bar{c})} d\lambda.$$

The thesis follows by setting $\alpha = \frac{\bar{c}}{2}$ and applying the expected value. □

14.6 Key Ideas to Remember

We summarize the most significant findings of the chapter and the fundamental concepts to be retained from an initial reading, while disregarding the more technical or secondary matters. As usual, if you have any doubt about what the following succinct statements mean, please review the corresponding section.

- Section 14.1: we introduce the concepts of *solution of an SDE on a set-up* $(W, \mathscr{F}_t)$ and of *solvability of an SDE* in the strong sense (i.e., with solutions adapted to the filtration generated by the initial datum and by W) and in the weak sense: in the latter case, not having the set-up fixed a priori, a solution is constituted by the *pair (X, W)*.
- Section 14.2: thanks to the regularizing effect of the Brownian motion and in contrast with what happens in the deterministic case, we can have existence and uniqueness of the solution of an SDE with a strongly irregular drift coefficient, even only measurable and bounded.
- Section 14.3: the solution transfer technique allows us to set the problem of solvability of an SDE in the canonical space of continuous trajectories: this is particularly useful for the study of weak solutions. The Yamada-Watanabe Theorem clarifies the relationship between the concepts of solvability in a weak and strong sense:
    (i) if an SDE is solvable in the strong sense then it is also solvable in the weak sense;

14.6 Key Ideas to Remember

(ii) if for an SDE there is uniqueness in the strong sense then there is also uniqueness in the weak sense;

(iii) if for an SDE there is solvability in the weak sense and uniqueness in the strong sense then there is solvability in the strong sense.

- Sections 14.4 and 14.5: under the "standard assumptions" of linear growth and Lipschitz continuity of the coefficients, we prove some integrability estimates that will be crucial in the study of strong solutions.

Main notations used or introduced in this chapter:

Symbol	Description	Page
$(W, \mathscr{F}_t)$	Set-up	264
$W = (W_t)_{t \in [t_0, T]}$	Brownian motion with initial point t_0	264
$\mathscr{F}^W$	Standard Brownian filtration	264
$X \in \text{SDE}(b, \sigma, W, \mathscr{F}_t)$	X is the solution of the SDE with coefficients b, σ related to $(W, \mathscr{F}_t)$	265
$\mathscr{F}^{Z,W}$	(Completed) filtration generated by $Z \in m\mathscr{F}_{t_0}$ and W	266
$\mathbf{\Omega}_n = C([0, T]; \mathbb{R}^n)$	Space of continuous n-dimensional trajectories	273
$\mathbf{X}_t(w) = w(t)$	Identity process on $\mathbf{\Omega}_n$	273
$(\mathscr{G}^n_t)_{t \in [0, T]}$	Filtration on $\mathbf{\Omega}_n$ generated by the identity process	273

Chapter 15
Feynman-Kac Formulas

> *I may never find all the answers*
> *I may never understand why*
> *I may never prove*
> *What I know to be true*
> *But I know that I still have to try*
>
> *Dream Theater, The spirit carries on*

Consider the SDE

$$dX_t = b(t, X_t)dt + \sigma(t, X_t)dW_t \qquad (15.0.1)$$

where W is a d-dimensional Brownian motion and

$$b = b(t, x) :]0, T[\times \mathbb{R}^N \longrightarrow \mathbb{R}^N, \qquad \sigma = \sigma(t, x) :]0, T[\times \mathbb{R}^N \longrightarrow \mathbb{R}^{N \times d}.$$

If there exists a solution $X^{t,x} = (X_s^{t,x})_{s \in [t,T]}$ to (15.0.1) with initial datum (t, x), then by Itô's formula, for any suitably smooth function u we have

$$u(s, X_s^{t,x}) = u(t, x) + \int_t^s (\partial_r + \mathscr{A}_r) u(r, X_r^{t,x}) dr$$
$$+ \int_t^s \nabla u(r, X_r^{t,x}) \sigma(r, X_r^{t,x}) dW_r, \qquad s \in [t, T], \qquad (15.0.2)$$

where

$$\mathscr{A}_t := \frac{1}{2} \sum_{i,j=1}^N c_{ij}(t, x) \partial_{x_i x_j} + \sum_{j=1}^N b_j(t, x) \partial_{x_j}, \qquad c := \sigma \sigma^*, \qquad (15.0.3)$$

is the so-called *characteristic operator of the SDE* (15.0.1) (see Definition 15.1.1).

The Feynman-Kac formulas offer a probabilistic framework for expressing solutions to partial differential equations (abbreviated as PDEs) that involve the operator $\mathscr{A}_t$. To fix ideas, suppose there exists a classical solution to the backward Cauchy problem

$$\begin{cases} (\partial_t + \mathscr{A}_t)u(t,x) = 0, & (t,x) \in \,]0,T[\,\times \mathbb{R}^N, \\ u(T,x) = \varphi(x), & x \in \mathbb{R}^N. \end{cases} \qquad (15.0.4)$$

Then, (15.0.2) reduces to

$$u(s, X_s^{t,x}) = u(t,x) + \int_t^s \nabla u(r, X_r^{t,x}) \sigma(r, X_r^{t,x}) dW_r, \qquad s \in [t,T],$$

and therefore the process $s \mapsto u(s, X_s^{t,x})$ is a local martingale: moreover, if $(u(s, X_s^{t,x}))_{s \in [t,T]}$ is a true martingale, by taking the expectation and using the final condition $u(T, \cdot) = \varphi$, we obtain

$$u(t,x) = E\left[u(T, X_T^{t,x})\right] = E\left[\varphi(X_T^{t,x})\right]. \qquad (15.0.5)$$

Formula (15.0.5) provides a representation of the solution of (15.0.4) in terms of the final datum φ: from an application standpoint, this formula can be readily implemented using Monte Carlo methods for numerical approximation of the solution; from a theoretical perspective, Eq. (15.0.5) provides a uniqueness result for the solution of problem (15.0.4).

In this chapter, we examine various variants and generalizations of formula (15.0.5), valid for second-order partial differential operators of elliptic and parabolic type.

15.1 Characteristic Operator of an SDE

Consider an SDE of the form (15.0.1) with coefficients $b, \sigma \in L^\infty_{\text{loc}}$ that satisfy the linear growth assumption (14.4.1). Suppose there exists a solution $X^{t,x} = (X_s^{t,x})_{s \in [t,T]}$ with initial datum (t,x). Then, given a function $\psi = \psi(x) \in bC^2(\mathbb{R}^N)$ (i.e., ψ has continuous and bounded derivatives up to the second order), by Itô's formula we have

$$E\left[\frac{\psi(X_s^{t,x}) - \psi(x)}{s-t}\right]$$

$$= E\left[\frac{1}{s-t}\int_t^s \mathscr{A}_r \psi(X_r^{t,x}) dr + \frac{1}{s-t}\int_t^s \nabla \psi(X_r^{t,x}) \sigma(r, X_r^{t,x}) dW_r\right] =$$

15.1 Characteristic Operator of an SDE

(since $|\nabla\psi(X_r^{t,x})\sigma(r, X_r^{t,x})| \leq c(1 + |X_r^{t,x}|) \in \mathbb{L}^2$ by the a priori integrability estimates of Theorem 14.5.2)

$$= E\left[\frac{1}{s-t}\int_t^s \mathscr{A}_r\psi(X_r^{t,x})dr\right] \xrightarrow[s-t \to 0^+]{} \mathscr{A}_t\psi(x)$$

where we used the dominated convergence theorem and the estimates of Theorem 14.5.2, to evaluate the limit: thus, we have

$$\frac{d}{ds}E\left[\psi(X_s^{t,x})\right]\bigg|_{s=t} = \mathscr{A}_t\psi(x). \tag{15.1.1}$$

This serves as the motivation for the following definition, which mirrors formula (2.5.5) for Markov processes.

Definition 15.1.1 (Characteristic Operator of an SDE) The operator $\mathscr{A}_t$ in (15.0.3) is called the *characteristic operator of the SDE* (15.0.1).

Remark 15.1.2 ([!]) Given $m \in \mathbb{R}^N$, consider the functions

$$\psi_i(x) := x_i, \qquad \psi_{ij}(x) := (x_i - m_i)(x_j - m_j), \qquad x \in \mathbb{R}^N, \; i, j = 1, \ldots, N,$$

and observe that

$$\mathscr{A}_t\psi_i(x) = b_i(t, x), \quad \mathscr{A}_t\psi_{ij}(x) = c_{ij}(t, x) + b_i(t, x)(x_j - m_j) + b_j(t, x)(x_i - m_i).$$

Formula (15.1.1) is valid with $\psi = \psi_i$ and $\psi = \psi_{ij}$: this can be proved using the same arguments as above since the linear growth hypothesis of the coefficients b, σ and the L^p estimates of Theorem 14.5.2 justify convergence and the martingale property of the stochastic integrals. Thus, we have

$$\frac{d}{ds}E\left[X_s^{t,x}\right]\bigg|_{s=t} = b(t, x), \tag{15.1.2}$$

$$\frac{d}{ds}E\left[(X_s^{t,x} - m)_i(X_s^{t,x} - m)_j\right]\bigg|_{s=t} = c_{ij}(t, x) + b_i(t, x)(x_j - m_j)$$
$$+ b_j(t, x)(x_i - m_i)$$

and in particular, for $m = x$,

$$\frac{d}{ds}E\left[(X_s^{t,x} - x)_i(X_s^{t,x} - x)_j\right]\bigg|_{s=t} = c_{ij}(t, x). \tag{15.1.3}$$

Based on formulas (15.1.2) and (15.1.3), the coefficients $b_i(t, x)$ and $c_{ij}(t, x)$ represent the *infinitesimal increments of expectation and covariance matrix of* $X^{t,x}$, in agreement with Remark 2.5.8.

Remark 15.1.3 Given $u \in C^{1,2}(\mathbb{R}^{N+1})$, by Itô's formula, the process

$$M_t := u(s, X_s^{t,x}) - \int_t^s (\partial_r + \mathscr{A}_r)u(r, X_r^{t,x})dr, \qquad s \geq t,$$

is a local martingale: this result is similar to Theorem 2.5.13 and shows how to "compensate" the process $s \mapsto u(s, X_s^{t,x})$ to obtain a (local) martingale. These similarities between Markov processes and solutions of SDEs are not coincidental: we will prove later (see Theorems 17.3.1 and 18.2.3) that, under suitable assumptions on the coefficients, the solution of an SDE is a diffusion.

15.2 Exit Time from a Bounded Domain

In this section, we provide some simple conditions that ensure that the first exit time of the solution of the SDE (15.0.1) from a bounded domain[1] D of $\mathbb{R}^N$, is absolutely integrable and therefore a.s. finite. We make the following

Assumption 15.2.1

(i) The coefficients of the SDE (15.0.1) are measurable and locally bounded, $b, \sigma \in L^\infty_{\text{loc}}([0, +\infty[\times \mathbb{R}^N)$;
(ii) for every $t \geq 0$ and $x \in D$ there exists a solution $X^{t,x}$ of (15.0.1) with initial condition $X_t^{t,x} = x$, on a set-up $(W, \mathscr{F}_t)$.

We denote by $\tau_{t,x}$ the first exit time of $X^{t,x}$ from D,

$$\tau_{t,x} = \inf\{s \geq t \mid X_s^{t,x} \notin D\},$$

and for simplicity, we write $X^{0,x} = X^x$ and $\tau_{0,x} = \tau_x$.

Proposition 15.2.2 *Assume that there exists a function* $f \in C^2(\mathbb{R}^N)$, *non-negative on D and such that*

$$\mathscr{A}_t f(x) \leq -1, \qquad t \geq 0, \ x \in D. \tag{15.2.1}$$

Then $E[\tau_x]$ is finite for every $x \in D$. In particular, such a function exists if for some $\lambda > 0$ and $i \in \{1, \ldots, N\}$ we have[2]

$$c_{ii}(t, \cdot) \geq \lambda, \qquad t \geq 0, \ x \in D. \tag{15.2.2}$$

---

[1] Open and connected set.
[2] Formula (15.2.2) is a non-total degeneracy condition on the matrix (c_{ij}) of the second-order coefficients of the characteristic operator $\mathscr{A}_t$ in (15.0.3): it is obviously satisfied if (c_{ij}) is uniformly positive definite.

15.2 Exit Time from a Bounded Domain

Proof For a fixed time t, by Itô's formula, we have

$$f(X^x_{t\wedge\tau_x}) = f(x) + \int_0^{t\wedge\tau_x} \mathscr{A}_s f(X^x_s)ds + \int_0^{t\wedge\tau_x} \nabla f(X^x_s)\sigma(s, X^x_s)dW_s.$$

Since ∇f and $\sigma(s, \cdot)$ are bounded on D for $s \leq t$, the stochastic integral has zero expectation and by (15.2.1) we have

$$E\left[f(X^x_{t\wedge\tau_x})\right] \leq f(x) - E\left[t \wedge \tau_x\right];$$

thus, since $f \geq 0$,

$$E\left[t \wedge \tau_x\right] \leq f(x).$$

Finally, taking the limit as $t \to \infty$, by Beppo Levi's theorem, we obtain

$$E\left[\tau_x\right] \leq f(x).$$

Now suppose that (15.2.2) holds and consider only the case $i = 1$: then it is enough to set

$$f(x) = \alpha(e^{\beta R} - e^{\beta x_1})$$

where α, β are suitable positive constants and R is large enough so that D is included in the Euclidean ball of radius R, centered at the origin. Indeed, f is non-negative on D and we have

$$\mathscr{A}_t f(x) = -\alpha e^{\beta x_1}\left(\frac{1}{2}c_{11}(t, x)\beta^2 + b_1(t, x)\beta\right)$$

$$\leq -\alpha\beta e^{-\beta R}\left(\frac{\lambda\beta}{2} - \|b\|_{L^\infty(D)}\right),$$

hence the thesis by choosing α, β suitably large. □

Remark 15.2.3 It is easy to determine a condition on the first-order terms, analogous to that of Proposition 15.2.2: if there exist $\lambda > 0$ and $i \in \{1, \ldots, N\}$ such that $b_i(t, \cdot) \geq \lambda$ or $b_1(t, x) \leq -\lambda$ on D for every $t \geq 0$ then $E[\tau_x]$ is finite. In fact, suppose for example that $b_1(t, x) \geq \lambda$: then applying Itô's formula to the function $f(x) = x_1$ we have

$$\left(X^x_{t\wedge\tau_x}\right)_1 = x_1 + \int_0^{t\wedge\tau_x} b_1(s, X^x_s)ds + \sum_{i=1}^d \int_0^{t\wedge\tau_x} \sigma_{1i}(s, X^x_s)dW^i_s,$$

and in expectation

$$E\left[\left(X^x_{t\wedge\tau_x}\right)_1\right] \geq x_1 + \lambda E\left[t \wedge \tau_x\right],$$

which proves the claim, in the limit as $t \to \infty$.

15.3 The Autonomous Case: The Dirichlet Problem

In this section, we consider the case where the coefficients $b = b(x)$ and $\sigma = \sigma(x)$ of the SDE (15.0.1) are independent of time and therefore denote $\mathscr{A}_t$ in (15.0.3) simply as $\mathscr{A}$. For many aspects, this condition is not restrictive since even problems with time dependence can be treated in this context by inserting time among the state variables as in the following Example 15.3.7. In addition to Assumption 15.2.1, we assume that $E[\tau_x]$ is finite for every $x \in D$, where D is a bounded domain.

The following result provides a representation formula (and, consequently, a uniqueness result) for the classical solutions of the *Dirichlet problem* for the elliptic-parabolic operator $\mathscr{A}$:

$$\begin{cases} \mathscr{A}u - au = f, & \text{in } D, \\ u|_{\partial D} = \varphi, \end{cases} \tag{15.3.1}$$

where f, a, φ are given functions. As previously stated, formula (15.3.2) serves as the foundation for Monte Carlo-type methods used in the numerical approximation of solutions to the Dirichlet problem (15.3.1).

Theorem 15.3.1 (Feynman-Kac Formula [!!]) *Let $f \in L^\infty(D)$, $\varphi \in C(\partial D)$ and $a \in C(D)$ such that $a \geq 0$. If $u \in C^2(D) \cap C(\bar{D})$ is a solution of the Dirichlet problem (15.3.1) then for every $x \in D$ we have*

$$u(x) = E\left[e^{-\int_0^{\tau_x} a(X^x_t)dt}\varphi(X^x_{\tau_x}) - \int_0^{\tau_x} e^{-\int_0^t a(X^x_s)ds} f(X^x_t)dt\right]. \tag{15.3.2}$$

Proof For $\varepsilon > 0$ sufficiently small, let D_ε be a domain such that

$$x \in D_\varepsilon, \qquad \bar{D}_\varepsilon \subseteq D, \qquad \text{dist}\,(\partial D_\varepsilon, \partial D) \leq \varepsilon.$$

Let τ_ε be the exit time of X^x from D_ε and observe that, being X^x continuous (Fig. 15.1),

$$\lim_{\varepsilon \to 0^+} \tau_\varepsilon = \tau_x.$$

15.3 The Autonomous Case: The Dirichlet Problem

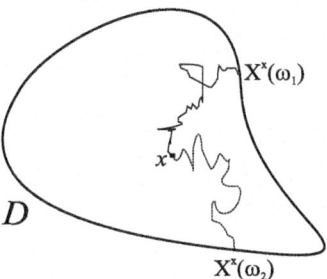

Fig. 15.1 The domain of a Dirichlet problem and two trajectories of the corresponding solution of the associated SDE

Let

$$Z_t = e^{-\int_0^t a(X_s^x)ds},$$

and note that, by hypothesis, $Z_t \in \,]0,1]$. Moreover, if $u_\varepsilon \in C_0^2(\mathbb{R}^N)$ is such that $u_\varepsilon = u$ on D_ε, by Itô's formula we have

$$d(Z_t u_\varepsilon(X_t^x)) = Z_t \left((\mathscr{A} u_\varepsilon - a u_\varepsilon)(X_t^x) dt + \nabla u_\varepsilon(X_t^x) \sigma(X_t^x) dW_t \right)$$

so that

$$Z_{\tau_\varepsilon} u(X_{\tau_\varepsilon}^x) = u(x) + \int_0^{\tau_\varepsilon} Z_t f(X_t^x) dt + \int_0^{\tau_\varepsilon} Z_t \nabla u(X_t^x) \sigma(X_t^x) dW_t.$$

Since ∇u and σ are bounded on D, in expectation we obtain

$$u(x) = E\left[Z_{\tau_\varepsilon} u(X_{\tau_\varepsilon}^x) - \int_0^{\tau_\varepsilon} Z_t f(X_t^x) dt \right].$$

Letting $\varepsilon \to 0^+$, we get the thesis by the dominated convergence: indeed, recalling that $Z_t \in \,]0,1]$, we have

$$\left| Z_{\tau_\varepsilon} u(X_{\tau_\varepsilon}^x) \right| \leq \|u\|_{L^\infty(D)}, \qquad \left| \int_0^{\tau_\varepsilon} Z_t f(X_t^x) dt \right| \leq \tau_x \|f\|_{L^\infty(D)},$$

and, by hypothesis, τ_x is absolutely integrable. □

Remark 15.3.2 The assumption $a \geq 0$ in Theorem 15.3.1 is essential: for example, the function

$$u(x,y) = \sin x \sin y$$

is a solution to the Dirichlet problem

$$\begin{cases} \frac{1}{2}\Delta u + u = 0, & \text{in } D =]0, 2\pi[\times]0, 2\pi[, \\ u|_{\partial D} = 0, \end{cases}$$

but does not satisfy (15.3.2).

Remark 15.3.3 (Maximum Principle) Under the assumptions of Theorem 15.3.1, from formula (15.3.2) it follows that if $f \geq 0$ then

$$u(x) \leq E\left[e^{-\int_0^{\tau_x} a(X_t^x)dt}\varphi(X_{\tau_x}^x)\right] \leq \max_{\partial D} \varphi.$$

Moreover, when $f = a = 0$, the following "maximum principle" holds:

$$\min_{\partial D} u \leq u(x) \leq \max_{\partial D} u.$$

Existence results for problem (15.3.1) are well known in the *uniformly elliptic* case: we recall the following classical theorem (see, for example, Theorem 6.13 in [53]).

Theorem 15.3.4 *Under the following assumptions*

(i) *$\mathscr{A}$ in (15.0.3) is a uniformly elliptic operator, i.e., there exists a constant $\lambda > 0$ such that*

$$\sum_{i,j=1}^{N} c_{ij}(x)\xi_i\xi_j \geq \lambda |\xi|^2, \qquad x \in D, \ \xi \in \mathbb{R}^N;$$

(ii) *the coefficients are Hölder continuous functions, $c_{ij}, b_j, a, f \in C^\alpha(D)$. Moreover, the functions c_{ij}, b_j, f are bounded and $a \geq 0$;*
(iii) *for each $y \in \partial D$ there exists[3] a Euclidean ball B contained in the complement of D and such that $y \in \bar{B}$;*
(iv) *$\varphi \in C(\partial D)$;*

there exists a classical solution $u \in C^{2+\alpha}(D) \cap C(\bar{D})$ of problem (15.3.1).

Now let us consider some significant examples.

Example 15.3.5 (Expectation of the Exit Time) If the problem

$$\begin{cases} \mathscr{A} u = -1, & \text{in } D, \\ u|_{\partial D} = 0, \end{cases}$$

has a solution, then by (15.3.2) we have $u(x) = E[\tau_x]$.

---

[3] This is a regularity condition for the boundary of D, that is satisfied if, for example, ∂D is a C^2-manifold.

15.3 The Autonomous Case: The Dirichlet Problem

Example 15.3.6 (Poisson Kernel) In the case $a = f = 0$, (15.3.2) is equivalent to a mean value formula. More precisely, let μ^x denote the distribution of the random variable $X^x_{\tau_x}$: then μ^x is a probability measure on ∂D and by (15.3.2) we have

$$u(x) = E\left[u(X^x_{\tau_x})\right] = \int_{\partial D} u(y) \mu^x(dy).$$

The law μ^x is usually called the *harmonic measure* of $\mathscr{A}$ on ∂D. If X^x is a Brownian motion with initial point $x \in \mathbb{R}^N$, then $\mathscr{A} = \frac{1}{2}\Delta$ and when $D = B(0, R)$ is the Euclidean ball of radius R, μ^x has a density (with respect to the surface measure) whose explicit expression is known: it corresponds to the so-called *Poisson kernel*.

$$\frac{1}{R\omega_N} \frac{R - |x|^2}{|x - y|^N},$$

where ω_N denotes the measure of the unit spherical surface in $\mathbb{R}^N$.

Example 15.3.7 (Heat Equation) Let W be a real Brownian motion. The process $X_t = (W_t, -t)$ is the solution of the SDE

$$\begin{cases} dX^1_t = dW_t, \\ dX^2_t = -dt, \end{cases}$$

and the corresponding characteristic operator

$$\mathscr{A} = \frac{1}{2} \partial_{x_1 x_1} - \partial_{x_2}$$

is the heat operator in $\mathbb{R}^2$.

Let us consider formula (15.3.2) on a rectangular domain

$$D =]a_1, b_1[\times]a_2, b_2[.$$

Examining the explicit expression of the trajectories of X (see Fig. 15.2), it is clear that the value $u(\bar{x}_1, \bar{x}_2)$ of a solution of the heat equation depends only on the values of u on the boundary part D contained in $\{x_2 < \bar{x}_2\}$. In general, the value of u in D depends only on the values of u on the *parabolic boundary* of D, defined by

$$\partial_p D = \partial D \setminus (\,]a_1, b_1[\times \{b_2\}).$$

This fact is consistent with the results on the Cauchy-Dirichlet problem of Sect. 20.1.1.

Fig. 15.2 Two paths traced by the solution of the SDE associated to a Cauchy-Dirichlet problem defined on a rectangular domain

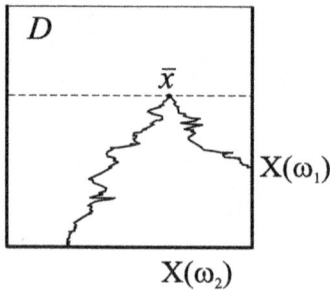

Example 15.3.8 (Method of Characteristics) If $\sigma = 0$ the characteristic operator is the first-order differential operator

$$\mathscr{A} = \sum_{i=1}^{N} b_i(x)\partial_{x_i}.$$

The corresponding SDE is actually deterministic and reduces to

$$X_t^x = x + \int_0^t b(X_s^x)ds,$$

that is, X is an integral curve of the vector field b:

$$\frac{d}{dt}X_t = b(X_t).$$

If the exit time of X from D is finite (cf. Remark 15.2.3) then we have the representation

$$u(x) = e^{-\int_0^{\tau_x} a(X_t^x)dt} \varphi(X_{\tau_x}^x) - \int_0^{\tau_x} e^{-\int_0^t a(X_s^x)ds} f(X_t^x)dt, \qquad (15.3.3)$$

for the solution of the problem

$$\begin{cases} \langle b, \nabla u \rangle - au = f, & \text{in } D, \\ u|_{\partial D} = \varphi. \end{cases}$$

Equation (15.3.3) is a particular case of the classic *method of characteristics* for the solution of first-order PDEs: for a full description of this method, we refer, for example, to Chapter 3.2 in [41].

As a particular example, let us consider the Cauchy problem in $\mathbb{R}^2$

$$\begin{cases} \partial_{x_1} u(x_1, x_2) - x_1 \partial_{x_2} u(x_1, x_2) = 0, & (x_1, x_2) \in D := \mathbb{R} \times \,]0, +\infty[, \\ u(x_1, 0) = \varphi(x_1), & x_1 \in \mathbb{R}. \end{cases}$$
(15.3.4)

In this case, $b(x_1, x_2) = (1, -x_1)$ and the correspondent "SDE" is

$$\begin{cases} \frac{d}{dt} X_{1,t} = 1, \\ \frac{d}{dt} X_{2,t} = -X_{1,t}. \end{cases}$$

Imposing the initial condition $X_0 = x \equiv (x_1, x_2) \in D$, we determine the solution

$$X_t^x = (X_{1,t}^x, X_{2,t}^x) = \left(x_1 + t, x_2 - tx_1 - \frac{t^2}{2} \right).$$

Imposing $X_{2,t}^x = 0$, we find the exit time from D of the trajectory X^x:

$$\tau_x = \sqrt{x_1^2 + x_2} - x_1.$$

Then $X_{\tau_x}^x = \left(\sqrt{x_1^2 + x_2}, 0 \right)$ is the exit point and, based on formula (15.3.3), the solution of the problem (15.3.4) is

$$u(x_1, x_2) = \varphi(X_{\tau_x}^x) = \varphi\left(\sqrt{x^2 + y} \right).$$

Note that, as in Example 2.5.12, the solution u inherits the regularity properties of φ and therefore in general the differential equation in (15.3.4) has to be understood in a distributional sense. From a probabilistic perspective, the transition law of the process X (which in this case is a Dirac delta distribution, i.e., $X_t^x \sim \delta_{(x_1+t, x_2-tx_1-t^2/2)}$) is the fundamental solution of the Cauchy problem (15.3.4).

15.4 The Evolutionary Case: The Cauchy Problem

Theorem 15.3.1 also has a parabolic counterpart, with a proof that is entirely analogous. Precisely, given the bounded domain D, we consider the cylinder

$$D_T = \,]0, T[\times D$$

and we denote by

$$\partial_p D_T := \partial D \setminus (\{0\} \times D)$$

the so-called *parabolic boundary of* D_T. The following theorem provides a representation formula for the classical solutions of the Cauchy-Dirichlet problem

$$\begin{cases} \mathscr{A}_t u - au + \partial_t u = f, & \text{in } D_T, \\ u|_{\partial_p D_T} = \varphi, \end{cases} \quad (15.4.1)$$

where f, a, φ are given functions.

Theorem 15.4.1 (Feynman-Kac Formula [!]) *Let* $f \in L^\infty(D_T)$, $\varphi \in C(\partial_p D_T)$ *and* $a \in C(D_T)$ *such that* $a_0 := \inf a$ *is finite. Under Assumption 15.2.1, if* $u \in C^2(D_T) \cap C(D_T \cup \partial_p D_T)$ *is a solution of problem* (15.4.1) *then, for any* $(t, x) \in D_T$, *we have*

$$u(t,x) = E\left[e^{-\int_t^{T \wedge \tau_{t,x}} a(s, X_s^{t,x})ds} \varphi(T \wedge \tau_{t,x}, X_{T \wedge \tau_{t,x}}^{t,x})\right.$$
$$\left. - \int_t^{T \wedge \tau_{t,x}} e^{-\int_t^s a(r, X_r^{t,x})dr} f(s, X_s^{t,x})ds\right]. \quad (15.4.2)$$

Remark 15.4.2 (Maximum Principle) Under the hypotheses of Theorem 15.4.1 and assuming $f = a = 0$, from formula (15.4.2) we deduce the following "maximum principle"

$$\min_{\partial_p D_T} u \leq u(x) \leq \max_{\partial_p D_T} u$$

which we will find, by analytical means, in Sect. 20.1.1.

We now prove a representation formula for the classical solution of the backward Cauchy problem

$$\begin{cases} \mathscr{A}_t u - au + \partial_t u = f, & \text{in } [0, T[\times \mathbb{R}^N, \\ u(T, \cdot) = \varphi, & \text{in } \mathbb{R}^N, \end{cases} \quad (15.4.3)$$

where $\mathscr{A}_t$ is the characteristic operator in (15.0.3) and f, a, φ are given functions. Chapter 20 is dedicated to a concise presentation of the main existence and uniqueness results for problem (15.4.3) in the case of uniformly parabolic operators with Hölder and bounded coefficients.

Since problem (15.4.3) is posed on an unbounded domain, it is necessary to introduce appropriate assumptions on the behavior at infinity of the coefficients.

15.4 The Evolutionary Case: The Cauchy Problem

Assumption 15.4.3

(i) The coefficients $b = b(t, x)$ and $\sigma = \sigma(t, x)$ are measurable functions, with at most linear growth in x uniformly in $t \in [0, T[$;
(ii) $a \in C([0, T[\times \mathbb{R}^N)$ with $\inf a =: a_0 > -\infty$.

Theorem 15.4.4 (Feynman-Kac Formula [!!]) *Suppose there exists a solution $u \in C^2([0, T[\times \mathbb{R}^N) \cap C([0, T] \times \mathbb{R}^N)$ of the Cauchy problem (15.4.3). Take Assumption 15.4.3 and at least one of the following conditions as given:*

(1) there exist two positive constants M, p such that

$$|u(t, x)| + |f(t, x)| \leq M(1 + |x|^p), \qquad (t, x) \in [0, T[\times \mathbb{R}^N; \qquad (15.4.4)$$

(2) the matrix σ is bounded and there exist two positive constants M and α, with α sufficiently small, such that

$$|u(t, x)| + |f(t, x)| \leq M e^{\alpha |x|^2}, \qquad (t, x) \in [0, T[\times \mathbb{R}^N. \qquad (15.4.5)$$

If the SDE (15.0.1) has a solution $X^{t,x}$ with initial datum $(t, x) \in [0, T[\times \mathbb{R}^N$ then the representation formula holds

$$u(t, x) = E\left[e^{-\int_t^T a(s, X_s^{t,x}) ds} \varphi(X_T^{t,x}) - \int_t^T e^{-\int_t^s a(r, X_r^{t,x}) dr} f(s, X_s^{t,x}) ds \right]. \qquad (15.4.6)$$

Proof Fix $(t, x) \in [0, T[\times \mathbb{R}^N$ and for simplicity, let $X = X^{t,x}$. If τ_R denotes the exit time of X from the Euclidean ball of radius R, by Theorem 15.4.1 we have

$$u(t, x) = E\left[e^{-\int_t^{T \wedge \tau_R} a(s, X_s) ds} u(T \wedge \tau_R, X_{T \wedge \tau_R}) \right.$$
$$\left. - \int_t^{T \wedge \tau_R} e^{-\int_t^s a(r, X_r) dr} f(s, X_s) ds \right]. \qquad (15.4.7)$$

Since

$$\lim_{R \to \infty} T \wedge \tau_R = T,$$

the thesis follows by taking the limit in R in (15.4.7) thanks to the dominated convergence theorem. In fact, we have pointwise convergence of the integrands and moreover, under condition 1), we have

$$e^{-\int_t^{T\wedge\tau_R} a(s,X_s)ds} \left|u(T \wedge \tau_R, X_{T\wedge\tau_R})\right| \leq M e^{|a_0|T}\left(1 + \bar{X}_T^p\right),$$

$$\left|\int_t^{T\wedge\tau_R} e^{-\int_t^s a(r,X_r)dr} f(s, X_s)ds\right| \leq T e^{|a_0|T} M \left(1 + \bar{X}_T^p\right),$$

where

$$\bar{X}_T = \sup_{0 \leq t \leq T} |X_t|$$

is absolutely integrable thanks to the a priori estimates of Theorem 14.5.2. Under condition (2), we proceed in a similar way using the exponential integrability estimate of Theorem 14.5.3. □

Remark 15.4.5 From the representation formula (15.4.6), it follows in particular the uniqueness of the solution of the Cauchy problem. As we will see in Sect. 20.1, the growth conditions (15.4.4) and (15.4.5) are necessary in order to select one among the solutions that are, in general, infinite.

15.5 Key Ideas to Remember

This chapter introduces several types of representation formulas for solutions to (Cauchy or Cauchy-Dirichlet) problems involving the characteristic operator of an SDE. Unsurprisingly, the definition of the characteristic operator closely mirrors that introduced in the context of Markov process theory. If you have any doubts about the following concise statements, please refer back to the relevant section.

- Section 15.1: we define the characteristic operator of an SDE: its coefficients represent the infinitesimal increments of expectation and covariance matrix of the solution to the associated SDE.
- Section 15.2: as a preliminary result, we provide simple conditions that ensure that the exit time from a bounded domain of the solution of an SDE is a.s. finite or even absolutely integrable.
- Section 15.3: using Itô's formula, it is almost immediate to obtain representation formulas for the classical solutions (assuming they exist) of the Dirichlet problem in terms of the expected value of the solution of the associated SDE. These formulas, known as Feynman-Kac formulas, have considerable theoretical and practical importance, which is illustrated with numerous examples.
- Section 15.4: we present the parabolic version of the Feynman-Kac formulas.

15.5 Key Ideas to Remember

Main notations used or introduced in this chapter:

Symbol	Description	Page
$\mathscr{A}_t$	Characteristic operator of an SDE	287
$\tau_{t,x}$	First exit time	290
$\partial_p D$	Parabolic boundary of the cylinder D	298
$\bar{X}_T = \sup_{0 \leq t \leq T} \lvert X_t \rvert$	Maximum process	300

Chapter 16
Linear Equations

Tant que nous sommes agités, nous pouvons être calmes[1]

Julien Green

In this chapter, we consider stochastic differential equations of the form

$$dX_t = (BX_t + b)dt + \sigma dW_t \qquad (16.0.1)$$

where $B \in \mathbb{R}^{N \times N}$, $b \in \mathbb{R}^N$, $\sigma \in \mathbb{R}^{N \times d}$, and W is a d-dimensional Brownian motion. Equation (16.0.1) is a particular case of (14.1.1) with coefficients $b(t, x) = Bx + b$ and $\sigma(t, x) = \sigma$ that are linear functions of the variable x (in fact, the diffusion coefficient is even constant) and therefore we say that (16.0.1) is a *linear SDE*. In this chapter, we exhibit the explicit expression of the solution and study the properties of its transition law, with particular attention to the absolutely continuous case, providing conditions for the existence of the transition density.

16.1 Solution and Transition Law of a Linear SDE

The following theorem provides the explicit expression of the solution of a linear SDE.

---

[1] As long as we are restless, we can be calm.

Theorem 16.1.1 *The solution $X^x = (X^x_t)_{t\geq 0}$ of (16.0.1) with initial datum $X^x_0 = x \in \mathbb{R}^N$ is given by*

$$X^x_t = e^{tB}\left(x + \int_0^t e^{-sB} b\, ds + \int_0^t e^{-sB}\sigma\, dW_s\right). \qquad (16.1.1)$$

The solution X^x is a Gaussian process: in particular, $X^x_t \sim \mathcal{N}_{m_t(x),\mathscr{C}_t}$ where

$$m_t(x) = e^{tB}\left(x + \int_0^t e^{-sB} b\, ds\right), \qquad \mathscr{C}_t = \int_0^t e^{sB}\sigma(e^{sB}\sigma)^*ds.$$

Proof To prove that X^x in (16.1.1) solves the SDE (16.0.1), it is sufficient to apply Itô's formula using the expression $X^x_t = e^{tB} Y^x_t$ where

$$dY^x_t = e^{-tB} b\, dt + e^{-tB}\sigma\, dW_t, \qquad Y^x_0 = x.$$

We now recall that, since Y^x is an Itô process with deterministic coefficients, by the multidimensional version of Example 11.1.9, we have

$$Y^x_t \sim \mathcal{N}_{\mu_t(x),C_t}, \qquad \mu_t(x) = x + \int_0^t e^{-sB} b\, ds, \qquad C_t = \int_0^t e^{-sB}\sigma\sigma^* e^{-sB^*} ds. \qquad (16.1.2)$$

The thesis follows easily from the fact that X^x is a linear transformation of Y^x. □

Remark 16.1.2 ([!]) The process

$$T \mapsto X^{t,x}_T := X^x_{T-t}, \qquad T \geq t,$$

solves the SDE (16.0.1) with initial datum (t, x). If the covariance matrix $\mathscr{C}_{T-t}$ is positive definite, then the random variable $X^{t,x}_T$ is absolutely continuous with Gaussian density $\Gamma(t, x; T, \cdot)$ given by

$$\Gamma(t,x;T,y) = \frac{1}{\sqrt{(2\pi)^N \det \mathscr{C}_{T-t}}} \exp\left(-\frac{1}{2}\langle \mathscr{C}^{-1}_{T-t}(y - m_{T-t}(x)), (y - m_{T-t}(x))\rangle\right).$$

By[2] Remark 2.5.10, Γ is a transition density of X in (16.0.1) and is the fundamental solution of the backward Kolmogorov operator $\mathscr{A}_t + \partial_t$ where

$$\mathscr{A}_t = \frac{1}{2}\sum_{i,j=1}^N c_{ij}\partial_{x_i x_j} + \langle Bx + b, \nabla\rangle, \qquad c := \sigma\sigma^*, \qquad (16.1.3)$$

is the characteristic operator of X.

---

[2] See also Theorem 17.3.1.

16.1 Solution and Transition Law of a Linear SDE

Example 16.1.3 (Langevin Equation [!]) Consider the SDE in $\mathbb{R}^2$

$$\begin{cases} dV_t = dW_t, \\ dX_t = V_t dt, \end{cases}$$

which is the simplified version of the Langevin equation [86] used in physics to describe the random motion of a particle in phase space: V_t and X_t represent the velocity and position of the particle at time t, respectively. Paul Langevin was the first, in 1908, to apply Newton's laws to the random Brownian motion studied by Einstein a few years earlier. Lemons [88] provides an interesting account of the approaches of Einstein and Langevin.

Referring to the general notation (16.0.1), we have $d = 1$, $N = 2$ and

$$B = \begin{pmatrix} 0 & 0 \\ 1 & 0 \end{pmatrix}, \qquad \sigma = \begin{pmatrix} 1 \\ 0 \end{pmatrix}. \tag{16.1.4}$$

Since $B^2 = 0$, the matrix B is nilpotent and

$$e^{tB} = I + tB = \begin{pmatrix} 1 & 0 \\ t & 1 \end{pmatrix}.$$

Moreover, setting $z = (v, x)$, we have

$$m_t(z) = e^{tB} z = (v, x + tv),$$

and

$$\mathscr{C}_t = \int_0^t e^{sB} \sigma \sigma^* e^{sB^*} ds = \int_0^t \begin{pmatrix} 1 & 0 \\ s & 1 \end{pmatrix} \begin{pmatrix} 1 & 0 \\ 0 & 0 \end{pmatrix} \begin{pmatrix} 1 & s \\ 0 & 1 \end{pmatrix} ds = \begin{pmatrix} t & \frac{t^2}{2} \\ \frac{t^2}{2} & \frac{t^3}{3} \end{pmatrix}. \tag{16.1.5}$$

Note that $\mathscr{C}_t$ is positive definite for every $t > 0$ and therefore (V, X) has transition density

$$\Gamma(t, z; T, \zeta) = \frac{\sqrt{3}}{\pi (T-t)^2} \exp\left(-\frac{1}{2} \langle \mathscr{C}_{T-t}^{-1} (\zeta - e^{(T-t)B} z), \zeta - e^{(T-t)B} z \rangle \right) \tag{16.1.6}$$

for $t < T$ and $z = (v, x)$, $\zeta = (\eta, \xi) \in \mathbb{R}^2$, where

$$\mathscr{C}_t^{-1} = \begin{pmatrix} \frac{4}{t} & -\frac{6}{t^2} \\ -\frac{6}{t^2} & \frac{12}{t^3} \end{pmatrix}.$$

Moreover, $(t, v, x) \mapsto \Gamma(t, v, x; T, \eta, \xi)$ is a fundamental solution of the backward Kolmogorov operator

$$\frac{1}{2}\partial_{vv} + v\partial_x + \partial_t \qquad (16.1.7)$$

and $(T, \eta, \xi) \mapsto \Gamma(t, v, x; T, \eta, \xi)$ is a fundamental solution of the forward Kolmogorov operator

$$\frac{1}{2}\partial_{\eta\eta} - \eta\partial_\xi - \partial_T. \qquad (16.1.8)$$

Operators in (16.1.7) and (16.1.8) are not uniformly parabolic because the matrix of the second-order part

$$\sigma\sigma^* = \begin{pmatrix} 1 & 0 \\ 0 & 0 \end{pmatrix}$$

is degenerate; nonetheless, like the classical heat equation operator, they have a Gaussian fundamental solution. Kolmogorov [70] was the first to exhibit the explicit expression (16.1.6) of the fundamental solution of (16.1.7) (see also the introduction of Hörmander's work [62]). In mathematical finance, the backward operator (16.1.7) is employed to evaluate some complex derivative instruments, notably including the so-called Asian options (see, for example, [8] and [112]).

Example 16.1.4 ([!]) In Example 16.1.3 we proved that, setting

$$X_t := \int_0^t W_s ds,$$

the pair (W, X) has a two-dimensional normal distribution with covariance matrix given in (16.1.5). It follows in particular that $X_t \sim \mathcal{N}_{0, \frac{t^3}{3}}$, confirming what we had already observed in Example 11.1.10.

Let us prove that *X is not a Markov process*. In Theorem 17.3.1 we will see that the pair (W, X), being a solution of the Langevin SDE, is a Markov process: Theorem 17.3.1 does not apply to X which is an Itô process but is not a solution of an SDE of the form (17.0.1). In fact, we have

$$E[X_T \mid \mathscr{F}_t] = X_t + E\left[\int_t^T W_s ds \mid \mathscr{F}_t\right] = X_t + (T-t)W_t \qquad (16.1.9)$$

since, by Itô's formula

$$d(tW_t) = W_t dt + t dW_t$$

16.1 Solution and Transition Law of a Linear SDE

namely

$$TW_T = tW_t + \int_t^T W_s ds + \int_t^T s dW_s$$

from which

$$E[TW_T \mid \mathscr{F}_t] = tW_t + E\left[\int_t^T W_s ds \mid \mathscr{F}_t\right] + E\left[\int_t^T s dW_s \mid \mathscr{F}_t\right]$$

and therefore

$$E\left[\int_t^T W_s ds \mid \mathscr{F}_t\right] = (T-t)W_t.$$

By (16.1.9), $E[X_T \mid \mathscr{F}_t]$ is a function not only of X_t but also of W_t: incidentally, this is a further confirmation of the Markov property of the pair (W, X). If X were a Markov process, then we should have[3]

$$E[X_T \mid X_t] = E[X_T \mid \mathscr{F}_t], \qquad t \leq T, \tag{16.1.10}$$

which combined with (16.1.9) would imply $W_t = f(X_t)$ a.s. for some $f \in m\mathscr{B}$. However, this is absurd: in fact, if $W_t = f(X_t)$ a.s. then $\mu_{W_t \mid X_t} = \delta_{f(X_t)}$ and this contrasts with the fact that (W_t, X_t) has a two-dimensional Gaussian density.

Remark 16.1.5 The results of this section extend to the case of linear SDEs of the type

$$dX_t = (b(t) + B(t)X_t)dt + \sigma(t)dW_t$$

where the matrices B, b and σ are measurable and bounded functions of time. In this case, the matrix exponential e^{tB} in the expression of the solution provided by Theorem 16.1.1 is replaced by the solution $\Phi(t)$ of the matrix Cauchy problem

$$\begin{cases} \Phi'(t) = B(t)\Phi(t), \\ \Phi(0) = I_N, \end{cases}$$

where I_N denotes the $N \times N$ identity matrix.

---

[3] Formula (16.1.10) must be interpreted according to Convention 4.2.5 in [113].

16.2 Controllability of Linear Systems and Absolute Continuity

We have seen that the solution X of the linear SDE (16.0.1) has a multi-normal transition law. Clearly, it is of particular interest when X admits a transition density and therefore the related Kolmogorov equations have a fundamental solution. In this section, we see that the non-degeneracy of the covariance matrix of X_t,

$$\mathscr{C}_t := \mathrm{cov}(X_t) = \int_0^t G_s G_s^* ds, \qquad G_t := e^{tB}\sigma, \tag{16.2.1}$$

can be characterized in terms of controllability of a system within the framework of optimal control theory (see, for example, [87] and [151]). We begin by introducing the following

Definition 16.2.1 The pair (B, σ) is *controllable* on $[0, T]$ if for every $x, y \in \mathbb{R}^N$ there exists a function $v \in C([0, T]; \mathbb{R}^d)$ such that the solution $\gamma \in C^1([0, T]; \mathbb{R}^N)$ of the problem

$$\begin{cases} \gamma'(t) = B\gamma(t) + \sigma v(t), & 0 < t < T, \\ \gamma(0) = x, \end{cases} \tag{16.2.2}$$

verifies the final condition $\gamma(T) = y$. We say that v is a *control* for (B, σ) on $[0, T]$.

Theorem 16.2.2 ([!]) *The matrix $\mathscr{C}_T$ in (16.2.1) is positive definite if and only if (B, σ) is controllable on $[0, T]$.*

Proof We preliminarily observe that $\mathscr{C}_t = e^{tB} C_t e^{tB^*}$, where

$$C_t = \int_0^t G_{-s} G_{-s}^* ds$$

is the covariance matrix in (16.1.2). Clearly, $\mathscr{C}_T > 0$ if and only if $C_T > 0$.

We suppose $C_T > 0$ and prove that (B, σ) is controllable on $[0, T]$. Consider the solution

$$\gamma(t) = e^{tB}\left(x + \int_0^t G_{-s} v(s) ds\right), \qquad t \in [0, T],$$

of the Cauchy problem (16.2.2). Given $y \in \mathbb{R}^N$, we have $\gamma(T) = y$ if and only if

$$\int_0^T G_{-s} v(s) ds = z := e^{-TB} y - x. \tag{16.2.3}$$

16.2 Controllability of Linear Systems and Absolute Continuity

Then it is easy to verify that a control is given explicitly by

$$v(s) = G^*_{-s} C_T^{-1} z, \qquad s \in [0, T]. \tag{16.2.4}$$

Conversely, assume that (B, σ) is controllable on $[0, T]$ and suppose, for contradiction, that C_T is degenerate, i.e., there exists $w \in \mathbb{R}^N \setminus \{0\}$ such that

$$\langle C_T w, w \rangle = 0.$$

Equivalently, we have

$$\int_0^T |w^* G_{-s}|^2 ds = 0$$

so that $w^* G_{-s} = 0$ for every $s \in [0, T]$ and therefore also

$$w^* \int_0^T G_{-s} v(s) ds = 0.$$

This contradicts (16.2.3), hence the controllability hypothesis, and concludes the proof. $\square$

Remark 16.2.3 The control v in (16.2.4) is *optimal* in the sense that it minimizes the "cost functional"

$$U(v) := \|v\|^2_{L^2([0,T])} = \int_0^T |v(t)|^2 dt.$$

This is a consequence of the Lagrange-Ljusternik theorem (cf., for example, [137]) which is the functional extension of the classical Lagrange multipliers theorem. More precisely, to minimize the functional U under the constraint (16.2.3), we consider the Lagrange functional

$$\mathscr{L}(v, \lambda) = \|v\|^2_{L^2([0,T])} - \lambda^* \left(\int_0^T G_{-t} v(t) dt - z \right),$$

where $\lambda \in \mathbb{R}^N$ is the Lagrange multiplier. Differentiating $\mathscr{L}$ in the Fréchet sense, we impose that v is a critical point for $\mathscr{L}$ and obtain

$$\partial_v \mathscr{L}(u) = 2 \int_0^T v(t)^* u(t) dt - \lambda^* \int_0^T G_{-t} u(t) dt = 0, \qquad u \in L^2([0, T]).$$

Then we find $v(s) = \frac{1}{2} G^*_{-s} \lambda$ with λ determined by the constraint (16.2.3), that is $\lambda = 2 C_T^{-1} z$, in agreement with (16.2.4).

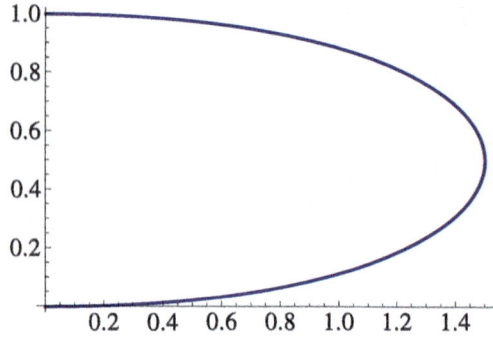

Fig. 16.1 Plot of the optimal trajectory $\gamma(t) = (6(t - t^2), 3t^3 - 2t^3)$, solution of problem (16.2.5) with initial condition $\gamma(0) = (0, 0)$ and final $\gamma(1) = (0, 1)$

Example 16.2.4 Let us resume Example 16.1.3 with the matrices B, σ as in (16.1.4). In this case, the control $v = v(t)$ has real values and the problem (16.2.2) becomes

$$\begin{cases} \gamma_1'(t) = v(t), \\ \gamma_2'(t) = \gamma_1(t), \\ \gamma(0) = (x_1, x_2). \end{cases} \tag{16.2.5}$$

The control acts directly only on the first component of γ but also affects the second component γ_2 through the second equation: by Theorem 16.2.2, (B, σ) is controllable on $[0, T]$ for every $T > 0$ with a control given explicitly by formula (16.2.4) (see Fig. 16.1).

16.3 Kalman Rank Condition

We provide a further operational criterion to verify the non-degeneracy of the covariance matrix $\mathscr{C}$.

Theorem 16.3.1 (Kalman Rank Condition) *The matrix $\mathscr{C}_T$ in (16.2.1) is positive definite for $T > 0$ if and only if the pair (B, σ) satisfies the following Kalman condition: the matrix of dimension $N \times (Nd)$, defined in blocks by*

$$\begin{pmatrix} \sigma & B\sigma & B^2\sigma & \cdots & B^{N-1}\sigma \end{pmatrix}, \tag{16.3.1}$$

has maximum rank, equal to N.

Proof Denote with

$$p(\lambda) := \det(B - \lambda I_N) = \lambda^N + a_1 \lambda^{N-1} + \cdots + a_{N-1}\lambda + a_N$$

16.3 Kalman Rank Condition

the characteristic polynomial of the matrix B: by the Cayley-Hamilton theorem, we have $p(B) = 0$. It follows that every power B^k, with $k \geq N$, is a linear combination of $I_N, B, \ldots, B^{N-1}$.

Now the matrix (16.3.1) does not have maximum rank if and only if there exists $w \in \mathbb{R}^N \setminus \{0\}$ such that

$$w^*\sigma = w^*B\sigma = \cdots = w^*B^{N-1}\sigma = 0. \tag{16.3.2}$$

Therefore, if the matrix (16.3.1) does not have maximum rank, by (16.3.2) and the Cayley-Hamilton theorem, we have

$$w^*B^k\sigma = 0, \qquad k \in \mathbb{N}_0,$$

from which also

$$w^*e^{tB}\sigma = 0, \qquad t \geq 0.$$

Consequently,

$$\langle \mathscr{C}_T w, w \rangle = \int_0^T |w^*e^{tB}\sigma|^2 dt = 0, \tag{16.3.3}$$

and $\mathscr{C}_T$ is degenerate for every $T > 0$.

Conversely, if $\mathscr{C}_T$ is degenerate then there exists $w \in \mathbb{R}^N \setminus \{0\}$ for which (16.3.3) holds and therefore

$$f(t) := w^*e^{tB}\sigma = 0, \qquad t \in [0, T].$$

By differentiating, we obtain

$$0 = \frac{d^k}{dt^k} f(t) \big|_{t=0} = w^*B^k\sigma, \qquad k \in \mathbb{N}_0,$$

and therefore, by (16.3.2), the matrix (16.3.1) does not have maximum rank. □

Remark 16.3.2 Since the Kalman condition does not depend on T, then $\mathscr{C}_T$ is positive definite for some $T > 0$ if and only if it is *for every* $T > 0$.

Example 16.3.3 In Example 16.1.3, we have

$$\sigma = \begin{pmatrix} 1 \\ 0 \end{pmatrix}, \qquad B\sigma = \begin{pmatrix} 0 & 0 \\ 1 & 0 \end{pmatrix} \begin{pmatrix} 1 \\ 0 \end{pmatrix} = \begin{pmatrix} 0 \\ 1 \end{pmatrix},$$

and thus $(\sigma \; B\sigma)$ is the 2×2 identity matrix which obviously satisfies the Kalman condition.

16.4 Hörmander's Condition

The non-degeneracy of the covariance matrix of a linear SDE can also be characterized in terms of a well-known condition in the context of partial differential equations. Consider the linear SDE (16.0.1) under the assumption that σ has rank d: then, up to a linear transformation, it is not restrictive to assume

$$\sigma = \begin{pmatrix} I_d \\ 0 \end{pmatrix}.$$

The corresponding Kolmogorov backward operator is

$$\mathscr{K} = \frac{1}{2}\Delta_d + \langle b + Bx, \nabla \rangle + \partial_t, \qquad (t, x) \in \mathbb{R}^{N+1}, \tag{16.4.1}$$

where Δ_d denotes the Laplace operator in the first d variables $x_1, \ldots, x_d$.

By convention, we identify a first-order differential operator on $\mathbb{R}^N$ of the type

$$Z := \sum_{i=1}^{N} \alpha_i(x) \partial_{x_i},$$

with the vector field of its coefficients and therefore also write

$$Z(x) = (\alpha_1(x), \ldots, \alpha_N(x)), \qquad x \in \mathbb{R}^N.$$

The *commutator* of two vector fields Z and U, with

$$U = \sum_{i=1}^{N} \beta_i \partial_{x_i},$$

is defined by

$$[Z, U] = ZU - UZ = \sum_{i=1}^{N} (Z\beta_i - U\alpha_i) \partial_{x_i}.$$

Hörmander's theorem [62] (see also Stroock [133] for a more recent treatment) stands as a remarkably broad theorem. Here, we revisit a specific version pertinent to the operator $\mathscr{K}$ in (16.4.1): this theorem states that $\mathscr{K}$ has a smooth fundamental solution if and only if, at every point $x \in \mathbb{R}^N$, the first-order operators (vector fields)

$$\partial_{x_1}, \ldots, \partial_{x_d}, \qquad Y := \langle Bx, \nabla \rangle,$$

16.4 Hörmander's Condition

together with their commutators of any order, span $\mathbb{R}^N$. This is the so-called *Hörmander's condition*. Note that $\partial_{x_1}, \ldots, \partial_{x_d}$ are the derivatives that appear in the second-order part of $\mathscr{K}$, corresponding to the directions of Brownian diffusion, while Y is the drift of the operator: therefore, essentially, the existence of the fundamental solution is equivalent to the fact that $\mathbb{R}^N$ is spanned at every point by the directional derivatives that appear in $\mathscr{K}$ as second derivatives and as drift, together with their commutators of any order.

Example 16.4.1

(i) If $d = N$ then $\mathscr{K}$ is a uniformly parabolic operator and Hörmander's condition is obviously satisfied, without resorting to the drift and commutators, since $\partial_{x_1}, \ldots, \partial_{x_N}$ form the canonical basis of $\mathbb{R}^N$.

(ii) In the case of the Langevin operator of Example 16.1.3, we have $Y = x_1 \partial_{x_2}$. Thus $\partial_{x_1} = (1, 0)$ together with the commutator

$$[\partial_{x_1}, Y] = \partial_{x_2} = (0, 1)$$

form the canonical basis of $\mathbb{R}^2$ and Hörmander's condition is satisfied.

(iii) Consider the Kolmogorov operator

$$\mathscr{K} = \frac{1}{2}\partial_{x_1 x_1} + x_1 \partial_{x_2} + x_2 \partial_{x_3} + \partial_t, \qquad (x_1, x_2, x_3) \in \mathbb{R}^3.$$

Here $N = 3$, $d = 1$ and $Y = x_1 \partial_{x_2} + x_2 \partial_{x_3}$: also in this case Hörmander's condition is satisfied since

$$\partial_{x_1}, \qquad [\partial_{x_1}, Y] = \partial_{x_2}, \qquad [[\partial_{x_1}, Y], Y] = \partial_{x_3},$$

form a basis of $\mathbb{R}^3$. This example can be considered a generalization of Langevin model in which, in addition to considering *position and velocity*, a third stochastic process is introduced that represents the *acceleration* of a particle and is defined as a real Brownian motion.

Theorem 16.4.2 *Kalman and Hörmander conditions are equivalent.*

Proof It is sufficient to note that, for $i = 1, \ldots, d$,

$$[\partial_{x_i}, Y] = \sum_{k=1}^{N} b_{ki} \partial_{x_k}$$

is the i-th column of matrix B. Moreover, $[[\partial_{x_i}, Y], Y]$ is the i-th column of matrix B^2 and an analogous representation holds for higher order commutators.

On the other hand, for $k = 1, \ldots, N$, the block $B^k \sigma$ in the Kalman matrix (16.3.1) is the $N \times d$ matrix whose columns are the first d columns of B^k. □

Building upon the research in [34, 85, 106, 119] and [114], a theory analogous to the classical treatment of uniformly parabolic equations has been developed for Kolmogorov equations *with variable coefficients* of the type $\partial_t + \mathscr{A}_t$ with $\mathscr{A}_t$ as in (16.1.3) and $\sigma = \sigma(t, x)$.

16.5 Examples and Applications

Linear SDEs are the basis of many important stochastic models: here we briefly present some examples.

Example 16.5.1 (Vasicek Model) One of the simplest and most famous stochastic models for the evolution of interest rates (also called short rates or short-term rates) was proposed by Vasicek [143]:

$$dr_t = \kappa(b - r_t)dt + \sigma dW_t.$$

Here W is a real Brownian motion, σ represents the volatility of the rate and the parameters κ, θ are called respectively "speed of mean reversion" and "long-term mean level". The particular form of the drift $\kappa(\theta - r_t)$, with $\kappa > 0$, is designed to capture the so-called "mean reversion" property, an essential characteristic of the interest rate that distinguishes it from other financial prices: unlike stock prices, for example, interest rates cannot rise indefinitely. This is because at very high levels they would hinder economic activity, leading to a decrease in interest rates. Consequently, interest rates move in a bounded range, showing a tendency to return to a long-term value, represented by the parameter θ in the model. As soon as r_t exceeds the level θ, the drift becomes negative and "pushes" r_t to decrease while on the contrary, if $r_t < \theta$, the drift is positive and tends to make r_t grow towards θ. The fact that r_t has a normal distribution makes the model very simple to use and allows for explicit formulas for more complex financial instruments, such as interest rate derivatives. Among various resources, [18] stands out as an excellent introductory text for interest rate modeling (Fig. 16.2).

Example 16.5.2 (Brownian Bridge) Given $b \in \mathbb{R}$, consider the one-dimensional SDE

$$dB_t = \frac{b - B_t}{1 - t}dt + dW_t$$

with solution

$$B_t = B_0(1 - t) + bt + (1 - t)\int_0^t \frac{dW_s}{1 - s}, \qquad 0 \leq t < 1.$$

16.5 Examples and Applications

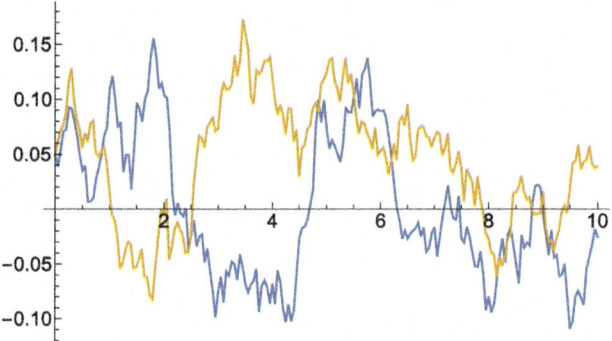

Fig. 16.2 Plot of two trajectories of the Vasicek process with parameters $\kappa = 1$, $X_0 = \theta = 5\%$ and $\sigma = 8\%$

We have

$$E[B_t] = B_0(1-t) + bt,$$

and, by Itô's isometry, we have

$$\mathrm{var}(B_t) = (1-t)^2 \int_0^t \frac{ds}{(1-s)^2} = t(1-t),$$

so that

$$\lim_{t \to 1^-} E[B_t] = b, \qquad \lim_{t \to 1^-} \mathrm{var}(B_t) = 0.$$

Let us prove that B_t converges to b for $t \to 1^-$ in L^2 norm:

$$E\left[(B_t - b)^2\right] = (1-t)^2(b - B_0)^2 - 2(1-t)^2(b - B_0) \underbrace{E\left[\int_0^t \frac{dW_s}{1-s}\right]}_{=0}$$

$$+ E\left[\left(\int_0^t \frac{dW_s}{1-s}\right)^2\right]$$

$$= (1-t)^2 \left((b - B_0)^2 + \int_0^t \frac{ds}{(1-s)^2}\right)$$

$$= (1-t)^2 \left((b - B_0)^2 + \frac{1}{1-t} - 1\right) \xrightarrow[t \to 1^-]{} 0.$$

The Brownian bridge is useful for modeling a system that starts at some level B_0 and is expected to reach level b at some future time, for example $t = 1$. In Fig. 16.3,

Fig. 16.3 Plot of four trajectories of a Brownian bridge

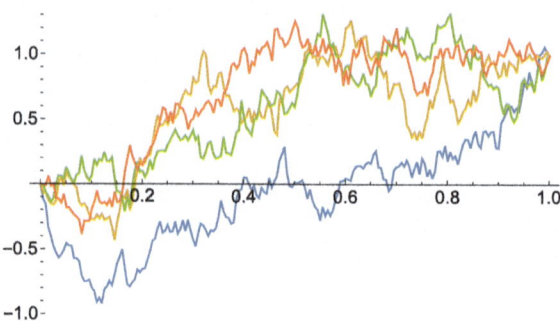

four trajectories of a Brownian bridge B with initial value $B_0 = 0$ and $B_1 = 1$ are shown.

Example 16.5.3 (Ornstein-Uhlenbeck [104]) The following system of equations for the motion of a particle extends the Langevin model by introducing an additional friction term:

$$\begin{cases} dX_t^1 = -\mu X_t^1 dt + \eta dW_t \\ dX_t^2 = X_t^1 dt. \end{cases}$$

Here W is a real Brownian motion, μ and η are the positive parameters of friction and diffusion. In matrix form

$$dX_t = BX_t dt + \sigma dW_t$$

with

$$B = \begin{pmatrix} -\mu & 0 \\ 1 & 0 \end{pmatrix}, \qquad \sigma = \begin{pmatrix} \eta \\ 0 \end{pmatrix}.$$

The validity of the Kalman condition is easily verified. Moreover, we have

$$B^n = \begin{pmatrix} (-\mu)^n & 0 \\ (-\mu)^{n-1} & 0 \end{pmatrix}, \qquad n \in \mathbb{N},$$

and

$$e^{tB} = I + \sum_{n=1}^{N} \frac{(tB)^n}{n!} = \begin{pmatrix} e^{-\mu t} & 0 \\ \frac{1-e^{-\mu t}}{\mu} & 1 \end{pmatrix}.$$

16.5 Examples and Applications

The solution X_t with initial datum $(x_1, x_2) \in \mathbb{R}^2$ is a two-dimensional Gaussian process with

$$E[X_t] = e^{tB}x = \begin{pmatrix} x_1 e^{-\mu t} \\ x_2 + \frac{x_1}{\mu}(1 - e^{-\mu t}) \end{pmatrix}$$

and

$$\mathscr{C}_t = \int_0^t e^{sB} \sigma \sigma^* e^{sB^*} ds$$

$$= \eta^2 \int_0^t \begin{pmatrix} e^{-\mu s} & 0 \\ \frac{1-e^{-\mu s}}{\mu} & 0 \end{pmatrix} \begin{pmatrix} e^{-\mu s} & \frac{1-e^{-\mu s}}{\mu} \\ 0 & 1 \end{pmatrix} ds$$

$$= y^2 \int_0^t \begin{pmatrix} e^{-2\mu s} & \frac{e^{-\mu s} - e^{-2\mu s}}{\mu} \\ \frac{e^{-\mu s} - e^{-2\mu s}}{\mu} & \left(\frac{1-e^{-\mu s}}{\mu}\right)^2 \end{pmatrix} ds$$

$$= y^2 \begin{pmatrix} \frac{1}{2\mu}(1 - e^{-2\mu t}) & \frac{1}{2\mu^2}(1 - 2e^{-\mu t} + e^{-2\mu t}) \\ \frac{1}{2\mu^2}(1 - 2e^{-\mu t} + e^{-2\mu t}) & \frac{1}{\mu^3}\left(\mu t + 2e^{-\mu t} - \frac{e^{-2\mu t} - 3}{2}\right) \end{pmatrix}.$$

Next, we present two examples of very popular SDEs frequently used in the field of mathematical finance. Although not linear SDEs of the form (16.0.1), these equations have an "affine structure" (in the sense of [36]) that allows to derive the expression of their CHF and density in terms of special functions.

Example 16.5.4 (CIR Model) The Cox-Ingersoll-Ross (CIR) model [29] is a variant of the Vasicek model of Example 16.5.1 in which the diffusion coefficient is a square root function: this implies that, unlike Vasicek, the solution (the interest rate) takes non-negative values. Specifically, we consider the following stochastic dynamics

$$dX_t = \kappa(\theta - X_t)dt + \sigma\sqrt{X_t}dW_t \tag{16.5.1}$$

where κ, θ, σ are positive parameters and W is a real Brownian motion. Using Itô's formula, we determine the CHF φ_{X_t} of X_t: first, we have

$$de^{i\eta X_t} = i\eta e^{i\eta X_t} dX_t - \frac{\eta^2}{2} e^{i\eta X_t} d\langle X \rangle_t$$

$$= e^{i\eta X_t} \left(i\eta\kappa(\theta - X_t) - \frac{(\eta\sigma)^2}{2} X_t \right) dt + i\eta\sigma e^{i\eta X_t} \sqrt{X_t} dW_t =$$

(putting $a(\eta) = i\eta\kappa\theta$, $b(\eta) = i\eta\kappa - \frac{(\eta\sigma)^2}{2}$ and $c(\eta, X_t) = i\eta\sigma e^{i\eta X_t}\sqrt{X_t}$)

$$= e^{i\eta X_t}(a(\eta) + b(\eta)X_t)dt + c(\eta, X_t)dW_t =$$

(exploiting the fact that $X_t e^{i\eta X_t} = -i\partial_\eta e^{i\eta X_t}$)

$$= (a(\eta) - ib(\eta)\partial_\eta)e^{i\eta X_t}dt + c(\eta, X_t)dW_t.$$

Applying the expected value and assuming $X_0 = x$, we have

$$\varphi_{X_t}(\eta) = e^{i\eta x} + \int_0^t \left(a(\eta) - ib(\eta)\partial_\eta\right)\varphi_{X_s}(\eta)ds.$$

Equivalently, the function $u(t, \eta) := \varphi_{X_t}(\eta)$ satisfies the following Cauchy problem for a first-order partial differential equation

$$\begin{cases} \partial_t u(t, \eta) = (a(\eta) - ib(\eta)\partial_\eta)\varphi_{X_t}(\eta), & t > 0, \ \eta \in \mathbb{R}, \\ u(0, \eta) = e^{i\eta x}. \end{cases}$$

This problem is solved using the method of characteristics of Example 15.3.8: setting

$$d(t) := \frac{2\kappa}{(1 - e^{-\kappa t})\sigma^2}, \qquad \lambda(t) := 2xe^{-\kappa t}d(t),$$

we obtain (Fig. 16.4)

$$\varphi_{X_t}(\eta) = \left(\frac{d(t)}{d(t) - i\eta}\right)^{\frac{\kappa}{2}} e^{\frac{i\eta\lambda(t)/2}{d(t) - i\eta}}.$$

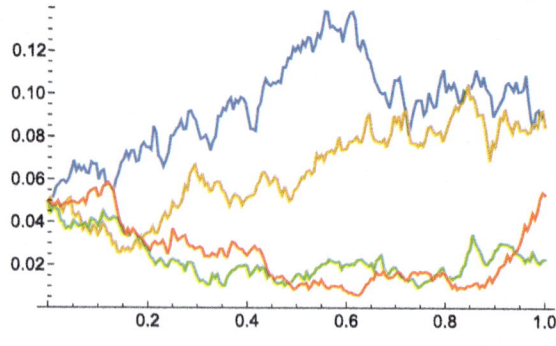

Fig. 16.4 Plot of four trajectories of the CIR process with parameters $X_0 = \theta = 5\%$, $\kappa = 1$ and $\sigma = 20\%$

16.5 Examples and Applications

Example 16.5.5 (CEV Model) The *constant elasticity of variance* (CEV) model has origins in physics and was introduced in mathematical finance by Cox [27, 28] to describe the dynamics of the price of a risky asset: the CEV equation is of the form

$$dX_t = \sigma X_t^\beta dW_t, \qquad (16.5.2)$$

with parameters $\sigma > 0$, $0 < \beta < 1$ and initial condition $X_0 = x \geq 0$.

We illustrate its peculiar characteristics here following the presentation in [105] (see also [32] and [33]): it is possible to construct a weak solution of (16.5.2) starting from the Kolmogorov equation, expressing the transition density[4] of the solution in terms of special functions. The process X has distinct properties in the two cases $\beta < \frac{1}{2}$ and $\beta \geq \frac{1}{2}$. To describe these properties, we first introduce the functions

$$\Gamma_\pm(t, x; T, y) = \frac{x^{\frac{1}{2}-2\beta}\sqrt{y} e^{-\frac{x^{2(1-\beta)}+y^{2(1-\beta)}}{2(1-\beta)^2\sigma^2(T-t)}}}{(1-\beta)\sigma^2(T-t)} \mathscr{I}_{\pm\frac{1}{2(1-\beta)}}\left(\frac{(xy)^{1-\beta}}{(1-\beta)^2\sigma^2(T-t)}\right),$$

where $\mathscr{I}_\nu(x)$ is the modified Bessel function of the first kind defined by

$$\mathscr{I}_\nu(x) = \left(\frac{x}{2}\right)^\nu \sum_{k=0}^\infty \frac{x^{2k}}{2^{2k} k! \Gamma_E(\nu+k+1)},$$

and Γ_E denotes the Euler Gamma. Both Γ_+ and Γ_- are fundamental solutions of $\partial_t + \mathscr{A}$ where $\mathscr{A}$ is the characteristic operator of X:

$$\mathscr{A} = \frac{\sigma^2 x^{2\beta}}{2} \partial_{xx}.$$

Precisely, we have

$$(\partial_t + \mathscr{A})\Gamma_\pm(t, x; T, y) = 0, \qquad \text{on }]0, T[\times \mathbb{R}_{>0},$$

---

[4] The transition density is constructed from the transformation

$$Y_t = \frac{X_t^{2(1-\beta)}}{\sigma^2(1-\beta)^2}$$

which leads (16.5.2) to the Bessel equation

$$dY_t = \delta dt + 2\sqrt{Y_t} dW_t \qquad (16.5.3)$$

with $\delta = \frac{1-2\beta}{1-\beta}$. Formula (16.5.3) is a particular case of (16.5.1).

and

$$\lim_{\substack{(t,x)\to(T,x_0)\\t<T}} \int_{\mathbb{R}_{>0}} \Gamma_{\pm}(t,x;T,y)\varphi(y)dy = \varphi(x_0), \qquad x_0 \in \mathbb{R}_{\geq 0},$$

for every continuous and bounded function φ.

The process X is non-negative and can take the value 0. If $\beta \geq \frac{1}{2}$ we say that 0 is an "absorbing" state since, if we denote by $\tau_x := \inf\{t \mid X_t = 0\}$ the first time when X reaches 0 starting from $X_0 = x$, then $X_t \equiv 0$ for $t \geq \tau_x$. The transition law of X is

$$p(t,x;T,H) = (1-a)\delta_0(H) + a\int_H \Gamma_+(t,x;T,y)dy, \qquad H \in \mathscr{B},$$

where

$$a := \int_0^{+\infty} \Gamma_+(t,x;T,y)dy < 1.$$

On the other hand, if $\beta < \frac{1}{2}$ then X reaches 0 but is "reflected": in this case Γ_- has an integral equal to one on $\mathbb{R}_{>0}$ and is the transition density of X.

In [33] and [61] it is proven that X is a strictly local martingale and for this reason it is not a good model for the price of a risky asset because it creates "arbitrage opportunities": in fact, if $\beta < \frac{1}{2}$, buying the asset at time τ_x at zero cost, there is a certain gain since the price later becomes positive. For this reason, in the CEV model introduced by Cox [27], the price is defined as the process obtained by stopping the solution X at time τ_x, that is

$$S_t := X_{t\wedge\tau_x}, \qquad t \geq 0.$$

In the financial interpretation, τ_x represents the default time of the risky asset. Delbaen and Shirakawa [33] show that S is a non-negative martingale for every $0 < \beta < 1$. The unstopped process X is instead used as a model for the dynamics of interest rates and volatility (or risk index, positive by definition) of financial assets, as in the famous CIR [29] and Heston [60] models. The CEV model (and its stochastic volatility counterpart, the popular SABR model [58] used in interest rate modeling) is an interesting example of a degenerate model because the infinitesimal generator is not uniformly elliptic and the law of the price process is not absolutely continuous with respect to the Lebesgue measure.

16.6 Key Ideas to Remember

We highlight the key outcomes of the chapter and the fundamental concepts to remember from an initial reading, omitting the more technical or peripheral matters. If any of the following brief statements are unclear, please refer back to the relevant section for clarification.

- Section 16.1: linear SDEs have explicit Gaussian solutions. A particularly interesting example is provided by the Langevin kinetic model whose solution admits a density although the diffusive coefficient of the SDE is degenerate.
- Sections 16.2, 16.3, and 16.4: the study of the absolute continuity of the solution of a linear SDE opens up interesting links with the theories of optimal control and PDEs. The fact that the covariance matrix of the solution of a linear SDE is positive definite is equivalent to the controllability of an appropriate linear system: in this regard, the Kalman condition provides a simple operational criterion. There is an additional equivalence with the Hörmander condition, which is well-known in the context of PDEs theory.
- Section 16.5: linear SDEs are the basis of classic stochastic models and find wide-ranging applications in various fields. In this section we present numerous examples of linear and non-linear SDEs used in mathematical finance and beyond.

Chapter 17
Strong Solutions

> *I spend many hours wandering the streets of Palermo, drinking strong black coffee and wondering what's wrong with me. I've made it - I'm the world's number one tennis player, yet I feel empty.*
>
> Andre Agassi [1]

We present classical results regarding the strong existence and pathwise uniqueness for SDEs. We maintain the general notations introduced in Chap. 14 and focus on the SDE

$$dX_t = b(t, X_t)dt + \sigma(t, X_t)dW_t \qquad (17.0.1)$$

where W is a d-dimensional Brownian motion and the coefficients

$$b = b(t, x) :]t_0, T[\times \mathbb{R}^N \longrightarrow \mathbb{R}^N, \qquad \sigma = \sigma(t, x) :]t_0, T[\times \mathbb{R}^N \longrightarrow \mathbb{R}^{N \times d}, \qquad (17.0.2)$$

satisfy the standard assumptions of Definition 14.4.1 for regularity (local Lipschitz continuity) and linear growth. Here $N, d \in \mathbb{N}$ and $0 \leq t_0 < T$ are fixed. We prove the following results:

- Theorem 17.1.1 on strong uniqueness;
- Theorem 17.2.1 on strong solvability and the flow property;
- Theorem 17.3.1 on the Markov property;
- Theorem 17.4.1 and Corollary 17.4.2 on estimates of dependence on the initial datum, regularity of trajectories, Feller property, and strong Markov property.

17.1 Uniqueness

Theorem 17.1.1 (Strong Uniqueness) *Assume the following hypothesis of local Lischitz continuity in x, uniform in t: for every $n \in \mathbb{N}$ there exists a constant κ_n such that*

$$|b(t,x) - b(t,y)| + |\sigma(t,x) - \sigma(t,y)| \leq \kappa_n |x-y|, \qquad (17.1.1)$$

for every $t \in [t_0, T]$ and $x, y \in \mathbb{R}^N$ such that $|x|, |y| \leq n$. Then for the SDE (17.0.1) with initial datum Z there is strong uniqueness according to Definition 14.1.11.

Proof Let X, Y be two solutions of the SDE (17.0.1) with initial datum Z, i.e. $X \in \text{SDE}(b, \sigma, W, \mathscr{F}_t)$ and $Y \in \text{SDE}(b, \sigma, W, \mathscr{G}_t)$. We use a localization argument[1] and set

$$\tau_n = \inf\{t \in [t_0, T] \mid |X_t| \vee |Y_t| \geq n\}, \qquad n \in \mathbb{N},$$

with the convention $\min \emptyset = T$. Note that $\tau_n = t_0$ on $(|Z| > n) \in \mathscr{F}_{t_0} \cap \mathscr{G}_{t_0}$. Since by hypothesis X, Y are adapted and a.s. continuous, τ_n is an increasing sequence of stopping times[2] with values in $[t_0, T]$, such that $\tau_n \nearrow T$ a.s. We set

$$b_n(t,x) = b(t,x) \mathbb{1}_{[t_0, \tau_n]}(t), \qquad \sigma_n(t,x) = \sigma(t,x) \mathbb{1}_{[t_0, \tau_n]}(t), \qquad n \in \mathbb{N}. \qquad (17.1.2)$$

The processes $X_{t \wedge \tau_n}, Y_{t \wedge \tau_n}$ satisfy almost surely the equation

$$X_{t \wedge \tau_n} - Y_{t \wedge \tau_n} = \int_{t_0}^{t \wedge \tau_n} (b(s, X_s) - b(s, Y_s)) \, ds + \int_{t_0}^{t \wedge \tau_n} (\sigma(s, X_s) - \sigma(s, Y_s)) \, dW_s$$

$$= \int_{t_0}^{t} \left(b_n(s, X_{s \wedge \tau_n}) - b_n(s, Y_{s \wedge \tau_n}) \right) ds$$

$$+ \int_{t_0}^{t} \left(\sigma_n(s, X_{s \wedge \tau_n}) - \sigma_n(s, Y_{s \wedge \tau_n}) \right) dW_s. \qquad (17.1.3)$$

Moreover, we have

$$\left| b_n(s, X_{s \wedge \tau_n}) - b_n(s, Y_{s \wedge \tau_n}) \right| = \left| b_n(s, X_{s \wedge \tau_n}) - b_n(s, Y_{s \wedge \tau_n}) \right| \mathbb{1}_{(|Z| \leq n)} \leq$$

---

[1] The localization argument is necessary even under the hypothesis of global Lischitz continuity because the idea is to apply Grönwall's lemma to the function

$$v(t) = E\left[\sup_{t_0 \leq s \leq t} |X_s - Y_s|^2 \right]$$

under the assumption that v is bounded.

[2] With respect to the filtration defined by $\mathscr{F}_t \vee \mathscr{G}_t := \sigma(\mathscr{F}_t \cup \mathscr{G}_t)$.

17.1 Uniqueness

(since $|X_{s\wedge\tau_n}|, |Y_{s\wedge\tau_n}| \leq n$ on $(|Z| \leq n)$ for $s \in [t_0, T]$)

$$\leq \kappa_n \left| X_{s\wedge\tau_n} - X_{s\wedge\tau_n} \right| \tag{17.1.4}$$

and a similar estimate is obtained with σ_n instead of b_n. Now let

$$v_n(t) = E\left[\sup_{t_0 \leq s \leq t} \left| X_{s\wedge\tau_n} - Y_{s\wedge\tau_n} \right|^2\right], \qquad t \in [t_0, T].$$

From (17.1.3) and (17.1.4), proceeding exactly as in the proof of estimate (14.4.5) with $p = 2$, we obtain

$$v_n(t) \leq \bar{c} \int_{t_0}^t v(s)\,ds, \qquad t \in [t_0, T],$$

for a positive constant $\bar{c} = \bar{c}(T, d, N, \kappa_n)$. Since X and Y are a.s. continuous and adapted (and therefore progressively measurable), Fubini's theorem ensures that v is a measurable function on $[t_0, T]$, that is, $v_n \in m\mathscr{B}$. Moreover, v_n is bounded, precisely $|v_n| \leq 4n^2$, by construction. From Grönwall's lemma, we obtain that $v_n \equiv 0$ and therefore

$$E\left[\sup_{t_0 \leq t \leq T} \left| X_{t\wedge\tau_n} - Y_{t\wedge\tau_n} \right|^2\right] = v_n(T) = 0.$$

Taking the limit as $n \to \infty$, by Beppo Levi's theorem, X and Y are indistinguishable on $[t_0, T]$. □

In the one-dimensional case, the following stronger result holds, which we report without proof (see, for example, Theorem 5.3.3 in [37] or Proposition 5.2.13 in [67]).

Theorem 17.1.2 (Yamada and Watanabe [149]) *In the case $N = d = 1$, there is strong uniqueness for the SDE* (17.0.1) *under the following conditions:*

$$|b(t, x) - b(t, y)| \leq k(|x - y|), \qquad |\sigma(t, x) - \sigma(t, y)| \leq h(|x - y|),$$
$$t \geq 0, x, y \in \mathbb{R}, \tag{17.1.5}$$

where

(i) *h is a strictly increasing function such that $h(0) = 0$ and for every $\varepsilon > 0$*

$$\int_0^\varepsilon \frac{1}{h^2(s)}\,ds = \infty; \tag{17.1.6}$$

(ii) k is a strictly increasing, concave function such that $k(0) = 0$ and for every $\varepsilon > 0$

$$\int_0^\varepsilon \frac{1}{k(s)} ds = \infty.$$

17.2 Existence

We are interested in studying the solvability in the strong sense, which, as seen in Sect. 14.1, requires that the solution is adapted to the standard filtration of the Brownian motion and the initial datum. As stated[3] in [124], the point where Itô's original theory of strong solutions of SDEs proves to be truly effective is the theory of flows, which plays an important role in many applications: in this regard, we indicate [82] as a reference monograph (additional valuable resources include [12, 47] and [51]).

Theorem 17.2.1 (Strong Solvability and Flow Property [!]) *Suppose that the coefficients b, σ satisfy the standard assumptions[4] (14.4.1) and (14.4.2) on $]t_0, T[\times \mathbb{R}^N$. Given a set-up $(W, \mathscr{F}_t)$, we have:*

(i) *for every $x \in \mathbb{R}^N$, there exists a strong solution $X^{t_0, x} \in SDE(b, \sigma, W, \mathscr{F}^W)$ with initial datum $X_{t_0}^{t_0, x} = x$. Moreover, for every $t \in [t_0, T]$ we have*

$$(x, \omega) \longmapsto \psi_{t_0, t}(x, \omega) := X_t^{t_0, x}(\omega) \in m(\mathscr{B}_N \otimes \mathscr{F}_t^W); \tag{17.2.1}$$

(ii) *for every $Z \in m\mathscr{F}_{t_0}$ the process $X^{t_0, Z}$ defined by*

$$X_t^{t_0, Z}(\omega) := \psi_{t_0, t}(Z(\omega), \omega), \qquad \omega \in \Omega, \ t \in [t_0, T], \tag{17.2.2}$$

is a strong solution of the SDE (17.0.1) (i.e. $X^{t_0, Z} \in SDE(b, \sigma, W, \mathscr{F}^{Z, W})$) with initial datum $X_{t_0}^{t_0, Z} = Z$;

(iii) *the flow property holds: for every $t \in [t_0, T[$, the processes $X^{t_0, Z}$ and $X^{t, X_t^{t_0, Z}}$ are indistinguishable on $[t, T]$, that is, almost surely*

$$X_s^{t_0, Z} = X_s^{t, X_t^{t_0, Z}} \quad \text{for every } s \in [t, T]. \tag{17.2.3}$$

---

[3] [124] page 136: "Where the 'strong' or 'pathwise' approach of Itô's original theory of SDEs really comes into its own is in the theory of flows. Flows are now very big business; and the martingale-problem approach, for all that is has other interesting things to say, cannot deal with them in any natural way."

[4] Actually, using a localization argument as in the proof of Theorem 17.1.1, it is sufficient to assume hypothesis (17.1.1) (local Lipschitz continuity) instead of (14.4.2).

17.2 Existence

Proof We divide the proof into several steps.

(1) We prove the existence of the solution of (17.0.1) on $[t_0, T]$ with deterministic initial datum $X_{t_0} = x \in \mathbb{R}^N$. We use the method of successive approximations and recursively define the sequence of Itô processes

$$X_t^{(0)} \equiv x,$$

$$X_t^{(n)} = x + \int_{t_0}^t b(s, X_s^{(n-1)}) ds + \int_{t_0}^t \sigma(s, X_s^{(n-1)}) dW_s, \qquad n \in \mathbb{N},$$
(17.2.4)

for $t \in [t_0, T]$. The sequence is well-defined and $X^{(n)}$ is adapted to $\mathscr{F}^W$ and a.s. continuous for every n. Moreover, an inductive argument[5] in n shows that $X_t^{(n)} = X_t^{(n)}(x, \omega) \in m(\mathscr{B}_N \otimes \mathscr{F}_t^W)$ for every $n \geq 0$ and $t \in [t_0, T]$.

We prove by induction the estimate

$$E\left[\sup_{t_0 \leq t \leq t_1} |X_t^{(n)} - X_t^{(n-1)}|^2\right] \leq \frac{c^n (t_1 - t_0)^n}{n!}, \qquad t_1 \in]t_0, T[, \ n \in \mathbb{N},$$
(17.2.5)

with $c = c(T, d, N, x, c_1, c_2) > 0$ where c_1, c_2 are the constants of the standard assumptions on coefficients. Let $n = 1$: by (14.4.4) we have

$$E\left[\sup_{t_0 \leq t \leq t_1} |X_t^{(1)} - X_t^{(0)}|^2\right] = E\left[\sup_{t_0 \leq t \leq t_1} \left|\int_{t_0}^t b(s, x) ds + \int_{t_0}^t \sigma(s, x) dW_s\right|^2\right]$$

$$\leq \bar{c}_1 (1 + |x|^2)(t_1 - t_0).$$

Supposing (17.2.5) true for n, let us prove it for $n + 1$: we have

$$E\left[\sup_{t_0 \leq t \leq t_1} |X_t^{(n+1)} - X_t^{(n)}|^2\right] = E\left[\sup_{t_0 \leq t \leq t_1} \left|\int_{t_0}^t \left(b(s, X_s^{(n)}) - b(s, X_s^{(n-1)})\right) ds \right.\right.$$

$$\left.\left. + \int_{t_0}^t \left(\sigma(s, X_s^{(n)}) - \sigma(s, X_s^{(n-1)})\right) dW_s\right|^2\right] \leq$$

(by (14.4.5))

$$\leq \bar{c}_2 \int_{t_0}^{t_1} E\left[\sup_{t_0 \leq r \leq s} |X_r^{(n)} - X_r^{(n-1)}|^2\right] ds \leq$$

---

[5] Measurability in (x, ω) is obvious for $n = 0$. Assuming the thesis true for $n - 1$, it is sufficient to approximate the integrand in (17.2.4) with simple processes and use Corollary 10.2.27, remembering that convergence in probability maintains the property of measurability.

(by inductive hypothesis, with $c = \bar{c}_2 \vee \bar{c}_1(1+|x|^2)$)

$$\leq c^{n+1} \int_{t_0}^{t_1} \frac{(s-t_0)^n}{n!} ds$$

and this proves (17.2.5).

Combining Markov's inequality with (17.2.5) we obtain

$$P\left(\sup_{t_0 \leq t \leq T} |X_t^{(n)} - X_t^{(n-1)}| \geq \frac{1}{2^n}\right) \leq 2^{2n} E\left[\sup_{t_0 \leq t \leq T} |X_t^{(n)} - X_t^{(n-1)}|^2\right]$$

$$\leq \frac{(4cT)^n}{n!}, \quad n \in \mathbb{N}.$$

Then, by Borel-Cantelli's Lemma 1.3.28 in [113] we have

$$P\left(\sup_{t_0 \leq t \leq T} |X_t^{(n)} - X_t^{(n-1)}| \geq \frac{1}{2^n} \text{ i.o.}\right) = 0$$

that is, for almost every $\omega \in \Omega$ there exists $n_\omega \in \mathbb{N}$ such that

$$\sup_{t_0 \leq t \leq T} |X_t^{(n)}(\omega) - X_t^{(n-1)}(\omega)| \leq \frac{1}{2^n}, \quad n \geq n_\omega.$$

Being

$$X_t^{(n)} = x + \sum_{k=1}^{n} (X_t^{(k)} - X_t^{(k-1)})$$

it follows that, almost surely, $X_t^{(n)}$ converges uniformly in $t \in [t_0, T]$ as $n \to +\infty$ to a limit that we denote by X_t: to express this fact, in symbols we write $X_t^{(n)} \rightrightarrows X_t$ a.s. Note that $X = (X_t)_{t \in [t_0, T]}$ is a.s. continuous (thanks to the uniform convergence) and adapted to $\mathscr{F}^W$: moreover, $X_t = X_t(x, \omega) \in m(\mathscr{B}_N \otimes \mathscr{F}_t^W)$ for each $t \in [t_0, T]$ because this measurability property holds for $X_t^{(n)}$ for each $n \in \mathbb{N}$.

By (14.4.1) and being X a.s. continuous, it is clear that condition (14.1.3) is satisfied. To verify that, almost surely, we have

$$X_t = x + \int_{t_0}^{t} b(s, X_s) ds + \int_{t_0}^{t} \sigma(s, X_s) dW_s, \quad t \in [t_0, T],$$

17.2 Existence

it is sufficient to observe that:

- by the Lipschitz property of b and σ uniform in t, it follows that $b(t, X_t^{(n)}) \rightrightarrows b(t, X_t)$ and $\sigma(t, X_t^{(n)}) \rightrightarrows \sigma(t, X_t)$ a.s., and therefore

$$\lim_{n \to +\infty} \int_{t_0}^t b(s, X_s^{(n)}) ds = \int_{t_0}^t b(s, X_s) ds \qquad \text{a.s.}$$

$$\lim_{n \to +\infty} \int_{t_0}^t \left| \sigma(s, X_s^{(n)}) - \sigma(s, X_s) \right|^2 ds = 0 \qquad \text{a.s.} \qquad (17.2.6)$$

- by Proposition 10.2.26, (17.2.6) implies that

$$\lim_{n \to +\infty} \int_{t_0}^t \sigma(s, X_s^{(n)}) dW_s = \int_{t_0}^t \sigma(s, X_s) dW_s \qquad \text{a.s.}$$

This concludes the proof of existence in the case of deterministic initial datum.
(2) Now consider the case of a random initial datum $Z \in m\mathscr{F}_{t_0}$. Let $f = f(x, \omega)$ be the function on $\mathbb{R}^N \times \Omega$ defined by

$$f(x, \cdot) := \sup_{t_0 \leq t \leq T} \left| X_t^{t_0, x} - x - \int_{t_0}^t b(s, X_s^{t_0, x}) ds - \int_{t_0}^t \sigma(s, X_s^{t_0, x}) dW_s \right|.$$

Note that $f \in m(\mathscr{B}_N \otimes \mathscr{F}_T^W)$ since $X_t^{t_0, \cdot} \in m(\mathscr{B}_N \otimes \mathscr{F}_t^W)$ for each $t \in [t_0, T]$. Moreover, for each $x \in \mathbb{R}^N$ we have $f(x, \cdot) = 0$ a.s. and therefore also $F(x) := E[f(x, \cdot)] = 0$. Then we have

$$0 = F(Z) = E[f(x, \cdot)]|_{x=Z} =$$

(by the freezing lemma in Theorem 4.2.10 in [113], since $Z \in m\mathscr{F}_{t_0}$, then $f \in m(\mathscr{B}_N \otimes \mathscr{F}_T^W)$ with $\mathscr{F}_{t_0}$ and $\mathscr{F}_t^W$ independent σ-algebras by Remark 14.1.4 and $f \geq 0$)

$$= E\left[f(Z, \cdot) \mid \mathscr{F}_{t_0} \right].$$

Applying the expected value we also have

$$E[f(Z, \cdot)] = 0$$

and therefore $X^{t_0, Z}$ in (17.2.2) is a solution of the SDE (17.0.1); actually, $X^{t_0, Z}$ is a strong solution because it is clearly adapted to $\mathscr{F}^{Z,W}$.

(3) For $t_0 \leq t \leq s \leq T$, with equalities holding almost surely, we have

$$X_s^{t_0,Z} = Z + \int_{t_0}^{s} b(r, X_r^{t_0,Z})dr + \int_{t_0}^{s} \sigma(r, X_r^{t_0,Z})dW_r$$

$$= Z + \int_{t_0}^{t} b(r, X_r^{t_0,Z})dr + \int_{t_0}^{t} \sigma(r, X_r^{t_0,Z})dW_r$$

$$+ \int_{t}^{s} b(r, X_r^{t_0,Z})dr + \int_{t}^{s} \sigma(r, X_r^{t_0,Z})dW_r$$

$$= X_t^{t_0,Z} + \int_{t}^{s} b(r, X_r^{t_0,Z})dr + \int_{t}^{s} \sigma(r, X_r^{t_0,Z})dW_r,$$

that is, $X^{t_0,Z}$ is a solution on $[t, T]$ of the SDE (17.0.1) with initial datum $X_t^{t_0,Z}$. On the other hand, as proven in point (2), also $X^{t,X_t^{t_0,Z}}$ is a solution of the same SDE. By uniqueness, the processes $X^{t_0,Z}$ and $X^{t,X_t^{t_0,Z}}$ are indistinguishable on $[t, T]$. This proves (17.2.3) and concludes the proof of the theorem.

□

17.3 Markov Property

In this section we show that, under suitable assumptions, the solution of an SDE is a continuous Markov process (i.e., a *diffusion*). Hereafter, we will refer systematically to the results of Sect. 2.5 concerning the characteristic operator of a Markov process.

Theorem 17.3.1 (Markov Property [!]) *Assume that the coefficients b, σ satisfy conditions (14.4.1) and (17.1.1) of linear growth and local Lipschitz continuity. If $X \in SDE(b, \sigma, W, \mathscr{F}_t)$ then X is a Markov process with transition law p where, for every $t_0 \leq t \leq s \leq T$ and $x \in \mathbb{R}^N$, $p = p(t, x; s, \cdot)$ is the law of the random variable $X_s^{t,x}$ that is, of the solution of the SDE with initial condition x at time t, evaluated at time s. Moreover, the characteristic operator of X is*

$$\mathscr{A}_t = \frac{1}{2}\sum_{i,j=1}^{N} c_{ij}(t, x)\partial_{x_i x_j} + \sum_{j=1}^{N} b_j(t, x)\partial_{x_i}, \qquad c_{ij} := (\sigma\sigma^*)_{ij}. \qquad (17.3.1)$$

Proof We observe that p is a transition law according to Definition 2.1.1. Indeed, we have:

(i) for every $x \in \mathbb{R}^N$, by definition, $p(t, x; s, \cdot)$ is a distribution such that $p(t, x; t, \cdot) = \delta_x$;

17.3 Markov Property

(ii) for every $H \in \mathscr{B}_N$

$$x \mapsto p(t,x;s,H) = E\left[\mathbb{1}_H\left(X_s^{t,x}\right)\right] \in m\mathscr{B}_N$$

thanks to the measurability property (17.2.1) and Fubini's theorem.

We prove that p is a transition law for X: according to Definition 2.1.1, we have to verify that

$$p(t, X_t; s, H) = P(X_s \in H \mid X_t), \qquad t_0 \le t \le s \le T, \ H \in \mathscr{B}_N.$$

Since, by uniqueness, X is indistinguishable from the solution $X^{t_0, X_{t_0}} \in$ SDE$(b, \sigma, W, \mathscr{F}_t^{X_{t_0}, W})$ constructed in Theorem 17.2.1, from the flow property (17.2.3) we have almost surely

$$X_s = X_s^{t, X_t} \quad \text{for every } s \in [t, T].$$

Therefore, we have

$$P(X_s \in H \mid X_t) \equiv E[\mathbb{1}_H(X_s) \mid X_t]$$

$$E\left[\mathbb{1}_H\left(X_s^{t, X_t}\right) \mid X_t\right] =$$

(by (4.2.7) in [113] of the freezing lemma, being $X_t \in m\mathscr{F}_t$ and therefore, by Remark 14.1.4, independent of $\mathscr{F}_s^W$ and $(x, \omega) \mapsto \mathbb{1}_H(X_s^{t,x}(\omega)) \in m(\mathscr{B}_N \otimes \mathscr{F}_s^W)$ thanks to (17.2.1))

$$= E\left[\mathbb{1}_H(X_s^{t,x})\right]|_{x=X_t} = p(t, X_t; s, H).$$

On the other hand, it is enough to repeat the previous steps, conditioning on $\mathscr{F}_t$ instead of X_t, to prove the Markov property

$$p(t, X_t; s, H) = P(X_s \in H \mid \mathscr{F}_t), \qquad 0 \le t_0 \le t \le s \le T, \ H \in \mathscr{B}_N.$$

Finally, the fact that $\mathscr{A}_t$ is the characteristic operator of X has been proved in Sect. 15.1 (in particular, compare (15.1.1) with definition (2.5.5)). □

Remark 17.3.2 Under the assumptions of Theorem 17.3.1, by the Markov property we have

$$E[\varphi(X_T) \mid \mathscr{F}_t] = u(t, X_t), \qquad \varphi \in b\mathscr{B},$$

where

$$u(t, x) := \int_{\mathbb{R}} p(t, x; T, dy)\varphi(y).$$

We recall that, by the results of Sects. 2.5.3 and 2.5.2, the transition law p is a solution of the Kolmogorov backward and forward equations, given respectively by

$$(\partial_t + \mathscr{A}_t)p(t,x;s,dy) = 0, \qquad (\partial_s - \mathscr{A}_s^*)p(t,x;s,dy) = 0, \qquad t_0 \leq t < s \leq T,$$

where $\mathscr{A}_s^*$ indicates the adjoint operator of $\mathscr{A}_t$ in (17.3.1), acting in the forward variable y.

17.3.1 Forward Kolmogorov Equation

The forward Kolmogorov equation of a diffusion X can be derived by a direct application of the Itô's formula. Under the assumptions of Theorem 17.3.1, we denote by $X^{t,x}$ the solution of the SDE (17.0.1) with initial condition $X_t^{t,x} = x$. Given a test function $\varphi \in C_0^\infty(\mathbb{R} \times \mathbb{R}^N)$, with compact support contained in $]t, T[\times \mathbb{R}^N$, by Itô's formula we have

$$0 = \varphi(T, X_T^{t,x}) - \varphi(t,x) = \int_t^T (\partial_s + \mathscr{A}_s)\varphi(s, X_s^{t,x})ds$$
$$+ \int_t^T \nabla \varphi(s, X_s^{t,x})\sigma(s, X_s^{t,x})dW_s$$

where $\mathscr{A}_t$ is the characteristic operator in (17.3.1). Applying the expected value and Fubini's theorem, we obtain

$$0 = E\left[\int_t^T (\partial_s + \mathscr{A}_s)\varphi(s, X_s^{t,x})ds\right]$$
$$= \int_t^T E\left[(\partial_s + \mathscr{A}_s)\varphi(s, X_s^{t,x})\right]ds = \int_t^T \int_{\mathbb{R}^N} (\partial_s + \mathscr{A}_s)\varphi(s,y)p(t,x;s,dy)ds \tag{17.3.2}$$

where $p(t,x;s,dy)$ denotes the law of the random variable $X_s^{t,x}$ which, by Theorem 17.3.1, is the transition law of the Markov process X.

By (17.3.2), for every $t \geq 0$ we have

$$\iint_{\mathbb{R}^{N+1}} (\partial_s + \mathscr{A}_s)\varphi(s,y)p(t,x;s,dy)ds = 0, \qquad \varphi \in C_0^\infty(]t, +\infty[\times\mathbb{R}^N),$$

and thus we recover the result of Sect. 2.5.3 according to which p is a distributional solution of the forward Kolmogorov equation

$$\left(\partial_s - \mathscr{A}_s^*\right) p(t,x;s,\cdot) = 0, \qquad s > t. \tag{17.3.3}$$

17.4 Continuous Dependence on Parameters

In particular, if p is absolutely continuous with density Γ, that is

$$p(t, x; t, H) = \int_H \Gamma(t, x; t, x)dx, \qquad H \in \mathscr{B}_N,$$

then $\Gamma(t, x; t, x)$ is a distributional solution of (17.3.3), that is

$$\iint_{\mathbb{R}^{N+1}} \Gamma(t, x; s, y)(\partial_s + \mathscr{A}_s)\varphi(t, x)dyds = 0, \qquad \varphi \in C_0^\infty(]t, +\infty[\times\mathbb{R}^N),$$

and we say that $(s, y) \mapsto \Gamma(t, x; s, y)$ is *fundamental solution of the forward operator* $\partial_s - \mathscr{A}_s^*$ *with pole in* (t, x).

17.4 Continuous Dependence on Parameters

Theorem 17.4.1 (Continuous Dependence Estimates on Parameters) *Under the standard assumptions* (14.4.1) *and* (14.4.2), *let* X^{t_0, Z_0} *and* X^{t_1, Z_1} *be solutions of the SDE* (17.0.1) *with initial data* (t_0, Z_0) *and* (t_1, Z_1), *respectively, with* $0 \leq t_0 \leq t_1 \leq t_2 \leq T$. *For every* $p \geq 2$ *there exists a positive constant* $c = c(T, d, N, p, c_1, c_2)$ *such that*

$$E\left[\sup_{t_2 \leq t, s \leq T} \left|X_t^{t_0, Z_0} - X_s^{t_1, Z_1}\right|^p\right] \leq cE\left[|Z_0 - Z_1|^p\right]$$
$$+ c\left(1 + E\left[|Z_1|^p\right]\right)\left(|t_1 - t_0|^{\frac{p}{2}} + |T - t_2|^{\frac{p}{2}}\right). \tag{17.4.1}$$

Proof By the elementary inequality (14.4.6) we have

$$E\left[\sup_{t_2 \leq t, s \leq T}\left|X_t^{t_0, Z_0} - X_s^{t_1, Z_1}\right|^p\right] \leq 3^{p-1}E\left[\sup_{t_2 \leq t \leq T}\left|X_t^{t_0, Z_0} - X_t^{t_0, Z_1}\right|^p\right]$$
$$+ 3^{p-1}E\left[\sup_{t_2 \leq t \leq T}\left|X_t^{t_0, Z_1} - X_t^{t_1, Z_1}\right|^p\right]$$
$$+ 3^{p-1}E\left[\sup_{t_2 \leq t, s \leq T}\left|X_t^{t_1, Z_1} - X_s^{t_1, Z_1}\right|^p\right]. \tag{17.4.2}$$

Again by (14.4.6) and (14.4.5) we have

$$v(t) := E\left[\sup_{t_0 \le s \le t} \left|X_s^{t_0,Z_0} - X_s^{t_0,Z_1}\right|^p\right]$$

$$\le 2^{p-1} E\left[|Z_0 - Z_1|^p\right] + 2^{p-1} \bar{c}_2 T^{\frac{p-2}{2}} \int_{t_0}^{t} v(s)ds,$$

and, by Grönwall's lemma,

$$E\left[\sup_{t_2 \le t \le T} \left|X_t^{t_0,Z_0} - X_t^{t_0,Z_1}\right|^p\right] \le v(T) \le cE\left[|Z_0 - Z_1|^p\right] \quad (17.4.3)$$

with c depending only on p, T and c_2.

On the other hand, by the flow property we have

$$E\left[\sup_{t_2 \le t \le T} \left|X_t^{t_0,Z_1} - X_t^{t_1,Z_1}\right|^p\right] = E\left[\sup_{t_2 \le t \le T} \left|X_t^{t_1, X_{t_1}^{t_0,Z_1}} - X_t^{t_1,Z_1}\right|^p\right] \le$$

(by (17.4.3))

$$\le cE\left[\left|X_{t_1}^{t_0,Z_1} - Z_1\right|^p\right] \le$$

(by (14.4.4))

$$\le c\bar{c}_1 |t_1 - t_0|^{\frac{p-2}{2}} \int_{t_0}^{t_1} \left(1 + E\left[\sup_{t_0 \le r \le s} |X_r^{t_0,Z_1}|^p\right]\right) ds \le$$

(by the L^p estimate (14.5.1), for a new constant $c = C(T, d, N, p, c_1, c_2)$)

$$\le c(1 + E\left[|Z_1|^p\right])|t_1 - t_0|^{\frac{p}{2}}.$$

We estimate the last term of (17.4.2) using a completely analogous approach, which concludes the proof. □

Corollary 17.4.2 (Feller and Strong Markov Properties) *Under the standard assumptions (14.4.1)–(14.4.2) and the usual conditions on the filtration, every $X \in$ SDE$(b, \sigma, W, \mathscr{F}_t)$ is a Feller process and satisfies the strong Markov property.*

Proof By Theorem 17.3.1, X is a Markov process with transition law $p = p(t, x; T, \cdot)$ where, for every $t, T \ge 0$ with $t \le T$ and $x \in \mathbb{R}^N$, $p(t, x; T, \cdot)$ is the law of the r.v. $X_T^{t,x}$. By (17.4.1) and Kolmogorov's continuity theorem (in the multidimensional version of Theorem 3.3.4), the process $(t, x, T) \mapsto X_T^{t,x}$ admits a modification $\widetilde{X}_T^{t,x}$ with locally α-Hölder continuous trajectories for every $\alpha \in [0, 1[$

17.4 Continuous Dependence on Parameters

with respect to the so-called "parabolic" distance: precisely, for every $\alpha \in [0, 1[$, $n \in \mathbb{N}$ and $\omega \in \Omega$ there exists $c_{\alpha,n,\omega} > 0$ such that

$$\left|\widetilde{X}_r^{t,x}(\omega) - \widetilde{X}_u^{s,y}(\omega)\right| \leq c_{\alpha,n,\omega} \left(|x-y| + |t-s|^{\frac{1}{2}} + |r-u|^{\frac{1}{2}}\right)^{\alpha},$$

for every $t, s, r, u \in [0, T]$ such that $t \leq r$, $s \leq u$, and for every $x, y \in \mathbb{R}^N$ such that $|x|, |y| \leq n$. Consequently, for every $\varphi \in bC(\mathbb{R}^N)$ and $h > 0$, the function

$$(t, x) \longmapsto \int_{\mathbb{R}^N} p(t, x; t+h, dy)\varphi(y) = E\left[\varphi(\widetilde{X}_{t+h}^{t,x})\right]$$

is continuous thanks to the dominated convergence theorem and this proves the Feller property. The strong Markov property follows from Theorem 7.1.2. □

Chapter 18
Weak Solutions

> *If someone were to ask me, as a philosopher, what should be learned in high school, I would answer: "first of all, only 'useless' things, ancient Greek, Latin, pure mathematics, and philosophy. Everything that is useless in life". The beauty is that by doing so, at the age of 18, you have a wealth of useless knowledge with which you can do everything. While with useful knowledge you can only do small things.*
>
> Agnes Heller

In this chapter, we present weak existence and uniqueness results for SDEs with coefficients

$$b = b(t,x) :]0, T[\times \mathbb{R}^N \longrightarrow \mathbb{R}^N, \qquad \sigma = \sigma(t,x) :]0, T[\times \mathbb{R}^N \longrightarrow \mathbb{R}^{N \times d}, \tag{18.0.1}$$

where $N, d \in \mathbb{N}$ and $T > 0$ are fixed. To this end, we describe what is known as the "martingale problem" due to Stroock and Varadhan [136]: this problem pertains to the construction of a distribution with respect to which the canonical process $\mathbf{X}$ is a semimartingale with drift $b(t, \mathbf{X}_t)$ and covariance matrix $(\sigma \sigma^*)(t, \mathbf{X}_t)$. The solution to the martingale problem, if it exists, is the law of the solution of the corresponding SDE: in fact, the martingale problem turns out to be equivalent to the weak solvability problem.

The analytical results on the fundamental solution of parabolic PDEs (cf. Chap. 20) provide a solution to the martingale problem under Hölder regularity and uniform ellipticity assumptions on the coefficients. Under these assumptions, we prove existence and uniqueness in the weak sense for SDEs, along with strong Markov, Feller, and other regularity properties of the trajectories of the solution. We also showcase broader findings from prominent mathematicians, including Skorokhod, Stroock, Varadhan, Krylov, Veretennikov and Zvonkin. In the last section, we prove a "regularization by noise" result that guarantees *strong* uniqueness for SDEs with bounded Hölder drift.

18.1 The Stroock-Varadhan Martingale Problem

Assume that the SDE with coefficients b, σ admits a weak solution (X, W) and denote as usual by $\mu_{X,W}$ its law. By Lemma 14.3.5, the canonical process $(\mathbf{X}, \mathbf{W})$ is also a solution of the SDE with coefficients b, σ on the space $(\mathbf{\Omega}_{N+d}, \mathscr{G}_T^{N+d}, \mu_{X,W})$ and consequently,[1] for each $i, j = 1, \ldots, N$, the processes

$$\mathbf{M}_t^i := \mathbf{X}_t^i - \int_0^t b_i(s, \mathbf{X}_s) ds, \tag{18.1.1}$$

$$\mathbf{M}_t^{ij} := \mathbf{M}_t^i \mathbf{M}_t^j - \int_0^t c_{ij}(s, \mathbf{X}_s) ds, \qquad (c_{ij}) := \sigma \sigma^*, \tag{18.1.2}$$

are local martingales with respect to the filtration $(\mathscr{G}_t^{N+d})_{t \in [0,T]}$ generated by $(\mathbf{X}, \mathbf{W})$.

Note that the Brownian motion $\mathbf{W}$ does not appear in the definitions (18.1.1) and (18.1.2) and, still denoting by $\mathbf{X}$ the identity process on $\mathbf{\Omega}_N$, one can verify that the processes formally defined as in (18.1.1) and (18.1.2) are local martingales on the space $(\mathbf{\Omega}_N, \mathscr{G}_T^N, \mu_X)$. This motivates the following

Definition 18.1.1 (Martingale Problem) A solution to the *martingale problem for b, σ* is a probability measure on the canonical space $(\mathbf{\Omega}_N, \mathscr{G}_T^N)$ such that the processes $\mathbf{M}^i, \mathbf{M}^{ij}$ in (18.1.1) and (18.1.2) are local martingales with respect to the filtration $\mathscr{G}_t^N$ generated by the identity process $\mathbf{X}$.

Remark 18.1.2 ([!!]) It is worth emphasizing that the martingale condition on the processes in (18.1.1) and (18.1.2) basically means that $\mathbf{X}$ *is a semimartingale with drift $b(t, \mathbf{X}_t)$ and covariation matrix* $\mathscr{C}_t := \big(c_{ij}(t, \mathbf{X}_t)\big)$.

If (X, W) is a solution of the SDE with coefficients b, σ then μ_X is a solution of the martingale problem for b, σ. We now show a result in the opposite direction

---

[1] Formula (18.1.1) follows from the fact that

$$\mathbf{M}_t = \mathbf{X}_0 + \int_0^t \sigma(s, \mathbf{X}_s) d\mathbf{W}_s;$$

then

$$\langle \mathbf{M}^i, \mathbf{M}^j \rangle_t = \int_0^t c_{ij}(s, \mathbf{X}_s) ds$$

is the covariation process of $\mathbf{M}$, leading to formula (18.1.2).

18.1 The Stroock-Varadhan Martingale Problem

that allows us to conclude that *the martingale problem and the weak solvability of an SDE are equivalent.*

Theorem 18.1.3 (Stroock and Varadhan) *If μ is a solution to the martingale problem for b, σ, then there exists a weak solution to the SDE with coefficients b, σ and initial law μ_0 defined by*

$$\mu_0(H) := \mu(\mathbf{X}_0 \in H), \qquad H \in \mathscr{B}_N.$$

Proof We provide the proof only in the scalar case $N = d = 1$ and refer, for example, to Section 5.4.B in [67] for the general case. The fact that μ is a solution to the martingale problem for b, σ, means that the process defined on $(\mathbf{\Omega}_N, \mathscr{G}_T^N, \mu)$ as in (18.1.1), that is

$$\mathbf{M}_t = \mathbf{X}_t - \int_0^t b(s, \mathbf{X}_s)ds, \tag{18.1.3}$$

is a local martingale with quadratic variation process $d\langle \mathbf{M} \rangle_t = \sigma^2(t, \mathbf{X}_t)dt$.

If $\sigma(t, x) \neq 0$ for every (t, x), the proof is very simple: in fact, the process

$$\mathbf{B}_t := \int_0^t \frac{1}{\sigma(s, \mathbf{X}_s)} d\mathbf{M}_s \tag{18.1.4}$$

is a local martingale with quadratic variation

$$\langle \mathbf{B} \rangle_t = \int_0^t \frac{1}{\sigma^2(s, \mathbf{X}_s)} d\langle \mathbf{M} \rangle_s = t.$$

Then, by Lévy's characterization Theorem 12.4.1, $\mathbf{B}$ is a Brownian motion and being $d\mathbf{B}_t = \sigma^{-1}(t, \mathbf{X}_t)d\mathbf{M}_t = \sigma^{-1}(t, \mathbf{X}_t)(d\mathbf{X}_t - b(t, \mathbf{X}_t)dt)$, we have

$$\int_0^t \sigma(s, \mathbf{X}_s)d\mathbf{B}_s = \mathbf{X}_t - \mathbf{X}_0 - \int_0^t b(s, \mathbf{X}_s)ds,$$

that is, $(\mathbf{X}, \mathbf{B})$ is a solution to the SDE with coefficients b, σ. Note that the solution $(\mathbf{X}, \mathbf{B})$ is defined on the space $(\mathbf{\Omega}_N, \mathscr{G}_T^N, \mu)$.

In the general case where σ can be zero, consider the space $(\mathbf{\Omega}_{N+d}, \mathscr{G}_T^{N+d}, \mu \otimes \mu_W)$ where μ_W is the Wiener measure and the canonical process $(\mathbf{X}, \mathbf{W})$ is such that $\mathbf{W}$ is a real Brownian motion (we recall that we are dealing only with the case $N = d = 1$). Let $J_t = \mathbb{1}_{(\sigma(t, \mathbf{X}_t) \neq 0)}$ and

$$\mathbf{B}_t = \int_0^t \frac{J_s}{\sigma(s, \mathbf{X}_s)} d\mathbf{M}_s + \int_0^t (1 - J_s)d\mathbf{W}_s.$$

Again, **B** is a real Brownian motion since it is a local martingale with quadratic variation

$$d\langle \mathbf{B}\rangle_t = \frac{J_t}{\sigma^2(t,\mathbf{X}_t)}d\langle \mathbf{M}\rangle_t + (1-J_t)d\langle \mathbf{W}\rangle_t + 2\frac{J_t(1-J_t)}{\sigma(t,\mathbf{X}_t)}d\langle \mathbf{M},\mathbf{W}\rangle_t = dt.$$

Furthermore, since $(1-J_t)\sigma(t,\mathbf{X}_t) = 0$, we have

$$\int_0^t \sigma(s,\mathbf{X}_s)d\mathbf{B}_s = \int_0^t J_s d\mathbf{M}_s = \mathbf{M}_t - \mathbf{M}_0 + \int_0^t (J_s - 1)d\mathbf{M}_s$$

$$= \mathbf{X}_t - \mathbf{X}_0 - \int_0^t b(s,\mathbf{X}_s)ds$$

where in the last step we used the fact that, by the Itô isometry,

$$E\left[\left(\int_0^t (J_s - 1)d\mathbf{M}_s\right)^2\right] = E\left[\int_0^t (J_s - 1)\sigma^2(s,\mathbf{X}_s)ds\right] = 0.$$

□

Remark 18.1.4 It is interesting to note in the previous proof that if $\sigma \neq 0$, i.e., in the non-degenerate case, the Brownian motion **B** is constructed as a functional of **X** and therefore the space Ω_N is sufficient to "support" the solution $(\mathbf{X},\mathbf{B})$ of the SDE. On the contrary, in the degenerate case where σ can be zero, the Brownian motion **W** comes into play to "guarantee sufficient randomness" to the system and it is therefore necessary to define the solution on the enlarged space Ω_{N+d}. This further explicates the difference between weak and strong solutions illustrated earlier in Remarks 14.1.7 and 14.3.7.

Remark 18.1.5 Stroock and Varadhan (cf. Theorem 6.2.3 in [136]) prove that, for the martingale problem, the equality of marginal distributions implies the equality of finite-dimensional distributions and therefore the uniqueness in law. Precisely, suppose that b, σ are measurable and bounded functions: if for every $t \in [0,T]$, $x \in \mathbb{R}^n$ and $\varphi \in bC(\mathbb{R}^n)$ we have

$$E^{\mu_1}[\varphi(\mathbf{X}_t)] = E^{\mu_2}[\varphi(\mathbf{X}_t)]$$

where μ_1, μ_2 are solutions of the martingale problem for b, σ with initial law δ_x, then there exists at most one solution of the martingale problem for b, σ with initial law δ_x. Hereafter, we will not use this result but will adopt a more analytical approach to prove weak uniqueness using existence theorems for the Kolmogorov equation associated with the SDE.

18.2 Equations with Hölder Coefficients

We consider an SDE with coefficients b, σ as in (18.0.1) and define the *diffusion matrix*

$$\mathscr{C} = (c_{ij}) := \sigma\sigma^*.$$

To specify the regularity conditions on the coefficients, we introduce the following

Notation 18.2.1 bC_T^α denotes the space of bounded, continuous functions on $]0, T[\times\mathbb{R}^n$, that are uniformly Hölder continuous in x with exponent $\alpha \in]0, 1]$. On bC_T^α, we consider the norm

$$[g]_\alpha := \sup_{]0,T[\times\mathbb{R}^n} |g| + \sup_{\substack{0<t<T \\ x\neq y}} \frac{|g(t,x) - g(t,y)|}{|x-y|^\alpha}. \qquad (18.2.1)$$

The elements of bC_T^α are continuous functions in (t, x), Hölder continuous in the spatial variable x, uniformly with respect to the time variable t. In fact, the continuity condition in t can be omitted and will only be assumed for the sake of simplifying the presentation.

In this section, we prove a weak existence and uniqueness result for SDE under the following

Assumption 18.2.2

(i) $c_{ij}, b_i \in bC_T^\alpha$ for some $\alpha \in]0, 1]$ and for each $i, j = 1, \ldots, N$;
(ii) the diffusion matrix $\mathscr{C}$ is uniformly positive definite: there exists a positive constant λ_0 such that

$$\frac{1}{\lambda_0}|\eta|^2 \leq \langle \mathscr{C}(t,x)\eta, \eta\rangle \leq \lambda_0|\eta|^2, \qquad (t,x) \in]0, T[\times\mathbb{R}^N, \, \eta \in \mathbb{R}^N. \qquad (18.2.2)$$

Theorem 18.2.3 ([!!]) *Under Assumption 18.2.2, for every distribution μ_0 on $\mathbb{R}^N$ there exists and is unique in law the weak solution (X, W) of the SDE*

$$dX_t = b(t, X_t)dt + \sigma(t, X_t)dW_t \qquad (18.2.3)$$

with initial law μ_0. Moreover:

(i) *X is a Feller and strong Markov process with characteristic operator*

$$\mathscr{A}_t := \frac{1}{2}\sum_{i,j=1}^N c_{ij}(t,x)\partial_{x_i x_j} + \sum_{i=1}^N b_i(t,x)\partial_{x_i}, \qquad (t,x) \in]0, T[\times\mathbb{R}^N.$$

(ii) X has a transition density $\Gamma(t,x;s,y)$ which is the fundamental solution[2] of $\partial_t + \mathscr{A}_t$;
(iii) X admits a modification with β-Hölder continuous trajectories for every $\beta < \frac{1}{2}$.

The proof of Theorem 18.2.3 is based on the existence results of the fundamental solution for parabolic PDEs of Theorem 18.2.6 below.

Notation 18.2.4 We denote by $C^{1,2}(]0,T[\times\mathbb{R}^N)$ the space of functions defined on $]0,T[\times\mathbb{R}^N$ that are continuously differentiable with respect to t and twice continuously differentiable with respect to x.

Definition 18.2.5 (Backward Cauchy Problem) A classical solution of the backward Cauchy problem for the operator $\partial_t + \mathscr{A}_t$ on $]0,s[\times\mathbb{R}^N$, is a function $u \in C^{1,2}(]0,s[\times\mathbb{R}^N) \cap C(]0,s] \times \mathbb{R}^N)$ such that

$$\begin{cases} \partial_t u(t,x) + \mathscr{A}_t u(t,x) = 0, & (t,x) \in]0,s[\times\mathbb{R}^N, \\ u(s,x) = \varphi(x), & x \in \mathbb{R}^N, \end{cases} \qquad (18.2.4)$$

where $\varphi \in C(\mathbb{R}^N)$ is the assigned *final datum*.

Section 20.3 is dedicated to the rather long and involved proof of the following result.[3]

Theorem 18.2.6 (Levi [89], Friedman [49]) *Under Assumption 18.2.2, there exists a continuous function $\Gamma = \Gamma(t,x;s,y)$, defined for $0 < t < s \leq T$ and $x,y \in \mathbb{R}^N$, such that:*

(i) *for every $s \in]0,T]$ and for every $\varphi \in bC(\mathbb{R}^N)$ the function defined by*

$$u(t,x) = \int_{\mathbb{R}^N} \Gamma(t,x;s,y)\varphi(y)dy, \qquad (t,x) \in]0,s[\times\mathbb{R}^N, \qquad (18.2.5)$$

and by $u(s,\cdot) = \varphi$, is a classical solution of the backward Cauchy problem on $]0,s[\times\mathbb{R}^N$ with final datum φ. We say that Γ is the fundamental solution of the operator $\partial_t + \mathscr{A}_t$ on $]0,T[\times\mathbb{R}^N$;

(ii) *the function*

$$p(t,x;s,H) := \int_H \Gamma(t,x;s,y)dy, \qquad 0 < t < s \leq T, \ x \in \mathbb{R}^N, \ H \in \mathscr{B}_N,$$

---

[2] See Theorem 18.2.6 for the definition of fundamental solution.
[3] In Sect. 20.3 we will prove an equivalent result, Theorem 20.2.5, which is the forward version of Theorem 18.2.6.

18.2 Equations with Hölder Coefficients

is a transition law,[4] enjoys the Feller property (cf. Definitions 2.1.1 and 2.1.10) and satisfies the Chapman-Kolmogorov equation (2.4.4);

(iii) for every $(s, y) \in \,]0, T] \times \mathbb{R}^N$, we have $\Gamma(\cdot, \cdot; s, y) \in C^{1,2}(]0, s[\times \mathbb{R}^N)$ and the following Gaussian estimates hold: there exist two positive constants λ, c that depend only on $T, N, \alpha, \lambda_0, [c_{ij}]_\alpha$ and $[b_i]_\alpha$, for which we have

$$\frac{1}{c} \mathbf{G}\left(\lambda^{-1}(s-t), x-y\right) \leq \Gamma(t, x; s, y) \leq c \, \mathbf{G}\left(\lambda(s-t), x-y\right),$$
(18.2.6)

$$\left|\partial_{x_i} \Gamma(t, x; s, y)\right| \leq \frac{c}{\sqrt{s-t}} \mathbf{G}\left(\lambda(s-t), x-y\right),$$

$$\left|\partial_{x_i x_j} \Gamma(t, x; s, y)\right| + \left|\partial_t \Gamma(t, x; s, y)\right| \leq \frac{c}{s-t} \mathbf{G}\left(\lambda(s-t), x-y\right)$$

for every $(t, x) \in \,]0, s[\times \mathbb{R}^N$, where $\mathbf{G}$ denotes the standard N-dimensional Gaussian function

$$\mathbf{G}(t, x) = \frac{1}{(2\pi t)^{\frac{N}{2}}} e^{-\frac{|x|^2}{2t}}, \qquad t > 0, \, x \in \mathbb{R}^N.$$

Proof of Theorem 18.2.3 It is a matter of combining Theorem 18.2.6 with a series of results proven earlier. We examine separately the existence and uniqueness:

Weak Solvability Let Γ be the fundamental solution on $]0, T[\times \mathbb{R}^N$ of the operator $\partial_t + \mathscr{A}_t$ as in Theorem 18.2.6. Due to the properties of Γ, in particular in Theorem 18.2.6-(ii), and the multidimensional version of Theorem 2.4.4, there exists a Markov process $X = (X_t)_{t \in [0,T]}$ that has transition density Γ and is such that $X_0 \sim \mu_0$. By Proposition 2.2.6, the identity process $\mathbf{X}$ is a Markov process on the canonical space $(\mathbf{\Omega}_N, \mathscr{G}_T^N, \mu_X)$ equipped with the filtration $(\mathscr{G}_t^N)_{t \in [0,T]}$ generated by $\mathbf{X}$.

We show that the law μ_X of the process X solves the martingale problem for b, σ and therefore, by Theorem 18.1.3, the SDE is solvable in the weak sense. We consider the functions

$$\psi_i(x) = x_i, \qquad \psi_{ij}(x) = x_i x_j, \qquad x \in \mathbb{R}^N, \, i, j = 1, \ldots, N,$$

for which we have

$$\mathscr{A}_t \psi_i(x) = b_i(t, x), \qquad \mathscr{A}_t \psi_{ij}(x) = c_{ij}(t, x) + b_i(t, x) x_j + b_j(t, x) x_i.$$

---

[4] In particular, according to Definition 2.1.1 of transition law, there exists

$$p(s, x; s, \cdot) := \lim_{t \to s^-} p(t, x; s, \cdot) = \delta_x$$

with the limit understood in the sense of weak convergence.

We observe that the boundedness hypothesis of the coefficients and the Gaussian estimate from above (18.2.6) guarantee that $\mathscr{A}_t \psi_i(\mathbf{X}_t), \mathscr{A}_t \psi_{ij}(\mathbf{X}_t) \in L^1([0, T] \times \Omega_N)$: then from Theorem 2.5.13 it follows that the processes

$$\mathbf{M}_t^i := \mathbf{X}_t^i - \int_0^t b_i(s, \mathbf{X}_s) ds,$$

$$Z_t^{ij} := \mathbf{X}_t^i \mathbf{X}_t^j - \int_0^t \left(c_{ij}(s, \mathbf{X}_s) + b_i(s, \mathbf{X}_s) \mathbf{X}_s^j + b_j(s, \mathbf{X}_s) \mathbf{X}_s^i \right) ds$$

are continuous martingales. To conclude, we prove that $\mathbf{M}^{ij}$ in (18.1.2) is indistinguishable from Z^{ij} or equivalently the process

$$Y_t^{ij} := \mathbf{M}_t^{ij} - Z_t^{ij} = \int_0^t \left(b_i(s, \mathbf{X}_s)(\mathbf{X}_s^j - \mathbf{X}_t^j) + b_j(s, \mathbf{X}_s)(\mathbf{X}_s^i - \mathbf{X}_t^i) \right)$$

$$+ \int_0^t b_i(s, \mathbf{X}_s) ds \int_0^t b_j(s, \mathbf{X}_s) ds,$$

is null. First of all, we have the equality

$$Y_t^{ij} = \int_0^t b_i(s, \mathbf{X}_s)(\mathbf{M}_s^j - \mathbf{M}_t^j) ds + \int_0^t b_j(s, \mathbf{X}_s)(\mathbf{M}_s^i - \mathbf{M}_t^i) ds$$

by the fact that

$$\int_0^t b_i(s, \mathbf{X}_s)(\mathbf{M}_s^j - \mathbf{M}_t^j) ds = \int_0^t b_i(s, \mathbf{X}_s)(\mathbf{X}_s^j - \mathbf{X}_t^j) ds$$

$$- \int_0^t b_i(s, \mathbf{X}_s) \left(\int_0^s b_j(r, \mathbf{X}_r) dr \right) ds$$

$$+ \int_0^t b_i(s, \mathbf{X}_s) ds \int_0^t b_j(s, \mathbf{X}_s) ds$$

and, by integration by parts, we have

$$\int_0^t b_i(s, \mathbf{X}_s) \left(\int_0^s b_j(r, \mathbf{X}_r) dr \right) ds = \int_0^t b_i(s, \mathbf{X}_s) ds \int_0^t b_j(s, \mathbf{X}_s) ds$$

$$- \int_0^t b_j(s, \mathbf{X}_s) \left(\int_0^s b_i(r, \mathbf{X}_r) dr \right) ds.$$

Moreover, we observe that

$$\int_0^t b_i(s, \mathbf{X}_s)(\mathbf{M}_s^j - \mathbf{M}_t^j) ds = - \int_0^t \left(\int_0^s b_i(r, \mathbf{X}_r) dr \right) d\mathbf{M}_s^j \qquad (18.2.7)$$

18.2 Equations with Hölder Coefficients

which is equivalent to the expression obtained from Itô's formula

$$d\left(\mathbf{M}_t^j \int_0^t b_j(s, \mathbf{X}_s)ds\right) = \mathbf{M}_t^j b_j(t, \mathbf{X}_t)dt + \left(\int_0^t b_j(s, \mathbf{X}_s)ds\right) d\mathbf{M}_t^j.$$

Formula (18.2.7) is an equality between a BV process and a continuous local martingale: by Theorem 9.3.6, both processes are null, so that $Y^{ij} = 0$. Then, by Theorem 18.1.3, **X** is a solution[5] of the SDE with coefficients b, σ and initial law μ_0 with respect to a Brownian motion **W**.

Uniqueness in Law and Main Properties Let us prove that if (X, W) is a weak solution of the SDE (18.2.3) on $[0, T]$ then X is a Markov process. Fixed $\varphi \in bC(\mathbb{R}^N)$, we consider the[6] solution u in (18.2.5) of the backward Cauchy problem (18.2.4). Note that u is a bounded function since, by the Gaussian estimate (18.2.6), we have

$$|u(t, x)| \le c\|\varphi\|_{L^\infty(\mathbb{R}^N)} \int_{\mathbb{R}^N} \mathbf{G}(\lambda(s - t), x - y)\, dy$$

$$= c\|\varphi\|_{L^\infty(\mathbb{R}^N)}, \qquad (t, x) \in \,]0, s] \times \mathbb{R}^N. \tag{18.2.8}$$

By Itô's formula, $u(t, X_t)$ is a local martingale and a bounded process by (18.2.8): therefore, $u(t, X_t)$ is a true martingale (cf. Remark 8.4.6-(v)) and we have

$$\varphi(X_s) = u(t, X_t) + \int_t^s \nabla_x u(r, X_r)\sigma(r, X_r)dW_r. \tag{18.2.9}$$

Conditioning (18.2.9) to $\mathscr{F}_t$, we obtain

$$E[\varphi(X_s) \mid \mathscr{F}_t] = u(t, X_t) = \int_{\mathbb{R}^N} \Gamma(t, X_t; s, y)\varphi(y)dy.$$

Given the arbitrariness of φ, it follows that X is a Markov process with transition density Γ: by Theorem 18.2.6-(ii) X is a Feller process and therefore also enjoys the strong Markov property by Theorem 7.1.2.

By Kolmogorov's continuity Theorem 3.3.4, the process X admits a modification with β-Hölder continuous trajectories for every $\beta < \frac{1}{2}$: indeed, for every $0 \le t < s \le T$ and $p > 0$, the following integral estimate holds

$$E\left[|X_t - X_s|^p\right] = E\left[E\left[|X_t - X_s|^p \mid X_t\right]\right]$$

---

[5] Possibly extending the canonical space to support also the Brownian motion **W** with respect to which to write the SDE, as in the proof of Theorem 18.1.3 and in the subsequent Remark 18.1.4.

[6] As we will see in Chap. 20, in general the Cauchy problem (18.2.4) admits more than one solution.

$$= E\left[\int_{\mathbb{R}^N} |X_t - y|^p \Gamma(t, X_t; s, y) dy\right] \le$$

(by the Gaussian estimate from above (18.2.6))

$$\le cE\left[\int_{\mathbb{R}^N} |X_t - y|^p \mathbf{G}\left(\lambda(s-t), X_t - y\right) dy\right] \le c(s-t)^{\frac{p}{2}}$$

where the last step is justified by the change of variable $z = \frac{X_t - y}{\sqrt{s-t}}$.

Finally, if (X^i, W^i) for $i = 1, 2$ are weak solutions of the SDE (18.2.3) on $[0, T]$, then, as just shown, Γ is a transition law for both X^1 and X^2. Therefore, if $X_0^1 \stackrel{d}{=} X_0^2$, i.e., if X^1, X^2 have the same initial law, then by Proposition 2.4.1 X^1, X^2 are equal in law.

To conclude, we observe that, under the assumption of uniform ellipticity (18.2.2), W^i is a functional of X^i: to fix ideas, in the case $d = 1$, from the SDE we obtain the explicit expression of such a functional as in (18.1.3) and (18.1.4). Then, from Corollary 10.2.28 we have the equality in law of (X^1, W^1) and (X^2, W^2). □

Remark 18.2.7 The latter part of the proof of Theorem 18.2.3 demonstrates a duality result, namely, that *the existence* of a fundamental solution of $\partial_t + \mathscr{A}_t$ implies *the uniqueness* in law of the solution (X, W) of the stochastic differential equation.

18.3 Other Results for the Martingale Problem

We present an existence and uniqueness result for weak solutions under significantly broader assumptions than those of Theorem 18.2.3.

Theorem 18.3.1 (Skorokhod [131], Stroock and Varadhan [136], Krylov [74, 75] [!!]) *Let μ_0 be a distribution on $\mathbb{R}^N$. Suppose that*

(i) *the coefficients b, σ are bounded, Borel measurable functions and at least one of the following assumptions holds:*
(ii) *$b(t, \cdot), \sigma(t, \cdot)$ are continuous functions for every $t \in [0, T]$;*
(iii) *condition (18.2.2) of uniform ellipticity holds.*

Then there exists a weak solution (X, W) of the SDE

$$dX_t = b(t, X_t)dt + \sigma(t, X_t)dW_t \qquad (18.3.1)$$

on $[0, T]$ with initial law μ_0. Moreover, if both assumptions (ii) and (iii) hold, then there is also uniqueness in the weak sense.

So as for Theorem 18.2.3, the proof of weak solvability hinges on the martingale problem, and therefore consists in the construction of the law of the solution. However, in the proof of Theorem 18.2.3, this probability distribution is defined by the fundamental solution of the backward Kolmogorov equation, whose existence is ensured by the classical results of the theory of PDEs. Conversely, Skorokhod's approach to proving Theorem 18.3.1 is more akin to the method employed in establishing the existence of strong solutions. It involves temporal discretization or smoothing of the coefficients to render the equation solvable, followed by taking the limit: the method of successive approximations employed in Theorem 17.2.1 is supplanted by an argument based on relative compactness or tightness (cf. Section 3.3.2 in [113]), in the space of distributions from which the weak convergence to a law is deduced; the latter is finally shown to be a solution to the martingale problem. For the details of the proof of Theorem 18.3.1, we refer to Section 2.6 in [77] and to Theorems 6.1.7 and 7.2.1 in [136].

Remark 18.3.2 (Bibliographic Note) The literature on the martingale problem is vast. We only mention some of the most recent contributions in which equations that do not satisfy the uniform ellipticity condition (18.2.2) are considered: [11, 44, 46, 95, 140] and [30].

18.4 Strong Uniqueness Through Regularization by Noise

The main result of the section is the following theorem that provides an example of "regularization by noise": it extends the results of Sect. 14.2 to the case of strong solutions.

Theorem 18.4.1 (Zvonkin [154], Veretennikov [144] [!!]) *Assume the following hypotheses:*

(i) *the drift coefficient is bounded and Hölder continuous, $b \in bC_T^\alpha$ for some $\alpha \in]0, 1]$;*
(ii) *the diffusion coefficient is bounded and Lipschitz continuous, $\sigma \in bC_T^1$;*
(iii) *condition (18.2.2) of uniform ellipticity holds.*

Then for the SDE

$$dX_t = b(t, X_t)dt + \sigma(t, X_t)dW_t \qquad (18.4.1)$$

there is existence and uniqueness in the strong sense.

Remark 18.4.2 ([!]) Theorem 18.4.1 illustrates the regularizing effect of noise, i.e., the diffusive part of the SDE: in the case of zero diffusion σ, the classic Example 14.2.1 by Peano shows that the Hölder continuity of the drift b is not sufficient to guarantee the uniqueness of the solution.

First, Zvonkin [154] proved the existence and uniqueness in the strong sense for SDEs in one dimension with $b \in L^\infty(]0, T[\times \mathbb{R})$ and $\sigma = 1$: Veretennikov [144] extended this result to the multidimensional case. Krylov and Röckner [78] showed that there is existence and uniqueness in the strong sense if $b \in L_{\text{loc}}^p$ with $p > N$ and Zhang [153] dealt with the case where the diffusion coefficient is not constant. In the recent work [23], Champagnat and Jabin study the existence and strong uniqueness for SDEs with irregular coefficients, without assuming the uniform ellipticity of the diffusion matrix, starting from suitable L^p estimates for the solutions of the associated Fokker-Planck equation. Finally, we point out the recent results in [57] on the approximation of solutions, under minimal regularity assumptions.

For the proof of Theorem 18.4.1 we follow Fedrizzi and Flandoli [43] who use the so-called *Itô-Tanaka trick* and the following

Proposition 18.4.3 *Under the assumptions of Theorem 18.2.6, let Γ be the fundamental solution of the Kolmogorov operator $\partial_t + \mathscr{A}_t$ on $]0, T[\times \mathbb{R}^N$. For every $\lambda \geq 1$, the vector-valued function in $\mathbb{R}^N$*

$$u_\lambda(t, x) := \int_t^T \int_{\mathbb{R}^N} e^{-\lambda(s-t)} \Gamma(t, x; s, y) b(s, y) dy ds, \qquad (t, x) \in]0, T] \times \mathbb{R}^N,$$

is a classical solution to the Cauchy problem

$$\begin{cases} (\partial_t + \mathscr{A}_t) u = \lambda u - b, & \text{in }]0, T[\times \mathbb{R}^N, \\ u(T, \cdot) = 0, & \text{in } \mathbb{R}^N. \end{cases}$$

Moreover, there exists a constant $c > 0$, which depends only on N, λ_0, T and the norms $[b_i]_\alpha$ and $[c_{ij}]_\alpha$ in (18.2.1), such that

$$|u_\lambda(t, x) - u_\lambda(t, y)| \leq c \frac{|x - y|}{\sqrt{\lambda}},$$
$$|\nabla_x u_\lambda(t, x) - \nabla_x u_\lambda(t, y)| \leq c|x - y|,$$
(18.4.2)

for every $t \in]0, T[$ and $x, y \in \mathbb{R}^N$, where $\nabla_x = (\partial_{x_1}, \ldots, \partial_{x_N})$.

Proposition 18.4.3 is a consequence of some estimates obtained in the proof of Theorem 18.2.6 on the existence of the fundamental solution: we refer to Sect. 20.3.5 for the proof and details.

Proof of Theorem 18.4.1 The existence of a weak solution is a consequence of Theorem 18.2.3: therefore, it remains to prove the strong uniqueness from which the thesis will follow thanks to the Yamada-Watanabe Theorem 14.3.6.

First, we present the Itô-Tanaka trick to transform the SDE into a new equation with a more regular drift. Let (X, W) be a solution of (18.4.1). By Proposition 18.4.3 and Itô's formula, we have[7]

---

[7] Here
$$(\nabla_x u_\lambda \cdot \sigma)_{ij} = \sum_{k=1}^N (\nabla_x u_\lambda)_{ik} \sigma_{kj}, \qquad i = 1, \ldots, N, \ j = 1, \ldots, d.$$

18.4 Strong Uniqueness Through Regularization by Noise

$$du_\lambda(t, X_t) = (\partial_t + \mathscr{A}_t)u_\lambda(t, X_t)dt + (\nabla_x u_\lambda \cdot \sigma)(t, X_t)dW_t$$
$$= (\lambda u_\lambda(t, X_t) - b(t, X_t))dt + (\nabla_x u_\lambda \cdot \sigma)(t, X_t)dW_t$$

or equivalently

$$\int_0^t b(s, X_s)ds = u_\lambda(0, X_0) - u_\lambda(t, X_t) + \lambda \int_0^t u_\lambda(s, X_s)ds$$
$$+ \int_0^t (\nabla_x u_\lambda \cdot \sigma)(s, X_s)dW_s. \quad (18.4.3)$$

Inserting (18.4.3) into (18.4.1), we obtain

$$X_t = X_0 + u_\lambda(0, X_0) - u_\lambda(t, X_t) + \lambda \int_0^t u_\lambda(s, X_s)ds + \int_0^t \sigma(s, X_s)dW_s$$
$$+ \int_0^t (\nabla_x u_\lambda \cdot \sigma)(s, X_s)dW_s. \quad (18.4.4)$$

In this way, the drift coefficient b is replaced by the more regular function u_λ: at this point, with some small adjustments, one can proceed as in the case of Lipschitz coefficients, using Grönwall's lemma to prove uniqueness. In fact, let X' be another solution of the SDE (18.4.1) related to the same Brownian motion W and let $Z := X - X'$. Writing also X' as in (18.4.4) and subtracting the two equations, we obtain

$$Z_t = -u_\lambda(t, X_t) + u_\lambda(t, X'_t) + \lambda \int_0^t (u_\lambda(s, X_s) - u_\lambda(s, X'_s))ds$$
$$+ \int_0^t \left(\sigma(s, X_s) - \sigma(s, X'_s)\right)dW_s$$
$$+ \int_0^t \left((\nabla_x u_\lambda \cdot \sigma)(s, X_s) - (\nabla_x u_\lambda \cdot \sigma)(s, X'_s)\right)dW_s.$$

By the elementary inequality (14.4.6) and Jensen and Burkholder inequalities (12.3.7), we have

$$\frac{1}{4}E\left[|Z_t|^2\right] \leq E\left[|u_\lambda(t, X_t) - u_\lambda(t, X'_t)|^2\right]$$
$$+ \lambda^2 TE\left[\int_0^t |u_\lambda(s, X_s) - u_\lambda(s, X'_s)|^2 ds\right]$$
$$+ E\left[\int_0^t |\sigma(s, X_s) - \sigma(s, X'_s)|^2 ds\right]$$
$$+ E\left[\int_0^t |(\nabla_x u_\lambda \cdot \sigma)(s, X_s) - (\nabla_x u_\lambda \cdot \sigma)(s, X'_s)|^2 ds\right] \leq$$

(by the estimates (18.4.2) of Proposition 18.4.3 with $\lambda \geq 1$ and the Lipschitz assumption of σ)

$$\leq \frac{c}{\lambda} E\left[|Z_t|^2\right] + c(1+\lambda) \int_0^t E\left[|Z_s|^2\right] ds,$$

for some positive constant c that depends only on N, λ_0, T and the norms $[b]_\alpha$ and $[\sigma]_1$. In other words, we have

$$\left(\frac{1}{4} - \frac{c}{\lambda}\right) E\left[|Z_t|^2\right] \leq c(1+\lambda) \int_0^t E\left[|Z_s|^2\right] ds.$$

Then, choosing λ suitably large, we get

$$E\left[|Z_t|^2\right] \leq \bar{c} \int_0^t E\left[|Z_s|^2\right] ds, \qquad t \in [0, T],$$

for a suitable positive constant $\bar{c}$. The thesis follows from Grönwall's lemma. □

Remark 18.4.4 Formula (18.4.4) can be used as in the proof of Theorem 17.4.1 to obtain the continuous dependence estimate (17.4.1) on the parameters. As a consequence of Kolmogorov's continuity Theorem 3.3.4, under the assumptions of Theorem 18.4.1 the solution of the SDE (18.4.1) with initial datum x at time t, admits a modification $(t, x, s) \mapsto X_s^{t,x}$ with locally α-Hölder continuous trajectories for every $\alpha \in [0, 1[$ with respect to the "parabolic" distance: precisely, for every $\alpha \in [0, 1[, n \in \mathbb{N}$ and $\omega \in \Omega$ there exists $c_{\alpha, n, \omega} > 0$ such that

$$\left|X_{s_1}^{t_1, x}(\omega) - X_{s_2}^{t_2, y}(\omega)\right| \leq c_{\alpha, n, \omega} \left(|x - y| + |t_1 - t_2|^{\frac{1}{2}} + |s_1 - s_2|^{\frac{1}{2}}\right)^\alpha,$$
(18.4.5)

for every $t_1, t_2, s_1, s_2 \in [0, T]$ such that $t_1 \leq s_1, t_2 \leq s_2$, and for every $x, y \in \mathbb{R}^N$ such that $|x|, |y| \leq n$.

18.5 Key Ideas to Remember

We summarize the most relevant results of the chapter. As usual, if you have any doubt about what the following succinct statements mean, please review the corresponding section.

- Section 18.1: through the Stroock-Varadhan martingale problem, the study of weak solvability of an SDE is reduced to the construction of a distribution (the law of the solution) on the canonical space that makes the processes in (18.1.1) and (18.1.2) martingales.

18.5 Key Ideas to Remember

- Sections 18.2 and 18.3: we exploit the analytical results on the existence of the fundamental solution of uniformly parabolic PDEs to solve the martingale problem. As a consequence, we prove existence, weak uniqueness, and Markov properties for SDEs with Hölder and bounded coefficients. The assumptions are further weakened in Theorem 18.3.1 whose proof is based on properties of relative compactness in the space of distributions.
- Section 18.4: we establish a "regularization by noise" result, ensuring *strong* uniqueness for SDEs with Hölder continuous and bounded drift, under a uniform ellipticity condition.

Main notations used or introduced in this chapter:

Symbol	Description	Page
$\Omega_n = C([0,T];\mathbb{R}^n)$	Space of continuous n-dimensional trajectories	273
$\mathbf{X}_t(w) = w(t)$	Identity process on Ω_n	273
$(\mathscr{G}_t^n)_{t\in[0,T]}$	Filtration on Ω_n generated by the identity process	273
bC_T^α	Continuous, bounded, and uniformly Hölder continuous functions in x	341
$[g]_\alpha$	Norm in bC_T^α	341
$C^{1,2}(]0,T[\times\mathbb{R}^N)$	Functions continuously differentiable w.r.t. t and twice continuously differentiable w.r.t x	342
$\mathscr{A}_t$	Characteristic operator	341
$\Gamma(t,x;s,y)$	Fundamental solution	342
$\mathbf{G}(t,x)$	Standard N-dimensional Gaussian	343

Chapter 19
Complements

> *The day a man realizes he cannot know everything is a day of mourning. Then comes the day when he is brushed by the suspicion that he will not be able to know many things; and finally that autumn afternoon when it will seem to him that he has never known too well what he thought he knew.*
>
> *Julien Green*

We offer a concise and relaxed exploration of various paths that the theory of stochastic differential equations has taken. At the end of each section, we include a bibliography, directing interested readers to further literature on the specific topics discussed.

19.1 Markovian Projection and Gyöngy's Lemma

Consider an Itô process of the form

$$dX_t = u_t dt + v_t dB_t \qquad (19.1.1)$$

where u is an N-dimensional process in $\mathbb{L}^1_{\text{loc}}$, v is $N \times d$-dimensional process in $\mathbb{L}^2_{\text{loc}}$ and B is a d-dimensional Brownian motion on the space $(\Omega, \mathscr{F}, P, \mathscr{F}_t)$. In general, at any time t, X_t may depend on the σ-algebra $\mathscr{F}_t$ (of information up to time t) in an extremely complicated way through the coefficients u_t and v_t. In this section, we present a result, known as Gyöngy's lemma, according to which there

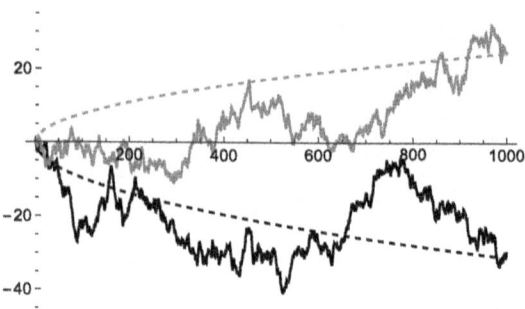

Fig. 19.1 Plot of the trajectories $t \mapsto W_t(\omega)$ (solid line) and $t \mapsto \widetilde{W}(\omega)$ (dashed line) of the processes of Remark 19.1.1, related to two outcomes $\omega = \omega_1$ (in black) and $\omega = \omega_2$ (in gray)

exists a diffusion Y, solution of an SDE of the type

$$dY_t = b(t, Y_t)dt + \sigma(t, Y_t)dW_t, \qquad (19.1.2)$$

which "mimicks" X in the sense that it has the same marginal distributions, i.e., it is such that $Y_t \stackrel{d}{=} X_t$ for each t. This result can be useful when one is interested in the law of X_t for a fixed time t and not in the entire law of the process X. Since the coefficients $b = b(t, y)$ and $\sigma = \sigma(t, y)$ in (19.1.2) are deterministic functions, by the results of the previous chapters, Y is a Markov process, sometimes called *Markovian projection* of X.

Remark 19.1.1 Processes with the same *one-dimensional* distributions can have very distinct properties: for example, we saw in Remark 4.1.5 that a Brownian motion W has the same one-dimensional distributions as the process $\widetilde{W}_t := \sqrt{t}W_1$. However, despite this equivalence, the two processes are inherently distinct in law, and their trajectories demonstrate entirely different properties, as illustrated in Fig. 19.1.

Theorem 19.1.2 (Gyöngy [56]) *Let X be an Itô process of the form (19.1.1) with the coefficients u, v being progressively measurable, bounded, and satisfying the uniform ellipticity condition*

$$\langle v_t v_t^* \eta, \eta \rangle \geq \lambda |\eta|^2, \qquad t \in [0, T], \ \eta \in \mathbb{R}^N,$$

for some positive constant λ. There exist two bounded and measurable functions

$$b : [0, T] \times \mathbb{R}^N \longrightarrow \mathbb{R}^N, \qquad \sigma : [0, T] \times \mathbb{R}^N \longrightarrow \mathbb{R}^{N \times N},$$

such that, setting $\mathscr{C} = \sigma\sigma^*$, we have[1]

$$b(t, X_t) = E[u_t \mid X_t], \qquad \mathscr{C}(t, X_t) = E\left[v_t v_t^* \mid X_t\right] \qquad (19.1.3)$$

---

[1] Formula (19.1.3) means that $b(t, \cdot)$ and $(\sigma\sigma^*)(t, \cdot)$ are respectively versions of the conditional expectation functions of u_t and $v_t v_t^*$ given X_t, according to Definition 4.2.16 in [113].

19.1 Markovian Projection and Gyöngy's Lemma

and the SDE (19.1.2) with coefficients b, σ admits a weak solution (Y, W) such that $Y_t \stackrel{d}{=} X_t$ for every $t \in [0, T]$.

Proof We only give a sketch of the proof. Let b and $\mathscr{C} = (c_{ij})$ be versions of the conditional expectation functions of u_t and $v_t v_t^*$ given X_t respectively, as in (19.1.3). Moreover, let $\sigma = \mathscr{C}^{\frac{1}{2}}$ be the positive definite square root of the positive definite matrix $\mathscr{C}$: the complete proof in [56] uses a regularization argument of the coefficients that allows to reduce to the case where $b_i(t, \cdot)$ and $c_{ij}(t, \cdot)$ are at least Hölder continuous functions so as to satisfy the hypotheses of Theorem 18.2.6 for the existence of a fundamental solution of the characteristic operator $\mathscr{A}_t + \partial_t$ where

$$\mathscr{A}_t := \frac{1}{2} \sum_{i,j=1}^N c_{ij}(t, x) \partial_{x_i x_j} + \sum_{i=1}^N b_i(t, x) \partial_{x_i}.$$

Hence, fixed $s \in]0, T]$ and $\varphi \in C_0^\infty(\mathbb{R}^N)$, consider the classical, bounded solution f of the backward Cauchy problem

$$\begin{cases} \partial_t f(t, x) + \mathscr{A}_t f(t, x) = 0, & (t, x) \in]0, s[\times \mathbb{R}^N, \\ f(s, x) = \varphi(x), & x \in \mathbb{R}^N. \end{cases}$$

By Itô's formula, we have

$$f(s, X_s) = f(0, X_0) + \frac{1}{2} \sum_{i,j=1}^N \int_0^s (v_t v_t^*)_{ij} \partial_{x_i x_j} f(t, X_t) dt$$
$$+ \int_0^s \left(u_t \nabla_x f(t, X_t) + \partial_t f(t, X_t) \right) dt + \int_0^s \nabla_x f(t, X_t) v_t dB_t$$
(19.1.4)

and taking the expectation[2]

$$E[f(s, X_s)] = E[f(0, X_0)] + \frac{1}{2} \sum_{i,j=1}^N \int_0^s E\left[(v_t v_t^*)_{ij} \partial_{x_i x_j} f(t, X_t) \right] dt$$
$$+ \int_0^s E[u_t \nabla_x f(t, X_t) + \partial_t f(t, X_t)] dt =$$

---

[2] Here we use a technical argument that relies on the analytical results of Chap. 20: the estimate of Corollary 20.2.7 guarantees that $\nabla_x f(t, X_t) v_t \in \mathbb{L}^2$ and therefore the stochastic integral in (19.1.4) has zero expectation.

(by the properties of conditional expectation)

$$= E\left[f(0, X_0)\right] + \frac{1}{2} \sum_{i,j=1}^{N} \int_0^s E\left[E\left[(v_t v_t^*)_{ij} \mid X_t\right] \partial_{x_i x_j} f(t, X_t)\right] dt$$

$$+ \int_0^s E\left[E\left[u_t \mid X_t\right] \nabla_x f(t, X_t) + \partial_t f(t, X_t)\right] dt =$$

(by (19.1.3))

$$= E\left[f(0, X_0)\right] + \int_0^s E\left[(\mathscr{A}_t f + \partial_t f)(t, X_t)\right] dt =$$

(being f a solution of the Cauchy problem)

$$= E\left[f(0, X_0)\right]. \tag{19.1.5}$$

On the other hand, by Theorem 18.3.1 there exists a weak solution (Y, W) of the SDE (19.1.2) with initial law equal to the law of X_0. By Itô's formula, the process $f(t, Y_t)$ is a martingale[3] and therefore, by (19.1.5) we have

$$E\left[\varphi(Y_s)\right] = E\left[f(s, Y_s)\right] = E\left[f(0, Y_0)\right] = E\left[f(0, X_0)\right] = E\left[f(s, X_s)\right]$$
$$= E\left[\varphi(X_s)\right]$$

so that $Y_s \stackrel{d}{=} X_s$, given the arbitrariness of φ. □

Remark 19.1.3 (Bibliographic Note) Markovian projection methods are widely used in mathematical finance for the calibration of local-stochastic volatility and interest rates models: in this regard, see, for example, [3], [83] and Section 11.5 in [55]. A version of Gyöngy's Theorem 19.1.2 that relaxes the hypotheses on the coefficients has been more recently proven by Brunick and Shreve [22].

19.2 Backward Stochastic Differential Equations

In the previous chapters, we examined SDEs with an assigned *initial* datum. However, in some applications, for example in stochastic optimal control theory or mathematical finance, problems arise where it is natural to assign a *final* condition:

---

[3] Precisely, by Itô's formula the process $f(t, Y_t)$ is a local martingale, but it is also a true martingale by the boundedness of the function f.

19.2 Backward Stochastic Differential Equations

in this case, we speak of backward SDEs (or BSDEs). The most elementary example is

$$\begin{cases} dY_t = 0, \\ Y_T = \eta. \end{cases} \qquad (19.2.1)$$

If the datum $\eta \in \mathbb{R}^N$ is not random, (19.2.1) is a simple ordinary differential equation (ODE) with constant solution $Y \equiv \eta$. The situation is completely different if we set the problem in a space $(\Omega, \mathscr{F}, P)$ on which a Brownian motion W is defined with standard filtration $\mathscr{F}^W$ and assume $\eta \in m\mathscr{F}_T^W$: in fact, to remain within the classical Itô calculus, we would like the solution Y to be an adapted process and therefore the constant solution equal to η is not acceptable. The first problem is therefore to correctly formulate the concept of a solution to a BSDE.

For each $\eta \in L^2(\Omega, \mathscr{F}_T^W, P)$, the *adapted* process that best (in L^2 norm) approximates the constant process equal to η is

$$Y_t := E\left[\eta \mid \mathscr{F}_t^W\right], \qquad t \in [0, T]. \qquad (19.2.2)$$

From this perspective, the process Y in (19.2.2) is the natural candidate to be a solution to the BSDE (19.2.1). Clearly, it is not necessarily the case that Y in (19.2.2) verifies the equation $dY_t = 0$. Indeed, since Y is a $\mathscr{F}^W$-square-integrable martingale, by the martingale representation Theorem 13.5.1 there exists a unique $Z \in \mathbb{L}^2$ such that

$$Y_t = Y_0 + \int_0^t Z_s dW_s = \underbrace{Y_0 + \int_0^T Z_s dW_s}_{=\eta} - \int_t^T Z_s dW_s.$$

This means that Y verifies the forward SDE

$$\begin{cases} dY_t = Z_t dW_t, \\ Y_0 = \eta - \int_0^T Z_s dW_s. \end{cases} \qquad (19.2.3)$$

Although it may not seem obvious, it is not difficult to prove that (Y, Z) is the only pair of processes in $\mathbb{L}^2$ that satisfies (19.2.3): in fact, if (19.2.3) were also satisfied by $(Y', Z') \in \mathbb{L}^2$, then, setting $A = Y - Y'$ and $B = Z - Z'$, we would have

$$\begin{cases} dA_t = B_t dW_t, \\ A_T = 0. \end{cases}$$

By Itô's formula, we have

$$dA_t^2 = 2A_t dA_t + d\langle A \rangle_t$$

and therefore

$$A_t = -\int_t^T 2A_s dA_s - \int_t^T B_s^2 ds$$

and

$$E\left[A_t^2 + \int_t^T B_s^2 ds\right] = E\left[\int_t^T 2A_s dA_s\right] = 0$$

where the last equality is due to the fact that A, and therefore also the stochastic integral, is a martingale. Based on what has just been proven, the following definition is well posed.

Definition 19.2.1 Let W be a Brownian motion on the space $(\Omega, \mathscr{F}, P)$ endowed with the standard filtration $\mathscr{F}^W$. We say that the pair $(Y, Z) \in \mathbb{L}^2$, unique solution of the SDE (19.2.3), is the *adapted solution* of the BSDE (19.2.1) with final datum $\eta \in L^2(\Omega, \mathscr{F}_T^W, P)$.

Note that by definition we have

$$\begin{cases} dY_t = Z_t dW_t, \\ Y_T = \eta. \end{cases}$$

In a similar way, more general backward equations of the form

$$\begin{cases} dY_t = f(t, Y_t, Z_t)dt + Z_t dW_t, \\ Y_T = \eta, \end{cases}$$

are studied. Under standard Lipschitz assumptions on the coefficient $f = f(t, y, z)$ in the variables (y, z), it is possible to prove the existence and uniqueness of the adapted solution (Y, Z): see, for example, Theorem 4.2, Chapter 1 in [93].

Often a BSDE is coupled with a forward SDE of the type

$$dX_t = b(t, X_t)dt + \sigma(t, X_t)dW_t.$$

Given $u = u(t, x) \in C^{1,2}([0, T[\times \mathbb{R}^N)$, applying Itô's formula to $Y_t := u(t, X_t)$ we obtain

$$dY_t = (\partial_t + \mathscr{A}_t)u(t, X_t)dt + Z_t dW_t$$

where $\mathscr{A}_t$ is the characteristic operator of X and

$$Z_t := (\nabla_x u)(t, X_t)\sigma(t, X_t).$$

19.3 Filtering and Stochastic Heat Equation

In particular, if u is a solution of the *quasi-linear* Cauchy problem

$$\begin{cases} (\partial_t + \mathscr{A}_t)u(t,x) = f(t,x,u(t,x), \nabla_x u(t,x)\sigma(t,x)) & (t,x) \in [0,T[\times \mathbb{R}^N, \\ u(T,x) = \varphi(x) & x \in \mathbb{R}^N, \end{cases} \qquad (19.2.4)$$

then (X, Y, Z) solves the forward-backward system of equations (FBSDE)

$$\begin{cases} dX_t = b(t, X_t)dt + \sigma(t, X_t)dW_t, \\ dY_t = f(t, X_t, Y_t, Z_t)dt + Z_t dW_t, \\ Y_T = \varphi(X_T). \end{cases} \qquad (19.2.5)$$

Under appropriate assumptions that guarantee the existence of a solution[4] of the problem (19.2.4), by construction we have

$$u(t,x) = Y_t^{t,x} \qquad (19.2.6)$$

where $Y^{t,x}$ is the solution of the FBSDE (19.2.5) with initial datum $X_t = x$. Formula (19.2.6) is a *nonlinear Feynman-Kac formula* that generalizes the classical representation formula of Sect. 15.4.

Remark 19.2.2 (Bibliographic Note) The main motivation for the study of BSDEs comes from the theory of optimal stochastic control, starting from the works [17] and [15]; some applications to mathematical finance are discussed in [39]. The earliest results about existence and the nonlinear Feynman-Kac representation come from [109], [117], and [2]. We point to the following books as essential references for the theory of backward equations: Ma and Yong [93], Yong and Zhou [150], Pardoux and Rascanu [110], and Zhang [152].

19.3 Filtering and Stochastic Heat Equation

In this section, we outline some basic ideas of the theory of stochastic filtering and, in a simple and explicit case, introduce the notion of *stochastic partial differential equation* (abbreviated as SPDE), which intervenes naturally in this type of problems.

Given (W, B) a standard two-dimensional Brownian motion, we consider the process

$$X_t^\sigma := \sigma W_t + \sqrt{1 - \sigma^2} B_t, \qquad \sigma \in [0, 1].$$

---

[4] Since it is a non-linear problem, the solution u is understood in a generalized sense, for example as a "viscosity solution" (see, for example, Theorem 2.1, Chap. 8 in [93]).

Suppose that X^σ represents a *signal* that is transmitted but not observable with precision due to some disturbance in the transmission: precisely, we assume that we can observe precisely W_t, called the *observation process*, while the Brownian motion B_t represents the *noise* in the transmission.

It is easy to verify that X^σ is a real Brownian motion for every $\sigma \in [0, 1]$. The problem of stochastic filtering consists in obtaining the best estimate of the signal X^σ based on the observation W: in fact, it is not difficult to prove that

$$\mu_{X_t^\sigma | \mathscr{F}_t^W} = \mathscr{N}_{\sigma W_t, (1-\sigma^2)t} \tag{19.3.1}$$

where $\mu_{X_t^\sigma | \mathscr{F}_t^W}$ denotes the conditional law of X_t^σ given the σ-algebra $\mathscr{F}_t^W$ of observations on W up to time t (here $\mathscr{F}^W$ is the standard filtration for W). To prove (19.3.1) it is enough to calculate the conditional characteristic function

$$\varphi_{X_t^\sigma | \mathscr{F}_t^W}(\eta) = E\left[e^{i\eta X_t^\sigma} \mid \mathscr{F}_t^W\right] = e^{i\eta \sigma W_t} E\left[e^{i\eta \sqrt{1-\sigma^2} B_t} \mid \mathscr{F}_t^W\right] =$$

(by independence of W and B)

$$= e^{i\eta \sigma W_t} E\left[e^{i\eta \sqrt{1-\sigma^2} B_t}\right]$$

which proves (19.3.1). We observe that in particular:

- when there is no noise, $\sigma = 1$, we have $X_t^\sigma = W_t$ and $\mu_{X_t^\sigma | \mathscr{F}_t^W} = \delta_{W_t}$, that is, the conditional law degenerates into a Dirac distribution;
- when there is no observation, $\sigma = 0$, then $X_t^\sigma = B_t$ and the conditioned law is obviously $\mu_{X_t^\sigma | \mathscr{F}_t^W} = \mathscr{N}_{0,t}$ with Gaussian density

$$\Gamma(t, x) = \frac{1}{\sqrt{2\pi t}} e^{-\frac{x^2}{2t}}, \qquad t > 0, \ x \in \mathbb{R}. \tag{19.3.2}$$

If $0 \leq \sigma < 1$ then X_t^σ has the following conditional density given $\mathscr{F}_t^W$:

$$p_t(x) = \Gamma((1-\sigma^2)t, x - \sigma W_t), \qquad t > 0, \ x \in \mathbb{R}. \tag{19.3.3}$$

If $\sigma > 0$, clearly the conditional density $p_t(x)$ is a stochastic process: from a practical standpoint, having the observation of W_t available and inserting it into (19.3.3), we obtain the expression of the law of X_t^σ estimated (or "filtered") based on such observation. Note that $p_t(x)$ is a Gaussian function with stochastic drift, dependent on the observation, and variance proportional to $1 - \sigma^2$. Figure 19.2 represents the plot of a simulation of the stochastic Gaussian density $p_t(x)$.

In analogy with the unconditioned case examined in Sects. 2.5.3 and 17.3.1, $p_t(x)$ is a solution of the Kolmogorov forward (Fokker-Planck) equation which in this case is a SPDE: in fact, recalling the expression (19.3.3) of $p_t(x)$ in terms of

19.3 Filtering and Stochastic Heat Equation

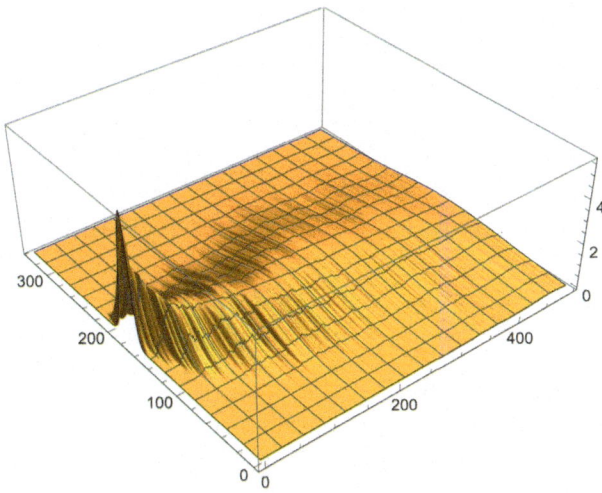

Fig. 19.2 Plot of a simulation of the fundamental solution $p_t(x)$ of the stochastic heat equation

$\Gamma = \Gamma(s, y)$ in (19.3.2), by Itô's formula we have

$$dp_t(x) = (1-\sigma^2)(\partial_s \Gamma)((1-\sigma^2)t, x-\sigma W_t)dt - \sigma(\partial_y \Gamma)((1-\sigma^2)t, x-\sigma W_t)dW_t$$
$$+ \frac{\sigma^2}{2}(\partial_{yy}\Gamma)((1-\sigma^2)t, x-\sigma W_t)dt =$$

(since Γ solves the forward heat equation $\partial_s \Gamma(s, y) = \frac{1}{2}\partial_{yy}\Gamma(s, y)$)

$$= \frac{1}{2}(\partial_{yy}\Gamma)((1-\sigma^2)t, x-\sigma W_t)dt - \sigma(\partial_y \Gamma)((1-\sigma^2)t, x-\sigma W_t)dW_t$$
$$= \frac{1}{2}\partial_{xx}p_t(x)dt - \sigma \partial_x p_t(x)dW_t.$$

In other words, the conditional density $p_t(x)$ is the fundamental solution of the *stochastic heat equation*

$$dp_t(x) = \frac{1}{2}\partial_{xx}p_t(x)dt - \sigma \partial_x p_t(x)dW_t$$

which, in the case $\sigma = 0$ where the observation is null, degenerates into the classical heat equation.

Remark 19.3.1 (Bibliographic Note) Among the numerous monographs on the theory of SPDEs, we particularly mention the books by Rozovskii [125], Kunita [82], Prévôt and Röckner [120], Kotelenez [73], Chow [24], Liu and Röckner [91], Lototsky and Rozovskii [92], and Pardoux [108]. For the study of stochastic filtering problems, we refer, for example, to Fujisaki et al. [52], Pardoux [107], Fristedt et al. [51], Elworthy et al. [40]. In [146] and [81], alternative approaches to the derivation of filtering SPDEs are proposed, based on arguments similar to those used for the proof of Feynman-Kac formulas.

19.4 Backward Stochastic Integral and Krylov's SPDE

In this section, we present an interesting result according to which, for a fixed time T, the solution of an SDE seen as a stochastic process varying over time and initial datum, i.e., $(t, x) \mapsto X_T^{t,x}$ in the usual notations of Chap. 15, solves a stochastic partial differential equation (SPDE) involving the characteristic operator of the SDE. The statement of this result and the formulation of the *Krylov's backward SPDE* (so called in Section 1.2.3 in [126]) requires the introduction of the *backward stochastic integral* in which the temporal structure of information, Brownian motion, and the related filtration are inverted.

Let W be a d-dimensional Brownian motion on the space $(\Omega, \mathscr{F}, P)$. For $t \in [0, T]$ we consider the (completed) σ-algebra of the increments of a Brownian motion between t and T, defined by

$$\hat{\mathscr{F}}_t := \sigma(\hat{\mathscr{G}}_t \cup \mathscr{N}), \qquad \hat{\mathscr{G}}_t := \sigma(W_s - W_t, s \in [t, T]). \tag{19.4.1}$$

Clearly, (19.4.1) defines a decreasing family of σ-algebras; we say that

$$\tilde{\mathscr{F}}_t := \hat{\mathscr{F}}_{T-t}, \qquad t \in [0, T],$$

is the *backward Brownian filtration*. It is straightforward to verify that the process

$$\tilde{W}_t := W_T - W_{T-t}, \qquad t \in [0, T],$$

is a Brownian motion on $(\Omega, \mathscr{F}, P, \tilde{\mathscr{F}}_t)$. The backward stochastic integral is defined as

$$\int_t^s u_r \star dW_r := \int_{T-s}^{T-t} u_{T-r} d\tilde{W}_r, \qquad 0 \le t \le s \le T, \tag{19.4.2}$$

under the assumptions on u for which the right-hand side of (19.4.2) is defined in the usual sense of Itô, i.e.,

(i) $t \mapsto u_{T-t}$ is $\tilde{\mathscr{F}}$-progressively measurable (thus $u_t \in m\hat{\mathscr{F}}_t$ for every $t \in [0, T]$);
(ii) $u \in L^2([0, T])$ a.s.

19.4 Backward Stochastic Integral and Krylov's SPDE

For practical purposes, according to Corollary 10.2.27, if u is continuous then the backward integral is the limit

$$\int_t^s u_r \star dW_r := \lim_{|\pi| \to 0^+} \sum_{k=1}^n u_{t_k} \left(W_{t_k} - W_{t_{k-1}} \right) \qquad (19.4.3)$$

in probability, where $\pi = \{t = t_0 < t_1 < \cdots < t_n = s\}$ denotes a partition of $[t, s]$: note, in particular, that unlike the usual Itô integral, the coefficient u in the sum in (19.4.3) is evaluated at the *right endpoint* of each interval of the partition and $u_{t_k} \in m\hat{\mathscr{F}}_{t_k}$ by hypothesis.

An N-dimensional backward Itô process is a process of the form

$$X_t = X_T + \int_t^T u_s ds + \int_t^T v_s \star dW_s, \qquad t \in [0, T],$$

also written in differential form as

$$-dX_t = u_t dt + v_t \star dW_t. \qquad (19.4.4)$$

We state the backward version of Itô's formula.

Theorem 19.4.1 (Backward Itô's Formula) *Let $F = F(t, x) \in C^{1,2}([0, T] \times \mathbb{R}^N)$ and let X be the process in (19.4.4). We have*

$$-dF(t, X_t)$$
$$= \left((\partial_t F)(t, X_t) + \frac{1}{2} \sum_{i,j=1}^N (v_t v_t^*)_{ij} (\partial_{x_i x_j} F)(t, X_t) + u_t (\nabla_x F)(t, X_t) \right) dt$$
$$+ \sum_{i=1}^N \sum_{j=1}^d (v_t)_{ij} (\partial_{x_i} F)(t, X_t) \star dW_t^j. \qquad (19.4.5)$$

The main result of the section is the following

Theorem 19.4.2 (Krylov's Backward SPDE) *Assume that $b, \sigma \in bC^3([0, T] \times \mathbb{R}^N)$ and let $s \mapsto X_s^{t,x}$ be the solution of the SDE*

$$dX_s^{t,x} = b(s, X_s^{t,x})ds + \sigma(s, X_s^{t,x})dW_s \qquad (19.4.6)$$

with initial condition $X_t^{t,x} = x$. Then the process $(t, x) \mapsto X_T^{t,x}$ solves the backward SPDE

$$\begin{cases} -dX_T^{t,x} = \mathscr{A}_t X_T^{t,x} dt + (\nabla_x X_T^{t,x})\sigma(t, x) \star dW_t, \\ X_T^{T,x} = x, \end{cases} \qquad (19.4.7)$$

where

$$\mathscr{A}_t = \frac{1}{2}\sum_{i,j=1}^{N} c_{ij}(t,x)\partial_{x_j x_i} + b_i(t,x)\partial_{x_i}, \qquad (c_{ij}) := \sigma\sigma^*,$$

is the characteristic operator of the SDE (19.4.6). The explicit expressions of the drift coefficient and the diffusion term in (19.4.7) are

$$\mathscr{A}_t X_T^{t,x} = \frac{1}{2}\sum_{i,j=1}^{N} c_{ij}(t,x)\partial_{x_j x_i} X_T^{t,x} + \sum_{i=1}^{N} b_i(t,x)\partial_{x_i} X_T^{t,x},$$

$$(\nabla_x X_T^{t,x})\sigma(t,x) \star dW_t = \sum_{i=1}^{N}\sum_{j=1}^{d} \sigma_{ij}(t,x)\partial_{x_i} X_T^{t,x} \star dW_t^j.$$

Proof We only give a sketch of the proof and refer to Proposition 5.3 in [126] for the details. To simplify the presentation, we only treat the one-dimensional and autonomous case, following the approach proposed in [145]. First of all, thanks to Kolmogorov's continuity theorem 3.3.4 and the L^p estimates of dependence on the initial datum, extending the results of Corollary 17.4.2, we have that, up to modifications, $x \mapsto X_T^{t,x}$ is sufficiently regular to support the derivatives that appear in the SPDE in the classical sense. We use the Taylor series expansion for functions of class $C^2(\mathbb{R})$:

$$f(\delta) - f(0) = \delta f'(0) + \frac{\delta^2}{2} f''(\lambda\delta), \qquad \lambda \in [0,1]. \tag{19.4.8}$$

Given a partition of $[t, T]$, we have

$$X_T^{t,x} - x = X_T^{t,x} - X_T^{T,x} = \sum_{k=1}^{n}\left(X_T^{t_{k-1},x} - X_T^{t_k,x}\right) =$$

(by the *flow property* of Theorem 17.2.1)

$$= \sum_{k=1}^{n}\left(X_T^{t_k, X_{t_k}^{t_{k-1},x}} - X_T^{t_k,x}\right) =$$

(by (19.4.8) with $f(\delta) = X_T^{t_k, x+\delta}$ and $\delta = \Delta_k X := X_{t_k}^{t_{k-1},x} - x$)

$$= \sum_{k=1}^{n}\left(\Delta_k X \partial_x X_T^{t_k,x} + \frac{(\Delta_k X)^2}{2}\partial_{xx} X_T^{t_k, x+\lambda_k \Delta_k X}\right) \tag{19.4.9}$$

19.4 Backward Stochastic Integral and Krylov's SPDE

with $\lambda_k = \lambda_k(\omega) \in [0, 1]$. Now we have

$$\Delta_k X = X_{t_k}^{t_{k-1},x} - x = \int_{t_{k-1}}^{t_k} b\left(X_s^{t_{k-1},x}\right) ds + \int_{t_{k-1}}^{t_k} \sigma\left(X_s^{t_{k-1},x}\right) dW_s.$$

Therefore, setting

$$\Delta_k t = t_k - t_{k-1}, \qquad \Delta_k W = W_{t_k} - W_{t_{k-1}}, \qquad \widetilde{\Delta}_k X = b(x)\Delta_k t + \sigma(x)\Delta_k W,$$

we have

$$\Delta_k X - \widetilde{\Delta}_k X = \int_{t_{k-1}}^{t_k} \left(b\left(X_s^{t_{k-1},x}\right) - b(x)\right) ds$$

$$+ \int_{t_{k-1}}^{t_k} \left(\sigma\left(X_s^{t_{k-1},x}\right) - \sigma(x)\right) dW_s = O(\Delta_k t),$$

$$\partial_{xx} X_T^{t_k,x+\lambda_k \Delta_k X} - \partial_{xx} X_T^{t_k,x} = O(\Delta_k t),$$

in $L^2(\Omega, P)$ norm or, more precisely,

$$E\left[|\Delta_k X - \widetilde{\Delta}_k X|^2 + \left|\partial_{xx} X_T^{t_k,x+\lambda_k \Delta_k X} - \partial_{xx} X_T^{t_k,x}\right|^2\right] \leq c(1+|x|^2)(\Delta_k t)^2$$

with the constant c depending only on T and the Lipschitz constants of b and σ. Hence, from (19.4.9) we obtain

$$X_T^{t,x} - x = \sum_{k=1}^n \left(\widetilde{\Delta}_k X \partial_x X_T^{t_k,x} + \frac{(\widetilde{\Delta}_k X)^2}{2} \partial_{xx} X_T^{t_k,x}\right) + O(\Delta_k t).$$

Note that $\partial_x X_T^{t_k,x}, \partial_{xx} X_T^{t_k,x} \in m\hat{\mathscr{F}}_{t_k}$; therefore, by (19.4.3), letting the partition mesh go to zero, we have

$$\sum_{k=1}^n \widetilde{\Delta}_k X \partial_x X_T^{t_k,x} \longrightarrow \int_t^T b(x) \partial_x X_T^{s,x} ds + \int_t^T \sigma(x) \partial_x X_T^{s,x} \star dW_s,$$

$$\sum_{k=1}^n (\widetilde{\Delta}_k X)^2 \partial_{xx} X_T^{t_k,x} \longrightarrow \int_t^T \sigma^2(x) \partial_{xx} X_T^{s,x} ds,$$

in norm $L^2(\Omega, P)$ and this concludes the proof. $\square$

Let us prove an interesting invariance property of Krylov's SPDE.

Corollary 19.4.3 *Given $F \in bC^2(\mathbb{R}^N)$ and X as in (19.4.6), let $V_T^{t,x} = F(X_T^{t,x})$. Then also $V_T^{t,x}$ satisfies the SPDE (19.4.7)*

$$-dV_T^{t,x} = \mathscr{A}_t V_T^{t,x} dt + (\nabla_x V_T^{t,x})\sigma(t,x) \star dW_t.$$

Proof To fix the ideas, let us start with the one-dimensional case: by the backward SPDE (19.4.7) and Itô's formula (19.4.5), we have

$$-dF(X_T^{t,x}) = \left(\frac{\sigma^2(t,x)}{2} F''(X_T^{t,x})(\partial_x X_T^{t,x})^2 + \frac{\sigma^2(t,x)}{2} F'(X_T^{t,x})\partial_{xx} X_T^{t,x}\right.$$
$$\left. + b(t,x) F'(X_T^{t,x})\partial_x X_T^{t,x}\right) dt$$
$$+ \sigma(t,x) F'(X_T^{t,x})\partial_x X_T^{t,x} \star dW_t =$$

(since $\partial_x V_T^{t,x} = F'(X_T^{t,x})\partial_x X_T^{t,x}$ and $\partial_{xx} V_T^{t,x} = F''(X_T^{t,x})(\partial_x X_T^{t,x})^2 + F'(X_T^{t,x})\partial_{xx} X_T^{t,x}$)

$$= \left(\frac{\sigma^2(t,x)}{2} \partial_{xx} V_T^{t,x} + b(t,x)\partial_x V_T^{t,x}\right) dt + \sigma(t,x)\partial_x V_T^{t,x} \star dW_t$$

which proves the thesis. The multidimensional case is analogous: first of all

$$\partial_{x_h} V_T^{t,x} = (\nabla_x F)(X_T^{t,x})\partial_{x_h} X_T^{t,x},$$

$$\partial_{x_h x_k} V_T^{t,x} = \sum_{i,j=1}^N (\partial_{x_i x_j} F)(X_T^{t,x})(\partial_{x_h} X_T^{t,x})_i (\partial_{x_k} X_T^{t,x})_j + (\nabla_x F)(X_T^{t,x})(\partial_{x_h x_k} X_T^{t,x}),$$

(19.4.10)

and by (19.4.7) and (19.4.5)

$$-dF(X_T^{t,x}) = \frac{1}{2}\sum_{i,j=1}^N \left((\nabla_x X_T^{t,x})\sigma(t,x)((\nabla_x X_T^{t,x})\sigma(t,x))^*\right)_{ij} (\partial_{x_i x_j} F)(X_T^{t,x}) dt$$
$$+ (\mathscr{A}_t X_T^{t,x})(\nabla_x F)(X_T^{t,x}) dt + (\nabla_x F)(X_T^{t,x})(\nabla_x X_T^{t,x})\sigma(t,x) \star dW_t =$$

(by (19.4.10))

$$= \mathscr{A}_t V_T^{t,x} dt + (\nabla_x V_T^{t,x})\sigma(t,x) \star dW_t.$$

which proves the thesis. □

19.4 Backward Stochastic Integral and Krylov's SPDE

Remark 19.4.4 (Bibliographic Note) The assumptions regarding the coefficients in Theorem 19.4.2 can be relaxed: in [126], Theorem 5.1, it is shown that if $b, \sigma \in bC^1([0, T] \times \mathbb{R}^N)$ then $(t, x) \mapsto X_T^{t,x}$ is a distributional solution of equation (19.4.7).

The material in this section is based on the original works of Krylov in [76], [79], [80] and Veretennikov [145]. The monographs [82] and [126] are now considered classical references for the study of SPDEs and stochastic flows.

Chapter 20
A Primer on Parabolic PDEs

Lo so
Del mondo e anche del resto
Lo so
Che tutto va in rovina
Ma di mattina
Quando la gente dorme
Col suo normale malumore
Mi può bastare un niente
Forse un piccolo bagliore
Un'aria già vissuta
Un paesaggio o che ne so

E sto bene
Io sto bene come uno quando sogna
Non lo so se mi conviene
Ma sto bene, che vergogna[1]

Giorgio Gaber

---

[1] *I know*
About the world and everything else
I know
That everything is crumbling
But in the morning
When people are asleep
With their usual bad mood
It can be enough for me
Maybe a small gleam
An already experienced air
A landscape or whatever

And I'm fine
I'm fine like someone when they dream
I don't know if it's good for me
But I'm fine, what a shame.

We provide a concise overview of fundamental results concerning the existence and uniqueness of solutions to the Cauchy problem for parabolic partial differential equations. The monographs by Friedman [49], Ladyzhenskaia et al. [84], Oleinik and Radkevic [103], although somewhat dated, are classic reference texts for a more complete and in-depth treatment.

We consider a second-order partial differential operator of the form

$$\mathscr{L} := \frac{1}{2} \sum_{i,j=1}^{N} c_{ij}(t,x) \partial_{x_i x_j} + \sum_{j=1}^{N} b_j(t,x) \partial_{x_j} + a(t,x) - \partial_t \qquad (20.0.1)$$

defined for (t,x) belonging to the strip

$$\mathscr{S}_T :=]0, T[\times \mathbb{R}^N, \qquad (20.0.2)$$

where $T > 0$ is fixed. We assume that the matrix of coefficients (c_{ij}) is symmetric and positive semidefinite: in this case, we say that $\mathscr{L}$ is a *forward parabolic operator*.

The interest in this type of operators is due to the fact that, as seen previously in Sects. 2.5 and 15.1,

$$\mathscr{A}_t := \frac{1}{2} \sum_{i,j=1}^{N} c_{ij}(t,x) \partial_{x_i x_j} + \sum_{j=1}^{N} \beta_j(t,x) \partial_{x_j}, \qquad (c_{ij}) := \sigma\sigma^*, \qquad (20.0.3)$$

is the characteristic operator of the SDE

$$dX_t = \beta(t, X_t)dt + \sigma(t, X_t)dW_t \qquad (20.0.4)$$

and the related forward Kolmogorov operator $\mathscr{L} = \mathscr{A}_t^* - \partial_t$ is, at least formally, of the form (20.0.1) with

$$b_j := -\beta_j + \sum_{i=1}^{N} \partial_{x_i} c_{ij}, \qquad a := -\sum_{i=1}^{N} \partial_{x_i} \beta_i + \frac{1}{2} \sum_{i,j=1}^{N} \partial_{x_i x_j} c_{ij}. \qquad (20.0.5)$$

Note that in a *forward* operator, the time derivative appears with a negative sign: as already mentioned in Sect. 2.5.2, this type of operators typically intervene in physics in the description of phenomena that evolve over time, such as heat diffusion in a body. On the other hand, every forward operator can be transformed into a parabolic *backward*[2] operator with the simple change of variables $s = T - t$: it follows that *all the results we prove in this chapter for forward operators admit an analogous backward formulation.*

---

[2] In which the time derivative appears with a positive sign.

20.1 Uniqueness: The Maximum Principle

We study the uniqueness of the solution to the Cauchy problem

$$\begin{cases} \mathscr{L}u(t,x) = 0, & (t,x) \in \mathscr{S}_T, \\ u(0,x) = \varphi(x), & x \in \mathbb{R}^N, \end{cases} \tag{20.1.1}$$

for the operator $\mathscr{L}$ in (20.0.1). A classic example due to Tychonoff [141] shows that the problem (20.1.1) for the heat operator admits infinite solutions: in fact, in addition to the identically zero solution, also the functions of the type

$$u_\alpha(t,x) := \sum_{k=0}^{\infty} \frac{x^{2k}}{(2k)!} \partial_t^k e^{-\frac{1}{t^\alpha}}, \qquad \alpha > 1, \tag{20.1.2}$$

are classical solutions to the Cauchy problem

$$\begin{cases} \frac{1}{2}\partial_{xx}u_\alpha - \partial_t u_\alpha = 0, & \text{in } \mathbb{R}_{>0} \times \mathbb{R}, \\ u_\alpha(0,\cdot) = 0, & \text{in } \mathbb{R}. \end{cases}$$

However, the solutions in (20.1.2) are in some sense "pathological": they oscillate, changing sign infinitely many times and have very rapid growth as $|x| \to \infty$. In light of Tychonoff's example, the study of the uniqueness of the solution of the problem (20.1.1) consists in determining suitable classes of functions, called *uniqueness classes for* $\mathscr{L}$, within which the solution, if it exists, is unique. In this section, we assume the following minimal hypotheses on the coefficients of $\mathscr{L}$ in (20.0.1):

Assumption 20.1.1

(i) For each $i,j = 1,\ldots,N$, the coefficients c_{ij}, b_i and a are real-valued measurable functions;
(ii) the matrix $\mathscr{C}(t,x) := (c_{ij}(t,x))$ is symmetric and positive semidefinite for every $(t,x) \in \mathscr{S}_T$;
(iii) the coefficient a is upper bounded: there exists $a_0 \in \mathbb{R}$ such that

$$a(t,x) \leq a_0, \qquad (t,x) \in \mathscr{S}_T.$$

We will prove that a uniqueness class is given by functions that do not grow too rapidly to infinity in the sense that they satisfy the estimate

$$|u(t,x)| \leq Ce^{C|x|^2}, \qquad (t,x) \in \mathscr{S}_T, \tag{20.1.3}$$

for some positive constant C. This result, contained in Theorem 20.1.8, is proven under very general conditions, namely Assumption 20.1.1 and the following

Assumption 20.1.2 There exists a constant M such that

$$|c_{ij}(t,x)| \leq M, \quad |b_i(t,x)| \leq M(1+|x|), \quad |a(t,x)| \leq M(1+|x|^2),$$
$$(t,x) \in \mathscr{S}_T, \ i,j = 1,\ldots,N.$$

It is possible to determine another uniqueness class by imposing other growth conditions on the coefficients.

Assumption 20.1.3 There exists a constant M such that

$$|c_{ij}(t,x)| \leq M(1+|x|^2), \quad |b_i(t,x)| \leq M(1+|x|), \quad |a(t,x)| \leq M,$$
$$(t,x) \in \mathscr{S}_T, \ i,j = 1,\ldots,N.$$

Theorem 20.1.10 shows that, under Assumptions 20.1.1 and 20.1.3, a uniqueness class is given by functions with at most polynomial growth, which satisfy an estimate of the type

$$|u(t,x)| \leq C(1+|x|^p), \qquad (t,x) \in \mathscr{S}_T, \tag{20.1.4}$$

for some positive constants C and p.

We explicitly note that the previous assumptions are so weak that they do not generally guarantee the *existence* of the solution.

20.1.1 Cauchy-Dirichlet Problem

In this section we study the operator $\mathscr{L}$ in (20.0.1) on a "cylinder" of the form

$$D_T =]0,T[\times D$$

where D is a *bounded* domain (open and connected set) of $\mathbb{R}^N$. We denote by ∂D the boundary of D and say that

$$\partial_p D_T := \underbrace{(\{0\} \times D)}_{\text{base}} \cup \underbrace{([0,T[\times \partial D)}_{\text{lateral boundary}}$$

is the *parabolic boundary* of D_T. As before, $C^{1,2}(D_T)$ is the space of functions defined on D_T, that are continuously differentiable with respect to t and twice continuously differentiable with respect to x.

20.1 Uniqueness: The Maximum Principle

Definition 20.1.4 (Cauchy-Dirichlet Problem) A classical solution of the Cauchy-Dirichlet problem for $\mathscr{L}$ on D_T is a function $u \in C^{1,2}(D_T) \cap C(D_T \cup \partial_p D_T)$ such that

$$\begin{cases} \mathscr{L}u = f, & \text{in } D_T, \\ u = \varphi, & \text{in } \partial_p D_T, \end{cases} \tag{20.1.5}$$

where $f \in C(D_T)$ and $\varphi \in C(\partial_p D_T)$ are given functions, called respectively *inhomogeneous term* and *boundary datum* of the problem.

The main result of the section, which subsequently leads to the uniqueness of the classical solution to problem (20.1.5) (cf. Corollary 20.1.6), is the following

Theorem 20.1.5 (Weak Maximum Principle) *Under Assumption 20.1.1, if $u \in C^{1,2}(D_T) \cap C(D_T \cup \partial_p D_T)$ is such that $\mathscr{L}u \geq 0$ in D_T and $u \leq 0$ on $\partial_p D_T$, then $u \leq 0$ on D_T.*

Proof First, we observe that it is not restrictive to take $a_0 < 0$ in Assumption 20.1.1. If it were not, it would be enough to prove the thesis for the function

$$u_\lambda(t,x) := e^{-\lambda t} u(t,x) \tag{20.1.6}$$

which satisfies

$$\mathscr{L}u_\lambda - \lambda u_\lambda = e^{-\lambda t} \mathscr{L}u, \tag{20.1.7}$$

choosing $\lambda > a_0$.

Now we proceed by contradiction. Denying the thesis, there would exist a point $(t,x) \in D_T$ such that $u(t,x) > 0$: in fact, we can also assume that

$$u(t,x) = \max_{[0,t] \times D} u.$$

It follows that

$$\mathscr{H}u(t,x) := (\partial_{x_i x_j} u(t,x)) \leq 0, \qquad \partial_{x_j} u(t,x) = 0, \qquad \partial_t u(t,x) \geq 0,$$

for every $j = 1, \ldots, N$. Then there exists a symmetric and positive semi-definite matrix $M = (m_{ij})$ such that

$$\mathscr{H}u(t,x) = -M^2 = \left(-\sum_{h=1}^{N} m_{ih} m_{jh}\right)_{i,j}$$

and therefore

$$\mathscr{L}u(t,x) = -\frac{1}{2}\sum_{i,j=1}^{N} c_{ij}(t,x) \sum_{h=1}^{N} m_{ih} m_{jh} + \sum_{j=1}^{N} b_j(t,x)\partial_{x_j} u(t,x)$$
$$+ a(t,x)u(t,x) - \partial_t u(t,x)$$
$$= -\frac{1}{2}\underbrace{\sum_{h=1}^{N}\sum_{i,j=1}^{N} c_{ij}(t,x)m_{ih}m_{jh}}_{\geq 0 \text{ since } \mathscr{C}=(c_{ij})\geq 0} + a(t,x)u(t,x) - \partial_t u(t,x)$$
$$\leq a(t,x)u(t,x) < 0,$$

and this contradicts the assumption $\mathscr{L}u \geq 0$ in D_T. $\square$

Corollary 20.1.6 (Comparison Principle) *Under Assumption 20.1.1, let $u, v \in C^{1,2}(D_T) \cap C(D_T \cup \partial_p D_T)$ be such that $\mathscr{L}u \leq \mathscr{L}v$ in D_T and $u \geq v$ on $\partial_p D_T$. Then $u \geq v$ in D_T. In particular, if it exists, the classical solution of the Cauchy-Dirichlet problem (20.1.5) is unique.*

Proof It is enough to apply the weak maximum principle to the function $v - u$. $\square$

The following useful result provides an estimate of the maximum of the solution of the Cauchy-Dirichlet problem (20.1.5) in terms of f and the boundary datum φ.

Theorem 20.1.7 *If the operator $\mathscr{L}$ satisfies Assumption 20.1.1 then for every $u \in C^{1,2}(D_T) \cap C(D_T \cup \partial_p D_T)$ we have*

$$\sup_{D_T} |u| \leq e^{a_0^+ T}\left(\sup_{\partial_p D_T} |u| + T \sup_{D_T} |\mathscr{L}u|\right), \qquad a_0^+ := \max\{0, a_0\}. \qquad (20.1.8)$$

Proof Consider first the case $a_0 \leq 0$ and therefore $a_0^+ = 0$. Suppose that u and $\mathscr{L}u$ are bounded respectively on $\partial_p D_T$ and D_T, otherwise there is nothing to prove. Letting

$$w(t) = \sup_{\partial_p D_T} |u| + t \sup_{D_T} |\mathscr{L}u|, \qquad t \in [0, T],$$

we have

$$\mathscr{L}w = aw - \sup_{D_T} |\mathscr{L}u| \leq \mathscr{L}u, \qquad \mathscr{L}(-w) = -aw + \sup_{D_T} |\mathscr{L}u| \geq \mathscr{L}u,$$

and $-w \leq u \leq w$ on $\partial_p D_T$. Then estimate (20.1.8) follows from the comparison principle, Corollary 20.1.6.

20.1 Uniqueness: The Maximum Principle

Let now $a_0 > 0$. Consider u_λ in (20.1.6) with $\lambda = a_0$: as just proved, we have

$$\sup_{D_T} |u_\lambda| \leq \sup_{\partial_p D_T} |u_\lambda| + T \sup_{D_T} |(\mathscr{L} - a_0)u_\lambda|.$$

Then, since $a_0 > 0$, we obtain

$$e^{-a_0 T} \sup_{D_T} |u| \leq \sup_{(t,x) \in D_T} |e^{-a_0 t} u(t,x)| \leq \sup_{\partial_p D_T} |u_\lambda| + T \sup_{D_T} |(\mathscr{L} - a_0)u_\lambda| \leq$$

(by (20.1.7))

$$\leq \sup_{(t,x) \in \partial_p D_T} |e^{-a_0 t} u(t,x)| + T \sup_{(t,x) \in D_T} |e^{-a_0 t} \mathscr{L}u(t,x)| \leq$$

(since $a_0 > 0$)

$$\leq \sup_{\partial_p D_T} |u| + T \sup_{D_T} |\mathscr{L}u|,$$

which proves the thesis. □

20.1.2 Cauchy Problem

We establish analogous results to those presented in the preceding section for the Cauchy problem (20.1.1).

Theorem 20.1.8 (Weak Maximum Principle) *Let Assumptions 20.1.1 and 20.1.2 be in force. If $u \in C^{1,2}(\mathscr{S}_T) \cap C([0, T[\times \mathbb{R}^N)$ is such that*

$$\begin{cases} \mathscr{L}u \leq 0, & \text{in } \mathscr{S}_T, \\ u(0, \cdot) \geq 0, & \text{in } \mathbb{R}^N, \end{cases} \qquad (20.1.9)$$

and verifies the estimate

$$u(t,x) \geq -Ce^{C|x|^2}, \qquad (t,x) \in [0, T[\times \mathbb{R}^N, \qquad (20.1.10)$$

for a positive constant C, then $u \geq 0$ in $[0, T[\times \mathbb{R}^N$. Consequently, there exists at most one classical solution of the Cauchy problem (20.1.1) that verifies the exponential growth estimate (20.1.3).

We explicitly note that Assumptions 20.1.1 and 20.1.2 are very mild, so as to include, for example, the case when $\mathscr{L}$ is a first-order operator. We first prove the following

Lemma 20.1.9 *Under Assumption 20.1.1, if $u \in C^{1,2}(\mathscr{S}_T) \cap C([0, T[\times \mathbb{R}^N)$ verifies (20.1.9) and is such that*

$$\liminf_{|x|\to\infty} \inf_{t\in]0,T[} u(t,x) \geq 0, \qquad (20.1.11)$$

then $u \geq 0$ on $[0, T[\times \mathbb{R}^N$.

Proof As in the proof of Theorem 20.1.5, it is not restrictive to assume $a_0 < 0$ so that, for every $\varepsilon > 0$, we have

$$\begin{cases} \mathscr{L}(u+\varepsilon) \leq 0, & \text{in } \mathscr{S}_T, \\ u(0,\cdot) + \varepsilon > 0, & \text{in } \mathbb{R}^N. \end{cases}$$

Fix $(t_0, x_0) \in \mathscr{S}_T$. Thanks to condition (20.1.11), there exists $R > |x_0|$ such that

$$u(t,x) + \varepsilon > 0, \qquad t \in]0, T[, \ |x| = R,$$

and from the weak maximum principle of Theorem 20.1.5, applied on the cylinder

$$D_T =]0, T[\times \{|x| < R\},$$

it follows that $u(t_0, x_0) + \varepsilon \geq 0$. Given the arbitrariness of ε, we also have $u(t_0, x_0) \geq 0$. □

Proof of Theorem 20.1.8 We prove that $u \geq 0$ on a strip $\mathscr{S}_{T_0}$ with $T_0 > 0$ that depends only on the constant M of Assumption 20.1.2 and on the constant C in (20.1.10): if necessary, we just need to apply this result repeatedly to prove the thesis on the strip $\mathscr{S}_T$.

First of all, to understand the general idea, we give the proof in the particular case of the heat operator

$$\mathscr{L} = \frac{1}{2}\Delta - \partial_t.$$

Fixed $\gamma > C$, let $T_0 = \frac{1}{4\gamma}$ and consider the function

$$v(t,x) := \frac{1}{(1-2\gamma t)^{\frac{N}{2}}} \exp\left(\frac{\gamma |x|^2}{1-2\gamma t}\right), \qquad (t,x) \in [0, T_0[\times \mathbb{R}^N,$$

20.1 Uniqueness: The Maximum Principle

such that

$$\mathscr{L}v(t, x) = 0 \quad \text{and} \quad v(t, x) \geq e^{\gamma |x|^2}.$$

From Lemma 20.1.9 we deduce that $u + \varepsilon v \geq 0$ for every $\varepsilon > 0$, which proves the thesis.

The general case is only technically more complicated and exploits Assumption 20.1.2 on the coefficients of the operator. Fixed $\gamma > C$ and two constants $\alpha, \beta \in \mathbb{R}$ that we will determine later, consider the function

$$v(t, x) = \exp\left(\frac{\gamma |x|^2}{1 - \alpha t} + \beta t\right), \quad 0 \leq t \leq \frac{1}{2\alpha}, \; x \in \mathbb{R}^N.$$

Since

$$\frac{\mathscr{L}v}{v} = \frac{2\gamma^2}{(1 - \alpha t)^2} \langle \mathscr{C}x, x \rangle + \frac{\gamma}{1 - \alpha t} \operatorname{tr} \mathscr{C} + \frac{2\gamma}{1 - \alpha t} \sum_{i=1}^{N} b_i x_i + a - \frac{\alpha \gamma |x|^2}{(1 - \alpha t)^2} - \beta,$$

by Assumption 20.1.2 it is possible to choose α, β large enough so that

$$\frac{\mathscr{L}v}{v} \leq 0. \tag{20.1.12}$$

Letting $w := \frac{u}{v}$, by condition (20.1.10), we have

$$\liminf_{|x| \to \infty} \left(\inf_{0 \leq t \leq \frac{1}{2\alpha}} w(t, x)\right) \geq 0,$$

and w satisfies the equation

$$\frac{1}{2} \sum_{i,j=1}^{N} c_{ij} \partial_{x_i x_j} w + \sum_{i=1}^{N} \hat{b}_i \partial_{x_i} w + \hat{a} w - \partial_t w = \frac{\mathscr{L}u}{v} \leq 0,$$

where

$$\hat{b}_i = b_i + \sum_{j=1}^{N} c_{ij} \frac{\partial_{x_j} v}{v}, \quad \hat{a} = \frac{\mathscr{L}v}{v}.$$

Since $\hat{a} \leq 0$ by (20.1.12), we can apply Lemma 20.1.9 to conclude that w (and thus also u) is non-negative. □

Theorem 20.1.10 (Weak Maximum Principle) *Assume Assumptions 20.1.1 and 20.1.3. If $u \in C^{1,2}(\mathscr{S}_T) \cap C([0, T[\times \mathbb{R}^N)$ satisfies (20.1.9) and the estimate*

$$u(t, x) \geq -C(1 + |x|^p), \qquad (t, x) \in [0, T[\times \mathbb{R}^N, \qquad (20.1.13)$$

for some positive constants C and p, then $u \geq 0$ in $[0, T[\times \mathbb{R}^N$. Consequently, there exists at most one classical solution of the Cauchy problem (20.1.1) that satisfies the polynomial growth estimate (20.1.4) at infinity.

Proof We only prove the case $a_0 < 0$. Consider the function

$$v(t, x) = e^{\alpha t} \left(\kappa t + |x|^2\right)^q$$

and verify that for every $q > 0$ it is possible to choose α, κ such that $\mathscr{L}v < 0$ on $\mathscr{S}_T$. Then for $p < 2q$ and for every $\varepsilon > 0$ we have $\mathscr{L}(u + \varepsilon v) < 0$ on $\mathscr{S}_T$ and, thanks to condition (20.1.13), we can apply Lemma 20.1.9 to deduce that $u + \varepsilon v \geq 0$ on $\mathscr{S}_T$. The thesis follows from the arbitrariness of ε. □

We now prove the analogue of Theorem 20.1.7: the following result provides estimates, in L^∞ norm, of the dependence of the solution in terms of the initial datum and the inhomogeneous term. These estimates play a crucial role, for example, in the proof of the stability of numerical schemes.

Theorem 20.1.11 *If the operator $\mathscr{L}$ satisfies Assumptions 20.1.1 and 20.1.2, then for every $u \in C^{1,2}(\mathscr{S}_T) \cap C([0, T[\times \mathbb{R}^N)$ that satisfies the exponential growth estimate (20.1.3) we have*

$$\sup_{[0,T[\times \mathbb{R}^N} |u| \leq e^{-a_0^+ T} \left(\sup_{\mathbb{R}^N} |u(0, \cdot)| + T \sup_{\mathscr{S}_T} |\mathscr{L}u| \right), \qquad a_0^+ := \max\{0, a_0\}.$$

Proof If $a_0 < 0$ then, let

$$w_\pm = \sup_{\mathbb{R}^N} |u(0, \cdot)| + t \sup_{\mathscr{S}_T} |\mathscr{L}u| \pm u,$$

we have

$$\begin{cases} \mathscr{L}w_\pm = a \sup |u(0, \cdot)| - \sup_{\mathscr{S}_T} |\mathscr{L}u| \pm \mathscr{L}u \leq 0, & \text{in } \mathscr{S}_T, \\ w_\pm(0, \cdot) \geq 0, & \text{in } \mathbb{R}^N, \end{cases}$$

and clearly $w_\pm$ satisfies the estimate (20.1.10). It follows from Theorem 20.1.8 that $w_\pm \geq 0$ in $\mathscr{S}_T$ and this proves the thesis. On the other hand, if $a_0 \geq 0$ then it is enough to proceed as in the proof of Theorem 20.1.7. □

20.2 Existence: The Fundamental Solution

In this section, we give an existence result of classical solutions of the Cauchy problem for the operator $\mathscr{L}$ in (20.0.1). The central concept in this regard is that of fundamental solution.

Definition 20.2.1 (Fundamental Solution) A fundamental solution for the operator $\mathscr{L}$ on $\mathscr{S}_T \equiv \,]0, T[\times\mathbb{R}^N$ is a function $\Gamma = \Gamma(t_0, x_0; t, x)$, with $0 \leq t_0 < t < T$ and $x_0, x \in \mathbb{R}^N$, such that for every $\varphi \in bC(\mathbb{R}^N)$ the function defined by

$$u(t, x) = \int_{\mathbb{R}^N} \varphi(x_0)\Gamma(t_0, x_0; t, x)dx_0, \qquad t_0 < t < T,\ x \in \mathbb{R}^N, \qquad (20.2.1)$$

and by $u(t_0, \cdot) = \varphi$, is a classical solution (i.e., $u \in C^{1,2}(]t_0, T[\times\mathbb{R}^N) \cap C([t_0, T[\times\mathbb{R}^N))$ of the Cauchy problem

$$\begin{cases} \mathscr{L}u = 0 & \text{in }]t_0, T[\times\mathbb{R}^N, \\ u(t_0, \cdot) = \varphi & \text{in } \mathbb{R}^N. \end{cases} \qquad (20.2.2)$$

A well-known technique for proving the existence of the fundamental solution is the *parametrix method* introduced by E.E. Levi in [89] and then developed by many other authors.[3] It is a fairly long and complex constructive procedure based on the following[4] Assumption 20.2.2 on the operator $\mathscr{L}$. We recall the definition of the space bC_T^α with the norm defined in (18.2.1): in particular, we emphasize that the functions in bC_T^α are Hölder continuous solely with respect to the spatial variables.

Assumption 20.2.2

(i) $c_{ij}, b_i, a \in bC_T^\alpha$ for some $\alpha \in\,]0, 1]$ and for each $i, j = 1, \ldots, N$;
(ii) the matrix $\mathscr{C} := (c_{ij})_{1 \leq i,j \leq N}$ is symmetric and satisfies the following *uniform parabolicity* condition: there exists a constant $\lambda_0 > 1$ such that

$$\frac{1}{\lambda_0}|\eta|^2 \leq \langle\mathscr{C}(t,x)\eta, \eta\rangle \leq \lambda_0|\eta|^2, \qquad (t,x) \in \mathscr{S}_T,\ \eta \in \mathbb{R}^N. \qquad (20.2.3)$$

For convenience, we assume λ_0 large enough so that $[c_{ij}]_\alpha, [b_i]_\alpha, [a]_\alpha \leq \lambda_0$ for each $i, j = 1, \ldots, N$.

---

[3] See, for example, the works of Pogorzelski [118] and Aronson [5] on the construction of the fundamental solution. The book by Friedman [50] is still a classic reference text for the parametrix method and the main source that inspired our presentation.

[4] It is possible to assume slightly weaker hypotheses: in this regard, see Section 6.4 in [50]. In particular, the continuity condition in time is only for convenience: the results of this section extend without difficulty to the case of coefficients that are measurable in t; in this case, the PDE is understood in an integro-differential sense, as in (20.2.5).

Remark 20.2.3 Let

$$\mathscr{A} := \frac{1}{2}\sum_{i,j=1}^{N} c_{ij}(t,x)\partial_{x_i x_j} + \sum_{j=1}^{N} b_j(t,x)\partial_{x_j} + a(t,x) \qquad (20.2.4)$$

so that $\mathscr{L} = \mathscr{A} - \partial_t$. Under Assumption 20.2.2, the following statements are equivalent:

(i) $u \in C^{1,2}(]t_0, T[\times \mathbb{R}^N)$ is a classical solution of the equation $\mathscr{L}u = 0$ on $]t_0, T[\times \mathbb{R}^N$;
(ii) $u \in C(]t_0, T[\times \mathbb{R}^N)$, is twice continuously differentiable with respect to x and satisfies the integro-differential equation

$$u(t,x) = u(t_1, x) + \int_{t_1}^{t} \mathscr{A}u(s,x)ds, \qquad t_0 < t_1 < t < T,\ x \in \mathbb{R}^N. \qquad (20.2.5)$$

In the following theorem, we consider the Cauchy problem with inhomogeneous term f satisfying growth and local Hölder continuity conditions.

Assumption 20.2.4 $f \in C(]t_0, T[\times \mathbb{R}^N)$ and there exists $\beta > 0$ such that:

(i)

$$|f(t,x)| \leq \frac{c_1 e^{c_2|x|^2}}{(t-t_0)^{1-\beta}}, \qquad (t,x) \in]t_0, T[\times \mathbb{R}^N, \qquad (20.2.6)$$

where c_1, c_2 are positive constants with $c_2 < \frac{1}{4\lambda_0(T-t_0)}$;
(ii) for every $n \in \mathbb{N}$, there exists a constant κ_n such that

$$|f(t,x) - f(t,y)| \leq \kappa_n \frac{|x-y|^\beta}{(t-t_0)^{1-\frac{\beta}{2}}}, \qquad t_0 < t < T,\ |x|, |y| \leq n. \qquad (20.2.7)$$

The main result of the chapter is the following

Theorem 20.2.5 (Fundamental Solution [!!!]) *Under Assumption 20.2.2, there exists a fundamental solution Γ for $\mathscr{A} - \partial_t$ in $\mathscr{S}_T$. Moreover:*

(i) $\Gamma = \Gamma(t_0, x_0; t, x)$ *is a continuous function of* (t_0, x_0, t, x) *for* $0 \leq t_0 < t < T$ *and* $x, x_0 \in \mathbb{R}^N$. *For every* $(t_0, x_0) \in [0, T[\times \mathbb{R}^N,\ \Gamma(t_0, x_0; \cdot, \cdot) \in C^{1,2}(]t_0, T[\times \mathbb{R}^N)$ *and the following Gaussian estimates hold: for every* $\lambda >$

20.2 Existence: The Fundamental Solution

λ_0, where λ_0 is the constant of Assumption 20.2.2, there exists a positive constant $c = c(T, N, \lambda, \lambda_0, \alpha)$ such that

$$\Gamma(t_0, x_0; t, x) \leq c\, \mathbf{G}\left(\lambda(t - t_0), x - x_0\right),$$
(20.2.8)

$$\left|\partial_{x_i} \Gamma(t_0, x_0; t, x)\right| \leq \frac{c}{\sqrt{t - t_0}} \mathbf{G}\left(\lambda(t - t_0), x - x_0\right),$$
(20.2.9)

$$\left|\partial_{x_i x_j} \Gamma(t_0, x_0; t, x)\right| + \left|\partial_t \Gamma(t_0, x_0; t, x)\right| \leq \frac{c}{t - t_0} \mathbf{G}\left(\lambda(t - t_0), x - x_0\right)$$
(20.2.10)

for every $(t, x) \in {]t_0, T[} \times \mathbb{R}^N$, where $\mathbf{G}$ is the Gaussian function in (20.3.1). Furthermore, there exist two positive constants $\bar{\lambda}, \bar{c}$, only dependent on T, N, λ_0, α, such that

$$\Gamma(t_0, x_0; t, x) \geq \bar{c}\, \mathbf{G}\left(\bar{\lambda}(t - t_0), x - x_0\right)$$
(20.2.11)

for every $(t, x) \in {]t_0, T[} \times \mathbb{R}^N$;

(ii) for every f satisfying Assumption 20.2.4 and $\varphi \in bC(\mathbb{R}^N)$, the function defined by

$$u(t, x) = \int_{\mathbb{R}^N} \varphi(x_0) \Gamma(t_0, x_0; t, x) dx_0$$
$$- \int_{t_0}^{t} \int_{\mathbb{R}^N} f(s, y) \Gamma(s, y; t, x) dy ds, \qquad t_0 < t < T,\ x \in \mathbb{R}^N,$$
(20.2.12)

and by $u(t_0, \cdot) = \varphi$, is a classical solution of the Cauchy problem

$$\begin{cases} \mathscr{L} u = f & \text{in }]t_0, T[\times \mathbb{R}^N, \\ u(t_0, \cdot) = \varphi & \text{in } \mathbb{R}^N. \end{cases}$$
(20.2.13)

Formula (20.2.12) is usually called[5] Duhamel's formula;

(iii) the Chapman-Kolmogorov equation holds

$$\Gamma(t_0, x_0; t, x)$$
$$= \int_{\mathbb{R}^N} \Gamma(t_0, x_0; s, y) \Gamma(s, y; t, x) dy, \qquad 0 \leq t_0 < s < t < T,\ x, x_0 \in \mathbb{R}^N;$$

---

[5] Duhamel's formula can be interpreted as a "forward version" of the Feynman-Kac formula (15.4.6).

(iv) *if the coefficient a is constant, then*

$$\int_{\mathbb{R}^N} \Gamma(t_0, x_0; t, x) dx_0 = e^{a(t-t_0)}, \qquad t \in]t_0, T[, \ x \in \mathbb{R}^N, \qquad (20.2.14)$$

and in particular, if $a \equiv 0$, *then* $\Gamma(t_0, \cdot; t, x)$ *is a density.*

The proof of Theorem 20.2.5 is deferred to Sect. 20.3 along with several preliminary results.

Notation 20.2.6 Let $\alpha \in]0, 1]$. We denote by $bC^\alpha(\mathbb{R}^N)$ the space of bounded, α-Hölder continuous functions on $\mathbb{R}^N$, equipped with the norm

$$\|\varphi\|_{bC^\alpha(\mathbb{R}^N)} := \sup_{\mathbb{R}^N} |\varphi| + \sup_{x \neq y} \frac{|\varphi(x) - \varphi(y)|}{|x - y|^\alpha}.$$

The following result shows that estimate (20.2.10) can be refined in the sense that the non-integrable singularity $\frac{1}{t-t_0}$ can be replaced by an integrable one when the initial datum is Hölder continuous.

Corollary 20.2.7 *Under the assumptions of Theorem 20.2.5, consider the solution* u *in* (20.2.12) *of the Cauchy problem* (20.2.13) *with* $a = f = 0$. *If* $\varphi \in bC^\delta(\mathbb{R}^N)$ *for some* $\delta > 0$, *then there exists a constant* c, *which depends only on* $T, N, \delta, \alpha, \lambda_0, [c_{ij}]_\alpha$ *and* $[b_i]_\alpha$, *such that*

$$|D_x^k u(t, x)| \leq \frac{c}{(t - t_0)^{\frac{k-\delta}{2}}} \|\varphi\|_{bC^\delta(\mathbb{R}^N)}, \qquad t > t_0, \ x \in \mathbb{R}^N, \ k = 0, 1, 2, \qquad (20.2.15)$$

where D_x^k *denotes a derivative of order* k *in the variables* $x_1, \ldots, x_N$.

Proof We give the proof for $k = 2$ as the other cases are analogous and simpler. Since

$$\int_{\mathbb{R}^N} \Gamma(t_0, x_0; t, x) dx_0 = 1, \qquad t_0 < t, \ x \in \mathbb{R}^N,$$

we have

$$0 = \partial_{x_i x_j} \int_{\mathbb{R}^N} \Gamma(t_0, x_0; t, x) dx_0 = \int_{\mathbb{R}^N} \partial_{x_i x_j} \Gamma(t_0, x_0; t, x) dx_0.$$

Hence

$$|\partial_{x_i x_j} u(t, x)| = \left| \int_{\mathbb{R}^N} \partial_{x_i x_j} \Gamma(t_0, x_0; t, x)(\varphi(x_0) - \varphi(x)) dy \right| \leq \qquad (20.2.16)$$

(by the triangle inequality and the Gaussian estimate (20.2.10))

$$\leq \frac{c}{t-t_0} \int_{\mathbb{R}^N} \mathbf{G}(\lambda(t-t_0), x-x_0)|\varphi(x_0) - \varphi(x)|dx_0 \leq \qquad (20.2.17)$$

(by the Hölder assumption on φ)

$$\leq \frac{c\|\varphi\|_{bC^\delta(\mathbb{R}^N)}}{(t-t_0)^{1-\frac{\delta}{2}}} \int_{\mathbb{R}^N} \left(\frac{|x-x_0|}{\sqrt{t-t_0}}\right)^\delta \mathbf{G}(\lambda(t-t_0), x-x_0)dx_0 \qquad (20.2.18)$$

and the conclusion follows thanks to the elementary estimates of Lemma 20.3.4. □

20.3 The Parametrix Method

This section is dedicated to the proof of Theorem 20.2.5. We consider $\mathscr{L}$ in (20.0.1) and assume that it verifies Assumption 20.2.2. The main idea of the parametrix method is to construct a fundamental solution through successive approximations; the first approximation term is referred to as the "parametrix", which is essentially the Gaussian fundamental solution of a heat operator obtained from $\mathscr{L}$ by freezing the coefficients in the spatial variables, while leaving the time variable free.

Notation 20.3.1 Given a constant $N \times N$, symmetric and positive definite matrix C, we set

$$\mathbf{G}(C, x) = \frac{1}{\sqrt{(2\pi)^N \det C}} e^{-\frac{1}{2}\langle C^{-1}x, x\rangle}, \qquad x \in \mathbb{R}^N.$$

Notice that

$$\frac{1}{2} \sum_{i,j=1}^N C_{ij} \partial_{x_i x_j} \mathbf{G}(tC, x) = \partial_t \mathbf{G}(tC, x), \qquad t > 0, \ x \in \mathbb{R}^N.$$

When C is the identity matrix, $C = I_N$, for simplicity we write

$$\mathbf{G}(t, x) \equiv \mathbf{G}(tI_N, x) = \frac{1}{(2\pi t)^{\frac{N}{2}}} e^{-\frac{|x|^2}{2t}}, \qquad t > 0, \ x \in \mathbb{R}^N, \qquad (20.3.1)$$

to indicate the usual standard Gaussian function, solution of the heat equation $\frac{1}{2}\Delta \mathbf{G}(t, x) = \partial_t \mathbf{G}(t, x)$.

Given $y \in \mathbb{R}^N$, we define the operator $\mathscr{L}_y$ as the result of computing the coefficients of $\mathscr{L}$ in y and removing terms of order lower than the second:

$$\mathscr{L}_y := \frac{1}{2} \sum_{i,j=1}^{N} c_{ij}(t, y) \partial_{x_i x_j} - \partial_t.$$

Operator $\mathscr{L}_y$ acts in the variables (t, x) and has coefficients that depend only on the time variable t, since y is fixed. Thanks to Assumption 20.2.2 and in particular to the fact that the matrix $\mathscr{C} = (c_{ij})$ is uniformly positive definite, we have that the fundamental solution of $\mathscr{L}_y$ has the following explicit expression

$$\Gamma_y(t_0, x_0; t, x) = \mathbf{G}(C_{t_0,t}(y), x - x_0), \qquad C_{t_0,t}(y) := \int_{t_0}^{t} \mathscr{C}(s, y) ds, \qquad (20.3.2)$$

for $0 \le t_0 < t < T$ and $x_0, x \in \mathbb{R}^N$.

We define the parametrix for $\mathscr{L}$ as

$$\mathbf{P}(t_0, x_0; t, x) := \Gamma_{x_0}(t_0, x_0; t, x), \qquad 0 \le t_0 < t < T, \ x_0, x \in \mathbb{R}^N. \qquad (20.3.3)$$

According to the parametrix method, the fundamental solution of $\mathscr{L}$ is sought in the form

$$\Gamma(t_0, x_0; t, x) = \mathbf{P}(t_0, x_0; t, x) + \int_{t_0}^{t} \int_{\mathbb{R}^N} \Phi(t_0, x_0; s, y) \mathbf{P}(s, y; t, x) dy ds$$

$$(20.3.4)$$

where Φ is an unknown function to be determined by imposing that[6] $\mathscr{L}\Gamma(t_0, x_0; t, x) = 0$. Formally, from (20.3.4) we have[7]

$$\mathscr{L}\Gamma(t_0, x_0; t, x) = \mathscr{L}\mathbf{P}(t_0, x_0; t, x) + \int_{t_0}^{t} \int_{\mathbb{R}^N} \Phi(t_0, x_0; s, y) \mathscr{L}\mathbf{P}(s, y; t, x) dy ds$$

$$- \Phi(t_0, x_0; t, x) \qquad (20.3.5)$$

---

[6] Remember that $\mathscr{L}$ acts in the variables (t, x).

[7] The last term in the right-hand side of (20.3.5) derives from applying ∂_t to integration limit of the outer integral in (20.3.4): we obtain

$$\int_{\mathbb{R}^N} \Phi(t_0, x_0; t, y) \mathbf{P}(t, y; t, x) dy = \Phi(t_0, x_0; t, x)$$

since formally $\mathbf{P}(t, y; t, x) dy = \delta_x(dy)$ where δ_x denotes the Dirac delta centered at x.

20.3 The Parametrix Method

which gives the equation for Φ

$$\Phi(t_0, x_0; t, x) = \mathscr{L}\mathbf{P}(t_0, x_0; t, x) + \int_{t_0}^{t} \int_{\mathbb{R}^N} \Phi(t_0, x_0; s, y) \mathscr{L}\mathbf{P}(s, y; t, x) dy ds \quad (20.3.6)$$

for $0 \leq t_0 < t < T$ and $x_0, x \in \mathbb{R}^N$. By successive approximations, we obtain

$$\Phi(t_0, x_0; t, x) = \sum_{k=1}^{\infty} (\mathscr{L}\mathbf{P})_k(t_0, x_0; t, x) \quad (20.3.7)$$

where

$$(\mathscr{L}\mathbf{P})_1(t_0, x_0; t, x) = \mathscr{L}\mathbf{P}(t_0, x_0; t, x),$$

$$(\mathscr{L}\mathbf{P})_{k+1}(t_0, x_0; t, x) = \int_{t_0}^{t} \int_{\mathbb{R}^N} (\mathscr{L}\mathbf{P})_k(t_0, x_0; s, y) \mathscr{L}\mathbf{P}(s, y; t, x) dy ds, \quad k \in \mathbb{N}. \quad (20.3.8)$$

In Sect. 20.3.2 we prove the following

Proposition 20.3.2 *The series in (20.3.7) converges and defines* $\Phi = \Phi(t_0, x_0; t, x)$ *which is a continuous function of* (t_0, x_0, t, x) *for* $0 \leq t_0 < t < T$ *and* $x, x_0 \in \mathbb{R}^N$, *and solves equation (20.3.6). Moreover, for every* $\lambda > \lambda_0$ *there exists a positive constant* $c = c(T, N, \lambda, \lambda_0)$ *such that*

$$|\Phi(t_0, x_0; t, x)| \leq \frac{c}{(t - t_0)^{1 - \frac{\alpha}{2}}} \mathbf{G}(\lambda(t - t_0), x - x_0), \quad (20.3.9)$$

$$|\Phi(t_0, x_0; t, x) - \Phi(t_0, x_0; t, y)| \leq \frac{c |x - y|^{\frac{\alpha}{2}}}{(t - t_0)^{1 - \frac{\alpha}{4}}} \Big(\mathbf{G}(\lambda(t - t_0), x - x_0) + \mathbf{G}(\lambda(t - t_0), y - x_0) \Big) \quad (20.3.10)$$

for every $0 \leq t_0 < t < T$ *and* $x, y, x_0 \in \mathbb{R}^N$.

20.3.1 Gaussian Estimates

In this section, we prove some preliminary estimates for Gaussian kernels.

Notation 20.3.3 We adopt Convention 14.4.3 to denote the dependence of constants. Moreover, for the sake of convenience, as we need to establish several estimates, we will use the symbol c to represent a generic constant whose value may vary from one line to another. When necessary, we will explicitly state the quantities on which c depends.

Lemma 20.3.4 *Let* $\mathbf{G}$ *be the Gaussian function in* (20.3.1). *For every* $p > 0$ *and* $\lambda_1 > \lambda_0$ *there exists a constant* $c = c(p, N, \lambda_1, \lambda_0)$ *such that*

$$\left(\frac{|x|}{\sqrt{t}}\right)^p \mathbf{G}(\lambda_0 t, x) \leq c\, \mathbf{G}(\lambda_1 t, x), \qquad t > 0,\ x \in \mathbb{R}^N.$$

Proof For simplicity, let $z = \frac{|x|}{\sqrt{t}}$, we have

$$z^p \mathbf{G}(\lambda_0 t, x) = \frac{z^p}{(2\pi \lambda_0 t)^{\frac{N}{2}}} \exp\left(-\frac{z^2}{2\lambda_0}\right) = \left(\frac{\lambda_1}{\lambda_0}\right)^N g(z) \mathbf{G}(\lambda_1 t, x)$$

where

$$g(z) := z^p e^{-\frac{\kappa z^2}{2}}, \qquad \kappa = \frac{1}{\lambda_0} - \frac{1}{\lambda_1} > 0, \qquad z \in \mathbb{R}_+,$$

reaches the global maximum in $z_0 = \sqrt{\frac{p}{\kappa}}$ where we have $g(z_0) = \left(\frac{p}{e\kappa}\right)^{\frac{p}{2}}$. $\square$

Lemma 20.3.5 *Consider* $\mathscr{L}$ *in* (20.0.1) *and assume that it verifies Assumption 20.2.2. For* $\mathbf{G}$ *and* Γ_y, *defined respectively in* (20.3.1) *and* (20.3.2), *we have*

$$\frac{1}{\lambda_0^N} \mathbf{G}\left(\frac{t-t_0}{\lambda_0}, x - x_0\right) \leq \Gamma_y(t_0, x_0; t, x) \leq \lambda_0^N \mathbf{G}\left(\lambda_0(t - t_0), x - x_0\right) \qquad (20.3.11)$$

for every $0 \leq t_0 < t < T$ *and* $x, x_0, y \in \mathbb{R}^N$, *where* λ_0 *is the constant of Assumption 20.2.2. Moreover, for every* $\lambda > \lambda_0$ *there exists a positive constant* $c = c(T, N, \lambda, \lambda_0)$ *such that*

$$\left|\partial_{x_i} \Gamma_y(t_0, x_0; t, x)\right| \leq \frac{c}{\sqrt{t - t_0}} \mathbf{G}\left(\lambda(t - t_0), x - x_0\right), \qquad (20.3.12)$$

$$\left|\partial_{x_i x_j} \Gamma_y(t_0, x_0; t, x)\right| \leq \frac{c}{t - t_0} \mathbf{G}\left(\lambda(t - t_0), x - x_0\right), \qquad (20.3.13)$$

$$\left|\partial_{x_i x_j x_k} \Gamma_y(t_0, x_0; t, x)\right| \leq \frac{c}{(t - t_0)^{3/2}} \mathbf{G}\left(\lambda(t - t_0), x - x_0\right), \qquad (20.3.14)$$

$$\left|\Gamma_y(t_0, x_0; t, x) - \Gamma_\eta(t_0, x_0; t, x)\right| \leq c|y - \eta|^\alpha \mathbf{G}\left(\lambda(t - t_0), x - x_0\right), \qquad (20.3.15)$$

20.3 The Parametrix Method

$$\left|\partial_{x_i}\Gamma_y(t_0, x_0; t, x) - \partial_{x_i}\Gamma_\eta(t_0, x_0; t, x)\right| \leq \frac{c|y-\eta|^\alpha}{\sqrt{t-t_0}} \mathbf{G}\left(\lambda(t-t_0), x-x_0\right),$$
(20.3.16)

$$\left|\partial_{x_ix_j}\Gamma_y(t_0, x_0; t, x) - \partial_{x_ix_j}\Gamma_\eta(t_0, x_0; t, x)\right| \leq \frac{c|y-\eta|^\alpha}{t-t_0} \mathbf{G}\left(\lambda(t-t_0), x-x_0\right),$$
(20.3.17)

for every $0 \leq t_0 < t < T$, $x, x_0, y, y_0 \in \mathbb{R}^N$ and $i, j, k = 1, \ldots, N$.

Proof By the definition of $C_{t_0,t}(y)$ in (20.3.2) and by the hypothesis of uniform parabolicity (20.2.3) we have

$$\frac{t-t_0}{\lambda_0}|y_0|^2 \leq \langle C_{t_0,t}(y)y_0, y_0\rangle \leq \lambda_0(t-t_0)|y_0|^2;$$
(20.3.18)

consequently, we have

$$\frac{|y_0|^2}{\lambda_0(t-t_0)} \leq \langle C_{t_0,t}^{-1}(y)y_0, y_0\rangle \leq \frac{\lambda_0|y_0|^2}{t-t_0}$$
(20.3.19)

and also

$$\left(\frac{t-t_0}{\lambda_0}\right)^N \leq \det C_{t_0,t}(y) \leq \lambda_0^N (t-t_0)^N.$$
(20.3.20)

Formula (20.3.19) follows from the fact that if A, B are symmetric and positive definite matrices, then the inequality between quadratic forms $A \leq B$ (i.e., $\langle Ay_0, y_0\rangle \leq \langle By_0, y_0\rangle$ for every $y_0 \in \mathbb{R}^N$) implies $B^{-1} \leq A^{-1}$. Formula (20.3.20) follows from the fact that the minimum and maximum eigenvalue of a symmetric matrix C are respectively $\min_{|y_0|=1}\langle Cy_0, y_0\rangle$ and $\max_{|y_0|=1}\langle Cy_0, y_0\rangle =: \|C\|$ where $\|C\|$ is the spectral norm of C. We note that (20.3.18)-(20.3.19) can be rewritten respectively in the form

$$\frac{t-t_0}{\lambda_0} \leq \|C_{t_0,t}(y)\| \leq \lambda_0(t-t_0), \qquad \frac{1}{\lambda_0(t-t_0)} \leq \|C_{t_0,t}^{-1}(y)\| \leq \frac{\lambda_0}{t-t_0}.$$
(20.3.21)

Estimates (20.3.11) then follow directly from the definition of $\Gamma_y(t_0, x_0; t, x)$. As for (20.3.12), letting $\nabla_x = (\partial_{x_1}, \ldots, \partial_{x_N})$, we have

$$\left|\nabla_x \Gamma_y(t_0, x_0; t, x)\right| = |C_{t_0,t}^{-1}(y)(x-x_0)|\Gamma_y(t_0, x_0; t, x)$$
$$\leq \|C_{t_0,t}^{-1}(y)\| \, |x-x_0|\Gamma_y(t_0, x_0; t, x) \leq$$

(by the second estimate in (20.3.21))

$$\leq \frac{\lambda_0}{\sqrt{t-t_0}} \left(\frac{|x-x_0|}{\sqrt{t-t_0}} \Gamma_y(t_0, x_0; t, x) \right) \leq$$

(by (20.3.11) and Lemma 20.3.4)

$$\leq \frac{c}{\sqrt{t-t_0}} G(\lambda(t-t_0), x-x_0).$$

Formulas (20.3.13) and (20.3.14) can be proven in a completely analogous way.

Using the explicit expression of Γ_y, (20.3.15) is a direct consequence of the following estimates:

$$\left| \frac{1}{\sqrt{\det C_{t_0,t}(y)}} - \frac{1}{\sqrt{\det C_{t_0,t}(\eta)}} \right| \leq \frac{c|y-\eta|^\alpha}{\sqrt{\det C_{t_0,t}(y)}}, \quad (20.3.22)$$

$$\left| e^{-\frac{1}{2}\langle C_{t_0,t}^{-1}(y)x,x\rangle} - e^{-\frac{1}{2}\langle C_{t_0,t}^{-1}(\eta)x,x\rangle} \right| \leq c|y-\eta|^\alpha e^{-\frac{|x|^2}{2\lambda(t-t_0)}}. \quad (20.3.23)$$

Regarding (20.3.22), we have

$$\left| \frac{1}{\sqrt{\det C_{t_0,t}(y)}} - \frac{1}{\sqrt{\det C_{t_0,t}(\eta)}} \right|$$

$$= \frac{1}{\sqrt{\det C_{t_0,t}(y)}} \frac{|\det C_{t_0,t}(y) - \det C_{t_0,t}(\eta)|}{\sqrt{\det C_{t_0,t}(\eta)} \left(\sqrt{\det C_{t_0,t}(y)} + \sqrt{\det C_{t_0,t}(\eta)} \right)} \leq$$

(by (20.3.20))

$$\leq \frac{\lambda_0^N}{\sqrt{\det C_{t_0,t}(y)}} \frac{|\det C_{t_0,t}(y) - \det C_{t_0,t}(\eta)|}{(t-t_0)^N}$$

$$= \frac{\lambda_0^N}{\sqrt{\det C_{t_0,t}(y)}} \left| \det \left(\frac{1}{t-t_0} C_{t_0,t}(y) \right) - \det \left(\frac{1}{t-t_0} C_{t_0,t}(y) \right) \right| \leq$$

(since $|\det A - \det B| \leq c\|A - B\|$ where $\|\cdot\|$ indicates the spectral norm and c is a constant that depends only on $\|A\|$, $\|B\|$ and the dimension of the matrices)

$$\leq \frac{c}{\sqrt{\det C_{t_0,t}(y)}} \left\| \frac{1}{t-t_0} \left(C_{t_0,t}(y) - C_{t_0,t}(\eta) \right) \right\|$$

20.3 The Parametrix Method

and (20.3.22) follows from Assumption 20.2.2, in particular from the Hölder condition on the coefficients c_{ij}. Regarding (20.3.23), by the mean value theorem and (20.3.19) we have

$$\left| e^{-\frac{1}{2}\langle C_{t_0,t}^{-1}(y)x,x\rangle} - e^{-\frac{1}{2}\langle C_{t_0,t}^{-1}(\eta)x,x\rangle} \right| \leq \left| \langle C_{t_0,t}^{-1}(y)x,x\rangle - \langle C_{t_0,t}^{-1}(\eta)x,x\rangle \right| e^{-\frac{|x|^2}{2\lambda_0(t-t_0)}}$$

$$\leq \| C_{t_0,t}^{-1}(y) - C_{t_0,t}^{-1}(\eta) \| \, |x|^2 e^{-\frac{|x|^2}{2\lambda_0(t-t_0)}} \leq$$

(by the identity $A^{-1} - B^{-1} = A^{-1}(B - A)B^{-1}$)

$$\leq c \| C_{t_0,t}^{-1}(y) \| \, \| C_{t_0,t}(y) - C_{t_0,t}(\eta) \| \, \| C_{t_0,t}^{-1}(\eta) \| \, |x|^2 e^{-\frac{|x|^2}{2\lambda_0(t-t_0)}} \leq$$

(by (20.3.21))

$$\leq c \left\| \frac{1}{t - t_0} \left(C_{t_0,t}(y) - C_{t_0,t}(\eta) \right) \right\| \frac{|x|^2}{t - t_0} e^{-\frac{|x|^2}{2\lambda_0(t-t_0)}} \leq$$

(by the assumption of Hölder continuity of the coefficients c_{ij} and by Lemma 20.3.4)

$$\leq c |y - \eta|^\alpha e^{-\frac{|x|^2}{2\lambda(t-t_0)}}$$

and this is sufficient to prove (20.3.23) and therefore (20.3.15).

The proof of the estimates (20.3.16) and (20.3.17) is analogous: for example, we have

$$\left| \nabla_x \Gamma_y(t_0, x_0; t, x) - \nabla_x \Gamma_\eta(t_0, x_0; t, x) \right|$$

$$= \left| C_{t_0,t}^{-1}(y)(x - x_0) \Gamma_y(t_0, x_0; t, x) - C_{t_0,t}^{-1}(\eta)(x - x_0) \Gamma_\eta(t_0, x_0; t, x) \right|$$

$$\leq \left| \left(C_{t_0,t}^{-1}(y) - C_{t_0,t}^{-1}(\eta) \right) (x - x_0) \right| \Gamma_y(t_0, x_0; t, x)$$

$$+ \left| C_{t_0,t}^{-1}(\eta)(x - x_0) \right| \left| \Gamma_y(t_0, x_0; t, x) - \Gamma_\eta(t_0, x_0; t, x) \right|$$

and the proof of (20.3.16) and (20.3.17) follows a similar line of reasoning as used previously. □

20.3.2 Proof of Proposition 20.3.2

Lemma 20.3.5 enables us to estimate the terms $(\mathscr{L}\mathbf{P})_k$ in (20.3.8) of the parametrix expansion.

Lemma 20.3.6 *For every $\lambda > \lambda_0$ there exists a positive constant $c = c(T, N, \lambda, \lambda_0)$ such that*

$$|(\mathscr{L}\mathbf{P})_k(t_0, x_0; t, x)| \leq \frac{\mathbf{m}_k}{(t-t_0)^{1-\frac{\alpha k}{2}}} \mathbf{G}(\lambda(t-t_0), x - x_0) \qquad (20.3.24)$$

for every $k \in \mathbb{N}$, $0 \leq t_0 < t < T$ and $x, x_0 \in \mathbb{R}^N$, where

$$\mathbf{m}_k = \frac{\left(c\Gamma_E\left(\frac{\alpha}{2}\right)\right)^k}{\Gamma_E\left(\frac{\alpha k}{2}\right)}$$

and Γ_E denotes the Euler Gamma function.

Proof First, we observe that by Assumption 20.2.2 we have

$$|c_{ij}(t, x) - c_{ij}(t, x_0)| \leq \lambda_0 |x - x_0|^\alpha, \qquad 0 \leq t < T, \ x, x_0 \in \mathbb{R}^N, \ i, j = 1, \ldots, N. \qquad (20.3.25)$$

For $k = 1$ we have

$$|\mathscr{L}\mathbf{P}(t_0, x_0; t, x)| = |(\mathscr{L} - \mathscr{L}_{x_0})\mathbf{P}(t_0, x_0; t, x)|$$

$$\leq \frac{1}{2} \sum_{i,j=1}^{N} |(c_{ij}(t, x) - c_{ij}(t, x_0)) \partial_{x_i x_j} \Gamma_{x_0}(t_0, x_0; t, x)|$$

$$+ \sum_{i=1}^{N} |b_i(t, x) \partial_{x_i} \Gamma_{x_0}(t_0, x_0; t, x)|$$

$$+ |a(t, x)| \Gamma_{x_0}(t_0, x_0; t, x).$$

The first term is the most delicate: by the estimates (20.3.25) and (20.3.13), for $\lambda' = \frac{\lambda_0 + \lambda}{2}$ we have

$$|(c_{ij}(t, x) - c_{ij}(t, x_0)) \partial_{x_i x_j} \Gamma_{x_0}(t_0, x_0; t, x)| \leq c \frac{|x - x_0|^\alpha}{t - t_0} \mathbf{G}(\lambda'(t-t_0), x - x_0) \leq$$

(by Lemma 20.3.4)

$$\leq \frac{c}{(t-t_0)^{1-\frac{\alpha}{2}}} \mathbf{G}(\lambda(t-t_0), x - x_0).$$

20.3 The Parametrix Method

The other terms are easily estimated using the boundedness hypothesis of the coefficients and estimate (20.3.12) of the first derivatives:

$$\left|b_i(t,x)\partial_{x_i}\Gamma_{x_0}(t_0,x_0;t,x)\right| + |a(t,x)|\Gamma_{x_0}(t_0,x_0;t,x)$$

$$\leq c\left(\frac{1}{\sqrt{t-t_0}}+1\right)G(\lambda(t-t_0),x-x_0).$$

This is sufficient to prove (20.3.24) in the case $k=1$.

Now we proceed by induction and, assuming the thesis is true for k, we prove it for $k+1$:

$$|(\mathscr{L}\mathbf{P})_{k+1}(t_0,x_0;t,x)| \leq \int_{t_0}^{t}\int_{\mathbb{R}^N}|(\mathscr{L}\mathbf{P})_k(t_0,x_0;s,y)|\,|\mathscr{L}\mathbf{P}(s,y;t,x)|\,dyds$$

$$\leq \int_{t_0}^{t}\frac{\mathbf{m}_k\mathbf{m}_1}{(s-t_0)^{1-\frac{\alpha k}{2}}(t-s)^{1-\frac{\alpha}{2}}}$$

$$\times \int_{\mathbb{R}^N}G(\lambda(s-t_0),y-x_0)G(\lambda(t-s),x-y)dyds =$$

(by the Chapman-Kolmogorov equation (2.4.4))

$$= G(\lambda(t-t_0),x-x_0)\int_{t_0}^{t}\frac{\mathbf{m}_k\mathbf{m}_1}{(s-t_0)^{1-\frac{\alpha k}{2}}(t-s)^{1-\frac{\alpha}{2}}}ds$$

and the thesis follows from the properties of Euler's Gamma function. □

Remark 20.3.7 The Chapman-Kolmogorov equation is a crucial tool in the parametrix method: it is proved by a direct calculation or, alternatively, as a consequence of the uniqueness result of Theorem 20.1.8. In fact, for $t_0 < s < t < T$ and $x, x_0, y \in \mathbb{R}^N$, we have that the functions $u_1(t,x) := G(t-t_0, x-x_0)$ and

$$u_2(t,x) = \int_{\mathbb{R}^N}G(s-t_0, y-x_0)G(t-s, x-y)dy$$

are both bounded solutions of the Cauchy problem

$$\begin{cases} \frac{1}{2}\Delta u - \partial_t u = 0 & \text{in }]s,T[\times\mathbb{R}^N, \\ u(s,y) = G(s-t_0, y-x_0) & \text{for } y \in \mathbb{R}^N, \end{cases}$$

and therefore they are equal.

Lemma 20.3.8 *Let $\kappa > 0$. Given $\kappa_1 \in]0, \kappa[$ there exists a positive constant c such that*

$$e^{-\kappa \frac{|\eta - x_0|^2}{t}} \leq c e^{-\kappa_1 \frac{|y - x_0|^2}{t}} \qquad (20.3.26)$$

for every $t > 0$ and $x_0, y, \eta \in \mathbb{R}^N$ such that $|y - \eta|^2 \leq t$.

Proof First of all, for every $\varepsilon > 0$ and $a, b \in \mathbb{R}$, the elementary inequalities hold

$$2|ab| \leq \varepsilon a^2 + \frac{b^2}{\varepsilon},$$

and

$$(a + b)^2 \leq (1 + \varepsilon) a^2 + \left(1 + \frac{1}{\varepsilon}\right) b^2.$$

Formula (20.3.26) follows from the fact that

$$\kappa_1 \frac{|y - x_0|^2}{t} - \kappa \frac{|\eta - x_0|^2}{t} \leq \kappa_1 \left(1 + \frac{1}{\varepsilon}\right) \frac{|y - \eta|^2}{t} + \frac{((1 + \varepsilon)\kappa_1 - \kappa)|\eta - x_0|^2}{t} \leq$$

(since $|y - \eta|^2 \leq t$ by hypothesis and for ε sufficiently small, being $\kappa_1 < \kappa$)

$$\kappa_1 \left(1 + \frac{1}{\varepsilon}\right).$$

□

Proof of Proposition 20.3.2 For every $\lambda > \lambda_0$ we have

$$|\Phi(t_0, x_0; t, x)| \leq \sum_{k=1}^{\infty} |(\mathscr{L}\mathbf{P})_k(t_0, x_0; t, x)| \leq$$

(by estimate (20.3.24))

$$\leq \sum_{k=1}^{\infty} \frac{\mathbf{m}_k}{(t - t_0)^{1 - \frac{\alpha k}{2}}} G(\lambda(t - t_0), x - x_0)$$

$$\leq \frac{c}{(t - t_0)^{1 - \frac{\alpha}{2}}} G(\lambda(t - t_0), x - x_0)$$

with $c = c(T, N, \lambda, \lambda_0)$ positive constant, since the power series $\sum_{k=1}^{\infty} \mathbf{m}_k r^{k-1}$ has infinite convergence radius. This proves (20.3.9). The convergence of the series is

20.3 The Parametrix Method

uniform in (t_0, x_0, t, x) if $t - t_0 \geq \delta > 0$, for every $\delta > 0$ sufficiently small, and consequently $\Phi(t_0, x_0; t, x)$ is a continuous function of (t_0, x_0, t, x) for $0 \leq t_0 < t < T$ and $x, x_0 \in \mathbb{R}^N$. Moreover, exchanging the signs of series and integral, we have

$$\int_{t_0}^{t} \int_{\mathbb{R}^N} \Phi(t_0, x_0; s, y) \mathscr{L}\mathbf{P}(s, y; t, x) dy ds$$

$$= \sum_{k=1}^{\infty} \int_{t_0}^{t} \int_{\mathbb{R}^N} (\mathscr{L}\mathbf{P})_k(t_0, x_0; s, y) \mathscr{L}\mathbf{P}(s, y; t, x) dy ds$$

$$= \sum_{k=2}^{\infty} (\mathscr{L}\mathbf{P})_k(t_0, x_0; t, x)$$

$$= \Phi(t_0, x_0; t, x) - \mathscr{L}\mathbf{P}(t_0, x_0; t, x)$$

and therefore Φ solves equation (20.3.6).

As for (20.3.10), we first prove the estimate

$$|\mathscr{L}\mathbf{P}(t_0, x_0; t, x) - \mathscr{L}\mathbf{P}(t_0, x_0; t, y)| \leq$$

$$\leq \frac{c |x-y|^{\alpha/2}}{(t-t_0)^{1-\alpha/4}} \left(\mathbf{G}(\lambda(t-t_0), x - x_0) + \mathbf{G}(\lambda(t-t_0), y - x_0) \right)$$

(20.3.27)

for every $\lambda > \lambda_0$, $0 \leq t_0 < t < T$ and $x, y, x_0 \in \mathbb{R}^N$, with $c = c(T, N, \lambda, \lambda_0) > 0$. Now, if $|x - y|^2 > t - t_0$ then (20.3.27) follows directly from (20.3.24) with $k = 1$.

To study the case $|x - y|^2 \leq t - t_0$, we observe that

$$\mathscr{L}\mathbf{P}(t_0, x_0; t, x) - \mathscr{L}\mathbf{P}(t_0, x_0; t, y)$$

$$= (\mathscr{L} - \mathscr{L}_{x_0})\mathbf{P}(t_0, x_0; t, x) - (\mathscr{L} - \mathscr{L}_{x_0})\mathbf{P}(t_0, x_0; t, y) = F_1 + F_2$$

where

$$F_1 = \frac{1}{2} \sum_{i,j=1}^{N} ((c_{ij}(t, x) - c_{ij}(t, x_0)) \partial_{x_i x_j} \mathbf{P}(t_0, x_0; t, x)$$

$$- (c_{ij}(t, y) - c_{ij}(t, x_0)) \partial_{y_i y_j} \mathbf{P}(t_0, x_0; t, y))$$

$$= \underbrace{\frac{1}{2} \sum_{i,j=1}^{N} (c_{ij}(t, x) - c_{ij}(t, y)) \partial_{x_i x_j} \mathbf{P}(t_0, x_0; t, x)}_{=: G_1}$$

$$+ \frac{1}{2} \underbrace{\sum_{i,j=1}^{N} (c_{ij}(t, y) - c_{ij}(t, x_0)) \left(\partial_{x_i x_j} \mathbf{P}(t_0, x_0; t, x) - \partial_{y_i y_j} \mathbf{P}(t_0, x_0; t, y)\right)}_{=:G_2},$$

$$F_2 = \sum_{j=1}^{N} \left(b_j(t, x) \partial_{x_j} \mathbf{P}(t_0, x_0; t, x) - b_j(t, y) \partial_{y_j} \mathbf{P}(t_0, x_0; t, y)\right)$$

$$+ a(t, x) \mathbf{P}(t_0, x_0; t, x) - a(t, y) \mathbf{P}(t_0, x_0; t, y).$$

Due to the Hölder continuity assumption of the coefficients and the Gaussian estimate (20.3.13), under the condition $|x - y|^2 \leq t - t_0$, we have

$$|G_1| \leq \frac{c |x - y|^\alpha}{t - t_0} \mathbf{G}(\lambda(t - t_0), x - x_0) \leq \frac{c |x - y|^{\frac{\alpha}{2}}}{(t - t_0)^{1-\frac{\alpha}{4}}} \mathbf{G}(\lambda(t - t_0), x - x_0).$$

Regarding G_2, we still use the Hölder continuity of the coefficients and combine the mean value theorem (with η belonging to the segment with endpoints x, y) with the Gaussian estimate (20.3.14) of the third derivatives: we obtain

$$|G_2| \leq |y - x_0|^\alpha \frac{c |x - y|}{(t - t_0)^{\frac{3}{2}}} \mathbf{G}\left(\frac{\lambda + \lambda_0}{2}(t - t_0), \eta - x_0\right) \leq$$

(since $|x - y|^2 \leq t - t_0$ and by Lemma 20.3.8)

$$\leq \frac{c |x - y|^{\frac{\alpha}{2}}}{(t - t_0)^{1+\frac{\alpha}{4}}} |y - x_0|^\alpha \mathbf{G}\left(\frac{\lambda + \lambda_0}{2}(t - t_0), y - x_0\right) \leq$$

(by Lemma 20.3.4)

$$\leq \frac{c |x - y|^{\frac{\alpha}{2}}}{(t - t_0)^{1-\frac{\alpha}{4}}} \mathbf{G}(\lambda(t - t_0), y - x_0).$$

A similar estimate holds for F_2, which can be proved using the Hölder continuity of the coefficients b_j and a. This concludes the proof of (20.3.27).

We now prove (20.3.10) using the fact that Φ solves equation (20.3.6), so we have

$$\Phi(t_0, x_0; t, x) - \Phi(t_0, x_0; t, y)$$
$$= \mathscr{L} \mathbf{P}(t_0, x_0; t, x) - \mathscr{L} \mathbf{P}(t_0, x_0; t, y)$$
$$+ \underbrace{\int_{t_0}^{t} \int_{\mathbb{R}^N} \Phi(t_0, x_0; s, \eta) \left(\mathscr{L} \mathbf{P}(s, \eta; t, x) - \mathscr{L} \mathbf{P}(s, \eta; t, y)\right) d\eta ds}_{=:I(t_0, x_0; t, x, y)}.$$

20.3 The Parametrix Method

Thanks to (20.3.27), it is sufficient to estimate the term $I(t_0, x_0; t, x, y)$: again by the estimates (20.3.9) and (20.3.27) we obtain

$$|I(t_0, x_0; t, x, y)| \leq \int_{t_0}^{t} \frac{c\,|x-y|^{\frac{\alpha}{2}}}{(s-t_0)^{1-\frac{\alpha}{2}}(t-s)^{1-\frac{\alpha}{4}}} \cdot$$
$$\cdot \int_{\mathbb{R}^N} \mathbf{G}(\lambda(s-t_0), \eta - x_0)(\mathbf{G}(\lambda(t-s), x - \eta)$$
$$+ \mathbf{G}(\lambda(t-s), y - \eta))d\eta ds =$$

(by the Chapman-Kolmogorov equation)

$$= \int_{t_0}^{t} \frac{c\,|x-y|^{\alpha/2}}{(s-t_0)^{1-\frac{\alpha}{2}}(t-s)^{1-\frac{\alpha}{4}}} ds \, (\mathbf{G}(\lambda(t-t_0), x - x_0) + \mathbf{G}(\lambda(t-t_0), y - x_0))$$

$$= \frac{c\,|x-y|^{\alpha/2}}{(t-t_0)^{1-\frac{3\alpha}{4}}} (\mathbf{G}(\lambda(t-t_0), x - x_0) + \mathbf{G}(\lambda(t-t_0), y - x_0))$$

given the general formula

$$\int_{t_0}^{t} \frac{1}{(s-t_0)^{\beta}(t-s)^{\gamma}} ds = \frac{\Gamma_E(1-\beta)\Gamma_E(1-\gamma)}{\Gamma_E(2-\beta-\gamma)}(t-t_0)^{1-\beta-\gamma} \qquad (20.3.28)$$

valid for every $\beta, \gamma < 1$. $\square$

20.3.3 Potential Estimates

Let Assumption 20.2.2 be in force and recall definition (20.3.3) of parametrix. In this section, we consider the so-called *potential*

$$V_f(t, x) := \int_{t_0}^{t} \int_{\mathbb{R}^N} f(s, y) \mathbf{P}(s, y; t, x) dy ds, \qquad (t, x) \in]t_0, T[\times \mathbb{R}^N,$$
$$(20.3.29)$$

where $f \in C(]t_0, T[\times \mathbb{R}^N)$ satisfies Assumption 20.2.4 of growth and local Hölder continuity. The main result of this section is the following

Proposition 20.3.9 *Definition* (20.3.29) *is well-posed and* $V_f \in C(]t_0, T[\times \mathbb{R}^N)$. *Moreover, for every* $i, j = 1, \ldots, N$ *there exist and are continuous on* $]t_0, T[\times \mathbb{R}^N$

the derivatives

$$\partial_{x_i} V_f(t,x) = \int_{t_0}^{t} \int_{\mathbb{R}^N} f(s,y) \partial_{x_i} \mathbf{P}(s,y;t,x) dy ds, \qquad (20.3.30)$$

$$\partial_{x_i x_j} V_f(t,x) = \int_{t_0}^{t} \int_{\mathbb{R}^N} f(s,y) \partial_{x_i x_j} \mathbf{P}(s,y;t,x) dy ds, \qquad (20.3.31)$$

$$\partial_t V_f(t,x) = f(t,x) + \int_{t_0}^{t} \int_{\mathbb{R}^N} f(s,y) \partial_t \mathbf{P}(s,y;t,x) dy ds. \qquad (20.3.32)$$

Proof Let

$$I(s;t,x) := \int_{\mathbb{R}^N} f(s,y) \Gamma_y(s,y;t,x) dy, \qquad t_0 \le s < t < T, \ x \in \mathbb{R}^N,$$

so that

$$V_f(t,x) = \int_{t_0}^{t} I(s;t,x) ds.$$

By estimate (20.3.11) and assumption (20.2.6), we have

$$|I(s;t,x)| \le \frac{c_1 \lambda_0^N}{(s-t_0)^{1-\beta}(2\pi\lambda_0(t-s))^{\frac{N}{2}}} \int_{\mathbb{R}^N} e^{c_2|y|^2 - \frac{|x-y|^2}{2\lambda_0(t-s)}} dy =$$

(by the change of variables $z = \frac{x-y}{\sqrt{2\lambda_0(t-s)}}$ and setting $c_0 = c_1 \lambda^N \pi^{-N/2}$)

$$= \frac{c_0}{(s-t_0)^{1-\beta}} \int_{\mathbb{R}^N} e^{c_2|x-z\sqrt{2\lambda_0(t-s)}|^2 - |z|^2} dz \le$$

(setting $\kappa = 1 - 4c_2 \lambda_0 T > 0$ by hypothesis)

$$\le \frac{c_0}{(s-t_0)^{1-\beta}} e^{2c_2|x|^2} \int_{\mathbb{R}^N} e^{-\kappa|z|^2} dz \le \frac{c e^{2c_2|x|^2}}{(s-t_0)^{1-\beta}} \qquad (20.3.33)$$

for some suitable positive constant $c = c(\lambda_0, T, N, c_1, c_2)$. It follows that the function $V_f \in C(]t_0, T[\times \mathbb{R}^N)$ is well-defined and

$$|V_f(t,x)| \le c(t-t_0)^{\beta} e^{2c_2|x|^2}, \qquad t_0 < t < T, \ x \in \mathbb{R}^N, \qquad (20.3.34)$$

with $\beta > 0$.

20.3 The Parametrix Method

Proof of (20.3.30) For $t_0 \leq s < t < T$ we have

$$\left|\partial_{x_i} I(s; t, x)\right| = \left|\int_{\mathbb{R}^N} f(s, y) \partial_{x_i} \mathbf{P}(s, y; t, x) dy\right| \leq$$

(proceeding as in the proof of (20.3.33), using estimate (20.3.12))

$$\leq \frac{c e^{2c_2 |x|^2}}{(s - t_0)^{1-\beta} \sqrt{t - s}}.$$

This is sufficient to prove (20.3.30) and moreover, by (20.3.28) we have

$$\left|\partial_{x_i} V_f(t, x)\right| \leq \frac{c e^{2c_2 |x|^2}}{(t - t_0)^{\frac{1}{2} - \beta}}, \qquad t_0 < t < T, \ x \in \mathbb{R}^N.$$

Proof of (20.3.31) The proof of the existence of the second order derivatives is more involved since repeating the previous argument using estimate (20.3.13) would result in a singular term of the type $\frac{1}{t-s}$ which is not integrable in the interval $[t_0, t]$. Proceeding carefully, it is possible to prove more precise and uniform estimates on $]t_0, T[\times D_n$ for each fixed $n \in \mathbb{N}$, where $D_n := \{|x| \leq n\}$.

Assume $x \in D_n$. First of all, for each $s < t$ we have

$$\partial_{x_i x_j} I(s; t, x) = \int_{\mathbb{R}^N} f(s, y) \partial_{x_i x_j} \mathbf{P}(s, y; t, x) dy = J(s; t, x) + H(s; t, x)$$

where

$$J(s; t, x) = \int_{D_{n+1}} f(s, y) \partial_{x_i x_j} \mathbf{P}(s, y; t, x) dy,$$

$$H(s; t, x) = \int_{\mathbb{R}^N \setminus D_{n+1}} f(s, y) \partial_{x_i x_j} \mathbf{P}(s, y; t, x) dy.$$

Decompose J into the sum of three terms, $J = J_1 + J_2 + J_3$, where[8]

$$J_1(s; t, x) = \int_{D_{n+1}} (f(s, y) - f(s, x)) \partial_{x_i x_j} \Gamma_y(s, y; t, x) dy,$$

$$J_2(s; t, x) = f(s, x) \int_{D_{n+1}} \left(\partial_{x_i x_j} \Gamma_y(s, y; t, x) - \left(\partial_{x_i x_j} \Gamma_\eta(s, y; t, x)\right)|_{\eta=x}\right) dy,$$

$$J_3(s; t, x) = f(s, x) \int_{D_{n+1}} \left(\partial_{x_i x_j} \Gamma_\eta(s, y; t, x)\right)|_{\eta=x} dy.$$

---

[8] For clarity, the term $\left(\partial_{x_i x_j} \Gamma_\eta(s, y; t, x)\right)|_{\eta=x}$ is obtained by first applying the derivatives $\partial_{x_i x_j} \Gamma_\eta(s, y; t, x)$, keeping η fixed, and then calculating the result obtained in $\eta = x$. Note that, under Assumption 20.2.2, $\Gamma_\eta(s, y; t, x)$ as a function of η is not differentiable.

By the local Hölder continuity of f, being $x, y \in D_{n+1}$, and estimate (20.3.13), we have

$$|J_1(s; t, x)| \leq \frac{c}{(s-t_0)^{1-\frac{\beta}{2}}} \int_{D_{n+1}} \frac{|x-y|^\beta}{t-s} \mathbf{G}\left(\lambda(t-s), x-y\right) dy \leq$$

(by Lemma 20.3.4)

$$\leq \frac{c}{(s-t_0)^{1-\frac{\beta}{2}}(t-s)^{1-\frac{\beta}{2}}} \int_{D_{n+1}} \mathbf{G}\left(2\lambda(t-s), x-y\right) dy \leq \frac{c}{(s-t_0)^{1-\frac{\beta}{2}}(t-s)^{1-\frac{\beta}{2}}},$$

with c positive constant that depends on κ_n in (20.2.7), as well as on T, N, λ and λ_0. Proceeding in a similar way, using (20.3.17) and (20.2.6), we have

$$|J_2(s; t, x)| \leq \frac{ce^{c_2|x|^2}}{(s-t_0)^{1-\beta}} \int_{D_{n+1}} \frac{|y-x|^\alpha}{t-s} \mathbf{G}\left(\lambda(t-s), x-y\right) dy$$

$$\leq \frac{ce^{c_2|x|^2}}{(s-t_0)^{1-\beta}(t-s)^{1-\frac{\alpha}{2}}}.$$

Now, we notice that

$$\partial_{x_i} \Gamma_\eta(s, y; t, x) = -\partial_{y_i} \Gamma_\eta(s, y; t, x)$$

and therefore

$$\int_{D_{n+1}} \left(\partial_{x_i x_j} \Gamma_\eta(s, y; t, x)\right)|_{\eta=x} dy = -\int_{D_{n+1}} \left(\partial_{y_i x_j} \Gamma_\eta(s, y; t, x)\right)|_{\eta=x} dy =$$

(by the divergence theorem, indicating with ν the external normal to D_{n+1} and with $d\sigma(y)$ the surface measure on the boundary ∂D_{n+1})

$$= -\int_{\partial D_{n+1}} \left(\partial_{x_j} \Gamma_\eta(s, y; t, x)\right)|_{\eta=x} \nu(y) d\sigma(y)$$

from which, again by (20.3.12) and (20.2.6), we obtain

$$|J_3(s; t, x)| \leq \frac{ce^{c_2|x|^2}}{(s-t_0)^{1-\beta}} \int_{\partial D_{n+1}} \frac{1}{\sqrt{t-s}} \mathbf{G}\left(\lambda(t-s), x-y\right) d\sigma(y)$$

$$\leq \frac{ce^{c_2|x|^2}}{(s-t_0)^{1-\beta}\sqrt{t-s}}.$$

20.3 The Parametrix Method

Finally, by (20.3.13) we have

$$|H(s; t, x)| \leq \int_{\mathbb{R}^N \setminus D_{n+1}} |f(s, y)| \frac{c}{t-s} G(\lambda(t-s), x-y) \, dy \leq$$

(being $|x - y| \geq 1$ since $|y| \geq n+1$ and $|x| \leq n$)

$$\leq c \int_{\mathbb{R}^N \setminus D_{n+1}} |f(s, y)| \frac{|x-y|^2}{t-s} G(\lambda(t-s), x-y) \, dy \leq$$

(by Lemma 20.3.4, with $\lambda' > \lambda$, and the assumption (20.2.6) on the growth of f)

$$\leq \frac{c}{(s-t_0)^{1-\beta}} \int_{\mathbb{R}^N} e^{c_2 |y|^2} G(\lambda'(t-s), x-y) \, dy \leq \frac{c e^{c|x|^2}}{(s-t_0)^{1-\beta}}$$

with $c > 0$, remembering that $c_2 < \frac{1}{4\lambda_0 T}$ by assumption and choosing $\lambda' - \lambda_0$ sufficiently small. In conclusion, we have proved that, for every $t_0 \leq s < t < T$ and $x \in D_n$, with $n \in \mathbb{N}$ fixed, there exists a constant c such that

$$|\partial_{x_i x_j} I(s; t, x)| = \left| \int_{\mathbb{R}^N} f(s, y) \partial_{x_i x_j} \mathbf{P}(s, y; t, x) dy \right| \leq \frac{c}{(s-t_0)^{1-\frac{\beta}{2}}(t-s)^{1-\frac{\gamma}{2}}}$$
(20.3.35)

where $\gamma = \alpha \wedge \beta$, from which also

$$|\partial_{x_i x_j} V_f(t, x)| \leq \frac{c}{(t-t_0)^{\frac{1}{2}-\frac{\beta}{2}-\frac{\gamma}{2}}}$$

thanks to (20.3.28). This concludes the proof of formula (20.3.31).

Proof of (20.3.32) First, we observe that

$$|\partial_t I(s; t, x)| = \left| \int_{\mathbb{R}^N} f(s, y) \partial_t \Gamma_y(s, y; t, x) dy \right| =$$

(since Γ_y is the fundamental solution of $\mathscr{L}_y$)

$$= \left| \int_{\mathbb{R}^N} f(s, y) \frac{1}{2} \sum_{i,j=1}^N c_{ij}(t, y) \partial_{x_i x_j} \Gamma_y(s, y; t, x) dy \right| \leq$$

(proceeding as in the proof of (20.3.35) and using the boundedness assumption on the coefficients)

$$\leq \frac{c}{(s-t_0)^{1-\beta}(t-s)^{1-\frac{\gamma}{2}}}. \tag{20.3.36}$$

for every $t_0 \leq s < t < T$ and $x \in D_n$, with $n \in \mathbb{N}$ fixed. Now, we have

$$\frac{V_f(t+h,x) - V_f(t,x)}{h} = \int_{t_0}^{t} \frac{I(s;t+h,x) - I(s;t,x)}{h} ds$$

$$+ \frac{1}{h}\int_{t}^{t+h} I(s;t+h,x)ds =: I_1(t,x) + I_2(t,x).$$

By the mean value theorem, there exists $\hat{t}_s \in [t, t+h]$ such that

$$I_1(t,x) = \int_{t_0}^{t} \partial_t I(s; \hat{t}_s, x) ds \xrightarrow[h \to 0]{} \int_{t_0}^{t} \partial_t I(s; t, x) ds$$

by the dominated convergence theorem thanks to estimate (20.3.36). As for I_2, we have

$$I_2(t,x) - f(t,x) = \frac{1}{h}\int_{t}^{t+h} (I(s;t+h,x) - f(s,x))\,ds$$

$$+ \frac{1}{h}\int_{t}^{t+h} (f(s,x) - f(t,x))ds$$

where the second integral on the right-hand side tends to zero as $h \to 0$ since f is continuous, while to estimate the first integral we assume $x \in D_n$ and proceed as in the proof of (20.3.31): specifically, we write

$$\frac{1}{h}\int_{t}^{t+h} (I(s;t+h,x) - f(s,x))\,ds$$

$$= \underbrace{\frac{1}{h}\int_{t}^{t+h}\int_{D_{n+1}} (f(s,y) - f(s,x))\Gamma_y(s,y;t+h,x)dy\,ds}_{=:J_1(t,x)}$$

$$+ \underbrace{\frac{1}{h}\int_{t}^{t+h}\int_{\mathbb{R}^N \setminus D_{n+1}} (f(s,y) - f(s,x))\Gamma_y(s,y;t+h,x)dy\,ds}_{=:J_2(t,x)}.$$

20.3 The Parametrix Method

Assuming $h > 0$ for simplicity: by the Hölder continuity of f and estimate (20.3.11) of Γ_y, we have

$$|J_1(t,x)| \le \frac{\lambda^N \kappa_{n+1}}{h} \int_t^{t+h} \int_{D_{n+1}} |x-y|^\beta \mathbf{G}(\lambda_0(t+h-s), x-y)\,dy ds \le$$

(by Lemma 20.3.4)

$$\le \frac{c}{h} \int_t^{t+h} (t+h-s)^{\frac{\beta}{2}} \underbrace{\int_{D_{n+1}} \mathbf{G}(\lambda_0(t+h-s), x-y)\,dy}_{\le 1}\,ds \xrightarrow[h\to 0^+]{} 0.$$

On the other hand, thanks to the growth assumption (20.2.6) on f and (20.3.11), it can be readily proved that

$$|J_2(t,x)| \le \frac{c}{h} \int_t^{t+h} \int_{|x-y|>1} e^{c_2|y|^2} \mathbf{G}(\lambda_0(t+h-s), x-y)\,dy ds \xrightarrow[h\to 0^+]{} 0.$$

This is enough to conclude the proof of the proposition.

□

20.3.4 Proof of Theorem 20.2.5

We divide the proof into several steps.

Step 1 By construction and the properties of Φ in Proposition 20.3.2, $\Gamma = \Gamma(t_0, x_0; t, x)$ in (20.3.4) is a continuous function of (t_0, x_0, t, x) for $0 \le t_0 < t < T$ and $x, x_0 \in \mathbb{R}^N$. We show that Γ is a solution of $\mathscr{L}$. Thanks to the estimates of Φ in Proposition 20.3.2, applying Proposition 20.3.9 we obtain

$$\partial_{x_i} \Gamma(t_0, x_0; t, x) = \partial_{x_i} \mathbf{P}(t_0, x_0; t, x) + \int_{t_0}^t \int_{\mathbb{R}^N} \Phi(t_0, x_0; s, y) \partial_{x_i} \mathbf{P}(s, y; t, x)\,dy ds,$$

$$\partial_{x_i x_j} \Gamma(t_0, x_0; t, x) = \partial_{x_i x_j} \mathbf{P}(t_0, x_0; t, x)$$
$$+ \int_{t_0}^t \int_{\mathbb{R}^N} \Phi(t_0, x_0; s, y) \partial_{x_i x_j} \mathbf{P}(s, y; t, x)\,dy ds,$$

$$\partial_t \Gamma(t_0, x_0; t, x) = \int_{t_0}^t \int_{\mathbb{R}^N} \Phi(t_0, x_0; s, y) \partial_t \mathbf{P}(s, y; t, x)\,dy ds + \Phi(t_0, x_0; t, x),$$

for $t_0 < t < T$, $x, x_0 \in \mathbb{R}^N$. Then we have

$$\mathscr{L}\Gamma(t_0, x_0; t, x) = \mathscr{L}\mathbf{P}(t_0, x_0; t, x) + \int_{t_0}^t \int_{\mathbb{R}^N} \Phi(t_0, x_0; s, y)\mathscr{L}\mathbf{P}(s, y; t, x)dyds$$
$$- \Phi(t_0, x_0; t, x)$$

from which we deduce that

$$\mathscr{L}\Gamma(t_0, x_0; t, x) = 0, \qquad 0 \leq t_0 < t < T, \ x, x_0 \in \mathbb{R}^N, \qquad (20.3.37)$$

since, by Proposition 20.3.2, Φ solves Eq. (20.3.6).

Step 2 We prove the upper Gaussian estimate (20.2.8). By using the definition (20.3.4) of Γ, we have

$$|\Gamma(t_0, x_0; t, x)| \leq \mathbf{P}(t_0, x_0; t, x) + \int_{t_0}^t \int_{\mathbb{R}^N} |\Phi(t_0, x_0; s, y)| \, \mathbf{P}(s, y; t, x)dyds \leq$$

(by (20.3.9) and (20.3.11))

$$\leq \lambda^N \mathbf{G}\left(\lambda(t - t_0), x - x_0\right)$$
$$+ \int_{t_0}^t \frac{c}{(s - t_0)^{1-\frac{\alpha}{2}}} \int_{\mathbb{R}^N} \mathbf{G}(\lambda(s - t_0), y - x_0)\mathbf{G}(\lambda(t - s), x - y)dyds =$$

(by the Chapman-Kolmogorov equation)

$$\leq \lambda^N \mathbf{G}\left(\lambda(t - t_0), x - x_0\right) + c(t - t_0)^{\frac{\alpha}{2}}\mathbf{G}(\lambda(t - t_0), x - x_0) \qquad (20.3.38)$$

and this proves, in particular, the upper bound (20.2.8). Formula (20.2.9) is proven in a completely analogous way.

Now, we prove (20.2.10). By repeating the proof of (20.3.35) with $\Phi(t_0, x_0; s, y)$ in place of $f(s, y)$ and using the estimates from Proposition 20.3.2, we establish the existence of a positive constant $c = c(T, N, \lambda, \lambda_b)$ such that

$$\left|\int_{\mathbb{R}^N} \Phi(t_0, x_0; s, y)\partial_{x_i x_j}\mathbf{P}(s, y; t, x)dy\right|$$
$$\leq \frac{c}{(s - t_0)^{1-\frac{\alpha}{4}}(t - s)^{1-\frac{\alpha}{4}}}\mathbf{G}(\lambda(t - t_0), x - x_0), \qquad t_0 \leq s < t < T, \ x, x_0 \in \mathbb{R}^N.$$
$$(20.3.39)$$

20.3 The Parametrix Method

Hence, by (20.3.4) and (20.3.31), we have

$$\left|\partial_{x_ix_j}\Gamma(t_0, x_0; t, x)\right| \le \left|\partial_{x_ix_j}\mathbf{P}(t_0, x_0; t, x)\right|$$
$$+ \left|\int_{t_0}^{t}\int_{\mathbb{R}^N} \Phi(t_0, x_0; s, y)\partial_{x_ix_j}\mathbf{P}(s, y; t, x)dyds\right| \le$$

(by (20.3.13) and (20.3.39))

$$\le c\left(\frac{1}{t-t_0} + \frac{1}{(t-t_0)^{1-\frac{\alpha}{2}}}\right)\mathbf{G}\left(\lambda(t-t_0), x-x_0\right).$$

Step 3 We prove that Γ is a fundamental solution of $\mathscr{L}$. Given $\varphi \in bC(\mathbb{R}^N)$, consider the function u in (20.2.1). Thanks to the estimates (20.2.8)–(20.2.10) we have

$$\mathscr{L}u(t,x) = \int_{\mathbb{R}^N} \varphi(\xi)\mathscr{L}\Gamma(t_0, \xi; t, x)d\xi = 0, \qquad 0 \le t_0 < t < T, \; x \in \mathbb{R}^N,$$

by (20.3.37). As for the initial datum, we have

$$u(t,x) = \underbrace{\int_{\mathbb{R}^N} \varphi(\xi)\mathbf{P}(t_0, \xi; t, x)d\xi}_{J(t,x)}$$
$$+ \underbrace{\int_{\mathbb{R}^N} \varphi(\xi) \int_{t_0}^{t}\int_{\mathbb{R}^N} \Phi(t_0, \xi; s, y)\mathbf{P}(s, y; t, x)dyds\, d\xi}_{H(t,x)}.$$

Now, for a fixed $x_0 \in \mathbb{R}^N$,

$$J(t,x) = \underbrace{\int_{\mathbb{R}^N} \varphi(\xi)\left(\Gamma_\xi(t_0, \xi; t, x) - \Gamma_{x_0}(t_0, \xi; t, x)\right)d\xi}_{J_1(t,x)}$$
$$+ \int_{\mathbb{R}^N} \varphi(\xi)\Gamma_{x_0}(t_0, \xi; t, x)d\xi$$

and, by (20.3.15), we have

$$|J_1(t,x)| \le c\int_{\mathbb{R}^N} |\varphi(\xi)||\xi - x_0|^\alpha \mathbf{G}\left(\lambda(t-t_0), x-\xi\right)d\xi \xrightarrow[(t,x)\to(t_0,x_0)]{} 0,$$

$$\int_{\mathbb{R}^N} \varphi(\xi)\Gamma_{x_0}(t_0, \xi; t, x)d\xi \xrightarrow[(t,x)\to(t_0,x_0)]{} \varphi(x_0).$$

Here we use the limit argument of Example 3.3.3 in [113]: in probabilistic terms, this correspond to the weak convergence of the normal distribution to the Dirac delta, as the variance tends to zero. On the other hand, by (20.3.38)

$$|H(t,x)| \leq c(t-t_0)^{\frac{\alpha}{2}} \int_{\mathbb{R}^N} \varphi(x_0) \mathbf{G}(\lambda(t-t_0), x-x_0) dx_0 \xrightarrow[(t,x) \to (t_0, \bar{x})]{} 0.$$

This proves that $u \in C([t_0, T[\times \mathbb{R}^N)$ and is therefore a classical solution of the Cauchy problem (20.2.2).

Step 4 We prove that u in (20.2.12) is a classical solution of the non-homogeneous Cauchy problem (20.2.13). We use the definition of Γ in (20.3.4) and focus on the term

$$\int_{t_0}^{t} \int_{\mathbb{R}^N} f(s,y) \Gamma(s,y;t,x) dy ds$$

$$= \int_{t_0}^{t} \int_{\mathbb{R}^N} f(s,y) \mathbf{P}(s,y;t,x) dy ds$$

$$+ \int_{t_0}^{t} \int_{\mathbb{R}^N} f(s,y) \int_{s}^{t} \int_{\mathbb{R}^N} \Phi(s,y;\tau,\eta) \mathbf{P}(\tau,\eta;t,x) d\eta d\tau dy ds =$$

(using notation (20.3.29), setting $\Phi(s,y;\tau,\eta) = 0$ for $\tau \leq s$ and exchanging the order of integration of the last integral)

$$= V_f(t,x) + V_F(t,x)$$

where

$$F(\tau, \eta) := \int_{t_0}^{\tau} \int_{\mathbb{R}^N} f(s,y) \Phi(s,y;\tau,\eta) dy ds.$$

We will soon prove that F satisfies Assumption 20.2.4 and it is therefore possible to apply Proposition 20.3.9 to V_f and V_F: we obtain

$$\mathscr{L}\left(V_f(t,x) + V_F(t,x)\right) = -f(t,x) - F(t,x)$$

$$+ \int_{t_0}^{t} \int_{\mathbb{R}^N} (f(s,y) + F(s,y)) \mathscr{L} \mathbf{P}(s,y;t,x) dy ds$$

$$= -f(t,x) + \int_{t_0}^{t} \int_{\mathbb{R}^N} f(s,y) I(s,y;t,x) dy ds$$

20.3 The Parametrix Method

where

$$I(s, y; t, x) := -\Phi(s, y; t, x) + \mathscr{L}\mathbf{P}(s, y; t, x)$$
$$+ \int_s^t \int_{\mathbb{R}^N} \Phi(s, y; \tau, \eta) \mathscr{L}\mathbf{P}(\tau, \eta; t, x) d\eta d\tau \equiv 0$$

by (20.3.6). This proves that

$$\mathscr{L}u(t, x) = f(t, x), \qquad 0 \le t_0 < t < T, \ x, x_0 \in \mathbb{R}^N.$$

Let us verify that F satisfies Assumption 20.2.4: by (20.3.9), the hypotheses on f and (20.3.28), we have

$$|F(\tau, \eta)| \le \int_{t_0}^{\tau} \int_{\mathbb{R}^N} \frac{c e^{c_2 |y|^2}}{(s-t_0)^{1-\frac{\beta}{2}} (\tau-s)^{1-\frac{\alpha}{2}}} \mathbf{G}(\lambda(\tau-s), \eta-y) dy ds$$
$$\le \frac{c}{(\tau-t_0)^{1-\frac{\alpha+\beta}{2}}} e^{c|\eta|^2}.$$

Moreover, by (20.3.10) we have

$$|F(\tau, \eta) - F(\tau, \eta')|$$
$$\le c|\eta - \eta'|^{\frac{\alpha}{2}} \int_{t_0}^{\tau} \int_{\mathbb{R}^N} \frac{e^{c_2 |y|^2}}{(s-t_0)^{1-\frac{\beta}{2}} (\tau-s)^{1-\frac{\alpha}{4}}}$$
$$\times \left(\mathbf{G}(\lambda(\tau-s), \eta-y) + \mathbf{G}(\lambda(\tau-s), \eta'-y) \right) dy ds$$
$$\le \frac{c|\eta - \eta'|^{\frac{\alpha}{2}}}{(\tau-t_0)^{1-\frac{\alpha+2\beta}{4}}} \left(e^{c|\eta|^2} + e^{c|\eta'|^2} \right).$$

Finally, using the upper bound (20.2.8) of Γ and proceeding as in the proof of estimate (20.3.34), we have that

$$\int_{t_0}^{t} \int_{\mathbb{R}^N} f(s, y) \Gamma(s, y; t, x) dy ds \xrightarrow[(t,x)\to(t_0,\bar{x})]{} 0,$$

for every $\bar{x} \in \mathbb{R}^N$. This concludes the proof that u in (20.2.12) is a classical solution of the non-homogeneous Cauchy problem (20.2.13).

Step 5 The Chapman-Kolmogorov equation and formula (20.2.14) can be proved as in Remark 20.3.7, as a consequence of the uniqueness result of Theorem 20.1.8.

In particular, as shown in the previous points, if a is constant, the functions

$$u_1(t,x) := e^{a(t-t_0)}, \qquad u_2(t,x) := \int_{\mathbb{R}^N} \Gamma(t_0, x_0; t, x) dx_0$$

are both bounded solutions (thanks to estimate (20.3.38)) of the Cauchy problem

$$\begin{cases} \mathscr{L}u = 0 & \text{in }]t_0, T[\times \mathbb{R}^N, \\ u(t_0, \cdot) = 1 & \text{in } \mathbb{R}^N, \end{cases}$$

and therefore coincide.

Step 6 As a last step, we prove the lower bound of Γ in (20.2.11). This is a non-trivial result, for which we adapt a technique introduced by D.G. Aronson that exploits some classical estimates of J. Nash: for further details, we also refer to Section 2 in [42]. Here, instead of Nash's estimates, we use other estimates derived directly from the parametrix method.

First, we prove that $\Gamma \geq 0$: for the sake of contradiction, if $\Gamma(t_0, x_0; t_1, x_1) < 0$ for certain $x_0, x_1 \in \mathbb{R}^N$ and $0 \leq t_0 < t_1 < T$, then by continuity we would have

$$\Gamma(t_0, y; t_1, x_1) < 0, \qquad |y - x_0| < r,$$

for a suitable $r > 0$. Consider $\varphi \in bC(\mathbb{R}^N)$ such that $\varphi(y) > 0$ for $|y - x_0| < r$ and $\varphi(y) \equiv 0$ for $|y - x_0| \geq r$: the function

$$u(t,x) := \int_{\mathbb{R}^N} \varphi(y) \Gamma(t_0, y; t, x) dy, \qquad t \in]t_0, T[, \; x \in \mathbb{R}^N,$$

is bounded thanks to estimate (20.3.38) of Γ, is such that $u(t_1, x_1) < 0$ and is a classical solution of the Cauchy problem (20.2.2). But this is absurd because it contradicts the maximum principle, Theorem 20.1.8.

Now we observe that for every $\lambda > 1$ we have

$$\mathbf{G}(\lambda t, x) \leq \mathbf{G}\left(\frac{t}{\lambda}, x\right)$$

if $|x| < c_\lambda \sqrt{t}$ where $c_\lambda = \sqrt{\frac{\lambda N}{\lambda^2 - 1} \log \lambda}$. Then, by definition (20.3.4) we have

$$\Gamma(t_0, x_0; t, x) \geq \mathbf{P}(t_0, x_0; t, x) - \left| \int_{t_0}^{t} \int_{\mathbb{R}^N} \Phi(t_0, x_0; s, y) \mathbf{P}(s, y; t, x) dy ds \right| \geq$$

$$\geq \frac{1}{\lambda^N} \mathbf{G}\left(\frac{t - t_0}{\lambda}, x - x_0\right) - c(t - t_0)^{\frac{\alpha}{2}} \mathbf{G}(\lambda(t - t_0), x - x_0) =$$

20.3 The Parametrix Method

(if $|x - x_0| \leq c_\lambda \sqrt{t - t_0}$)

$$\geq \left(\lambda^{-N} - c(t - t_0)^{\frac{\alpha}{2}}\right) \mathbf{G}\left(\frac{t - t_0}{\lambda}, x - x_0\right)$$

$$\geq \frac{1}{2\lambda^N} \mathbf{G}\left(\frac{t - t_0}{\lambda}, x - x_0\right) \qquad (20.3.40)$$

if $0 < t - t_0 \leq T_\lambda := \left(2c\lambda^N\right)^{-\frac{2}{\alpha}} \wedge T$.

Given $x, x_0 \in \mathbb{R}^N$ and $0 \leq t_0 < t < T$, let $m \in \mathbb{N}$ be the integer part of

$$\max\left\{\frac{4|x - x_0|^2}{c_\lambda^2(t - t_0)}, \frac{T}{T_\lambda}\right\}.$$

We set

$$t_k = t_0 + k\frac{t - t_0}{m + 1}, \qquad x_k = x_0 + k\frac{x - x_0}{m + 1}, \qquad k = 1, \ldots, m,$$

and observe that, thanks to the choice of m, we have

$$t_{k+1} - t_k = \frac{t - t_0}{m + 1} \leq \frac{T}{m + 1} \leq T_\lambda. \qquad (20.3.41)$$

Moreover, if $y_k \in D(x_k, r) := \{y \in \mathbb{R}^N \mid |x_k - y| < r\}$ for each $k = 1, \ldots, m$ then, choosing $r = \frac{c_\lambda}{4}\sqrt{\frac{t-t_0}{m+1}}$, we have

$$|y_{k+1} - y_k| \leq 2r + |x_{k+1} - x_k| = 2r + \frac{|x - x_0|}{m + 1} \leq 2r + \frac{c_\lambda}{2}\sqrt{\frac{t - t_0}{m + 1}}$$

$$= c_\lambda\sqrt{\frac{t - t_0}{m + 1}} \qquad (20.3.42)$$

$$= c_\lambda\sqrt{t_{k+1} - t_k}. \qquad (20.3.43)$$

Applying the Chapman-Kolmogorov equation repeatedly, we have

$$\Gamma(t_0, x_0; t, x) = \int_{\mathbb{R}^{Nm}} \Gamma(t_0, x_0; t_1, y_1)$$

$$\times \prod_{k=1}^{m-1} \Gamma(t_k, y_k; t_{k+1}, y_{k+1}) \Gamma(t_m, y_m; t, x) dy_1 \ldots dy_m \geq$$

(using the fact that $\Gamma \geq 0$)

$$\geq \int_{\mathbb{R}^{Nm}} \Gamma(t_0, x_0; t_1, y_1)$$
$$\times \prod_{k=1}^{m-1} \mathbb{1}_{D(x_k,r)}(y_k)\Gamma(t_k, y_k; t_{k+1}, y_{k+1})\mathbb{1}_{D(x_m,r)}(y_m)\Gamma(t_m, y_m; t, x)dy_1\ldots dy_m \geq$$

(since, by (20.3.41) and (20.3.43), estimate (20.3.40) holds)

$$\geq \frac{1}{(2\lambda^N)^{m+1}} \int_{\mathbb{R}^{Nm}} \mathbf{G}\left(\frac{t-t_0}{\lambda(m+1)}, y_1 - x_0\right) \cdot$$
$$\cdot \prod_{k=1}^{m-1} \mathbb{1}_{D(x_k,r)}(y_k)\mathbf{G}\left(\frac{t-t_0}{\lambda(m+1)}, y_{k+1} - y_k\right)$$
$$\times \mathbb{1}_{D(x_m,r)}(y_m)\mathbf{G}\left(\frac{t-t_0}{\lambda(m+1)}, x - y_m\right) dy_1\ldots dy_m \geq$$

(denoting by ω_N the volume of the unit ball in $\mathbb{R}^N$, by (20.3.42))

$$\geq \frac{1}{(2\lambda^N)^{m+1}} \left(\omega_N r^N\right)^m \left(\frac{\lambda(m+1)}{2\pi(t-t_0)}\right)^{\frac{N}{2}(m+1)} \exp\left(-\frac{\lambda c_\lambda^2}{2}(m+1)\right).$$

It follows that there exists a constant $c = c(N, T, \alpha, \lambda, \lambda_0)$ such that

$$\Gamma(t_0, x_0; t, x) \geq \frac{1}{c(t-t_0)^{\frac{N}{2}}} e^{-cm}$$

and by the choice of m, this is enough to prove the thesis and conclude the proof of Theorem 20.2.5.

20.3.5 Proof of Proposition 18.4.3

For consistency with the notations of this chapter, we state and prove Proposition 18.4.3 in its *forward* version.

Proposition 20.3.10 *Under Assumption 20.2.2, let Γ be the fundamental solution of the operator $\mathscr{A} - \partial_t$ on $\mathscr{S}_T$ with $\mathscr{A}$ in (20.2.4). For every $\lambda \geq 1$, the vector-valued*

20.3 The Parametrix Method

function

$$u_\lambda(t, x) := \int_0^t e^{-\lambda(t-t_0)}$$
$$\times \int_{\mathbb{R}^N} b(t_0, x_0)\Gamma(t_0, x_0; t, x) dx_0 dt_0, \qquad (t, x) \in [0, T[\times \mathbb{R}^N,$$

is a classical solution of the Cauchy problem

$$\begin{cases} (\partial_t + \mathscr{A}_t)u = \lambda u - b, & \text{in } \mathscr{S}_T, \\ u(0, \cdot) = 0, & \text{in } \mathbb{R}^N. \end{cases}$$

Moreover, there exists a constant $c > 0$, which depends only on N, λ_0 and T, such that

$$|u_\lambda(t, x) - u_\lambda(t, y)| \leq \frac{c}{\sqrt{\lambda}}|x - y|, \tag{20.3.44}$$

$$|\nabla_x u_\lambda(t, x) - \nabla_x u_\lambda(t, y)| \leq c|x - y|, \tag{20.3.45}$$

for every $t \in]0, T[$ and $x, y \in \mathbb{R}^N$, where $\nabla_x = (\partial_{x_1}, \ldots, \partial_{x_N})$.

Proof We use the representation (20.3.4) of the fundamental solution provided by the parametrix method:

$$u_\lambda(t, x) = \int_0^t e^{-\lambda(t-t_0)}(I_b(t_0; t, x) + J_b(t_0; t, x))dt_0$$

where

$$I_b(t_0; t, x) := \int_{\mathbb{R}^N} b(t_0, x_0)\mathbf{P}(t_0, x_0; t, x) dx_0,$$

$$J_b(t_0; t, x) := \int_{\mathbb{R}^N} b(t_0, x_0) \underbrace{\int_{t_0}^t \int_{\mathbb{R}^N} \Phi(t_0, x_0; s, y)\mathbf{P}(s, y; t, x) dy ds}_{=:R(t_0, x_0; t, x)} dx_0,$$

$$\tag{20.3.46}$$

with Φ defined in (20.3.7). Since b is bounded, by (20.3.12) we have

$$|I_b(t_0; t, x) - I_b(t_0; t, y)| \leq c\frac{|x - y|}{\sqrt{t - t_0}}, \qquad x, y \in \mathbb{R}^N.$$

A similar result holds for J_b: in fact, by (20.3.9) and by the mean value theorem and estimate (20.3.12), for $\lambda_1 > \lambda_0$ we have

$$|R(t_0, x_0; t, x) - R(t_0, x_0; t, y)|$$
$$\leq c \int_{t_0}^t \frac{|x-y|}{(s-t_0)^{1-\frac{\alpha}{2}}\sqrt{t-s}}$$
$$\times \int_{\mathbb{R}^N} \mathbf{G}(\lambda_1(t-s), \bar{x}-y)\mathbf{G}(\lambda_1(s-t_0), y-x_0) dy ds =$$

(integrating and using (20.3.28))

$$= c \frac{|x-y|}{(t-t_0)^{\frac{1-\alpha}{2}}} \mathbf{G}(\lambda_1(t-t_0), \bar{x}-x_0) \qquad (20.3.47)$$

Plugging estimate (20.3.47) into (20.3.46) and being b bounded, we obtain

$$|J_b(t_0; t, x) - J_b(t_0; t, y)| \leq c \frac{|x-y|}{(t-t_0)^{\frac{1-\alpha}{2}}} \qquad x, y \in \mathbb{R}^N.$$

Hence, we have

$$|u_\lambda(t, x) - u_\lambda(t, y)| \leq c|x-y| \int_0^t \frac{e^{-\lambda(t-t_0)}}{\sqrt{t-t_0}} dt_0$$

which yields (20.3.44). The proof of (20.3.45) is analogous and is based on the arguments also used for the proof of Proposition 20.3.9. □

20.4 Key Ideas to Remember

The chapter is structured into two parts, focusing on the study of uniqueness and existence for the parabolic Cauchy problem, respectively.

- Section 20.1: uniqueness is proven under very general assumptions (cf. Assumption 20.1.1, (20.1.2), and 20.1.3). The main results are the maximum and comparison principles. Uniqueness classes for the Cauchy problem are given by functions that do not grow too rapidly at infinity.
- Sections 20.2 and 20.3: we present the classic *parametrix method* for the construction of the fundamental solution of a uniformly parabolic operator with bounded coefficients that are Hölder continuous in the spatial variable. This is a fairly long and complex technique based on suitable estimates involving Gaussian functions and on the study of singular integrals. The fundamental Theorem 20.2.5 provides, in addition to existence and the property of being a

20.4 Key Ideas to Remember

density, also a comparison between the fundamental solution and the Gaussian function, the Chapman-Kolmogorov property, and Duhamel's formula for the solution of the non-homogeneous Cauchy problem.

Main notations used or introduced in this chapter:

Symbol	Description	Page
$\mathscr{L}$	Forward parabolic operator	370
$\mathscr{S}_T := \,]0, T[\, \times \mathbb{R}^N$	Strip in $\mathbb{R}^{N+1}$	370
Γ	Fundamental solution	379
$\mathbf{G}$	Standard Gaussian function	383
$\mathbf{P}$	Parametrix	384
bC_T^α	Bounded, uniformly α-Hölder continuous (w.r.t. x) functions on $\mathscr{S}_T$	341
$[g]_\alpha$	Norm in bC_T^α	341
$bC^\alpha(\mathbb{R}^N)$	Bounded, α-Hölder continuous functions on $\mathbb{R}^N$	382
$\|\varphi\|_{bC^\alpha(\mathbb{R}^N)}$	Norm in $bC^\alpha(\mathbb{R}^N)$	382

References

1. Agassi, A.: Open: An Autobiography. Einaudi (2011)
2. Antonelli, F.: Backward-forward stochastic differential equations. Ann. Appl. Probab. **3**, 777–793 (1993)
3. Antonov, A., Misirpashaev, T., Piterbarg, V.: Markovian projection on a Heston model. J. Comput. Finance **13**, 23–47 (2009)
4. Applebaum, D.: Lévy Processes and Stochastic Calculus, vol. 93 of Cambridge Studies in Advanced Mathematics. Cambridge University Press, Cambridge (2004)
5. Aronson, D.G.: The fundamental solution of a linear parabolic equation containing a small parameter. Illinois J. Math. **3**, 580–619 (1959)
6. Baldi, P.: Stochastic Calculus. Universitext, Springer, Cham (2017). An introduction through theory and exercises
7. Barlow, M.T.: One-dimensional stochastic differential equations with no strong solution. J. London Math. Soc. (2) **26**, 335–347 (1982)
8. Barucci, E., Polidoro, S., Vespri, V.: Some results on partial differential equations and Asian options. Math. Models Methods Appl. Sci. **11**, 475–497 (2001)
9. Bass, R.F.: Stochastic Processes, vol. 33 of Cambridge Series in Statistical and Probabilistic Mathematics. Cambridge University Press, Cambridge (2011)
10. Bass, R.F.: Real Analysis for Graduate Students (2013). Available at http://bass.math.uconn.edu/real.html
11. Bass, R.F., Perkins, E.: A new technique for proving uniqueness for martingale problems. Astérisque, 47–53 (2009), (2010)
12. Baudoin, F.: An Introduction to the Geometry of Stochastic Flows. Imperial College Press, London (2004)
13. Baudoin, F.: Diffusion Processes and Stochastic Calculus. EMS Textbooks in Mathematics. European Mathematical Society (EMS), Zürich (2014)
14. Beiglböck, M., Schachermayer, W., Veliyev, B.: A short proof of the Doob-Meyer theorem. Stoch. Process. Appl. **122**, 1204–1209 (2012)
15. Bensoussan, A.: Stochastic maximum principle for distributed parameter systems. J. Franklin Inst. **315**, 387–406 (1983)
16. Billingsley, P.: Convergence of Probability Measures. Wiley Series in Probability and Statistics: Probability and Statistics, second edn. John Wiley & Sons, New York (1999). A Wiley-Interscience Publication
17. Bismut, J.-M.: Théorie probabiliste du contrôle des diffusions. Mem. Amer. Math. Soc. **4**, xiii+130 (1976)

18. Bjork, T.: Arbitrage Theory in Continuous Time, 2nd edn. Oxford University Press, Oxford (2004)
19. Black, F., Scholes, M.: The pricing of options and corporate liabilities. J. Polit. Econ. **81**, 637–654 (1973)
20. Blumenthal, R.M., Getoor, R.K.: Markov Processes and Potential Theory. Pure and Applied Mathematics, vol. 29. Academic Press, New York-London (1968)
21. Brémaud, P.: Point Processes and Queues. Springer, New York (1981). Martingale dynamics, Springer Series in Statistics
22. Brunick, G., Shreve, S.: Mimicking an Itô process by a solution of a stochastic differential equation. Ann. Appl. Probab. **23**, 1584–1628 (2013)
23. Champagnat, N., Jabin, P.-E.: Strong solutions to stochastic differential equations with rough coefficients. Ann. Probab. **46**, 1498–1541 (2018)
24. Chow, P.-L.: Stochastic Partial Differential Equations, second edn. Advances in Applied Mathematics. CRC Press, Boca Raton, FL (2015)
25. Chung, K.L., Doob, J.L.: Fields, optionality and measurability. Amer. J. Math. **87**, 397–424 (1965)
26. Courrège, P.: Générateur infinitésimal d'un semi-groupe de convolution sur $\mathbb{R}^n$, et formule de Lévy-Khinchine. Bull. Sci. Math. (2) **88**, 3–30 (1964)
27. Cox, J.C.: Notes on Option Pricing I: Constant Elasticity of Variance Diffusion. Working Paper, Stanford University, Stanford CA (1975)
28. Cox, J.C.: The constant elasticity of variance option pricing model. J. Portfolio Manag. **23**, 15–17 (1997)
29. Cox, J.C., Ingersoll, J.E., Ross, S.A.: The relation between forward prices and futures prices. J. Financ. Econ. **9**, 321–346 (1981)
30. Criens, D., Pfaffelhuber, P., Schmidt, T.: The martingale problem method revisited. Electron. J. Probab. **28** (2023), 1–46
31. Davie, A.M.: Uniqueness of solutions of stochastic differential equations. Int. Math. Res. Not. IMRN **2007**, Art. ID rnm124, 26 (2007)
32. Davydov, D., Linetsky, V.: Pricing and hedging path-dependent options under the CEV process. Manag. Sci. **47**, 949–965 (2001)
33. Delbaen, F., Shirakawa, H.: A note on option pricing for the constant elasticity of variance model. Asia-Pac. Financ. Mark. **9**, 85–99 (2002)
34. Di Francesco, M., Pascucci, A.: On a class of degenerate parabolic equations of Kolmogorov type. AMRX Appl. Math. Res. Express **3**, 77–116 (2005)
35. Doob, J.L.: Stochastic Processes. John Wiley & Sons/Chapman & Hall, New York/London (1953)
36. Duffie, D., Filipović, D., Schachermayer, W.: Affine processes and applications in finance. Ann. Appl. Probab. **13**, 984–1053 (2003)
37. Durrett, R.: Stochastic Calculus. Probability and Stochastics Series. CRC Press, Boca Raton, FL (1996). A practical introduction
38. Durrett, R.: Probability: Theory and Examples, vol. 49 of Cambridge Series in Statistical and Probabilistic Mathematics. Cambridge University Press, Cambridge (2019). Available at https://services.math.duke.edu/~rtd/PTE/pte.html
39. El Karoui, N., Peng, S., Quenez, M.C.: Backward stochastic differential equations in finance. Math. Finance **7**, 1–71 (1997)
40. Elworthy, K.D., Le Jan, Y., Li, X.-M.: The Geometry of Filtering. Frontiers in Mathematics. Birkhäuser Verlag, Basel (2010)
41. Evans, L.C.: Partial Differential Equations, second edn., vol. 19 of Graduate Studies in Mathematics. American Mathematical Society, Providence, RI (2010)
42. Fabes, E.B., Stroock, D.W.: A new proof of Moser's parabolic Harnack inequality using the old ideas of Nash. Arch. Rational Mech. Anal. **96**, 327–338 (1986)
43. Fedrizzi, E., Flandoli, F.: Pathwise uniqueness and continuous dependence of SDEs with non-regular drift. Stochastics **83**, 241–257 (2011)

44. Feehan, P.M.N., Pop, C.A.: On the martingale problem for degenerate-parabolic partial differential operators with unbounded coefficients and a mimicking theorem for Itô processes. Trans. Amer. Math. Soc. **367**, 7565–7593 (2015)
45. Feller, W.: Zur Theorie der stochastischen Prozesse. Math. Ann. **113**, 113–160 (1937)
46. Figalli, A.: Existence and uniqueness of martingale solutions for SDEs with rough or degenerate coefficients. J. Funct. Anal. **254**, 109–153 (2008)
47. Flandoli, F.: Regularity Theory and Stochastic Flows for Parabolic SPDEs, vol. 9 of Stochastics Monographs. Gordon and Breach Science Publishers, Yverdon (1995)
48. Flandoli, F.: Random Perturbation of PDEs and Fluid Dynamic Models, vol. 2015 of Lecture Notes in Mathematics. Springer, Heidelberg (2011). Lectures from the 40th Probability Summer School held in Saint-Flour, 2010, École d'Été de Probabilités de Saint-Flour. [Saint-Flour Probability Summer School]
49. Friedman, A.: Partial Differential Equations of Parabolic Type. Prentice-Hall, Englewood Cliffs, NJ (1964)
50. Friedman, A.: Stochastic Differential Equations and Applications. Dover Publications, Mineola, NY (2006). Two volumes bound as one, Reprint of the 1975 and 1976 original published in two volumes
51. Fristedt, B., Jain, N., Krylov, N.: Filtering and Prediction: A Primer, vol. 38 of Student Mathematical Library. American Mathematical Society, Providence, RI (2007)
52. Fujisaki, M., Kallianpur, G., Kunita, H.: Stochastic differential equations for the non linear filtering problem. Osaka J. Math. **9**, 19–40 (1972)
53. Gilbarg, D., Trudinger, N.S.: Elliptic Partial Differential Equations of Second Order, second edn., vol. 224 of Grundlehren der mathematischen Wissenschaften [Fundamental Principles of Mathematical Sciences]. Springer, Berlin (1983)
54. Goodfellow, I., Bengio, Y., Courville, A.: Deep Learning. MIT Press (2016). Available at http://www.deeplearningbook.org
55. Guyon, J., Henry-Labordère, P.: Nonlinear Option Pricing. Chapman & Hall/CRC Financial Mathematics Series. CRC Press, Boca Raton, FL (2014)
56. Gyöngy, I.: Mimicking the one-dimensional marginal distributions of processes having an Itô differential. Probab. Theory Relat. Fields **71**, 501–516 (1986)
57. Gyöngy, I., Krylov, N.V.: Existence of strong solutions for Itô's stochastic equations via approximations: revisited. Stoch. Partial Differ. Equations Anal. Comput. **10**, 693–719 (2022)
58. Hagan, P.S., Kumar, D., Lesniewski, A., Woodward, D.E.: Managing smile risk. Wilmott Magazine, September, 84–108 (2002)
59. Halmos, P.R.: Measure Theory. D. Van Nostrand Company, New York, NY (1950)
60. Heston, S.: A closed-form solution for options with stochastic volatility with applications to bond and currency options. Rev. Financ. Stud. **6**, 327–343 (1993)
61. Heston, S.L., Loewenstein, M., Willard, G.A.: Options and bubbles. Rev. Financ. Stud. **20**(2), 359–390 (2007)
62. Hörmander, L.: Hypoelliptic second order differential equations. Acta Math. **119**, 147–171 (1967)
63. Ikeda, N., Watanabe, S.: Stochastic Differential Equations and Diffusion Processes, vol. 24 of North-Holland Mathematical Library. North-Holland Publishing Co./Kodansha, Amsterdam/Tokyo (1981)
64. Itô, K., Watanabe, S.: Introduction to stochastic differential equations. In: Proceedings of the International Symposium on Stochastic Differential Equations (Res. Inst. Math. Sci., Kyoto Univ., Kyoto, 1976), pp. i–xxx. Wiley, New York (1978)
65. Jacod, J., Shiryaev, A.N.: Limit Theorems for Stochastic Processes, second edn., vol. 288 of Grundlehren der Mathematischen Wissenschaften [Fundamental Principles of Mathematical Sciences]. Springer, Berlin (2003)
66. Kallenberg, O.: Foundations of Modern Probability, second edn. Probability and its Applications (New York). Springer, New York (2002)
67. Karatzas, I., Shreve, S.E.: Brownian Motion and Stochastic Calculus, second edn., vol. 113 of Graduate Texts in Mathematics. Springer, New York (1991)

68. Klenke, A.: Probability Theory, second edn. Universitext. Springer, London (2014). A comprehensive course
69. Kolmogorov, A.N.: Über die analytischen Methoden in der Wahrscheinlichkeitsrechnung. Math. Ann. **104**, 415–458 (1931)
70. Kolmogorov, A.N.: Selected Works of A. N. Kolmogorov. Vol. III. Kluwer Academic Publishers Group, Dordrecht (1993). Edited by A. N. Shiryayev
71. Kolokoltsov, V.N.: Markov Processes, Semigroups and Generators, vol. 38 of De Gruyter Studies in Mathematics. Walter de Gruyter & Co., Berlin (2011)
72. Komlós, J.: A generalization of a problem of Steinhaus. Acta Math. Acad. Sci. Hungar. **18**, 217–229 (1967)
73. Kotelenez, P.: Stochastic Ordinary and Stochastic Partial Differential Equations, vol. 58 of Stochastic Modelling and Applied Probability. Springer, New York (2008). Transition from microscopic to macroscopic equations
74. Krylov, N.V.: Itô's stochastic integral equations. Teor. Verojatnost. i Primenen **14**, 340–348 (1969)
75. Krylov, N.V.: Correction to the paper "Itô's stochastic integral equations" (Teor. Verojatnost. i Primenen. **14**, 340–348 (1969)). Teor. Verojatnost. i Primenen. **17**, 392–393 (1972)
76. Krylov, N.V.: The selection of a Markov process from a Markov system of processes, and the construction of quasidiffusion processes. Izv. Akad. Nauk SSSR Ser. Mat. **37**, 691–708 (1973)
77. Krylov, N.V.: Controlled Diffusion Processes, vol. 14 of Stochastic Modelling and Applied Probability. Springer, Berlin (2009). Translated from the 1977 Russian original by A. B. Aries, Reprint of the 1980 edition
78. Krylov, N.V., Röckner, M.: Strong solutions of stochastic equations with singular time dependent drift. Probab. Theory Related Fields **131**, 154–196 (2005)
79. Krylov, N.V., Rozovsky, B.L.: On the first integrals and Liouville equations for diffusion processes. In: Stochastic Differential Systems (Visegrád, 1980), pp. 117–125, vol. 36 of Lecture Notes in Control and Information Sci. Springer, Berlin (1981)
80. Krylov, N.V., Rozovsky, B.L.: Characteristics of second-order degenerate parabolic Itô equations. Trudy Sem. Petrovsk. **8**, 153–168 (1982)
81. Krylov, N.V., Zatezalo, A.: A direct approach to deriving filtering equations for diffusion processes. Appl. Math. Optim. **42**, 315–332 (2000)
82. Kunita, H.: Stochastic Flows and Stochastic Differential Equations, vol. 24 of Cambridge Studies in Advanced Mathematics. Cambridge University Press, Cambridge (1997). Reprint of the 1990 original
83. Lacker, D., Shkolnikov, M., Zhang, J.: Inverting the Markovian projection, with an application to local stochastic volatility models. Ann. Probab. **48**, 2189–2211 (2020)
84. Ladyzhenskaia, O.A., Solonnikov, V.A., Ural'tseva, N.N.: Linear and Quasilinear Equations of Parabolic Type. Translations of Mathematical Monographs, vol. 23. American Mathematical Society, Providence, RI (1968). Translated from the Russian by S. Smith
85. Lanconelli, E., Polidoro, S.: On a class of hypoelliptic evolution operators. Rend. Sem. Mat. Univ. Politec. Torino **52**, 29–63 (1994)
86. Langevin, P.: Sur la théorie du mouvement Brownien. C.R. Acad. Sci. Paris **146**, 530–532 (1908)
87. Lee, E.B., Markus, L.: Foundations of Optimal Control Theory, second edn. Robert E. Krieger Publishing Co., Melbourne, FL (1986)
88. Lemons, D.S.: An Introduction to Stochastic Processes in Physics. Johns Hopkins University Press, Baltimore, MD (2002). Containing "On the theory of Brownian motion" by Paul Langevin, translated by Anthony Gythiel
89. Levi, E.E.: Sulle equazioni lineari totalmente ellittiche alle derivate parziali. Rend. Circ. Mat. Palermo **24**, 275–317 (1907)
90. Liptser, R.S., Shiryaev, A.N.: Statistics of Random Processes. I, expanded edn., vol. 5 of Applications of Mathematics (New York). Springer, Berlin (2001). General theory, Translated from the 1974 Russian original by A. B. Aries, Stochastic Modelling and Applied Probability

References

91. Liu, W., Röckner, M.: Stochastic Partial Differential Equations: An Introduction. Universitext. Springer, Cham (2015)
92. Lototsky, S.V., Rozovsky, B.L.: Stochastic Partial Differential Equations. Universitext. Springer, Cham (2017)
93. Ma, J., Yong, J.: Forward-backward Stochastic Differential Equations and Their Applications, vol. 1702 of Lecture Notes in Mathematics. Springer, Berlin (1999)
94. Mazliak, L., Shafer, G.: The Splendors and Miseries of Martingales - Their History from the Casino to Mathematics. Trends in the History of Science. Birkhäuser, Cham (2022)
95. Menozzi, S.: Parametrix techniques and martingale problems for some degenerate Kolmogorov equations. Electron. Commun. Probab. **16**, 234–250 (2011)
96. Meyer, P.-A.: Probability and Potentials. Blaisdell Publishing Co. Ginn, Waltham (1966)
97. Meyer, P.A.: Stochastic processes from 1950 to the present. J. Électron. Hist. Probab. Stat. **5**, 42 (2009). Translated from the French [MR1796860] by Jeanine Sedjro
98. Mörters, P., Peres, Y.: Brownian Motion, vol. 30 of Cambridge Series in Statistical and Probabilistic Mathematics. Cambridge University Press, Cambridge (2010). With an appendix by Oded Schramm and Wendelin Werner
99. Mumford, D.: The dawning of the age of stochasticity. Atti Accad. Naz. Lincei Cl. Sci. Fis. Mat. Natur. Rend. Lincei (9) Mat. Appl. **11**, 107–125 (2000). Mathematics towards the third millennium (Rome, 1999)
100. Novikov, A.A.: A certain identity for stochastic integrals. Teor. Verojatnost. i Primenen. **17**, 761–765 (1972)
101. Nualart, D.: The Malliavin Calculus and Related Topics, second edn. Probability and its Applications (New York). Springer, Berlin (2006)
102. Oksendal, B.: Stochastic Differential Equations, fifth edn. Universitext. Springer, Berlin (1998). An introduction with applications
103. Oleinik, O.A., Radkevic, E.V.: Second Order Equations with Nonnegative Characteristic Form. Plenum Press, New York (1973). Translated from the Russian by Paul C. Fife
104. Ornstein, L.S., Uhlenbeck, G.E.: On the theory of the Brownian motion. Phys. Rev. **36**, 823–841 (1930)
105. Pagliarani, S., Pascucci, A.: The exact Taylor formula of the implied volatility. Finance Stoch. **21**, 661–718 (2017)
106. Pagliarani, S., Pascucci, A., Pignotti, M.: Intrinsic Taylor formula for Kolmogorov-type homogeneous groups. J. Math. Anal. Appl. **435**, 1054–1087 (2016)
107. Pardoux, E.: Stochastic partial differential equations and filtering of diffusion processes. Stochastics **3**, 127–167 (1979)
108. Pardoux, E.: Stochastic Partial Differential Equations. SpringerBriefs in Mathematics. Springer, Cham (2021). An introduction
109. Pardoux, E., Peng, S.G.: Adapted solution of a backward stochastic differential equation. Syst. Control Lett. **14**, 55–61 (1990)
110. Pardoux, E., Rascanu, A.: Stochastic Differential Equations, Backward SDEs, Partial Differential Equations, vol. 69 of Stochastic Modelling and Applied Probability. Springer, Cham (2014)
111. Pascucci, A.: Calcolo stocastico per la finanza, vol. 33 of Unitext. Springer, Milano (2008)
112. Pascucci, A.: PDE and Martingale Methods in Option Pricing, vol. 2 of Bocconi & Springer Series. Springer/Bocconi University Press, Milan (2011)
113. Pascucci, A.: Probability Theory. Volume 1 - Random Variables and Distributions. Unitext. Springer, Milan (2024)
114. Pascucci, A., Pesce, A.: Sobolev embeddings for kinetic Fokker-Planck equations. J. Funct. Anal. **286**, Paper No. 110344, 40 (2024)
115. Pascucci, A., Runggaldier, W.J.: Financial Mathematics, vol. 59 of Unitext. Springer, Milan (2012). Theory and problems for multi-period models, Translated and extended version of the 2009 Italian original
116. Paulos, J.A.: A Mathematician Reads the Newspaper. Basic Books, New York (2013). Paperback edition of the 1995 original with a new preface

117. Peng, S.G.: A nonlinear Feynman-Kac formula and applications. In: Control Theory, Stochastic Analysis and Applications (Hangzhou, 1991), pp. 173–184. World Sci. Publ., River Edge, NJ (1991)
118. Pogorzelski, W.: Étude de la solution fondamentale de l'équation parabolique. Ricerche Mat. **5**, 25–57 (1956)
119. Polidoro, S.: Uniqueness and representation theorems for solutions of Kolmogorov-Fokker-Planck equations. Rend. Mat. Appl. (7) **15**, 535–560 (1995)
120. Prévôt, C., Röckner, M.: A Concise Course on Stochastic Partial Differential Equations, vol. 1905 of Lecture Notes in Mathematics. Springer, Berlin (2007)
121. Protter, P.E.: Stochastic Integration and Differential Equations, Second edn., vol. 21 of Stochastic Modelling and Applied Probability. Springer, Berlin (2005). Version 2.1, Corrected third printing
122. Rasmussen, C.E., Williams, C.K.I.: Gaussian Processes for Machine Learning. MIT Press (2006). Available at http://www.gaussianprocess.org/gpml/
123. Revuz, D., Yor, M.: Continuous Martingales and Brownian Motion, third edn., vol. 293 of Grundlehren der Mathematischen Wissenschaften [Fundamental Principles of Mathematical Sciences]. Springer, Berlin (1999)
124. Rogers, L.C.G., Williams, D.: Diffusions, Markov Processes, and Martingales. Vol. 2, Cambridge Mathematical Library. Cambridge University Press, Cambridge (2000). Itô calculus, Reprint of the second (1994) edition
125. Rozovsky, B.L.: Stochastic Evolution Systems, vol. 35 of Mathematics and its Applications (Soviet Series). Kluwer Academic Publishers Group, Dordrecht (1990). Linear theory and applications to nonlinear filtering, Translated from the Russian by A. Yarkho
126. Rozovsky, B.L., Lototsky, S.V.: Stochastic Evolution Systems, vol. 89 of Probability Theory and Stochastic Modelling. Springer, Cham (2018). Linear theory and applications to nonlinear filtering
127. Salsburg, D.: The Lady Tasting Tea: How Statistics Revolutionized Science in the Twentieth Century. Henry Holt and Company (2002)
128. Schilling, R.L.: Sobolev embedding for stochastic processes. Expo. Math. **18**, 239–242 (2000)
129. Schilling, R.L.: Brownian Motion—A Guide to Random Processes and Stochastic Calculus. De Gruyter Textbook, De Gruyter, Berlin (2021). With a chapter on simulation by Björn Böttcher, Third edition [of 2962168]
130. Shaposhnikov, A., Wresch, L.: Pathwise vs. path-by-path uniqueness, preprint, arXiv:2001.02869 (2020)
131. Skorokhod, A.V.: Studies in the Theory of Random Processes. Translated from the Russian by Scripta Technica, Inc., Mineola, NY: Dover Publications, reprint of the 1965 edition ed., 2017
132. Stroock, D.W.: Markov Processes from K. Itô's Perspective, vol. 155 of Annals of Mathematics Studies. Princeton University Press, Princeton, NJ (2003)
133. Stroock, D.W.: Partial Differential Equations for Probabilists, vol. 112 of Cambridge Studies in Advanced Mathematics. Cambridge University Press, Cambridge (2012). Paperback edition of the 2008 original
134. Stroock, D.W., Varadhan, S.R.S.: Diffusion processes with continuous coefficients. I. Comm. Pure Appl. Math. **22**, 345–400 (1969)
135. Stroock, D.W., Varadhan, S.R.S.: Diffusion processes with continuous coefficients. II. Comm. Pure Appl. Math. **22**, 479–530 (1969)
136. Stroock, D.W., Varadhan, S.R.S.: Multidimensional Diffusion Processes. Classics in Mathematics. Springer, Berlin (2006). Reprint of the 1997 edition
137. Struwe, M.: Variational Methods, fourth edn., vol. 34 of Ergebnisse der Mathematik und ihrer Grenzgebiete. 3. Folge. A Series of Modern Surveys in Mathematics [Results in Mathematics and Related Areas. 3rd Series. A Series of Modern Surveys in Mathematics]. Springer, Berlin (2008). Applications to nonlinear partial differential equations and Hamiltonian systems
138. Taira, K.: Semigroups, Boundary Value Problems and Markov Processes, second edn. Springer Monographs in Mathematics. Springer, Heidelberg (2014)

139. Tanaka, H.: Note on continuous additive functionals of the 1-dimensional Brownian path. Z. Wahrscheinlichkeitstheorie Verw. Gebiete **1**, 251–257 (1962/1963)
140. Trevisan, D.: Well-posedness of multidimensional diffusion processes with weakly differentiable coefficients. Electron. J. Probab. **21**, Paper No. 22, 41 (2016)
141. Tychonoff, A.: Théorèmes d'unicité pour l'equation de la chaleur. Math. Sbornik **42**, 199–216 (1935)
142. van Casteren, J.A.: Markov Processes, Feller Semigroups and Evolution Equations, vol. 12 of Series on Concrete and Applicable Mathematics. World Scientific Publishing, Hackensack (2011)
143. Vasicek, O.: An equilibrium characterization of the term structure. J. Financ. Econ. **5**, 177–188 (1977)
144. Veretennikov, A.Y.: Strong solutions and explicit formulas for solutions of stochastic integral equations. Mat. Sb. (N.S.) **111**(153), 434–452, 480 (1980)
145. Veretennikov, A.Y.: "Inverse diffusion" and direct derivation of stochastic Liouville equations. Mat. Zametki **33**, 773–779 (1983)
146. Veretennikov, A.Y.: On backward filtering equations for SDE systems (direct approach). In: Stochastic Partial Differential Equations (Edinburgh, 1994), pp. 304–311, vol. 216 of London Math. Soc. Lecture Note Ser. Cambridge Univ. Press, Cambridge (1995)
147. Vespri, V.: Le anime della matematica. Da Pitagora alle intelligenze artificiali. Diarkos editore, Santarcangelo di Romagna (2023)
148. Williams, D.: Probability with Martingales. Cambridge Mathematical Textbooks. Cambridge University Press, Cambridge (1991)
149. Yamada, T., Watanabe, S.: On the uniqueness of solutions of stochastic differential equations. J. Math. Kyoto Univ. **11**, 155–167 (1971)
150. Yong, J., Zhou, X.Y.: Stochastic Controls, vol. 43 of Applications of Mathematics (New York). Springer, New York (1999). Hamiltonian systems and HJB equations
151. Zabczyk, J.: Mathematical Control Theory—An Introduction, Systems & Control: Foundations & Applications. Birkhäuser/Springer, Cham (2020). Second edition [of 2348543]
152. Zhang, J.: Backward Stochastic Differential Equations, vol. 86 of Probability Theory and Stochastic Modelling. Springer, New York (2017). From linear to fully nonlinear theory
153. Zhang, X.: Stochastic homeomorphism flows of SDEs with singular drifts and Sobolev diffusion coefficients. Electron. J. Probab. **16**, 38, 1096–1116 (2011)
154. Zvonkin, A.K.: A transformation of the phase space of a diffusion process that will remove the drift. Mat. Sb. (N.S.) **93**(135), 129–149, 152 (1974)

Index

Symbols
$\mathbb{L}^2$, 176
$\mathbb{L}^2_{B,\text{loc}}$, 184
$\mathbb{L}^2_B$, 184
$\mathbb{L}^2_{S,\text{loc}}$, 202
$\mathbb{L}^2_{\text{loc}}$, 195
$\mathcal{F}^X$, 111
$\mathcal{F}_\infty$, 109
$\mathcal{F}_\tau$, 119
$\mathcal{G}^X$, 14
σ-algebra
    completion, 11
bC_T^α, 341
$\mathcal{M}^{c,2}$, 141
$\mathcal{M}^{c,\text{loc}}$, 143

A
Almost everywhere (a.e.), xix
Almost surely (a.s.), xix
A priori estimate
    L^p, 281
    exponential, 283
Arg max, xii
Aronson, D.G., 406
Assumptions
    standard for SDE, 277

B
Bachelier, L., 71
Black&Scholes, 247
Blumenthal, O., 114, 117
Boundary
    parabolic, 298
Bounded variation (BV), 152
Brownian bridge, 314
Brownian motion, 41, 71
    canonical, 117
    correlated, 228, 238, 239
    with drift, 161
    Feller property, 75
    finite-dimensional densities, 76
    geometric, 277
    Lévy characterization, 238
    Markov property, 75
    multidimensional, 227
    with random initial value, 144
Burkholder-Davis-Gundy, 216, 219

C
Càdlàg, 88
Canonical version
    of a continuous process, 63
    of a Markov process, 33
    of a process, 12
Cauchy-Schwarz, 167
Change of drift, 244
Chapman-Kolmogorov, 37, 381, 391
Characteristic exponent, 116
Characteristics, 296
Coefficient
    diffusion, 205, 264
    drift, 264
Commutator, 312
Completion, 11
Condition
    Hörmander, 312

Kalman, 310
Novikov, 250
Constant elasticity of variance, 319
Continuity in mean, 182
Continuous dependence on parameters, 333
Controllability, 308
Control theory, 308
Courrège, P., 45
Covariation process, 166, 178, 186, 198
Cox-Ingersoll-Ross (CIR), 317

D

Decomposition
    Doob, 17, 165
Delta
    Kronecker, 228
Density
    transition, 28
Differential notation, 156, 204
Diffusion, 55
Dirichlet problem, 292
Distribution
    of a stochastic process, 4
    transition, 25
        homogeneous, 27
Doob, J.L., 17, 103, 135, 165
Drift, 205, 264
    change of, 244
Duhamel, J.M.C., 381
Durrett, R., 71
Dyadic
    partition, 67
    rationals, 67, 133

E

Einstein, A., 71, 305
Enlargement
    filtrations, 110
    standard, 111
Equation
    Chapman-Kolmogorov, 37
    Fokker-Planck, 51, 52, 360
    heat, 295
        backward, 50
        forward, 49
    Kolmogorov
        backward, 47
        forward, 49, 51, 52, 332
    Langevin, 305
    stochastic differential, 263
    Volterra, 269

Estimates
    Gaussian, 381, 385
    potential, 395
Euler Gamma, 390
Exit time, 290

F

Feller, W., 29, 124
Feynman-Kac, 287, 292, 298, 299
    non-linear, 359
Filtering, 359
Filtration, 14
    Brownian
        backward, 362
    complete, 107
    enlargement, 110
        standard, 111
    $\mathcal{G}^X$, 14
    generated, 14
    right-continuous, 107
    standard, 111
    usual conditions, 107
Finite-dimensional
    cylinder, 3
    distributions, 4
Flandoli, F., 348
Fokker-Planck, 51
Formula
    Black&Scholes, 247
    Duhamel, 381
    Feynman-Kac, 287, 292, 298, 299
        non-linear, 359
    Itô, 209, 210
        backward, 363
        for Brownian motion, 211
        for Brownian motion correlated, 239
        for continuous semimartingales, 233
        deterministic, 156
        for Itô processes, 214, 234
    Lévy-Khintchine, 116
Forward-backward system of equations
    (FBSDE), 359
Friedman, A., 342
Function
    of bounded variation, 152
    BV, 152
    càdlàg, 88
    Euler Gamma, 390
    Gaussian, 383
        standard, 383
    indicator, xi

G
Girsanov, I.V., 251, 253
Grönwall, T.H., 281
Gyöngy, I., 353

H
Hörmander, L., 306, 312, 313
Hilbert-Schmidt, 231

I
Independent increments, 34
Inequality
　Burkholder-Davis-Gundy, 216, 219
　Doob's maximal, 103, 104, 135
Infinitesimal generator, 42
Inhomogeneous term, 373
Integral
　Itô, 175
　Lebesgue-Stieltjes, 158
　Riemann-Stieltjes, 151, 154, 200
Intensity, 86
　stochastic, 90
Isometry
　Itô, 178, 186, 199, 232
Itô
　formula, 209, 210
　　for Brownian motion, 211, 239
　　for continuous semimartingales, 233
　　for Itô processes, 214, 234
　integral, 175
　isometry, 178, 186, 199, 232
　process, 204
　　multidimensional, 231
　　process with deterministic coefficients, 215
Itô-Tanaka, 348

K
Kalman, R.E., 310
Kernel
　Poisson, 295
Kolmogorov, A.N., 11, 23, 64, 306
Kolmogorov equation
　backward, 47
　forward, 51, 332
Komlós, J., 168
Kronecker, L., 228
Krylov, N.V., 346, 363

L
Langevin, P., 305
Laplace, P.S., 129

Law
　0-1 of Blumenthal, 114, 117
　of a continuous process, 63
　iterated logarithm, 74
　of a stochastic process, 4
　transition, 25
　　Gaussian, 28, 40
　　homogeneous, 27
　　linear SDE, 303
　　Poisson, 27, 39
Lebesgue-Stieltjes, 158
Lemma
　Grönwall, 281
　Gyöngy, 353
　Komlós, 168
　upcrossing, 105
Levi E.E., 342, 379
Lévy P., 114, 238
Lévy-Khintchine, 116
Linear system
　controllability, 308

M
Markov
　process, 30
　　finite-dimensional laws, 36
　property, 30
　　extended, 32
Markov, A., 25, 30, 123, 330
Markovian projection, 354
Martingale, 15, 338
　Brownian, 77, 257
　càdlàg, 139
　discrete, 15
　exponential, 77, 212, 235, 244
　local, 143
　problem, 338
　quadratic, 77, 234
　stopped, 143
　sub-, 17
　super-, 17
　uniformly square-integrable, 146
Matrix
　covariation, 166, 232
Mean reversion, 314
Measurability
　progressive, 118
Measure
　harmonic, 295
　Lebesgue-Stieltjes, 158
　Wiener, 76
Mesh, 152

Method
  characteristics, 296
  parametrix, 379, 383
Model
  CEV, 319
  CIR, 317
  Vasicek, 314
Modification, 8
Mumford, D.B., vii

N
Nash, J., 406
Norm
  Hilbert-Schmidt, 231
  spectral, 387
Novikov, A., 250

O
Operator
  adjoint, 52
  characteristic, 42
    an SDE, 288
  elliptic-parabolic, 46
  Kolmogorov
    backward, 342
    forward, 370
  Laplace, 49, 129
  local, 44
  parabolic, 371
  pseudo-differential, 116
  symbol, 116
  translation, 128
Option, 245
  Asian, 306
Optional sampling, 102, 137, 148
Ornstein-Uhlenbeck, 316

P
Parabolic
  boundary, 298, 372
  distance, 335
  PDE, 369
Parametrix, 379, 383
Partial differential equation (PDE), 288
  parabolic, 369
Partition, 151
  dyadic, 133
Peano's brush, 269
Poisson, 27, 84, 295
  characteristic exponent, 87
  kernel, 295
  transition law, 39
Positive part, xii
Potential, 395
Predictable, 17
Principle
  comparison, 374
  Duhamel, 381
  maximum, 44, 294, 298, 371, 373
    weak, 375, 378
  reflection, 126
Problem
  Cauchy, 371
    backward, 342
    classical solution, 342
    quasi-linear, 359
  Cauchy-Dirichlet, 372
  Dirichlet, 292
  martingale, 338
Processes
  absolutely integrable, 15
  adapted, 14
  Brownian motion, 71
  BV, 160
  càdlàg, 88
  canonical version, 12, 14, 63
  CEV, 319
  CIR, 317
  continuous, 59
    canonical version, 63
    law, 63
  covariation, 166, 178, 186, 198
  diffusion, 55
  equal in law, 8
  Feller, 29
  Gaussian, 5, 12
  increasing, 160
  indistinguishable, 9
  Itô, 204
    backward, 363
    with deterministic coefficients, 215
    multidimensional, 231
  Lévy, 114
  Markov, 25, 30, 306
  martingale, 15
  maximum, 127
  measurable, 8
  modifications, 8
  Poisson, 40, 84, 88
    compensated, 90
    compound, 87
    with stochastic intensity, 90
  predictable, 17
  progressively measurable, 118
  quadratic variation, 165, 205, 209, 220

Index 425

reflected, 126
simple, 177, 189
square root, 317
stochastic, 1–3
  discrete, 2
  law, 4
  real, 2
stopped, 100
with independent increments, 34
Progressively measurable, 118
Property
  Feller, 29, 124
    for SDE, 334
    strong, 41
  flow, 326, 364
  Markov, 30
    extended, 32
    for SDE, 330
    strong, 123
  martingale, 15
  semigroup, 41
  strong Markov, 124
    homogeneous case, 129
    for SDE, 334
Pseudo-differential operator, 116

Q
Quadratic variation, 165, 205, 209, 220

R
Random variable (r.v.), xix
Reflected process, 126
Reflection principle, 126
Regularization by noise, 347
Representation of Brownian martingales, 257
Riemann-Stieltjes, 151, 154
Risk-neutral valuation, 245

S
Scalar product, xii
Semigroup, 41
Semimartingale, 161, 202
  BV, 163
  continuous
    uniqueness of decomposition, 164
Set-up, 264
Shreve, S., 356
Skorokhod, A.V., 63, 346
Solution
  distributional, 52
  fundamental, 50, 333, 342, 379

of an SDE, 265
strong (of an SDE), 266
transfer of, 273
weak (of an SDE), 272
Solvability of an SDE, 267
Space
  of trajectories, 1
  Polish, 61
  probability
    complete, 9
  Skorokhod, 63
  trajectories, 2
  Wiener, 76
Stochastic differential equation (SDE), 263
  backward, 356
  forward-backward, 359
  linear, 303
  solution, 265
  solvability, 267
  standard assumptions, 277
  strong solution, 266
  uniqueness, 268
  weak solution, 272
Stochastic partial differential equation (SPDE), 359
  heat, 359
  Krylov's, 363
Stopping time, 107
  discrete, 97
Stroock, D.W., 337, 339
Sub-martingale, 17
Super-martingale, 17
Symbol
  Kronecker, 228
  of an operator, 116

T
Tanaka, H., 267, 268
Tanaka's example, 267, 268
Theorem
  Courrège, 45
  Doob's decomposition, 17, 165
  Girsanov, 251, 253
  Kolmogorov's continuity, 64, 65
  Kolmogorov's extension, 11, 22
  Lévy's characterization, 238
  optional sampling, 102, 137, 148
  representation of Brownian martingales, 257
  Skorokhod, 346
  Stroock and Varadhan, 339
  Yamada-Watanabe, 274, 325

Time
    exit, 98, 108, 290
        from a closed set, 109
        from an open set, 108
    stopping, 107
        discrete, 97
Trajectory, 2, 4
Transfer of solutions, 273
Tychonoff, A.N., 371

U
Uniqueness
    class, 371
    for an SDE, 268
    strong for SDE, 324
Upcrossing, 105
Usual conditions, 107

V
Varadhan, S.R.S., 337, 339
Variation
    first, 152
    quadratic, 161
Vasicek, O., 314
Vector field, 312
Veretennikov, A.Y., 270, 347
Version
    canonical
        of a continuous process, 63
        of a Markov process, 33
        of a process, 12
    continuous, 60
Vespri, V., vii

W
Watanabe, S., 274
Wiener, N., 71, 76

Y
Yamada, T., 274

Z
Zvonkin, A.K., 270, 347

SPRINGER NATURE

GPSR Compliance

The European Union's (EU) General Product Safety Regulation (GPSR) is a set of rules that requires consumer products to be safe and our obligations to ensure this.

If you have any concerns about our products, you can contact us on ProductSafety@springernature.com

In case Publisher is established outside the EU, the EU authorized representative is:

Springer Nature Customer Service Center GmbH
Europaplatz 3
69115 Heidelberg, Germany

The manufacturer's authorised representative in the EU is Springer Nature Customer Service Centre GmbH, Europaplatz 3, 69115 Heidelberg, Germany. If you have any concerns regarding our products, please contact ProductSafety@springernature.com

Printed and bound by CPI Group (UK) Ltd, Croydon, CR0 4YY

27/03/2026

02079527-0001